Ecological Management of Agricultural Weeds

Concerns over environmental and human health impacts of conventional weed management practices, herbicide resistance in weeds, and rising costs of crop production and protection have led agricultural producers and scientists in many countries to seek strategies that take greater advantage of ecological processes. This book provides principles and practices for ecologically based weed management in a wide range of temperate and tropical farming systems. After examining weed life histories and processes determining the assembly of weed communities, the authors describe how tillage and cultivation practices, manipulations of soil conditions, competitive cultivars, crop diversification, grazing livestock, arthropod and microbial biocontrol agents, and other factors can be used to reduce weed germination, growth, competitive ability, reproduction, and dispersal. Special attention is given to the evolutionary challenges that weeds pose and the roles that farmers can play in the development of new weed management strategies.

MATT LIEBMAN is an Associate Professor in the Department of Agronomy at Iowa State University. He is the co-editor of *Weed Management in Agro-Ecosystems: Ecological Approaches* (1988).

CHARLES L. MOHLER is a Senior Research Associate in the Department of Crop and Soil Science at Cornell University. He is an associate editor of the journal *Weed Science*.

CHARLES P. STAVER is a project co-coordinator at CATIE (Center for Teaching and Research in Tropical Agriculture), Nicaragua, where he works on integrated pest management.

Ecological Management of Agricultural Weeds

Written and Edited by

MATT LIEBMAN
Iowa State University

CHARLES L. MOHLER
Cornell University

CHARLES P. STAVER
Centro Agronómico Tropical de Investigación y Enseñanza, Managua, Nicaragua

CAMBRIDGE UNIVERSITY PRESS
Cambridge, New York, Melbourne, Madrid, Cape Town, Singapore, São Paulo

Cambridge University Press
The Edinburgh Building, Cambridge CB2 8RU, UK

Published in the United States of America by Cambridge University Press, New York

www.cambridge.org
Information on this title: www.cambridge.org/9780521560689

© Cambridge University Press 2001

This publication is in copyright. Subject to statutory exception
and to the provisions of relevant collective licensing agreements,
no reproduction of any part may take place without the written
permission of Cambridge University Press.

First published 2001
This digitally printed version 2007

A catalogue record for this publication is available from the British Library

Library of Congress Cataloguing in Publication data

Liebman, Matt.
  Ecological management of agricultural weeds / written and edited by Matt Liebman,
Charles L. Mohler, Charles P. Staver.
    p. cm.
  Includes bibliographical references (p.).
  ISBN 0 521 56068 3
    1. Weeds – Biological control. 2. Weeds – Ecology. 3. Agricultural ecology. 4. Tillage.
  5. Agricultural systems. I. Mohler, Charles L., 1947– II. Staver, Charles P., 1949– III. Title.

SB611.5.L54 2001
632′.5 – dc21   00-068869

ISBN 978-0-521-56068-9 hardback
ISBN 978-0-521-03787-7 paperback

Contents

Preface ix

1 Weed management: a need for ecological approaches 1
MATT LIEBMAN
Introduction 1
Weed management objectives 2
Herbicide sales and use 5
Unintended impacts of herbicide use 8
Weed management and farm profitability 19
Transitions to ecological weed management 24
Summary 30

2 Weed life history: identifying vulnerabilities 40
CHARLES L. MOHLER
Weeds from an ecological perspective 40
The life history of weeds 42
Dormancy and germination 46
Survival of weed seeds in the soil 53
Hazards of establishment 60
Vegetative growth and crop–weed competition 62
Survival after emergence 67
Life span and seed production 71
Dispersal of seeds and ramets 75
Conclusions 84

3 Knowledge, science, and practice in ecological weed management: farmer–extensionist–scientist interactions 99
CHARLES P. STAVER
Introduction 99
Knowledge and technology for weed management: an historical perspective 100

Contrasting perspectives of farmers, extensionists, and scientists on weeds 104
Weed patchiness and uncertainty: the challenge to improving weed management 109
Three approaches to farmer management of weed patchiness and uncertainty 113
Adaptive management and farmer–extensionist–scientist interactions 117
Participatory learning for ecological weed management: a proposal 118
Farmers, extensionists, and scientists learning together: four examples 127
A concluding note 133

4 Mechanical management of weeds 139
CHARLES L. MOHLER

Introduction 139
Tillage: pros and cons 140
Mechanical management of perennial weeds 141
Effects of tillage on weed seedling density 151
Basic principles of mechanical weeding 169
Machinery for mechanical weeding 173
Comparison of chemical and mechanical weed management 190
Directions for future research 192

5 Weeds and the soil environment 210
MATT LIEBMAN AND CHARLES L. MOHLER

Introduction 210
Temperature management 212
Water management 215
Fertility management 220
Crop residue management 229
Toward the integration of weed and soil management 250

6 Enhancing the competitive ability of crops 269
CHARLES L. MOHLER

Introduction 269
Crop density 270
Crop spatial arrangement 281
Crop genotype 287
Phenology 297
Conclusions 305

7 Crop diversification for weed management 322
MATT LIEBMAN AND CHARLES P. STAVER

Introduction 322
Crop diversity in conventional, traditional, and organic farming systems 323
Principles guiding crop diversification for weed management 325
Crop rotation 326
Intercropping 336

Agroforestry 351
Obstacles and opportunities in the use of crop diversification for weed management 363

8 Managing weeds with insects and pathogens 375
MATT LIEBMAN

Introduction 375
Conservation of resident herbivores and pathogens 377
Inoculative releases of control agents 380
Inundative releases of control agents 391
The integration of multiple stress factors 398
Moving ahead with weed biocontrol 400

9 Livestock grazing for weed management 409
CHARLES P. STAVER

Introduction 409
Matching grazing strategies with weed problems 410
Weed control through herbivory in short-cycle crops 420
Aftermath and fallow grazing for weed control 421
Grazing for weed control in tree crops 423
Grazing for weed control in pastures and rangelands 427
Research directions 435

10 Weed evolution and community structure 444
CHARLES L. MOHLER

Introduction 444
Formation and management of weed communities 445
Human-dominated ecosystems as an evolutionary context 454
Origins of weeds 455
Weed genecology 463
Managing the adaptation of weed populations 474
Controlling the spread of new weeds 481
Conclusions 484

11 Weed management: the broader context 494
CHARLES L. MOHLER, MATT LIEBMAN, AND CHARLES P. STAVER

Introduction 494
If ecological weed management is effective, why do farmers rely heavily on herbicides? 495
Feeding a growing human population 502
Developing an environment for research on ecological weed management 506
Implementing ecological weed management 510

Taxonomic index 519
Subject index 525

Preface

Of the many books that have been written about weed management, most have focused on the use of herbicides. This volume is different. Instead of providing information about chemical weed control technologies, the emphasis here is on weed management procedures that rely on manipulations of ecological conditions and relationships. By focusing on ecologically based methods of management, we have been able to provide in-depth treatment of subjects that most weed science books treat only briefly.

Although the reader will find much information on the ecology of weeds here, the primary purpose of the book is not to explain weed ecology. Rather, our intent is to elucidate the role of ecological principles in weed management. We believe that ecology can provide a theoretical basis for weed science, much as physics provides a theoretical basis for engineering and biology acts as the theoretical basis for medicine. Accordingly, throughout this book we show ways in which insights into ecological processes provide explanations for the successes and failures of weed management and avenues for developing better management strategies.

This volume could be used as a textbook for an advanced course in weed management, but it was not written primarily for that purpose. Rather, we have attempted to offer the reader a critical analysis and synthesis of the literature on ecological weed management and relevant aspects of weed ecology. Several goals motivated this review process. First, we wanted to identify clearly the principles that underlie ecological management practices. Second, we wanted to assess the strengths and weaknesses of specific weed management tactics in different cropping systems. Third, we sought to identify the current gaps in understanding of ecological approaches to weed management. As we wrote, we regularly asked ourselves, "What are the interesting research questions relating to this subject, and how could they be answered?" Fourth, we wanted to point out possible new roles for weed scientists within the context of dynamically changing agricultural systems. Finally, we sought to develop the argument that ecological weed management can greatly reduce herbicide use through the creation of agricultural systems that suppress weeds and resist their impacts.

We recognize that the latter point is likely to be controversial. Some controversy is desirable, however, for spurring discussion of the issues involved. In any case, we have attempted to be fully honest in disclosing our agenda.

Science, and particularly an applied discipline like weed science, has important effects on society. Those effects depend on which topics scientists choose to pursue and which they choose to ignore. The volume of work on ecological weed management is increasing rapidly due to rising public demand for environmentally friendly agricultural systems and food products, increasing environmental regulation of agriculture by governments, and changing priorities for public funding. Simultaneously, the increasing industrialization of farm production makes herbicides appear more essential than ever to many farmers and weed scientists. These conflicting pressures on the weed science community need to be confronted and addressed with a maximum of clarity and collegiality and a minimum of acrimony. The ways in which weed scientists resolve this tension will largely determine the fate of weed science as a discipline. We hope our book contributes perspectives that are useful during that process.

This book was conceived and created as an integrated work. The scope and organization of the book were decided at the outset, and we have striven to create unity in tone and perspective throughout. Every draft of each chapter received detailed scrutiny and comment from the other authors/editors. This developed consistency in style and allowed each successive chapter to build on concepts and information presented in previous chapters. Nevertheless, the essential ideas in any particular chapter were generated primarily by one or two of us, and it seemed desirable to indicate that fact with chapter bylines. Despite the identification of authorship on the chapters, we hope that readers will view this as a whole book rather than as a compilation of papers on assorted topics.

A work of this scope cannot be accomplished without the help and support of many people. We are especially indebted to the many colleagues who provided critical reviews of various parts of the manuscript. These include Carol Baskin, Susan Boyetchko, Robert Bugg, Douglas Buhler, Brian Caldwell, John Cardina, Nancy Creamer, Moacyr Dias-Filho, Francis Drummond, Michael Duffy, Frank Forcella, Eric Gallandt, Monica Geber, Carol Greiner, Vern Grubinger, Robert Hartzler, Jeff Herrick, Wayne Honeycutt, John Ikerd, Nicholas Jordan, Peter Marks, Diane Mayerfeld, Milton E. McGiffen Jr., Catrin

Meir, Stephen Moss, Kristen Nelson, Stewart Smith, Marty Strange, James Sumberg, John Teasdale, Mark Vellend, and William Vorley. Any errors, however, are solely the responsibility of the authors. We also received ideas, information, or help with technical questions from Doug Derksen, Elizabeth Dyck, Sana Gardescu, Stephen Moss, and Jacob Weiner. Loden Mohler prepared the line drawings in Chapter 4. Frank Forcella generously provided the data for Figure 10.1. CLM was partially supported while writing this book by Hatch funds (Regional Project NE-92, NY(C)-183458) from the Cornell Agricultural Experiment Station. Finally, we thank our families for their patience and support during the long process of preparing this book: Laura Merrick, Chan Liebman, Marika Liebman, Carol Mohler, Ariel Mohler, Loden Mohler, Jan Salick, Carla Staver, and Benjamin Staver.

MATT LIEBMAN

1

Weed management: a need for ecological approaches

Introduction

Agriculture is the process of managing plant communities to obtain useful materials from the small set of species we call crops. Weeds comprise the "other" set of plant species found in agroecosystems. Although they are not intentionally sown, weed species are well adapted to environments dominated by humans and have been associated with crop production since the origins of agriculture (Harlan, 1992, pp. 83–99).

The ecological role of weeds can be seen in very different ways, depending on one's perspective. Most commonly, weeds are perceived as unwanted intruders into agroecosystems that compete for limited resources, reduce crop yields, and force the use of large amounts of human labor and technology to prevent even greater crop losses. In developing countries, farmers may spend 25 to 120 days hand-weeding a hectare of cropland (Akobundu, 1991), yet still lose a quarter of the potential yield to weed competition (Parker & Fryer, 1975). In the USA, where farmers annually spend $6 billion on herbicides, tillage, and cultivation for weed control (Chandler, 1991), crop losses due to weed infestation currently exceed $4 billion per year (Bridges & Anderson, 1992).

At the other end of the spectrum, weeds can be viewed as valuable agroecosystem components that provide services complementing those obtained from crops. In India (Alstrom, 1990, pp. 25–9) and Mexico (Bye, 1981; Mapes, Basurto & Bye, 1997), farmers consume *Amaranthus*, *Brassica*, and *Chenopodium* species as nutritious foods before crop species are ready to harvest. In western Rajasthan, yields of sesame and pearl millet can be increased by allowing the crops to grow in association with the leguminous weed *Indigofera cordifolia* (Bhandari & Sen, 1979). Certain weeds may limit insect damage to crops by interfering with pest movement or by providing habitat for natural enemies of pests (Andow, 1988; Nentwig, Frank & Lethmayer, 1998). Weed species can reduce soil erosion (Weil, 1982), serve as important sources of fodder and medicine (Datta & Banerjee, 1979; Chacon & Gliessman, 1982), and provide habitat for game birds and other desirable wildlife species (Sotherton, Rands & Moreby, 1985; Sotherton, Boatman & Rands, 1989). These types of beneficial effects indicate that weeds are not just agricultural pests, but can also play beneficial roles in agroecosystems.

In this chapter, we outline the objectives of weed management systems and then discuss how weeds are managed conventionally. We follow with a discussion of why alternatives to conventional management strategies are needed. Finally, we suggest how a broad range of ecological processes and farming practices might be exploited to manage weeds more effectively, while better protecting human health and environmental quality, and potentially increasing farm profitability. In subsequent chapters, we will examine these ecological processes and farming practices in more detail.

Weed management objectives

From the standpoint of crop protection, weed management has three principal objectives:

(1) *Weed density should be reduced to tolerable levels.* Experimental studies with a range of species indicate that the relationship between crop yield loss and weed density can be described by a rectangular hyperbola (Cousens, 1985; Weaver, Smits & Tan, 1987; Norris, 1992; Blackshaw, 1993; Knezevic, Weise & Swanton, 1994; Chikoye, Weise & Swanton, 1995). The specific parameters of this relationship change with differences in weather and soil conditions, species combinations, and other factors (Mortensen & Coble, 1989; Bauer *et al.*, 1991; Lindquist *et al.*, 1996), but, in general, reductions in weed density reduce crop yield loss (Figure 1.1a). Although the relationship shown in Figure 1.1a might argue for total elimination of weeds from crops, eradication efforts may be excessively expensive, incur unacceptable environmental damage, and deprive farmers and others of the ecological services certain weeds provide. Thus, with the exceptions of particularly noxious or invasive species, weed management rather than eradication is desirable.

(2) *The amount of damage that a given density of weeds inflicts on an associated crop should be reduced* (Figure 1.1b). The negative effect of weeds on crops can be limited not only by reducing weed density, but also by minimizing the resource consumption, growth, and competitive ability of each surviving weed (Mortensen, Dieleman & Johnson, 1998). This can be accomplished

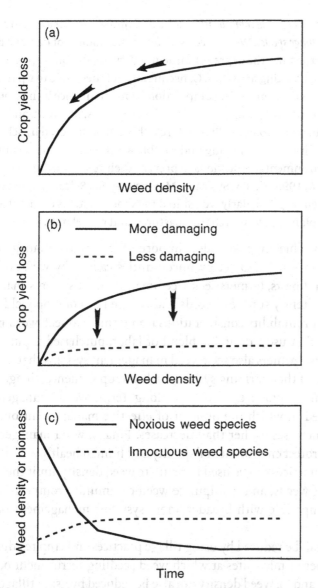

Figure 1.1 Three objectives of weed management: (a) reducing weed density to decrease crop yield loss; (b) reducing the amount of damage a given density of weeds inflicts on a crop; and (c) shifting the composition of weed communities from undesirable to desirable species.

by (i) delaying weed emergence relative to crop emergence (Cousens *et al.*, 1987; Blackshaw, 1993; Chikoye, Wiese & Swanton, 1995), (ii) increasing the proportion of available resources captured by crops (Berkowitz, 1988), and (iii) damaging, but not necessarily killing, weeds with chemical, mechanical, or biological agents (Kropff, Lotz & Weaver, 1993).

(3) *The composition of weed communities should be shifted toward less aggressive, easier-to-manage species.* Weed species differ in the amount of damage they inflict on crops and the degree of difficulty they impose on crop management and harvesting activities. Consequently, it is desirable to tip the balance of weed community composition from dominance by noxious species toward a preponderance of species that crops, livestock, and farmers can better tolerate (Figure 1.1c). This can be achieved by selectively and directly suppressing undesirable weed species while manipulating environmental conditions to prevent their re-establishment (Staver *et al.*, 1995; Sheley, Svejcar & Maxwell, 1996). Selective vegetation management is particularly well suited to agroecosystems dominated by perennial plants, such as orchards, pastures, and rangelands.

Other, broader objectives are also important for weed management systems. Because farming is beset by uncertainties caused by variations in prices, weather, and pests, farmers seek weed management systems that predictably and consistently suppress weeds and reduce risks of crop yield loss. Convenience and profitability considerations lead farmers to seek weed management systems that use a desirable blend of labor, purchased inputs, and management skills. Farmers also seek weed management systems that fit well with other aspects of their farming system, such as crop sequence, tillage, and residue management practices. Over the long term, weed management systems are needed in which the number of effective management options holds steady or increases, rather than decreases. Finally, weed management systems need to protect environmental quality and human health.

What specific practices can be used to regulate weed density, limit the competitive impact of weeds, and manipulate weed community composition in ways that are compatible with broader, more systemic management objectives?

Weed density can be reduced by using tillage practices and crop residues to restrict the number of microsites at which weed seedling recruitment occurs (see Chapters 4, 5, and 7). Weed density can also be reduced by using tillage and cultivation tools (see Chapter 4), biological control agents (see Chapter 8), grazing livestock (see Chapter 9), and herbicides to kill or displace weed seeds, vegetative propagules, seedlings, and mature plants. Monitoring and decision-making are key components of managing weed density, and the development and implementation of procedures for doing so are discussed in Chapter 3.

Weed competitive ability can be reduced by killing early-emerging cohorts of weeds with herbicides or cultivation tools (see Chapter 4) and by choosing particular crop densities, spatial arrangements, and genotypes to enhance crop resource capture and competitive ability (see Chapter 6). Sequences and mixtures of different crops can also be used to preempt resources from weeds (see Chapter 7). Allelochemicals released from live crops and crop residues (see Chapters 5, 6, and 7), biological control agents (see Chapter 8), grazing livestock (see Chapter 9), and herbicides may be used to damage weeds and improve crop performance.

Desirable shifts in weed species composition can be promoted by tillage practices (see Chapter 4), grazing practices (see Chapter 9), and manipulations of soil conditions (see Chapter 5) and crop canopy characteristics (see Chapters 6 and 7). Selective herbicides can also be applied to alter weed species composition.

Currently, herbicides are the primary method for managing weeds in industrialized countries and are becoming more widely used in developing countries. Although we do not believe that they should be excluded from the weed management tool kit, we have given them relatively little attention in this book. There are four reasons for our orientation.

First, a large amount of information about herbicides and their effects on weeds and crops already exists, whereas much less information is available about other management tactics. We hope this book contributes to the closure of that information gap. Second, we believe that, over time, heavy reliance on herbicides reduces their efficacy by selecting for resistant or tolerant weed species and genotypes. To maintain the effectiveness of herbicides as weed management tools, weeds should be exposed to them as infrequently as possible. Third, we believe that certain herbicides can jeopardize environmental quality and human health. To minimize the potential for damage, effective weed management systems that are less reliant on herbicides are needed. Finally, herbicides constitute a rising proportion of crop value at a time when farmers are challenged by serious economic pressures. To promote farm profitability, there is an important need to develop effective weed management strategies that maximize opportunities for farmers to reduce input costs and increase the value of the crop and livestock products they sell.

We examine these points in more detail in the following sections.

Herbicide sales and use

Herbicides dominate the world market for pesticides and pervade the production of staple crops. Worldwide in 1997, $16.9 billion was spent for 1.0 billion kg of herbicide active ingredients, compared with $11.6 billion for 0.7 billion kg of insecticides and $6.0 billion for 0.2 billion kg of fungicides (Aspelin & Grube, 1999). Global herbicide sales are greatest for materials used for maize, soybean, wheat, and rice (Figure 1.2).

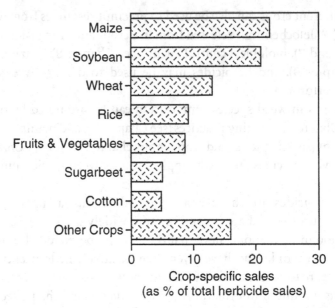

Figure 1.2 Global sales of herbicides in 1985 for the world's major crops. Data are expressed as percentages of total herbicide sales. (After Jutsum, 1988.)

In the USA, herbicide application to agricultural land has risen nearly fourfold since 1966 (National Research Council, 1989, p. 45), and now exceeds 200 million kg of active ingredients annually (Aspelin & Grube, 1999). Herbicides used for maize, soybean, wheat, cotton, and sorghum account for most pesticides applied to American cropland (Aspelin & Grube, 1999; United States Department of Agriculture, 1999a) (Table 1.1).

Herbicide use is also intensifying in many developing countries. In India, herbicide use increased more than 350% from 1971 to 1987, primarily for wheat and rice production (Alstrom, 1990, pp. 167–8). From 1987 to 1992, herbicide sales in South Asia and East Asia grew about 4% per year (Pingali & Gerpacio, 1997). By the early 1990s, herbicides were applied to half of the area planted with rice in the Philippines (Naylor, 1994) and more than 40% of the land planted with wheat in Punjab and Haranya, the two states that account for a third of India's total wheat production (Gianessi & Puffer, 1993). Sales and application of herbicides and other pesticides are also expanding in many regions of Latin America and certain areas of Africa (Repetto & Baliga, 1996, pp. 3–8).

Multiple factors promote the use of herbicides as primary tools for weed management. Herbicides can markedly reduce labor requirements for weed management in both mechanized (Gunsolus & Buhler, 1999) and nonmechanized (Posner & Crawford, 1991) farming systems. Consequently, herbicides

Table 1.1. *Estimated applications of pesticides[a] used in greatest quantities for crop production in the USA in 1987 and 1997*

Pesticide	Use	Active ingredients (millions of kg)	
		Applied in 1987	Applied in 1997
Atrazine	herbicide	32–35	34–37
Metolachlor	herbicide	20–23	29–31
Metam sodium	fumigant (broad-spectrum biocide)	2–4	24–26
Methyl bromide	fumigant (broad-spectrum biocide)	no data	17–20
Glyphosate	herbicide	3–4	15–17
Dichloropropene	fumigant (broad-spectrum biocide)	14–16	15–17
Acetochlor	herbicide	0	14–16
2,4-D	herbicide	13–15	13–15
Pendimethalin	herbicide	5–6	11–13
Trifluralin	herbicide	11–14	10–11
Cyanazine	herbicide	10–11	8–10
Alachlor	herbicide	25–27	6–7
Copper hydroxide	fungicide	0.4–0.9	4–6
Chlorpyrifos	insecticide	3–4	4–6
Chlorothanil	fungicide	2–3	3–4
Dicamba	herbicide	2–3	3–5
Mancozeb	fungicide	2–3	3–5
EPTC	herbicide	8–10	3–5
Terbufos	insecticide	4–5	3–4
Dimethenamid	herbicide	no data	3–4
Bentazon	herbicide	3–4	3–4
Propanil	herbicide	3–5	3–4
Simazine	herbicide	1–2	2–3
MCPA	herbicide	2–3	2–3
Chloropicrin	fumigant (broad-spectrum biocide)	no data	2–3

Note:
[a] Excluded from this list are pesticidal uses of sulfur (22–34 million kg in 1997) and petroleum oils and distillates (30–34 million kg in 1997).
Source: Aspelin & Grube (1999).

are commonly used or becoming more widespread in regions where rising agricultural wages have reduced the cost-effectiveness of hand-weeding (Naylor, 1994; Pingali & Gerpacio, 1997) or mechanical cultivation (Miranowski & Carlson, 1993). Tractor-powered cultivation equipment greatly reduces manual labor requirements for weeding, but may be less consistently successful than herbicides in reducing weed density and protecting crop yield (Hartzler et al., 1993). The cost-effectiveness and timeliness of cultivation can be particularly problematic on large farms with low crop diversity (Gunsolus & Buhler, 1999). Additionally, herbicide use is favored by the adoption of reduced and zero tillage practices (Johnson, 1994) and by the use of direct-seeding techniques in place of transplanting, as in the case of rice (Naylor, 1994).

Public and private institutions also play an important role in promoting herbicide use. In developing countries, herbicide use is encouraged by national and international organizations that provide technical advice and loans to farmers (Alstrom, 1990, p. 169; Pretty, 1995, pp. 26–57) and by government subsidies for herbicides and other pesticides, which lower their cost to farmers (Repetto, 1985). Throughout the world, advertising emphasizes chemical solutions to weed problems. Agrichemical companies spent an estimated $32 million for herbicide advertising in printed media in the USA in 1994 (Benbrook, 1996, p. 165), and herbicide advertisements on radio and television are also common.

A concentration of scientific research upon herbicides has strongly contributed to their importance as weed management tools in both industrialized and developing countries (Alstrom, 1990, pp. 162–5; Wyse, 1992). Abernathy & Bridges (1994) and Benbrook (1996, p. 163) surveyed weed science publications cited in *Weed Abstracts* and the *Agricola* database between 1970 and 1994 and reported that more than two-thirds of the articles focused on various aspects of herbicides and their application. Although some research focused on weed biology and ecology, only a small fraction of articles addressed components of alternative weed management strategies, such as tillage, cultivation, crop rotation, cover crops, mulches, and biological control.

Technical and social factors that favor the dominance of herbicides over other approaches for weed management are discussed in more detail in Chapter 11. Here we will review some of the unintended impacts of herbicide use that are leading a growing number of farmers, scientists, and policy makers to seek alternatives to heavy reliance on herbicide technology.

Unintended impacts of herbicide use

Herbicide resistance in weeds and herbicide product development

Reappraisal of herbicide technology has been driven, in part, by the detection of herbicide resistance in a growing number of weed species. Herbicide resistance is an evolved condition whereby exposure of a weed population to a herbicide leads to a predominance of genotypes that can survive and grow when treated with herbicide concentrations that are normally fatal in untreated populations. Before 1980, herbicide resistance was observed in only a few weed species and was generally limited to triazine compounds

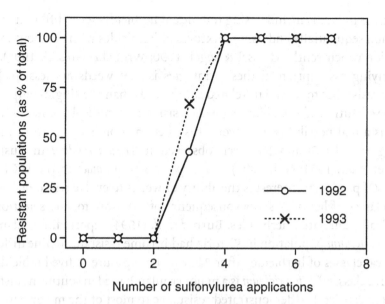

Figure 1.3 Relationship between the number of sulfonylurea herbicide applications made to individual fields and the percentage of *Lolium rigidum* populations with detectable resistance to sulfonylurea compounds. Plant collections were made in Western Australia in 1992 and 1993. (After Gill, 1995.)

(Warwick, 1991; Holt, 1992). Since that time, however, herbicide resistance has been reported for 145 weed species in 45 countries throughout the world (Heap, 1999). Herbicide resistance is appearing in additional weed species at a rate equal to that observed for insecticide and acaricide resistance in arthropod pests (Holt & LeBaron, 1990), and weed biotypes now exist with resistance to one or more herbicides in at least 16 different chemical classes, including the arsenical, aryloxyphenoxyproprionate, benzonitrile, bipyridilium, chloroacetamide, cyclohexanedione, dinitroaniline, dithiocarbamate, imidazolinone, phenoxy, substituted urea, sulfonylurea, triazine, and uracil compounds (Heap, 1999).

Under field conditions in which the same herbicide or chemical class of herbicides is applied repeatedly, herbicide resistance may evolve in four to five years (Holt, 1992). As shown in Figure 1.3, resistance to sulfonylurea herbicides was detected in all populations of the grass weed *Lolium rigidum* collected from Western Australia wheat fields that had been treated with those compounds only four times (Gill, 1995). Evolved resistance to glyphosate, which was thought unlikely to occur, was reported in 1998 for a *L. rigidum* population collected from an Australian orchard that had been treated with glyphosate two or three times a year for 15 years (Powles *et al.*, 1998).

Suggested strategies for preventing or delaying the evolution of herbicide resistance in weeds include using individual herbicides with different modes of action sequentially and using mixtures of herbicides with different modes of action concurrently (Gressel & Segel, 1990; Wrubel & Gressel, 1994). The underlying assumption in these strategies is that weeds are less likely to evolve resistance to several unrelated compounds than to a single compound.

The evolution of weed biotypes with resistance to multiple classes of herbicides is a real possibility, however. This phenomenon is common in insects (Georghiou, 1986) and has been observed in *Lolium rigidum* in Australia (Burnet et al., 1994; Gill, 1995) and *Alopecurus myosuroides* in the UK (Holt, 1992). Of particular interest is the ability of weeds to evolve resistance to distinct classes of herbicides as a consequence of exposure to, and selection by, chemically unrelated herbicides. Burnet et al. (1994) reported, for example, that a *L. rigidum* population in Victoria had become resistant to nine different chemical classes of herbicides after 21 years of exposure to five herbicides in only five classes. *Lolium rigidum* is a major cropland weed in southern Australia and, as a species, has demonstrated resistance to most of the major herbicide chemistries used there (Powles et al., 1997).

Increasing costs of research, development, and registration are reducing the rate at which new herbicides are introduced into the marketplace. The cost to a company of developing and registering a pesticide product increased from $1.2 million in 1956 to an estimated $70 million in 1991 (Holt & LeBaron, 1990; Leng, 1991). Concomitantly, the chances of a newly discovered chemical becoming a legally registered product have decreased greatly; Holt & LeBaron (1990) cited the odds as 1 in 1000 in 1956, compared with 1 in 18 000 in 1984. Increased costs of toxicological testing and legal work associated with the regulatory process are also leading many agrichemical firms to not seek re-registration for the use of herbicides in crops that occupy only small areas, e.g., vegetables and fruits (Anonymous, 1989).

Partly as a consequence of rising costs for discovering, developing, and registering new herbicides, agrichemical firms have merged with seed and biotechnology companies to produce new crop varieties with resistance to existing herbicides, especially glyphosate, glufosinate, bromoxynil, and sulfonylurea, cyclohexanedione, and imidazolinone compounds (Duke, 1999). Many of these varieties have been produced using recombinant DNA technologies. Worldwide in 1999, herbicide-resistant, transgenic varieties of soybean, maize, cotton, rapeseed, and other crops were planted on 28 million ha (Ferber, 1999). The broadscale deployment of these and other genetically engineered crops has been met with controversy in Europe, Japan, the USA, and elsewhere because of environmental and consumer concerns. Thus, the extent to which herbicide-resistant crops will be used in the future is uncertain.

If herbicide-resistant crops are accepted and used widely in coming years, herbicide resistance in weeds will remain a concern, since herbicides used with these crops will exert the same types of selection pressures that they do in herbicide-tolerant, non-genetically engineered crops. Shifts in weed community composition toward species pre-adapted to tolerate herbicides applied to herbicide-resistant crops are also possible (Owen, 1997). In addition, transfer of herbicide resistance from crops to related weed species through pollen movement may create new herbicide-resistant weed populations (Snow & Morán-Palma, 1997; Seefeldt et al., 1998), which would have to be controlled by different herbicides or other means.

The combination of herbicide resistance in an increasing number of weed species, slower introduction of new herbicides, and withdrawal of older herbicides means that farmers are likely to have fewer chemical control options within the next several decades. For this reason, alternative weed management strategies that make full use of nonchemical tactics need to be developed.

Herbicides and water quality

Since the 1980s there has been increasing recognition that herbicides, applied in the course of normal farming practices, have contaminated surface and ground water in many agricultural regions (Barbash et al., 1999; Larson, Gilliom & Capel, 1999; United States Geological Survey, 1999). Among the herbicides detected most frequently in drinking-water sources, there are a number of compounds classified as probable (e.g., acetochlor), likely (e.g., alachlor), and possible (e.g., atrazine, cyanazine, metolachlor, and simazine) carcinogens (United States Environmental Protection Agency, 1999). Several herbicides contaminating drinking-water sources are also under scrutiny as possible disrupters of human immune, endocrine, and reproductive systems (see section "Acute and chronic effects of herbicides on human health" below). The effects of low-level exposure to herbicides are poorly understood, but there is considerable popular and regulatory concern over contamination of drinking-water sources.

Herbicide contamination of the Mississippi River drainage basin has been particularly well documented (United States Geological Survey, 1999). The 12 states that drain to the Mississippi River contain about 65% of the harvested cropland in the USA, and fields of maize, soybean, sorghum, rice, wheat, and cotton are dominant features of the region's landscape (United States Department of Agriculture, 1999b). The Mississippi River basin receives the majority of herbicides applied in the USA; during the late 1980s, more than 125 000 metric tons of herbicide active ingredients were applied annually to cropland in the watershed (Gianessi & Puffer, 1991; Goolsby, Battaglin & Thurman, 1993).

About 18 million people rely on the Mississippi River and its tributaries as their primary source of drinking water (Goolsby, Coupe & Markovchick, 1991). Public water systems serving that population are required to take at least four samples each year to measure concentrations of pollutants, including certain herbicides, for which the US Environmental Protection Agency (1996) has set legally enforceable safety standards called maximum contaminant levels. A public water system is out of compliance with the federal Safe Drinking Water Act of 1986 if the yearly average concentration of a pollutant exceeds its maximum contaminant level, or if a pollutant's concentration in any one quarterly sample is more than four times higher than its maximum contaminant level.

For several herbicides currently lacking legally enforceable standards, the US Environmental Protection Agency (1996) has specified health advisory levels, which are maximum chemical concentrations that may be consumed in drinking water over an average human lifetime with minimal risk that they will cause "adverse non-carcinogenic effects." Health advisory levels can eventually become enforceable standards. Both maximum contaminant and health advisory levels have been established only for individual compounds; standards have not been set for mixtures of herbicides and other chemicals, including metabolites of herbicides (Goolsby, Battaglin & Thurman, 1993).

After application to cropland in the midwestern USA, herbicides not degraded or bound to soil are detected in surface water in pulses corresponding to late spring and summer rainfall (Thurman *et al.*, 1991). In 1991, the US Geological Survey detected atrazine, which is widely used for weed control in maize and sorghum, in each of 146 water samples collected at eight locations throughout the Mississippi River basin (Goolsby, Coupe & Markovchick, 1991). More than 75% of the samples also contained other herbicides used in maize, soybean, and sorghum production: alachlor, metolachlor, cyanazine, and simazine. Between April and July 1991, atrazine concentrations exceeded the US Environmental Protection Agency's maximum contaminant level of 3 μg L^{-1} for 6 to 9 weeks at sites in the Illinois, Mississippi, Missouri, Platte, and White Rivers (Figure 1.4). In those same rivers, cyanazine concentrations exceeded the US Environmental Protection Agency's health advisory level of 1 μg L^{-1} for 7 to 14 weeks. Alachlor concentrations exceeded the agency's maximum contaminant level of 2 μg L^{-1} for 1 to 3 weeks in the Illinois, Platte, and White Rivers.

In a review of data from 12 studies of herbicide concentrations in finished tap water and raw drinking-water sources (rivers and reservoirs) in the

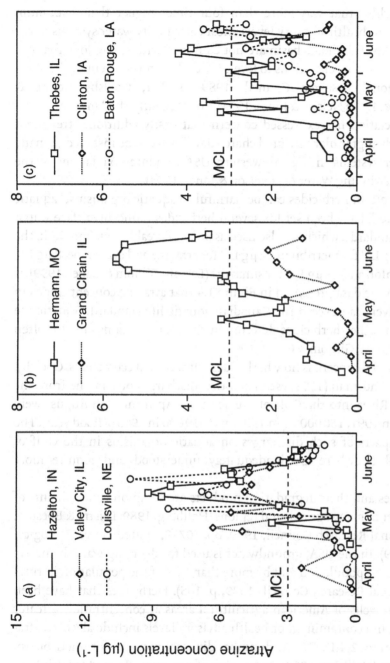

Figure 1.4 Atrazine concentrations during April through June 1991 in the White River at Hazelton, IN (a); the Illinois River at Valley City, IL (a); the Platte River at Louisville, NE (a); the Missouri River at Hermann, MO (b); the Ohio River at Grand Chain, IL (b); and the Mississippi River at Thebes, IL, Clinton, IA, and Baton Rouge, LA (c). The maximum contaminant level (MCL) set by the US Environmental Protection Agency for atrazine in drinking water is 3 µg L^{-1}. (After Goolsby, Coupe & Markovchick, 1991.)

American maize belt, Nelson & Jones (1994) noted that a substantial proportion of sampled locations had at least one measurement of atrazine, cyanazine, or alachlor that was more than four times higher than maximum contaminant or health advisory levels. Most community water systems in the Mississippi River drainage basin are not equipped with technology that can reduce herbicide concentrations to levels lower than government health standards (National Research Council, 1989, p. 101; Goolsby, Coupe & Markovchick, 1991; Nelson & Jones, 1994). Consequently, the American Water Works Association has expressed concern that costly additional treatment systems, such as granular activated charcoal, will have to be installed in many public water systems in the midwestern USA to address violations of the federal Safe Drinking Water Act (Nelson & Jones, 1994).

Because certain herbicides can be harmful to aquatic organisms, "aquatic life guidelines" have been set for several herbicides found in surface water. Canadian standards, which are also used as nonenforceable benchmarks in the USA, are 1 μg L^{-1} for metribuzin, 2 μg L^{-1} for atrazine and cyanazine, 8 μg L^{-1} for metolachlor, and 10 μg L^{-1} for simazine (Larson, Gilliom & Capel, 1999). It is clear from the data presented in Figure 1.4 that atrazine concentrations in American rivers can exceed the Canadian aquatic life standard. Aquatic life standards for other herbicides detected in rivers and streams are also often exceeded (Larson, Gilliom & Capel, 1999).

An additional concern is how herbicides affect coastal ecosystems. Goolsby, Battaglin & Thurman (1993) estimated that discharges of atrazine from the Mississippi River into the Gulf of Mexico from April through August were 296 000 kg in 1991, 160 000 kg in 1992, and 539 000 in 1993 (a flood year). The possible impacts of such discharges on aquatic organisms in the Gulf of Mexico and elsewhere are inadequately understood and require more research.

Herbicides and their degradation products are common contaminants of groundwater in many agricultural regions (Hallberg, 1989; Leistra & Boesten, 1989; National Research Council, 1989, pp. 107–9; United States Geological Survey, 1999). In the USA, groundwater is used for drinking water by nearly half of the total population and by more than 95% of the population in rural areas (National Research Council, 1989, p. 105). Herbicides that have been measured in wells of American agricultural areas at concentrations greater than maximum contaminant or health advisory levels include alachlor, atrazine, cyanazine, 2,4-D, DCPA, dicamba, dinoseb, metolachlor, metribuzin, and simazine (Hallberg, 1989). In a survey of private wells used for drinking water in Ohio, Indiana, Illinois, West Virginia, and Kentucky, Richards et al. (1996) detected chloroacetamide and triazine herbicides in 9.7% and 4.9% of the 12 362 samples tested; maximum contaminant levels for alachlor and atrazine were exceeded in 1.1% and 0.1% of the samples, respectively. Two large-scale, multistate investigations of herbicides in American wells and springs detected at least one of seven targeted compounds in 35% to 40% of the sites sampled, although maximum contaminant or health advisory levels were exceeded at fewer than 0.1% of the sites (Barbash et al., 1999).

Although concentrations of individual herbicides in American groundwater rarely exceed existing regulatory standards, important concerns remain concerning health risks. Detection of one herbicide in groundwater at an individual site is often accompanied by the detection of others (Barbash et al., 1999), but little is known about the health-related impacts of exposure to multiple herbicides, or to herbicides in combination with nitrates, which are also common water contaminants. Breakdown products of herbicides are generally found in well water more frequently and at higher concentrations than the corresponding parent compounds (Kolpin, Thurman & Goolsby, 1996), but little is known about their possible effects on human health. Health-based standards for breakdown product concentrations in groundwater generally do not exist.

Herbicide drift

Herbicides can contaminate off-target sites by moving in air as well as in water. Generally, herbicide drift from tractor-mounted sprayers is about 5% to 10% of the material applied, with most off-site deposition occurring within 20 m of field edges (Freemark & Boutin, 1995). However, depending on meteorological conditions, application equipment, and physical characteristics of herbicide products, spray drift concentrations of 0.02% to 2% of application rates may occur at distances as great as 400 m from application sites (Fletcher et al., 1996).

The implications of aerial movement of herbicides are especially problematic for highly phytotoxic chemicals, such as sulfonylurea and imidiazolinone compounds. Although these compounds may have low mammalian toxicity, their drift onto nontarget crops and wild land areas, even at low concentrations, may greatly alter plant performance, particularly reproduction. Fletcher et al. (1996) found that flower and seed production by rapeseed, soybean, sunflower, and *Polygonum persicaria* could be reduced by exposure to chlorsulfuron at rates from 0.1% to 0.8% of those recommended for field applications to cereal crops. For certain combinations of plant species, chlorsulfuron rates, and application times, reproductive damage occurred even when effects on vegetative growth were minimal. For example, chlorsulfuron treatment of rapeseed (at 9.2×10^{-5} kg a.i. ha^{-1}) and soybean (at 1.8×10^{-4} kg a.i. ha^{-1}) during anthesis reduced seed yield 92% and 99%, respectively, compared with untreated plants, whereas height was reduced only 12% and 8%. Similarly, treatment of cherry trees with low rates of chlorsulfuron reduced fruit yield but created little or no foliar damage (Fletcher, Pfleeger & Ratsch, 1993).

Other herbicides do not necessarily have such potent effects at low concentrations. Rapeseed and soybean were unaffected by applications of atrazine, glyphosate, and 2,4-D at rates and stages of plant development at which chlorsulfuron suppressed reproduction (Fletcher et al., 1996). None the less, the experiments with chlorsulfuron indicate that low doses of certain compounds can profoundly affect plant reproduction, and the results emphasize the potential for serious off-target damage due to herbicide drift. Currently, data concerning the impacts of chlorsulfuron and other herbicides on nontarget plant reproduction are not required for product registration in the USA (Fletcher et al., 1996).

Acute and chronic effects of herbicides on human health

Although much remains to be learned about the acute and chronic health impacts of herbicide use, public health reports and epidemiological studies indicate that certain herbicides can be responsible for direct, unintentional poisoning and may be associated with increased incidence of cancer and other disorders. Farmers, farm families, and agricultural workers are exposed to herbicides at higher concentrations than the general public and consequently may be subjected to greater health risks. Health issues relating to exposure to herbicides and other pesticides are particularly important in developing countries, where safe use is difficult because of unavailable or prohibitively expensive protective equipment, inadequate and poorly enforced safety standards, poor labeling, illiteracy, and insufficient knowledge of hazards by handlers and applicators (Pimentel et al., 1992; Repetto & Baliga, 1996, pp. 9–16).

Acute symptoms of pesticide poisoning include headache, skin and eye irritation, fatigue, dizziness, nausea, cramping, fever, diarrhea, and difficulty in breathing (Stone et al., 1988). Most incidents of pesticide poisoning go unreported (Jeyaratnum, 1990), but it is conservatively estimated that one million serious accidental pesticide poisonings occur throughout the world each year (World Health Organization, 1990, p. 86). Pesticide poisonings of farmers and agricultural workers occur in industrialized countries, such as the USA (Stone et al., 1988), but are more frequent in developing countries (Repetto & Baliga, 1996, pp. 9–16).

Public health data from Costa Rica suggest that herbicides may contribute to a significant portion of acute pesticide poisonings in developing countries. Hilje *et al.* (1992, p. 79) reported that bipyridilium, chloroacetamide, dinitroaniline, phenoxy, picolinic acid, substituted urea, and triazine herbicides accounted for 19% of the 787 pesticide poisonings registered in 1984 by the Costa Rican National Poison Control Center. Similarly, Dinham (1993, p. 105) noted that various herbicides were responsible for 22% of the acute pesticide poisonings in the region of Limón, Costa Rica, in the first six months of 1990. Hilje *et al.* (1992, p. 79) stated that the actual number of pesticide poisonings in Costa Rica is higher than that reported to government agencies, but that available data accurately reflect the percentage of poisonings attributable to different types of pesticides.

Chronic health effects of chemical exposure can include cancer and disorders of the immune, endocrine, neurological, and reproductive systems. Unambiguous cause-and-effect relationships are often difficult to establish for these types of health problems because a long lag period typically exists between exposure to causative agents and presentation of clinical symptoms, and because exposure to other chemicals or behaviors such as smoking may be contributing factors. Epidemiological studies can be conducted, however, to determine patterns of risk associated with exposure to herbicides and other pesticides.

Thirty-nine herbicide active ingredients are classified by the US Environmental Protection Agency (1999) as probable, likely, or possible carcinogens, and a number of epidemiological studies have examined possible links between herbicides and cancer in human populations. Significant correlations between herbicide use and several types of cancer were noted by Stokes & Brace (1988) in a study of cancer deaths in 1497 nonmetropolitan counties in the USA. The percentage of land area treated with herbicides in each county was significantly correlated with the incidence of genital, lymphatic, hematopoietic, and digestive system cancers. Herbicide use had no relationship with urinary system cancers, however, and was negatively correlated with respiratory system cancers. On Saskatchewan farms of less than 400 ha, death of male farmers due to non-Hodgkin's lymphoma (NHL) rose significantly with increasing numbers of hectares sprayed with herbicides (Blair, 1990; Wigle *et al.*, 1990). No significant relationship was found on farms of more than 400 ha, where farmers may have been less likely to apply herbicides personally or may have used aircraft for applications.

Hoar *et al.* (1986) reported that the incidence of NHL among men in Kansas increased significantly with the number of days per year that they used herbicides; men who used herbicides more than 20 days per year had a six-fold higher chance of contracting NHL than did nonfarmers or farmers not using herbicides. Increased risk of NHL was specifically associated with use of phenoxy herbicides, especially 2,4-D, which is widely used in field crop production in Kansas. Exposure to phenoxy herbicides has been linked to increased risks of NHL, Hodgkin's lymphoma, and soft-tissue sarcoma in a number of other studies (Hardell & Sandstrom, 1979; Hardell et al., 1981; Blair, 1990), although reviews of the subject have concluded that no consistent cause-and-effect pattern exists (Smith & Bates, 1989; Ibrahim et al., 1991).

In addition to concerns about possible links to various cancers, concerns also exist about potential effects of herbicide exposure on other aspects of human health. Repetto & Baliga (1996, pp. 17–49) noted that three widely used herbicides – atrazine, 2,4-D, and paraquat – are immunotoxic to laboratory animals whose immune systems are similar to that of humans, and they suggested that exposure to these and other pesticides may increase human susceptibility to infectious diseases and certain types of cancer because of immune system suppression. They noted, however, that the epidemiological studies necessary to test that hypothesis have not been conducted. The herbicides alachlor, atrazine, 2,4-D, metribuzin, and trifluralin have been identified as potentially disruptive to the human endocrine system (Colborn, vom Saal & Soto, 1993), but how actual exposure through agricultural use affects endocrine function is unknown. Public health data from Minnesota suggest that exposure to 2,4-D and MCPA significantly increased the rate of birth defects in offspring of pesticide applicators and members of the general population in areas with high application rates (Garry et al., 1996). However, exposure to 2,4-D and MCPA was confounded with exposure to a number of fungicides, making it impossible to draw firm conclusions about the reproductive system effects of specific compounds.

Because manipulative experiments with human subjects and possible toxins are unethical, uncertainty about the chronic health effects of herbicides will continue. How should this uncertainty be dealt with? Many proponents of herbicide use do not find available data sufficiently compelling to assume that herbicides pose important human health risks. Opponents believe there is adequate evidence that they do, particularly in developing countries. We suggest that it is prudent to err on the side of safety by minimizing herbicide exposure and toxicity. Greater safety could be obtained by producing and distributing superior application and protective equipment, and by developing new herbicides whose chemistries limit their persistence, mobility, and toxicity to nontarget organisms, including people. The development of effective nonchemical weed management strategies would address the problem at its source and is the focus of this book.

Weed management and farm profitability

An additional factor motivating the development of ecologically based weed management strategies is the need to increase farm profitability. In both industrialized and developing countries, the economic viability of many farmers has been challenged as input costs rise faster than the market values of the crops they produce. Weed management strategies that make better use of ecological processes may improve profitability by reducing production costs and helping farmers produce crops and livestock that are worth more in the marketplace.

The cost–price squeeze

The cost–price squeeze confronting farmers in the USA is exemplified by the maize–soybean cropping system used in much of Iowa, where a total of 9.3 million ha was planted with the two crops in 1998 (United States Department of Agriculture, 1999c). Average yields of maize and soybean in Iowa rose 28% and 24%, respectively, from 1972–80 to 1990–98 (Figure 1.5a). For those same periods, average non-land production costs in constant dollars fell 37% for maize and 31% for soybean (Figure 1.5b). Costs for maize and soybean herbicides, in constant dollars, decreased 9% and 13%, respectively.

Increases in yields and reductions in production costs would seem to bode well for profitability, but prices fell precipitously for both crops. Between 1972–80 and 1990–98, the average price of a metric ton of maize, in constant dollars, decreased 60%; soybean price dropped 62% (Figure 1.5c). Consequently, gross returns declined 47% for maize and 52% for soybean (Figure 1.5d). Returns over non-land costs also declined sharply. For maize, average returns in constant dollars dropped from $396 per hectare in 1972–80 to $153 per hectare in 1990–98, a 61% decline; for soybean, average returns dropped from $530 to $182 per hectare, a 66% decline (Figure 1.5e).

For many Iowa farmers, reductions in returns per unit of cropland have reinforced the importance of herbicides within the production process. Herbicides accounted for 7% of non-land production costs for maize in 1972–80, but 11% in 1990–98; for soybean, the proportion of non-land costs spent on herbicides rose from 12% to 15% (Figure 1.5f). As discussed in Chapter 11, these increases reflect, in part, the greater land area farmers must harvest to maintain farm-derived income, the shift toward hired applications of agricultural chemicals to cover more hectares, and the limited time available for weed management and other farming activities when farmers add nonfarm jobs to their existing responsibilities.

A cost–price squeeze also confronts farmers in developing countries. Beets

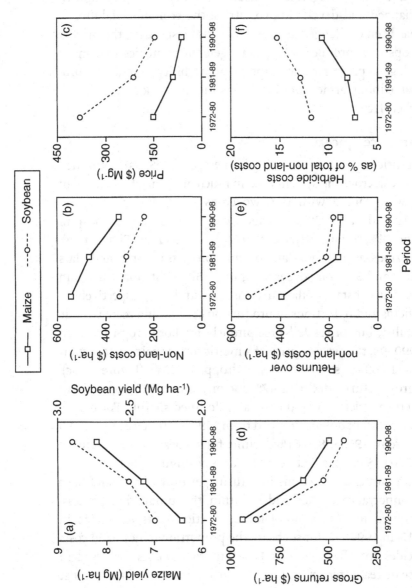

Figure 1.5 Economic characteristics of maize and soybean production in Iowa, 1972–98: (a) yields, (b) non-land production costs, (c) prices, (d) gross returns, (e) returns above non-land production costs, and (f) herbicide costs as percentages of total non-land production costs. Prices, costs, and returns have been adjusted for inflation using the Consumer Price Index of the US Bureau of Labor Statistics (base period: 1982–4). Production costs are for machinery, seeds, pesticides, fertilizers, and labor. Gross returns have been calculated as the product of state average yields (Mg ha^{-1}) and prices ($ Mg^{-1}). Sources: Duffy & Vontalge (1998 and previous years), and M. D. Duffy, Iowa State University, personal communication (2000).

(1990, pp. 81–128) noted three important contributing factors: (i) low domestic prices for farm products because of government policies that keep food inexpensive for urban consumers; (ii) low international prices for export crops because of surplus production; and (iii) high costs for farm production inputs, most of which are imported. In constant dollars, prices for agricultural commodities produced in developing countries, including maize, wheat, rice, cacao, coffee, palm oil, rubber, sugar, and tea, fell about 50% in the international marketplace from 1980–82 to 1990–92 (Pretty, 1995, pp. 54–5). The net result of these factors, combined with the volatility of international commodity prices, is economic insecurity for many farmers in developing countries.

Strategies to lower production costs

Reduction of production costs is one strategy for increasing farm profitability. The possible importance of this strategy is illustrated by results from three studies sponsored by South Dakota State University comparing conventional and alternative (organic) cropping systems. "Study 1" and "Study 2" were conducted from 1986 to 1992 on replicated field plots at a university research farm (Smolik, Dobbs & Rickerl, 1995); "Study 3" was a paired comparison from 1985 to 1992 of two commercial South Dakota farms, one managed conventionally and the other managed organically, without the use of synthetic fertilizers, herbicides, and other pesticides (Dobbs & Smolik, 1996).

In Study 1, a conventional maize–soybean–wheat rotation was compared with an alternative maize–oat + alfalfa–alfalfa–soybean rotation; in Study 2, a conventional wheat–barley–soybean rotation was compared with an alternative wheat–oat + clover–clover–soybean rotation (a mixture of red and sweet clovers was sown). The conventional systems used commercial fertilizers and moldboard plowing; the alternative systems relied on legumes and manure as sources of fertility and used only surface tillage practices. Weeds were controlled with herbicides and cultivation in the conventional systems, whereas in the alternative systems, weeds were controlled without herbicides, but with additional cultivation operations. Inclusion of the forage legume crops (alfalfa and clovers) was also considered to make a positive contribution toward weed control in the alternative systems.

In Study 1, average maize yield per unit area was higher in the conventional than the alternative system, but average soybean yield was similar in the two systems (Smolik et al., 1993). Soybean and wheat yields were also similar between the conventional and alternative systems of Study 2 (Smolik et al., 1993). Gross income, which was calculated using crop sales prices (without premiums for organic products) and relevant government subsidy

Table 1.2. *Economic results from three comparisons of cropping systems conducted in South Dakota*

Study, location, date, management system, and crop rotation	Gross income ($ ha^{-1} yr^{-1})	Costs other than land and management ($ ha^{-1} yr^{-1})	Net income over costs other than land and management ($ ha^{-1} yr^{-1})
Study 1: Research station plots, 1986–92			
Alternative system (maize–oat + alfalfa–alfalfa–soybean)	378	222	156
Conventional system (maize–soybean–wheat)	373	252	121
Study 2: Research station plots, 1986–92			
Alternative system (wheat–oat + clover–clover–soybean)	249	156	93
Conventional system (wheat–barley–soybean)	314	218	96
Study 3: On-farm comparison, 1985–92			
Alternative system (maize–small grain + alfalfa–alfalfa–soybean)	405	220	185
Conventional system (maize–soybean)	561	304	257

Sources: Adapted from Smolik, Dobbs & Rickerl (1995) and Dobbs & Smolik (1996).

payments, was essentially the same for the conventional and alternative systems of Study 1, but 26% higher for the conventional system than for the alternative system of Study 2 (Table 1.2). In contrast, production costs (including labor, but excluding land and management time) were 12% and 28% lower in the alternative systems of Studies 1 and 2, respectively, than in the corresponding conventional systems (Table 1.2). Comparisons of herbicide and cultivation practices similar to those used in the conventional and alternative systems indicated that weed control in the latter was less costly. Compared to the conventional systems, weed control costs (including labor) in the alternative systems were 13% lower for maize, 13% to 28% lower for soybean, and 89% lower for wheat (Smolik *et al.*, 1993).

Largely because of lower production costs, the alternative systems of both Studies 1 and 2 were economically competitive with the conventional systems. In Study 1, average net income over all costs except land and management (i.e., planning, organizing, marketing) was 29% higher for the alternative system; in Study 2, average net income was 3% higher for the conventional system (Table 1.2).

Different results were obtained from the on-farm comparison. The conventional farm in Study 3 used a two-year maize–soybean rotation and applied synthetic fertilizers and pesticides, whereas the alternative (organic) farm used a four-year rotation (maize–small grain + alfalfa–alfalfa–soybean) and did not apply synthetic fertilizers and pesticides. Livestock production, which occurred on both farms, was not evaluated in this study, and premium prices for the organic crops were not included in the baseline analyses. Various forms of government farm support payments were included in the analyses.

Over the eight-year study period, earnings from crop production on both farms in Study 3 were considered "respectable for the area" (Dobbs & Smolik, 1996), but the conventional farm was more profitable than the alternative farm (Table 1.2). Crop production costs were lower on the alternative farm (Table 1.2), but did not overcome the effects of lower crop yields and a smaller percentage of the cropland being used for maize and soybean, which were more profitable than small grains and alfalfa during the study period. Although weed densities in maize and soybean were higher on the organic farm, results from research station experiments led the investigators to conclude that weeds were unlikely to have influenced yields substantially on either the conventional or alternative farm (Dobbs & Smolik, 1996).

Results from these studies indicate that cutting costs for weed management and other farming activities can increase the profitability of crop production under certain circumstances. However, a favorable outcome is not guaranteed. The success or failure of cost-cutting strategies depends on crop choices, impacts of government farm programs, skills and knowledge of farm operators, site-specific soil, weather, and pest conditions, and other factors (Welsh, 1999). More research is needed to increase or maintain crop yields while reducing production costs. This might be achieved, in part, by focusing on ecological processes within farming systems that can reduce requirements for cultivation, herbicides, and other external inputs.

Strategies to increase crop value

In addition to reducing production costs, farm profitability may be improved by increasing prices received for crop and livestock products. Organic farming is one option for adding value to farm products that has become increasingly popular in recent years. During the mid to late 1990s, retail sales of organic products in the USA and Europe rose 20% to 30% annually (Tate, 1994; Burros, 1997; Welsh, 1999). In 1997, the American and European markets for organic products were each estimated to be between $4 and $5 billion, while sales in Japan were estimated to be $2 billion (Welsh, 1999). Geier (1998) predicted the worldwide market for organic foods will reach $100 billion by 2010.

Price premiums paid to American farmers for organic products can be substantial: more than 100% above conventional prices for broccoli in California (Franco, 1989); 33% to 38% higher for apples in California (Swezey *et al.*, 1994); and 45% to 119% higher for maize, 95% to 223% higher for soybean, 28% to 94% higher for oat, and 46% to 74% higher for wheat in South Dakota (Dobbs & Pourier, 1999). Substantial price premiums are also paid in many European countries for organic products (Padel & Lampkin, 1994) and are a major factor driving organic production of certain tropical products, such as coffee.

In a review of data collected by land-grant universities comparing conventional and organic grain and forage production in the midwestern USA (including the South Dakota studies cited previously), Welsh (1999) concluded that organic systems that receive price premiums at existing market levels can consistently match or exceed the profitability of most common conventional systems. Welsh (1999) also found that "break-even" premium levels required for organic systems to match the profitability of conventional systems were lower than the average premiums available during the 1990s. In contrast, Klonsky & Livingston (1994) compared the economic performance of different tomato-based cropping systems in California and found that net returns from a conventionally managed two-year rotation were greater than returns from an organically managed four-year system, even when available organic premiums were included in the analysis.

These and other studies indicate that organic production is a viable means of increasing farm profitability for some, but not all, farmers. Owing to a long-standing paucity of weed research relevant to organic farming systems, strategies to improve weed management are among the top research priorities of organic farmers (Organic Farming Research Foundation, 1998). Ecological approaches for weed management need to be fully developed to achieve the agronomic and economic potentials of organic farming systems.

Transitions to ecological weed management

Characteristics and benefits of a systemic approach

Concerns over pesticide resistance, environmental and health hazards of pesticides, and declining profitability are not unique to weed management. Such problems have been recognized for several decades in the management of arthropod pests and plant pathogens and led in the 1960s and 1970s to development of the concept of integrated pest management (IPM). The IPM concept has never been fully implemented for managing weeds, but its usefulness in managing insects, mites, and plant pathogens calls attention to its potential for weeds.

As described by Bottrell (1979), IPM involves the concerted use of multiple tactics to suppress and kill pests and reduce crop damage to economically acceptable levels. Emphasis is placed on modifying habitat characteristics to reduce pest densities and promote crop health, conserving and releasing beneficial organisms that attack pests, and planting pest-resistant cultivars. Pesticides are used in IPM systems as therapeutic tools only when preventive practices fail to provide adequate control. If pesticide applications are deemed necessary, selective materials are applied in a manner that poses minimal risks to human health and the environment. A key component of IPM systems is timely farmer decision-making based on knowledge of (i) crop, pest, and natural enemy biology; (ii) pest abundance and distribution; (iii) impacts of environmental factors and farming practices on crop–pest–natural-enemy interactions; (iv) cost and income implications of different management options; and (v) human health and environmental impacts of different management options.

Multitactic, ecologically based, information-intensive pest management strategies are desirable for several reasons (Bottrell & Weil, 1995; Lewis et al., 1997). First, effective pest control can result from tactics whose individual impacts are weak, but whose cumulative impacts are strong. Second, risks of crop failure or serious loss can be reduced when the burden of crop protection is distributed across many tactics, and when information is available to allow rapid adjustments in management strategies. Third, the rate at which pests adapt or evolve resistance to a given management tactic can be decreased when the frequency of their exposure to that tactic is reduced. Fourth, environmental disruptions and threats to human health can be minimized as pesticide inputs are reduced. Finally, reductions in operating costs and increases in profitability can result from lowering the need for purchased inputs through better use of locally generated materials and site-specific knowledge.

How can transitions be made from conventional weed management systems toward more sustainable, ecologically based systems? Bird et al. (1990) and MacRae et al. (1990) have described a general model that we find relevant for weed management in industrialized countries. It involves passage from heavy reliance on conventional herbicides through stages of *improved efficiency* of herbicide use, *substitution* of more benign inputs and practices for conventional herbicides, and finally *system-level redesign* to manipulate multiple ecological interactions, facilitate decision-making, and minimize reliance on purchased, nonrenewable inputs. Alternatively, for farmers in developing countries who are not yet fully reliant on herbicides, transitions toward ecological weed management systems may involve substantial agroecosystem redesign but lack intermediate stages of improved herbicide efficiency and input substitution. None the less, system redesign in developing countries is still likely to emphasize the improved use of multiple tactics, local biological resources, on-farm labor and knowledge, and skills for timely monitoring and decision-making.

Increased efficiency in the use of conventional inputs

Because weed populations are often distributed patchily throughout fields with many areas having low densities (Mortensen, Johnson & Young, 1993; Cardina, Johnson & Sparrow, 1997), increased efficiency in herbicide use can be achieved by treating weed populations only where and when their densities warrant it (Mortensen, Dieleman & Johnson, 1998). Johnson, Mortensen and Martin (1995) mapped weeds in commercial Nebraska maize fields in which pre-emergence herbicides were applied only in a band over crop rows and found that, on average, 71% of the intra-row area was free of broadleaf weeds and 94% free of grass weeds; 30% of the area between crop rows was free of broadleaf species and 72% free of grasses. They concluded that if herbicides were applied only where weeds were present or exceeded a threshold density, large reductions in herbicide use would be possible.

Recent advances in real-time sensing technologies may soon allow spraying weeds with post-emergence herbicides on the scale of individual plants; remote sensing and geographic information systems already allow herbicide applications to be made on the scale of small sections of fields (Hanson, Robert & Bauer, 1995; Mortensen *et al.*, 1995; Mortensen, Dieleman & Johnson, 1998). Backpack sprayers and wick applicators are also suitable for locally targeted use of herbicides. Accurate predictions of the timing and location of weed emergence based on better knowledge of seed bank dynamics may also allow farmers to avoid unneeded herbicide applications. If the costs of monitoring weeds and applying herbicides at specific locations are lower than broadcast, prophylactic applications, then direct cost savings and greater returns will be possible at the farm level.

Substitution of benign inputs

The input substitution approach can involve replacement of conventional herbicides with new synthetic materials (Zoschke, 1994) or microbial products (see Chapter 8) that have shorter residual periods, less mobility, and lower toxicity to humans and other nontarget organisms. Input substitution can also involve partial or complete replacement of herbicides with mechanical controls; a common method is the combination of cultivation between crop rows with herbicide application in narrow bands over crop rows (Buhler, Gunsolus & Ralston, 1992; Eadie *et al.*, 1992; Mt. Pleasant, Burt & Frisch, 1994). By reducing reliance on more toxic materials, the input substitution approach can reduce environmental and health hazards. It can also, in the case of herbicide banding, decrease production costs and increase returns (Mulder & Doll, 1993). To be used most effectively and reliably, the input substitution approach requires monitoring weed populations and their responses to management tactics. Like the improved efficiency approach, it is responsive rather than proactive, and tends to maintain a farmer's dependence on externally derived curative solutions and inputs (MacRae *et al.*, 1990).

Agroecosystem redesign

The agroecosystem redesign approach is characteristic of ecological weed management and involves a shift from linear, one-to-one relationships between target weeds and a particular weed management tactic, to webs of relationships between weeds, multiple weed management tactics, and other farming practices (Liebman & Gallandt, 1997). Emphasis is placed on preventing weed problems and reducing requirements for purchased inputs through better use of ecological factors that stress and kill weeds. Emphasis is also placed on integrating weed management activities with other farming practices that maintain soil productivity and crop health, minimize the impacts of other pests and unfavorable weather, and reduce financial risks (Swanton & Murphy, 1996).

Exner, Thompson & Thompson (1996) have described an Iowa crop and livestock farm that uses an agroecosystem approach to manage weeds successfully. Most of the 125-ha farm is in a five-year rotation sequence (maize–soybean–maize–oat + clover-and-forage-grasses–hay) that challenges weeds with varying patterns of soil disturbance and resource competition. A rye cover crop, planted between the maize and soybean phases of the rotation, is used to allelopathically suppress weed emergence and growth. The ridge tillage system used for maize and soybean production kills weeds within the crop row at planting, but also minimizes soil disturbance and stimulation of weed germination before the crops are sown. A rotary hoe and an inter-row cultivator designed for high-residue conditions are also used to control weeds mechanically. Crops are planted at higher than conventional rates to increase crop competitive ability.

The owner–operators of this farm are willing to apply post-emergence herbicides if other tactics fail to provide sufficient weed control, but generally have not needed to do so. They monitor weeds in their fields, tinker with and improve machinery and other components of their farming system, conduct field day tours, exchange information with other farmers, and participate in collaborative research projects with scientists at Iowa State University and other institutions (Chapter 3) (Harp, 1996; Thompson, Thompson & Thompson, 1998). Experiments conducted on the farm have shown that its weed management system protects maize and soybean from weed competition as effectively as conventional herbicide-based systems (Exner, Thompson & Thompson, 1996; Thompson, Thompson & Thompson, 1998). Production costs on the farm are lower than for conventional farms in the area, but crop yields are as high (National Research Council, 1989, pp. 308–23).

Combinations of ecologically based weed management tactics are also used effectively by certain farmers in the Canadian prairie provinces and the American northern plains states. Matheson *et al.* (1991) and Hilander (1997) described crop and livestock farms in the region that range from 400 to 1100 ha and operate profitably with little or no herbicide use. Insect herbivores and mixed stocking of sheep with cattle are used to suppress perennial weeds in grazing lands. In arable fields, weeds are managed through the use of higher seeding rates, competitive crop varieties, pre- and post-emergence cultivation, and crop rotation. Various annual crops (e.g., wheat, barley, oat, lentil, pea, flax, buckwheat, and sunflower) are grown in sequence or in mixture with perennial or biennial forage legumes (e.g., alfalfa and sweet clover). In some cases, short-duration legumes (e.g., pea and lentil) are included in rotations as green manures. Where possible, fall-sown crops are alternated with spring-sown crops. As discussed in Chapter 7, sequences and mixtures of diverse crops can help to prevent the proliferation of adapted weed species by challenging them with complex sets of stress and mortality factors.

Additional options have been proposed for ecological weed management in the Canadian prairies and American northern plains. Derksen, Blackshaw & Boyetchko (1996) suggested that increased use of residue-conserving tillage techniques in concert with moderate use of herbicides may improve habitat for insects, fungi, and bacteria that attack weed seeds and seedlings. The investigators noted that there is considerable potential to minimize herbicide use in conservation tillage systems through improved use of crop rotations, competitive cultivars, and crop densities and fertilizer placement strategies that enhance crop competitive ability against weeds.

Lightfoot *et al.* (1989) described the use of the agroecosystem redesign approach by a group of farmers and scientists in the Philippines seeking to manage the perennial grass *Imperata cylindrica*. To begin the process, group meetings and individual farm visits were used to facilitate discussions and identify impacts of different cropping systems, soil fertility practices, burning and cultivation regimes, and various socioeconomic issues, such as land tenure, cash requirements, labor availability, and family health. These discussions led to recognition of several interrelated factors that favored the proliferation of *I. cylindrica*: (i) there was minimal soil cover on many fields because of residue burning and intensive tillage; (ii) the lack of soil cover was exacerbated by low fertility due to continuous cropping and erosion; and (iii) *I. cylindrica* seeds blew in from surrounding fallow areas, and germinated easily and grew well in bare soil.

In further discussions, plowing and herbicides were deemed inappropriate options for dealing with *I. cylindrica* because cash reserves were insufficient to hire draught animals or purchase chemical inputs. Labor availability was also identified as a key constraint. Several of the farmers had observed, however, that the weed was effectively suppressed when shaded by vigorously growing vines.

Farmers in this group then visited field experiments and demonstration sites where rapidly growing legume cover crops were being tested by research and extension workers for their ability to improve soil fertility and provide erosion protection. The farmers discussed the weed-suppression potential of the various legume species they saw and chose several (*Pueraria*, *Centrosema*, and *Desmodium* spp.) with which to conduct trials on their own farms. Seven months after the initial discussions, 31 farmers in the group had begun experiments testing legume cover crops for soil improvement and *I. cylindrica* suppression. The farmers, with some assistance from researchers, made measurements of weed and cover crop performance, and visited experiments on other farms. The scientists collected information on labor inputs. All the information collected was discussed by group participants.

The story is incomplete, in that Lightfoot *et al.* (1989) did not describe how introduction of cover crops ultimately affected *I. cylindrica* management. What does emerge, however, is that by participating in the problem-solving process as partners, both the farmers and scientists improved their capacity for decision-making and future interactions. The farmers gained a better idea of how their farming systems functioned, what they wanted, and what their options were for achieving their goals. The scientists were better able to produce and refine a relevant research and extension agenda. These types of interactions are as important in industrial countries as they are in developing countries and are examined more thoroughly in Chapter 3.

Summary

Much of the last half-century of weed science and weed management technology has been directed, implicitly and explicitly, at weed eradication. Is this a realistic possibility in most arable fields, pastures, and rangelands? We believe it is not, given the ecological similarities between weeds and the crops they infest, the dispersal ability of many weed species, and the capacity of most weeds to adapt rapidly to selection pressures imposed upon them (see Chapter 10). Conventional efforts to eradicate weeds with herbicides have reduced weed competition and improved farm labor efficiency, but have also incurred substantial costs, including environmental pollution, threats to human health, and growing dependence on purchased inputs. New approaches are needed to manage weeds effectively while minimizing or eliminating such costs.

This chapter has introduced the concept of weed management systems that are less reliant on herbicides and more reliant on ecological processes, such as resource competition, allelopathy, herbivory, disease, seed and seedling responses to soil disturbance, and succession. We call this concept ecological weed management. Ecological weed management does not exclude the use of herbicides, but minimizes their use through the creation of weed-suppressive agricultural systems. Like conventional management systems, ecological weed management will not eliminate weeds. However, as discussed in later chapters, it has the potential to effectively reduce weed density, limit weed competitive ability, and prevent undesirable shifts in weed community composition, while lowering the use of nonrenewable resources, minimizing threats to human health and the environment, and providing a net benefit to local and national economies.

In contrast to chemically based approaches, ecological weed management has no shortlist of prepackaged, broadly applicable remedies. Instead, it relies on biological information, multiple tactical options, farmer decision-making, and careful adaptation of general design principles to site-specific conditions. Farmers clearly assume a larger burden of responsibility for insuring success when using ecological rather than chemical weed management systems. On the other hand, the benefits of using an ecologically based approach may include more durable weed suppression, cleaner air and water, and less damage to nontarget organisms. Ecological farming may also promote greater farm profits, through cost reductions and price premiums, and healthier rural communities, through practices that are especially well suited to farms that are family-owned and operated.

The development of ecological weed management systems is in its infancy.

As noted in the following chapters, many important research questions remain to be answered. In addition, changes in educational modes and government policies are required if ecological weed management is to be implemented on a broad scale. None the less, we believe that knowledge about ecological weed management and opportunities to apply that knowledge in farm fields are sufficiently advanced to justify increased use and further development of ecological management methods. We hope this book provides the reader with some of the information necessary to proceed.

REFERENCES

Abernathy, J. R., & Bridges, D. C. (1994). Research priority dynamics in weed science. *Weed Technology*, **8**, 396–9.

Akobundu, I. O. (1991). Weeds in human affairs in sub-Saharan Africa: implications for sustainable food production. *Weed Technology*, **5**, 680–90.

Alstrom, S. (1990). *Fundamentals of Weed Management in Hot Climate Peasant Agriculture*. Uppsala, Sweden: Swedish University of Agricultural Sciences.

Andow, D. A. (1988). Management of weeds for insect manipulation in agroecosystems. In *Weed Management in Agroecosystems: Ecological Approaches*, ed. M. A. Altieri & M. Liebman, pp. 265–301. Boca Raton, FL: CRC Press.

Anonymous (1989). Prepare for registration losses. *American Vegetable Grower*, **37**(10), 24–6.

Aspelin, A. L., & Grube, A. H. (1999). *Pesticide Industry Sales and Usage: 1996 and 1997 Market Estimates*. Office of Prevention, Pesticides, and Toxic Substances, publication no. 733–R–99–001. Washington, DC: US Environmental Protection Agency.

Barbash, J. E., Thelin, G. P., Kolpin, D. W., & Gilliom, R. J. (1999). *Distribution of Major Herbicides in Ground Water of the United States*. Water Resources Investigations Report 98–4245. Sacramento, CA: United States Geological Survey.

Bauer, T. A., Mortensen, D. A., Wicks, G. A., Hayden, T. A., & Martin, A. R. (1991). Environmental variability associated with economic thresholds for soybeans. *Weed Science*, **39**, 564–9.

Beets, W. C. (1990). *Raising and Sustaining Productivity of Smallholder Farming Systems in the Tropics*. Alkmaar, The Netherlands: AgBe Publishing.

Benbrook, C. (1996). *Pest Management at the Crossroads*. Mount Vernon, NY: Consumers Union.

Berkowitz, A. R. (1988). Competition for resources in weed–crop mixtures. In *Weed Management in Agroecosystems: Ecological Approaches*, ed. M. A. Altieri & M. Liebman, pp. 89–120. Boca Raton, FL: CRC Press.

Bhandari, D. C., & Sen, D. N. (1979). Agroecosystem analysis of the Indian arid zone. 1. *Indigofera cordifolia* Heyne ex Roth. as a weed. *Agroecosystems*, **5**, 257–62.

Bird, G. W., Edens, T., Drummond, F., & Groden, E. (1990). Design of pest management systems for sustainable agriculture. In *Sustainable Agriculture in Temperate Zones*, ed. C. A. Francis, L. D. King & C. B. Flora, pp. 55–110. New York: John Wiley.

Blackshaw, R. E. (1993). Downy brome (*Bromus tectorum*) density and relative time of emergence affect interference in winter wheat (*Triticum aestivum*). *Weed Science*, **41**, 551–6.

Blair, A. (1990). Herbicides and non-Hodgkin's lymphoma: new evidence from a study of Saskatchewan farmers. *Journal of the National Cancer Institute*, **82**, 544–5.

Bottrell, D. G. (1979). *Integrated Pest Management*. Washington, DC: Council on Environmental Quality.

Bottrell, D. G., & Weil, R. R. (1995). Protecting crops and the environment: striving for durability. In *Agriculture and Environment: Bridging Food Production and Environmental Protection in Developing Countries*, ed. A. S. R. Juo & R. D. Freed, pp. 55–74. Madison, WI: American Society of Agronomy.

Bridges, D. C., & Anderson, R. L. (1992). Crop losses due to weeds in the United States – by state. In *Crop Losses Due to Weeds in the United States, 1992*, ed. D. C. Bridges, pp. 1–60. Champaign, IL: Weed Science Society of America.

Buhler, D. D., Gunsolus, J. L., & Ralston, D. R. (1992). Integrated weed management techniques to reduce herbicide inputs in soybean. *Agronomy Journal*, **84**, 973–8.

Burnet, M. W. M., Hart, Q., Holtum, J. A. M., & Powles, S. B. (1994). Resistance to nine herbicide classes in a population of rigid ryegrass (*Lolium rigidum*). *Weed Science*, **42**, 369–77.

Burros, M. (1997). US to subject organic foods, long ignored, to federal rules. *The New York Times*, 15 December 1997, A1 & A14.

Bye, R. A. Jr. (1981). Quelites – ethnoecology of edible greens – past, present, and future. *Journal of Ethnobiology*, **1**, 109–23.

Cardina, J., Johnson, G. A., & Sparrow, D. H. (1997). The nature and consequence of weed spatial distribution. *Weed Science*, **45**, 364–73.

Chacon, J. C., & Gliessman, S. R. (1982). Use of the "non-weed" concept in traditional tropical agroecosystems of south-eastern Mexico. *Agroecosystems*, **8**, 1–11.

Chandler, J. M. (1991). Estimated losses of crops to weeds. In *CRC Handbook of Pest Management in Agriculture*, vol. 1, ed. D. Pimentel, pp. 53–65. Boca Raton, FL: CRC Press.

Chikoye, D., Weise, S. F., & Swanton, C. J. (1995). Influence of common ragweed (*Ambrosia artemisiifolia*) time of emergence and density on white bean (*Phaseolus vulgaris*). *Weed Science*, **43**, 375–80.

Colborn, T., vom Saal, F. S., & Soto, A. M. (1993). Developmental effects of endocrine-disrupting chemicals in wildlife and humans. *Environmental Health Perspectives*, **101**, 378–84.

Cousens, R. (1985). A simple model relating crop yield loss to weed density. *Annals of Applied Biology*, **107**, 239–52.

Cousens, R., Brain, P., O'Donovan, J. T., & O'Sullivan, P. A. (1987). The use of biologically realistic equations to describe the effects of weed density and relative time of emergence on crop yield. *Weed Science*, **35**, 720–5.

Datta, S. C., & Banerjee, A. K. (1979). Useful weeds of West Bengal rice fields. *Economic Botany*, **32**, 297–310.

Derksen, D. A., Blackshaw, R. E., & Boyetchko, S. M. (1996). Sustainability, conservation tillage and weeds in Canada. *Canadian Journal of Plant Science*, **76**, 651–9.

Dinham, B. (1993). *The Pesticide Hazard: A Global Health and Environmental Audit*. London: Zed Books.

Dobbs, T. L., & Pourier, J. L. (1999). Organic price premiums for grains and beans remain high. *Economics Commentator*, no. 397, 5 April 1999. Brookings, SD: South Dakota State University.

Dobbs, T. L., & Smolik, J. D. (1996). Productivity and profitability of conventional and alternative farming systems: a long-term on-farm paired comparison. *Journal of Sustainable Agriculture*, **9**, 63–79.

Duffy, M., & Vontalge, A. (1998 and previous years). *Estimated Costs of Crop Production in Iowa*, Bulletin no. Fm-1712. Ames, IA: Iowa State University Extension.

Duke, S. O. (1999). Weed management: implications of herbicide resistant crops. In *Ecological Effects of Pest Resistance Genes in Managed Ecosystems*, ed. P. L. Traynor & J. H. Westwood, pp. 21–5. Blacksburg, VA: Information Systems for Biotechnology, Virginia Polytechnic Institute and State University.

Eadie, A. G., Swanton, C. J., Shaw, J. E., & Anderson, G. W. (1992). Banded herbicide applications and cultivation in a modified no-till corn (*Zea mays*) system. *Weed Technology*, **6**, 535–42.

Exner, D. N., Thompson, R. L., & Thompson, S. N. (1996). Practical experience and on-farm research with weed management in an Iowa ridge tillage-based system. *Journal of Production Agriculture*, **9**, 496–500.

Ferber, D. (1999). GM crops in the cross hairs. *Science*, **286**, 1662–6.

Fletcher, J. S., Pfleeger, T. G., & Ratsch, H. C. (1993). Potential environmental risks associated with the new sulfonylurea herbicides. *Environmental Science and Technology*, **27**, 2250–2.

Fletcher, J. S., Pfleeger, T. G., Ratsch, H. C., & Hayes, R. (1996). Potential impact of low levels of chlorsulfuron and other herbicides on growth and yield of nontarget plants. *Environmental Toxicology and Chemistry*, **15**, 1189–96.

Franco, J. (1989). An analysis of the California market for organically grown produce. *American Journal of Alternative Agriculture*, **4**, 22–7.

Freemark, K., & Boutin, C. (1995). Impacts of agricultural herbicide use on terrestrial wildlife in temperate landscapes: a review with special reference to North America. *Agriculture, Ecosystems and Environment*, **52**, 67–91.

Garry, V. F., Schreinemachers, D., Harkins, M. E., & Griffith, J. (1996). Pesticide appliers, biocides, and birth defects in rural Minnesota. *Environmental Health Perspectives*, **104**, 394–9.

Geier, B. (1998). The organic market: opportunities and challenges. *ILEIA Newsletter*, **14**(4), 6–7. Leusden, The Netherlands: Centre for Research and Information on Low-External-Input and Sustainable Agriculture.

Georghiou, G. P. (1986). The magnitude of the resistance problem. In *Pesticide Resistance: Strategies and Tactics for Management*, ed. E. H. Glass, pp. 14–43. Washington, DC: National Academy Press.

Gianessi, L. P., & Puffer, C. M. (1991). *Herbicide Use in the United States*. Washington, DC: Resources for the Future.

Gianessi, L. P., & Puffer, C. M. (1993). Herbicide-resistant weeds may threaten wheat production in India. *Resources*, **111**, 17–22. Washington, DC: Resources for the Future.

Gill, G. S. (1995). Development of herbicide resistance in annual ryegrass populations (*Lolium rigidum* Gaud.) in the cropping belt of Western Australia. *Australian Journal of Experimental Agriculture*, **35**, 67–72.

Goolsby, D. A., Battaglin, W. A., & Thurman, E. M. (1993). *Occurrence and Transport of*

Agricultural Chemicals in the Mississippi River Basin, July through August 1993, US Geological Survey Circular 1120–C. Washington, DC: US Government Printing Office.

Goolsby, D. A., Coupe, R. C., & Markovchick, D. J. (1991). *Distribution of Selected Herbicides and Nitrate in the Mississippi River and its Major Tributaries, April through June 1991*, Water Resources Investigations Report 91–4163. Denver, CO: US Geological Survey.

Gressel, J., & Segel, L. A. (1990). Modelling the effectiveness of herbicide rotations and mixtures as strategies to delay or preclude resistance. *Weed Technology*, **4**, 186–98.

Gunsolus, J. L., & Buhler, D. D. (1999). A risk management perspective on integrated weed management. *Journal of Crop Production*, **2**, 167–87.

Hallberg, G. R. (1989). Pesticide pollution of groundwater in the humid United States. *Agriculture, Ecosystems and Environment*, **26**, 299–367.

Hanson, L. D., Robert, P. C., & Bauer, M. (1995). Mapping wild oats infestations using digital imagery for site-specific management. In *Site-Specific Management for Agricultural Systems*, ed. P. Robert, R. H. Rust & W. E. Larson, pp. 495–503. Madison, WI: American Society of Agronomy.

Hardell, L., & Sandstrom, A. (1979). Case-control study: soft-tissue sarcomas and exposure to phenoxyacetic acids or chlorophenols. *British Journal of Cancer*, **39**, 711–17.

Hardell, L., Eriksson, M., Lenner, P., & Lundgren, E. (1981). Malignant lymphoma and exposure to chemicals, especially organic solvents, chlorophenols, and phenoxy acids: a case-control study. *British Journal of Cancer*, **43**, 169–76.

Harlan, J. R. (1992). *Crops and Man*, 2nd edn. Madison, WI: American Society of Agronomy.

Harp, A. (1996). Iowa, USA: an effective partnership between the Practical Farmers of Iowa and Iowa State University. In *New Partnerships for Sustainable Agriculture*, ed. L. A. Thrupp, pp. 127–36. Washington, DC: World Resources Institute.

Hartzler, R. G., Van Kooten, B. D., Stoltenberg, D. E., Hall, E. M., & Fawcett, R. S. (1993). On-farm evaluation of mechanical and chemical weed management practices in corn (*Zea mays*). *Weed Technology*, **7**, 1001–4.

Heap, I. (1999). *International Survey of Herbicide Resistant Weeds*. Internet: http://www.weedscience.com.

Hilander, S. K. (ed.) (1997). *Weeds as Teachers: Many Little Hammers Weed Management*. Helena, MT: Alternative Energy Resources Organization.

Hilje, L., Castillo, L. E., Thrupp, L., & Wesselling, I. (1992). *El Uso de los Plaguicidas en Costa Rica*. San José, Costa Rica: Heliconia, Editorial Universidad Estatal a Distancia.

Hoar, S. K., Blair, A., Holmes, F. F., Boysen, C. D., Robel, R. J., Hoover, R., & Fraumeni, J. F. (1986). Agricultural herbicide use and risk of lymphoma and soft-tissue sarcoma. *Journal of the American Medical Association*, **256**, 1141–7.

Holt, J. S. (1992). History of identification of herbicide-resistant weeds. *Weed Technology*, **6**, 615–20.

Holt, J. S., & LeBaron, H. M. (1990). Significance and distribution of herbicide resistance. *Weed Technology*, **4**, 141–9.

Ibrahim, M. A., Bond, G. G., Burke, T. A., Cole, P., Dost, F. N., Enterline, P. E., Gough, M., Greenberg, R. S., Halperin, W. E., McConnell, E., Munro, I. C., Swenberg, J. A., Zahm, S. H. & Graham, J. D. (1991). Weight of the evidence on the human carcinogenicity of 2,4-D. *Experimental Health Perspectives*, **96**, 213–22.

Jeyaratnum, J. (1990). Acute pesticide poisoning: a major global health problem. *World Health Statistics Quarterly*, **43**, 139–44.

Johnson, G. A., Mortensen, D. A., & Martin, A. R. (1995). A simulation of herbicide use based on weed spatial distribution. *Weed Research*, **35**, 197–205.

Johnson, R. R. (1994). Influence of no-till on soybean cultural practices. *Journal of Production Agriculture*, **7**, 43–9.

Jutsum, A. R. (1988). Commercial application of biological control: status and prospects. *Philosophical Transactions of the Royal Society of London, Series B*, **318**, 357–73.

Klonsky, K., & Livingston, P. (1994). Alternative systems aim to reduce inputs, maintain profits. *California Agriculture*, **48**, 34–42.

Knezevic, S. Z., Weise, S. F., & Swanton, C. J. (1994). Interference of redroot pigweed (*Amaranthus retroflexus*) in corn (*Zea mays*). *Weed Science*, **42**, 568–73.

Kolpin, D. W., Thurman, E. M., & Goolsby, D. A. (1996). Occurrence of selected pesticides and their metabolites in near-surface aquifers of the midwestern United States. *Environmental Science and Technology*, **30**, 335–40.

Kropff, M. J., Lotz, L. A. P., & Weaver, S. E. (1993). Practical applications. In *Modelling Crop–Weed Interactions*, ed. M. J. Kropff & H. H. van Laar, pp. 149–67. Wallingford, UK: CAB International.

Larson, S. J., Gilliom, R. J., & Capel, P. D. (1999). *Pesticides in Streams of the United States: Initial Results from the National Water Quality Assessment Program*, Water Resources Investigations Report 98–4222. Sacramento, CA: US Geological Survey.

Leistra, M., & Boesten, J. J. T. I. (1989). Pesticide contamination of groundwater in western Europe. *Agriculture, Ecosystems and Environment*, **26**, 369–89.

Leng, M. L. (1991). Consequences of reregistration on existing pesticides. In *Regulation of Agrochemicals: A Driving Force in Their Evolution*, ed. G. J. Marco, R. M. Hollingworth & J. R. Plimmer, pp. 27–44. Washington, DC: American Chemical Society.

Lewis, W. J., van Lenteren, J. C., Phatak, S. C., & Tumlinson, J. H. (1997). A total system approach to sustainable pest management. *Proceedings of the National Academy of Sciences, USA*, **94**, 12243–8.

Liebman, M., & Gallandt, E. R. (1997). Many little hammers: ecological management of crop–weed interactions. In *Agricultural Ecology*, ed. L. E. Jackson, pp. 291–343. San Diego, CA: Academic Press.

Lightfoot, C., De Guia O. Jr., Aliman, A., & Ocado, F. (1989). Systems diagrams to help farmers decide in on-farm research. In *Farmer First: Farmer Innovation and Agricultural Research*, ed. R. Chambers, A. Pacey & L. A. Thrupp, pp. 93–100. London: Intermediate Technology Publications.

Lindquist, J. L., Mortensen, D. A., Clay, S. A., Schemnk, R., Kells, J. J., Howatt, K., & Westra, P. (1996). Stability of coefficients in the corn yield loss–velvetleaf density relationship across the north central US. *Weed Science*, **44**, 309–13.

MacRae, R. J., Hill, S. B., Henning, J., & Bentley, A. J. (1990). Policies, programs and regulations to support the transition to sustainable agriculture in Canada. *American Journal of Alternative Agriculture*, **5**, 76–92.

Mapes, C., Basurto, F., & Bye, R. (1997). Ethnobotany of quintonil: knowledge, use, and management of edible greens *Amaranthus* spp. (Amaranthaceae) in the Sierra Norte del Puebla, Mexico. *Economic Botany*, **51**, 293–306.

Matheson, N., Rusmore, B., Sims, J. R., Spengler, M., & Michalson, E. L. (1991). *Cereal–Legume Cropping Systems: Nine Farm Case Studies in the Dryland Northern Plains, Canadian Prairies, and Intermountain Northwest.* Helena, MT: Alternative Energy Resources Organization.

Miranowski, J., & Carlson, G. (1993). Agricultural resource economics: an overview. In *Agricultural and Environmental Resource Economics*, ed. G. Carlson, D. Zilberman & J. Miranowski, pp. 3–27. New York: Oxford University Press.

Mortensen, D. A., & Coble, H. D. (1989). The influence of soil water content on common cocklebur (*Xanthium stramonium*) interference in soybeans (*Glycine max*). *Weed Science*, 37, 76–83.

Mortensen, D. A., Dieleman, J. A. & Johnson, G. A. (1998). Weed spatial variation and weed management. In *Integrated Weed and Soil Management*, ed. J. L. Hatfield, D. D. Buhler & B. A. Stewart, pp. 293–309. Chelsea, MI: Ann Arbor Press.

Mortensen, D. A., Johnson, G. A., & Young, L. J. (1993). Weed distribution in agricultural fields. In *Soil-Specific Crop Management*, ed. P. Robert, R. H. Rust & W. E. Larson, pp. 113–24. Madison, WI: American Society of Agronomy.

Mortensen, D. A., Johnson, G. A., Wyse, D. Y., & Martin, A. R. (1995). Managing spatially variable weed populations. In *Site-Specific Management for Agricultural Systems*, ed. P. Robert, R. H. Rust & W. E. Larson, pp. 397–415. Madison, WI: American Society of Agronomy.

Mt. Pleasant, J., Burt, R. F., & Frisch, J. C. (1994). Integrating mechanical and chemical weed management in corn (*Zea mays*). *Weed Technology*, 8, 217–23.

Mulder, T. A., & Doll, J. D. (1993). Integrating reduced herbicide use with mechanical weeding in corn (*Zea mays*). *Weed Technology*, 7, 382–9.

National Research Council (1989). *Alternative Agriculture.* Washington, DC: National Academy Press.

Naylor, R. (1994). Herbicide use in Asian rice production. *World Development*, 22, 55–70.

Nelson, H., & Jones, R. D. (1994). Potential regulatory problems associated with atrazine, cyanazine, and alachlor in surface water source drinking water. *Weed Technology*, 8, 852–61.

Nentwig, W., Frank, T., & Lethmayer, C. (1998). Sown weed strips: artificial ecological compensation areas as an important tool in conservation biological control. In *Conservation Biological Control*, ed. P. Barbosa, pp. 133–53. San Diego, CA: Academic Press.

Norris, R. F. (1992). Case history for weed competition/population ecology: barnyardgrass (*Echinochloa crus-galli*) in sugarbeets (*Beta vulgaris*). *Weed Technology*, 6, 220–7.

Organic Farming Research Foundation (1998). *Third Biennial National Organic Farmers' Survey.* Santa Cruz, CA: Organic Farming Research Foundation.

Owen, M. D. K. (1997). North American developments in herbicide tolerant crops. In *Proceedings of the 1997 Brighton Crop Protection Conference: Weeds*, pp. 955–63. Farnham, UK: British Crop Protection Council.

Padel, S., & Lampkin, N. H. (1994). Farm-level performance of organic farming systems: an overview. In *The Economics of Organic Farming: An International Perspective*, ed. N. H. Lampkin & S. Padel, pp. 201–19. Wallingford, UK: CAB International.

Parker, C., & Fryer, J. D. (1975). Weed control problems causing major reductions in world food supplies. *FAO Plant Protection Bulletin*, 23(3/4), 83–95.

Pimentel, D., Acquay, H., Biltonen, M., Rice, P., Silva, M., Nelson, J., Lipner, V., Giordano, S., Horowitz, A. & D'Amore, M. (1992). Environmental and economic costs of pesticide use. *Bioscience*, **42**, 750–60.

Pingali, P. L., & Gerpacio, R. V. (1997). *Towards Reduced Pesticide Use for Cereal Crops in Asia*. Economics Working Paper 97–04. Mexico, DF: Centro Internacional de Mejoramiento de Maíz y Trigo.

Posner, J. L., & Crawford, E. W. (1991). An agroeconomic analysis from a farming systems research perspective: weed control in rainfed lowland rice in Senegal. *Experimental Agriculture*, **27**, 231–41.

Powles, S. B., Lorraine-Colwill, D. F., Dellow, J. J., & Preston, C. (1998). Evolved resistance to glyphosate in rigid ryegrass (*Lolium rigidum*) in Australia. *Weed Science*, **46**, 604–7.

Powles, S. B., Preston, C., Bryan, I. B., & Jutsum, A. R. (1997). Herbicide resistance: impacts and management. *Advances in Agronomy*, **58**, 57–93.

Pretty, J. (1995). *Regenerating Agriculture: Policies and Practice for Sustainability and Self-Reliance*. Washington, DC: Joseph Henry Press.

Repetto, R. (1985). *Paying the Price: Pesticide Subsidies in Developing Countries*. Washington, DC: World Resources Institute.

Repetto, R., & Baliga, S. S. (1996). *Pesticides and the Immune System: The Public Health Risks*. Washington, DC: World Resources Institute.

Richards, R. P., Baker, D. B., Creamer, N. L., Kramer, J. W., Ewing, D. E., Merryfield, B. J., & Wallrabstein, L. K. (1996). Well water quality, well vulnerability, and agricultural contamination in the midwestern United States. *Journal of Environmental Quality*, **25**, 389–402.

Seefeldt, S., Zemetra, R., Yound, F., & Jones, S. (1998). Production of herbicide-resistant jointed goatgrass (*Aegilops cylindrica*) × wheat (*Triticum aestivum*) hybrids in the field by natural hybridization. *Weed Science*, **46**, 632–4.

Sheley, R. L., Svejcar, T. J., & Maxwell, B. D. (1996). A theoretical framework for developing successional weed management strategies on rangeland. *Weed Technology*, **10**, 766–73.

Smith, A. H., & Bates, M. N. (1989). Epidemiological studies of cancer and pesticide exposure. In *Carcinogenicity and Pesticides: Principles, Issues, and Relationships*, ed. N. N. Ragsdale & R. E. Menzer, pp. 207–22. Washington, DC: American Chemical Society.

Smolik, J. D., Dobbs, T. L., & Rickerl, D. H. (1995). The relative sustainability of alternative, conventional, and reduced-till farming systems. *American Journal of Alternative Agriculture*, **10**, 25–35.

Smolik, J. D., Dobbs, T. L., Rickerl, D. H., Wrage, L. J., Buchenau, G. W., & Machacek, T. A. (1993). *Agronomic, Economic, and Ecological Relationships in Alternative (Organic), Conventional, and Reduced-Till Farming Systems*, Bulletin 718. Brookings, SD: South Dakota Agricultural Experiment Station.

Snow, A., & Morán-Palma, P. (1997). Commercialization of transgenic plants: potential ecological risks. *Bioscience*, **47**, 86–96.

Sotherton, N. W., Boatman, N. D., & Rands, M. R. W. (1989). The "conservation headland" experiment in cereal ecosystems. *The Entomologist*, **108**, 135–43.

Sotherton, N. W., Rands, M. R. W., & Moreby, S. J. (1985). Comparison of herbicide treated and untreated headlands on the survival of game and wildlife. *1985 British Crop Protection Conference: Weeds*, **3**, 991–8.

Staver, C., Aguilar, A., Aguilar, V., & Somarriba, S. (1995). Selective weeding. *ILEIA Newsletter*, **11**(3), 22–3. Leusden, The Netherlands: Centre for Research and Information on Low-External-Input and Sustainable Agriculture.

Stokes, C. S., & Brace, K. D. (1988). Agricultural chemical use and cancer mortality in selected rural counties in the USA. *Journal of Rural Studies*, **4**, 239–47.

Stone, J. F., Eichner, M. L., Kim, C., & Koehler, K. (1988). Relationship between clothing and pesticide poisoning: symptoms among Iowa farmers. *Journal of Environmental Health*, **50**, 210–15.

Swanton, C. J., & Murphy, S. D. (1996). Weed science beyond the weeds: the role of integrated weed management (IWM) in agroecosystem health. *Weed Science*, **44**, 437–45.

Swezey, S. L., Rider, J., Werner, M. R., Buchanan, M., Allison, J., & Gliessman, S. R. (1994). Granny Smith conversions to organic show early success. *California Agriculture*, **48**(6), 36–44.

Tate, W. B. (1994). The development of the organic industry and market: an international perspective. In *The Economics of Organic Farming: An International Perspective*, ed. N. H. Lampkin & S. Padel, pp. 11–25. Wallingford, UK: CAB International.

Thompson, R., Thompson, S., & Thompson, R. (1998). *Alternatives in Agriculture: 1998 Report*. Boone, IA: Thompson On-Farm Research.

Thurman, E. M., Goolsby, D. A., Meyer, M. T., & Kolpin, D. W. (1991). Herbicides in surface waters of the midwestern United States: the effect of the spring flush. *Environmental Science and Technology*, **25**, 1794–6.

United States Department of Agriculture (1999a). *Agricultural Chemical Usage: 1998 Field Crops Summary*, Publication Ag Ch 1–99. Washington, DC: US Department of Agriculture National Agricultural Statistics Service.

United States Department of Agriculture (1999b). *1997 Census of Agriculture*. Washington, DC: US Department of Agriculture National Agricultural Statistics Service.

United States Department of Agriculture. (1999c). *Acreage*, Publication Cr Pr 2–5. Washington, DC: US Department of Agriculture National Agricultural Statistics Service.

United States Environmental Protection Agency (1996). *Drinking Water Regulations and Health Advisories*. Washington, DC: US Environmental Protection Agency, Office of Water.

United States Environmental Protection Agency (1999). *Pesticidal Chemicals Classified as Known, Probable or Possible Human Carcinogens*. Washington, DC: US Environmental Protection Agency, Office of Pesticide Programs.

United States Geological Survey (1999). *The Quality of Our Nation's Waters: Nutrients and Pesticides*, Circular 1225. Washington, DC: US Department of Interior and US Geological Survey.

Warwick, S. I. (1991). Herbicide resistance in weedy plants: physiology and population biology. *Annual Review of Ecology and Systematics*, **22**, 95–114.

Weaver, S. E., Smits, N., & Tan, C. S. (1987). Estimating yield losses of tomato (*Lycopersicon esculentum*) caused by nightshade (*Solanum* spp.) interference. *Weed Science*, **35**, 163–8.

Weil, R. R. (1982). Maize–weed competition and soil erosion in unweeded maize. *Tropical Agriculture*, **59**, 207–13.

Welsh, R. (1999). *The Economics of Organic Grain and Soybean Production in the Midwestern United States.* Greenbelt, MD: Henry A. Wallace Institute for Alternative Agriculture.

Wigle, D. T., Semenciw, R. M., Wilkins, K., Riedel, D., Ritter, L., Morrison, H. I., & Mao, Y. (1990). Mortality study of Canadian male farm operators: non-Hodgkin's lymphoma mortality and agricultural practices in Saskatchewan. *Journal of the National Cancer Institute*, **82**, 575–82.

World Health Organization (1990) *Public Health Impacts of Pesticides Used in Agriculture.* Geneva, Switzerland: World Health Organization.

Wrubel, R. P., & Gressel, J. (1994). Are herbicide mixtures useful for delaying the rapid evolution of resistance? A case study. *Weed Technology*, **8**, 635–48.

Wyse, D. L. (1992). Future of weed science research. *Weed Technology*, **6**, 162–5.

Zoschke, A. (1994). Toward reduced herbicide rates and adapted weed management. *Weed Technology*, **8**, 376–86.

CHARLES L. MOHLER

2

Weed life history: identifying vulnerabilities

Weeds from an ecological perspective

Weeds share certain ecological characteristics that distinguish them from other plants. Specifically, *weeds are plants that are especially successful at colonizing disturbed, but potentially productive, sites and at maintaining their abundance under conditions of repeated disturbance.* That is, weeds are the plants that thrive where soil and climate are favorable to plant growth, but disturbance frequently reduces competition among plants to low levels. Unlike previous conceptions of weediness (Baker, 1965; Harlan & de Wet, 1965; Buchholtz et al., 1967), this ecologically based definition lacks reference to humans and human disturbance. The species people refer to as weeds mostly existed prior to human disturbance, and the repertoire of behaviors that makes them invasive and persistent in human-dominated habitats largely evolved independently of human society. Nevertheless, as discussed in Chapter 10, human activities selectively modify weed characteristics such that weeds are becoming better adapted to human disturbance regimes.

The subcategory of weeds dealt with in this book consists of the weeds of agriculture – specifically, the plants that colonize and increase in the disturbances created by farming. These are sometimes termed *agrestal* weeds, as distinguished from the *ruderal* weeds of roadsides, waste piles, and other non-agricultural disturbances (Baker, 1965). Agricultural weeds share certain life-history characteristics that adapt them for life on farms (Table 2.1). The thesis of this chapter is that *understanding life-history characteristics provides insights into how weed management practices work and how they can be improved.* In particular, *differences between weeds and crops in germination characteristics, seed size, growth rate, and susceptibility of different life stages to stress provide weed management options.*

Relative to most ecosystems, agricultural fields are not stressful environ-

Table 2.1. *Ecological characteristics of agricultural weeds and crops*

Character	Weed	Crop
Maximum relative growth rate (g g^{-1}d^{-1})	Very high	High
Early growth rate (g d^{-1})	Low	High
Shade tolerance	Low	Low
Tolerance of nutrient stress	Low	Low
Nutrient uptake rate	Very high	High
Seed size	Mostly small	Mostly large
Size at establishment	Mostly small	Mostly large
Reproductive rate	High	Varies with crop
Seasonal innate seed dormancy	Frequent	Very rare
Germination in response to tillage related cues[a]	Common	Rare
Seed longevity in soil	Often long	Usually short
Dispersal	Mostly by humans	By humans

Notes:
[a] Light, fluctuating temperature, nitrate.

ments for plants: to get high productivity from crops, the grower reduces stress through seedbed preparation, fertilization, irrigation, and artificial drainage. Moreover, in annual cropping systems, resources greatly exceed the needs of both crop and weeds for several weeks after the crop is planted, and during this period competition has a negligible effect on seedling establishment. The species that do well in these conditions, namely agricultural weeds, prosper because they have very high maximum relative growth rates (see section "Vegetative growth and crop–weed competition" below). This allows them to grow large rapidly and occupy space before resources are monopolized by crops and any ruderal species that happen to be present. The very high relative growth rates of agricultural weeds are coincident with inefficient resource use. Weeds are more susceptible to the negative effects of shade than are species commonly found in less disturbed conditions (Fenner, 1978). Weeds typically accumulate higher concentrations of mineral nutrients than crop species when nutrients are plentiful, but often suffer greater relative declines in growth than crops when nutrients are in short supply (Vengris, Colby & Drake, 1955; Alkämper, 1976). Inherent physiological trade-offs appear to prevent plants from fully adapting to both high and low light levels (Givnish, 1988), or to both high and low nutrient availability (Schläpfer & Ryser, 1996). Agricultural weeds are at one extreme of these adaptive continua.

Because agricultural weeds establish primarily in conditions of low competition, only minimal provisioning of offspring by the mother plant is required. Hence, weed seeds usually weigh only a few milligrams or less (Table

2.2, below). Small seed size allows for production of many seeds by mature individuals. This facilitates colonization of new sites. Moreover, a high reproductive rate is necessary to compensate for high mortality caused by (i) repeated disturbance during the growing season, and (ii) the environmental unpredictability created by crop rotation and variation in weather. Weeds avoid some unpredictability via dormancy mechanisms and germination cues that allow synchronization of establishment with favorable conditions. They also spread risks across years with different environmental conditions by means of perennation and seed banks. Although all these characteristics allow agricultural weeds to prosper in farm fields, they also provide opportunities for weed management.

Each of the properties of agricultural weeds mentioned above is discussed further in the following sections, with a focus on how the nature of weeds indicates their vulnerability to control. The following discussion focuses on broad patterns and generalities regarding various sorts of weeds. Naturally, exceptions exist for each of these generalizations. To avoid undue digression, however, these exceptions are usually not discussed explicitly. Hopefully, understanding of the usual properties shared by many weed species will also clarify the functional significance of the exceptional properties of unusual species.

The life history of weeds

Weeds progress through a series of stages in the life cycle: germination, establishment, growth, reproduction, dispersal, and dormancy. Management tactics generally apply to a particular stage. Moreover, differences in the behavior of species in each stage lead to differences in susceptibility to control by a particular approach. Thus, life history is an organizing principle for the integration of weed management tactics.

Although agricultural weeds commonly share many ecological attributes, they are by no means a homogeneous group of species. In particular, four broad categories of life history can be distinguished (Table 2.2). Annual weeds grow from germination to reproduction within a single growing season. With few exceptions, their seeds persist in the soil for at least a few years and in many cases for decades.

Stationary perennials live from two to several growing seasons (biennials are included in this group). Because they generally do not rejuvenate via vegetative reproduction, they eventually die. As with the annuals, their seeds usually persist in the soil for at least a few years, and often much longer. Under favorable conditions some of these species may set seed the year of

Table 2.2. Four types of weed life-history strategies

Character	Annuals	Stationary perennials	Wandering perennials	Woody perennials
Vegetative life span	<1 year	2 to a few years	Long, indefinite	Long
Vegetative propagation	No	Accidental	Yes	Some species
Usual seed persistence	Years to decades	Years to decades	A few years	Months to years
Energy allocated to seed production	High	Medium high	Medium low	Low
Seed size[a]	Mostly small	Mostly small	Mostly small	Mostly large
Usual mode of establishment	Seeds	Seeds	Vegetative propagules	Seeds
Main dispersal modes	With soil, manure	With soil, wind, feces, crop seed	With soil	Birds, wind
Position in succession	Year 1 (2 in gaps)	Year 1 to 5 (10)	After year 1	Middle
Taxonomy	Monocot and dicot	Mostly dicot	Monocot & dicot	Mostly dicot
Crop types	Annual	Forage, annual	All	Orchard, pasture, swidden, no-till
Examples	*Chenopodium album*	*Rumex crispus*	*Imperata cylindrica*	*Lantana camara*
	Setaria faberi	*Poa annua*[b]	*Convolvulus arvensis*	*Toxicodendron radicans*

Notes:

[a] Seed size ranges and medians from an analysis of 39 annual, 18 stationary perennial, and 16 wandering perennial British weed species of arable land and well-drained grassland reported in Salisbury (1961) were 0.02–35 (1.1), 0.2–3 (1.2) and 0.13–14 (0.7) mg, respectively, indicating no difference in seed size between the three categories.

[b] Because *Poa annua* sets seed during its first season of life, it can behave as an annual in an annual cropping systems. However, if left undisturbed it usually lives at least two seasons, and often sets more seed the second year (Law, Bradshaw & Putwain, 1977).

establishment, but often they require longer to mature. Most stationary perennials are broadleaf species. Perennial bunchgrasses are intermediate between stationary and wandering perennials in that most can rejuvenate indefinitely, but have limited capacity for spreading vegetatively.

Because wandering perennials reproduce by vegetative spread and fragmentation, the life span of a genetic individual is indefinite and potentially very long. Although most of these species produce seeds, most reproduction is by vegetative propagation. Seeds of many wandering perennials persist for more than one year, but relatively few have great seed longevity; consequently, wandering perennials are rarely well represented in the seed bank.

Woody weeds are perennials that develop persistent shoot structures. Although some species spread clonally, most reproduce primarily by seed. Their seeds are often relatively large, short-lived, and dispersed by wind or birds. As explained later, they are problems in long-lived crops like orchards and permanent pastures, and are increasingly problematic in no-till planted annual crops (J. Cardina, personal communication). Woody weeds with a vining growth habit (lianas) are often the most difficult to control because they can sprawl laterally and can rapidly reach an orchard canopy by using the crop trees for support.

The contrasting life-history characters of the four groups express ecological rather than physiological trade-offs. For example, propagation by rhizomes is probably not physiologically related to lack of persistence in the seed bank. Basically, adaptation to different stages of ecological succession has grouped characteristics into suites, thereby forming four ecologically distinct types of weed species.

Following a severe disturbance like tillage, annuals predominate because they can survive the disturbance event in a physiologically dormant state as seeds. The stationary perennials are similarly tied to establishment shortly after tillage. However, because they persist in a vegetative state for a longer period, their allocation of resources to roots is greater, and consequently, their seedling growth rate tends to be lower. Thus, in the first year after disturbance, annuals often predominate even if stationary perennials are abundant in the seed bank. However, because stationary perennials start growth with greater reserves during the second season of life compared with newly germinated annuals they are better able to compete with established perennial crops after the first year. Consequently, they are particularly common in hay fields.

In pre-agricultural landscapes, wandering perennials were probably found primarily after the first year of regrowth in fertile, disturbed locations like areas where animals congregated or flood deposited soil along streams. Today, wandering perennial agricultural weeds like *Elytrigia repens* and *Cirsium arvense* are commonly found in well-vegetated, abandoned fields, roadsides, and even disturbed forest. To spread under highly competitive conditions in closed vegetation, they were selected to substantially provision vegetative propagules. Seeds probably served primarily for the risky enterprise of long-distance dispersal.

The advent of tillage greatly changed conditions of life for wandering perennials. Tillage separates daughter plants from the parent and spreads them within and between fields. Simultaneously, tillage removes the competing vegetation. This puts spreading perennials in the advantageous position of having well-provisioned propagules establishing with relatively little competitive pressure. Consequently, many of the world's worst weeds are wandering perennials (Holm *et al.*, 1977). Essentially, these species have characteristics that evolved in response to conditions quite different from present-day agriculture, but they were fortuitously pre-adapted to thrive under moderate tillage. However, deep and frequently repeated tillage is often detrimental to these species (Chapter 4).

Woody perennial weeds are primarily problems in orchards and pastures. Although they help restore soil and eliminate pests and disease during the regenerative phase of shifting agricultural systems (swidden), they can also reduce crop productivity during the cropping phase (Staver, 1991). Woody perennials are poorly adapted to cropping systems with annual tillage for two reasons. First, only a few of these species form persistent seed banks, and consequently, synchronizing establishment with an annual crop is rarely possible. Second, wood is an energetically expensive way to hold up a plant relative to fiber and turgor pressure. Consequently, woody plants grow more slowly than herbs when young (Grime & Hunt, 1975), and as young plants they are rarely competitive with herbaceous annual crops. Because tillage prevents woody weeds from surviving more than one year, they are poorly adapted to most annual cropping systems. Nevertheless, their long life span, tall stature, and vigorous resprouting after cutting and browsing make them serious weeds in tree crops and pastures, and a significant nuisance in no-till planted annual crops.

Parasitic weeds constitute an additional life-history category not included in Table 2.2. To the extent that they behave like other vascular plants (e.g., seed production, seed dispersal), the principles discussed in this book apply to them as well as to weeds that affect the crop primarily through competition for resources. To the extent that they behave more like pathogens (e.g., germination in the presence of host roots, source of nutrition), their study is a specialized field beyond the scope of this book. Several recent books treat the

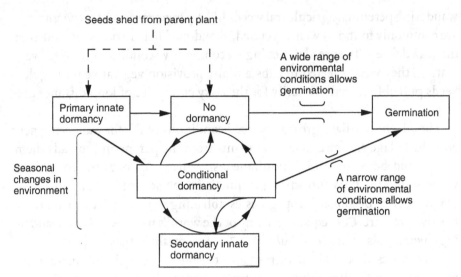

Figure 2.1 Dormancy/germination states of weed seeds. (Redrawn from Egley (1995) based on the concepts of Baskin & Baskin (1985, 1998a).)

ecology and management of parasitic weeds (Musselman, 1987; Pieterse, Verkleij & ter Borg, 1994; Hosmani, 1995).

Dormancy and germination

Seasonal and aseasonal germination

Weed seeds often undergo several changes in dormancy state between seed shed and germination (Figure 2.1) (Baskin & Baskin, 1985). These changes represent an adaptive response to the problem of immobility: a seed has little control over where it lands, but through dormancy response to environment, it can choose when to germinate. When first shed from the parent plant, seeds may lack dormancy and be ready to germinate if environmental conditions are favorable. This is commonly the case for those winter annuals like *Galium aparine* that commonly shed seeds in mid to late summer (Håkansson, 1983). Seeds of these species need to be ready for immediate germination since winter annuals usually do best when they establish early in the autumn.

Alternatively, seeds may have innate dormancy when shed (primary innate dormancy). Innate dormancy may be due to impermeable (hard) seed coats, chemical germination inhibitors in the seed coat or embryo, a cold or heat requirement, or other physiological mechanisms (Povilaitis, 1956; Baskin & Baskin, 1985; Taylorson, 1987). Such mechanisms are found in most weedy species of the temperate zone (Baskin & Baskin, 1988) and provide means for matching the period of germination to weather conditions that are suitable for establishment and growth of the plant. With time, seed coats break down, chemical inhibitors are leached away, and cold or heat requirements are satisfied by winter or summer temperatures, depending on the species. The seed then becomes capable of germination.

Nondormant seeds still may not germinate, however, if environmental conditions are unfavorable. Frequently, seeds remain in a quiescent state until appropriate temperatures, water, light, and other germination cues indicate that conditions are favorable for germination and establishment. For some species, seeds that can not germinate because appropriate conditions are lacking may enter a secondary state of innate dormancy (e.g., *Ambrosia artemisiifolia* – Baskin & Baskin, 1980; *Arabidopsis thaliana* – Baskin & Baskin, 1983). In that state, the seed must undergo another period of chilling, heating, leaching, etc., before germination is again possible. The transition into (and out of) innate dormancy is gradual: the seed passes through a series of conditional dormancy states in which the range of environmental conditions that trigger immediate germination becomes increasingly narrow (Baskin & Baskin, 1998a, pp. 50–64). Seeds of many species may cycle between innate dormancy and non- or conditional dormancy for several years before the environment happens to favor germination in an appropriate season.

Due to these dormancy processes, most weed species germinate at particular times of year. For example, Chepil (1946) observed the timing of emergence of 59 species in Saskatchewan and grouped the species into five categories. The categories of peak emergence were (i) early spring (e.g., *Bromus tectorum*, *Chenopodium album*, *Plantago major*), (ii) mid spring (e.g., *Setaria viridis*, *Cirsium arvense*), (iii) summer (e.g., *Amaranthus retroflexus*, *Capsella bursa-pastoris*, *Portulaca oleracea*), (iv) autumn (e.g., *Sophia multifida*, *Lepidium perfoliatum*), and (v) no consistent period of peak emergence (e.g., *Taraxacum officinale*, *Sinapis arvensis*, *Medicago lupulina*). Other authors have found similar variation in the emergence times of temperate weeds (Figure 2.2) (Lawson, Waister & Stephens, 1974; Roberts & Neilson, 1980; Roberts & Potter, 1980; Håkansson, 1983; Roberts, 1984), although some have also found species with bimodal germination in spring and fall (e.g., *Veronica hederifolia* in Figure 2.2).

Several points can be made regarding these studies. First, most "winter annual" agricultural weeds are only facultatively tied to autumn germination. Only a few have secondary dormancy mechanisms that prevent spring germination (e.g., winter annual races of *Arabidopsis thaliana*, Baskin & Baskin, 1983), and so most agricultural weeds that show a winter annual phenology are found in spring-sown crops as well (Hald, 1999). Second, little overlap in

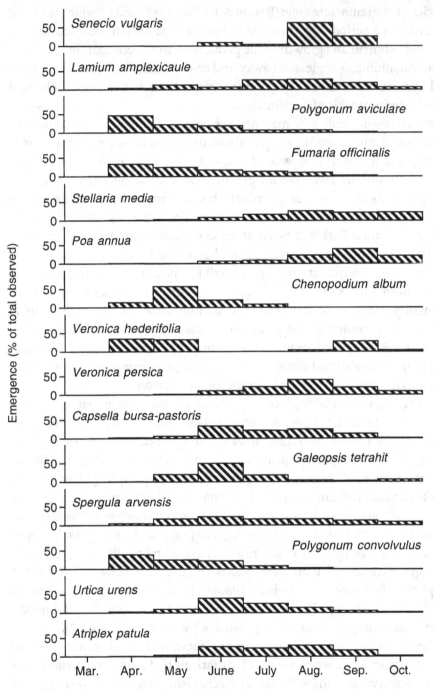

Figure 2.2 Seasonality of emergence in 15 weed species (from Lawson, Waister & Stephens, 1974, experiment I). Data are percent of total counts observed three weeks after soil disturbances performed at monthly intervals in the southern United Kingdom.

the time of germination may occur between species characteristic of early versus late spring (e.g., Figure 2.2). For some species, early and late spring are apparently very different seasons. Third, although most species have times of the year in which germination is most probable, at least some seeds germinate over a wide range of other seasons. This occurs because the dormancy mechanisms are less than perfect in matching germination to one particular season. However, that variability provides adaptation to environmental unpredictability. Finally, the species that have a very broad season of germination tend to be small-statured, rapidly maturing plants, with broad-amplitude temperature tolerance. Such species are well adapted to gardens and mixed high-intensity vegetable farms where the cropping conditions are unpredictable, and therefore, early catastrophic mortality is frequent.

A large comparative study by Forcella *et al.* (1997) illustrates the importance of understanding how dormancy affects the seasonality of seedling emergence. By examining percentage emergence of several weed species in relation to environmental conditions at 22 site and year combinations, they identified soil temperature and moisture thresholds that induced secondary dormancy in *Setaria faberi*, *S. viridis*, *Polygonum convolvulus*, *P. pennsylvanicum*, and *Amaranthus* spp. (mostly *A. retroflexus*). Once the threshold for one of these species has been passed during a growing season, the seeds are induced into secondary dormancy. Few individuals are likely to emerge after that date, and a grower can modify management plans accordingly.

Probably the principal utility of understanding periodicity of weed seed germination is that it allows disruption of weed life cycles. Two approaches are mentioned here and discussed in depth in Chapters 4, 6, and 7. First, if the time of germination is known for the dominant weeds in the seed bank, crop planting dates can be adjusted so that either (i) the crop emerges before the weeds and thereby obtains a competitive advantage, or (ii) weeds are allowed to germinate and are then destroyed during seedbed preparation. Second, by rotating between crops with radically different planting dates, a grower can block the establishment and reproduction of particular groups of weeds in any given year. Thus, for example, in a fall-sown grain crop, spring germinating weeds will either remain dormant or, if they do germinate, suffer heavy competition from the already well-established crop. Those that do not germinate are subjected to another year of mortality risk as seeds, reducing weed pressure on later crops. Alternation of early spring grains or vegetables with late spring- or summer-planted soybean or vegetable crops may be equally effective in disrupting weed life cycles.

Why tillage promotes the germination of weed seeds

Tillage promotes germination of most agricultural weeds, provided the soil disturbance comes at a time of year when the seeds are not innately dormant. Agricultural weeds have adapted to respond to cues associated with soil disturbance because their small seedlings make them poor competitors early in life. Vigorous, well-established plants are unlikely to be present immediately after soil disturbance, and hence weedy species have been selected for germination under conditions that indicate soil disturbance. Relative to undisturbed soil with established vegetation, recently tilled ground tends to be warmer, have higher diurnal temperature fluctuations, higher nitrate concentration, and better aeration (Gebhardt et al., 1985; Cox et al., 1990; Dou, Fox & Toth, 1995). Perhaps most importantly, when tillage or natural processes stir soil, exposure to light prompts seed germination (Sauer & Struik, 1964).

Each of these factors promotes the germination of some common agricultural weeds (Table 2.3). For example, *Rumex crispus*, *Chenopodium album*, and *Panicum dichotomiflorum* have a higher percentage germination when exposed to fluctuating temperatures than when exposed to a constant temperature with the same mean (Henson, 1970; Totterdell & Roberts, 1980) or when separately tested against the two temperature extremes (Fausey & Renner, 1997). *Amaranthus retroflexus* germinates best at 30–40 °C (McWilliams, Landers & Mahlstede, 1968; Weaver & Thomas, 1986), a soil temperature that is unlikely to occur under the shade of established vegetation.

Tillage and other soil disturbances stimulate decomposition of organic matter and nitrification of the ammonium released by decomposition. The presence of nitrate thus indicates not only enhanced availability of mineral nutrients, but also the elimination of competing vegetation. In any case, nitrate indicates favorable growing conditions, and germination of several weed species, including *Chenopodium album* and *Plantago lanceolata*, increases in response to elevated nitrate concentrations (Williams & Harper, 1965; Pons, 1989).

Gas exchange in the soil during tillage probably prompts germination of many weed species. Although oxygen concentration influences germination (Edwards, 1969; Popay & Roberts, 1970; Brennan et al., 1978), oxygen levels near the soil surface are rarely low enough to directly inhibit germination, except when the soil is saturated with water (Egley, 1995). Several studies have shown, however, that flushing the soil with air substantially increased germination or emergence of buried seeds (Wesson & Wareing, 1969b; Holm, 1972; Benvenuti & Macchia, 1995). Since flushing with nitrogen also enhanced ger-

Table 2.3. *Factors associated with tillage that have been shown to promote the germination of weed seeds*

Factor	Species	+Factor (%)[a,b]	−Factor (%)[a,b]	Reference
Light	*Alopecurus myosuroides*	86	0	Froud-Williams (1985)
	Amaranthus retroflexus	98	14	Kigel (1994)
	Brassica arvensis	78	53	Povilaitis (1956)
	Datura ferox	96	1	Scopel, Ballaré & Sánchez (1991)
	Lolium multiflorum	95	82	Schafer & Chilcote (1970)
	Poa annua	89	1	Froud-Williams (1985)
	Portulaca oleracea	28	12	Povilaitis (1956)
Alternating temperature	*Poa annua*	92	47	Froud-Williams (1985)
	Rumex crispus	100	0	Totterdell & Roberts (1980)
	Sonchus arvensis	57	3	Håkansson & Wallgren (1972)
	Sorghum halepense	20	7	Ghersa, Benech Arnold & Martinez-Ghersa (1992)
	Stellaria media	93	47	Roberts & Lockett (1975)
Nitrate	*Chenopodium album*	92	55	Williams & Harper (1965)
	Erysimum cheiranthoides	89	57	Steinbauer & Grigsby (1957a)
	Plantago lanceolata	48	25	Pons (1989)
	Plantago major	93	3	Steinbauer & Grigsby (1957b)

Notes:
[a] +Factor: seeds germinated in light, in alternating temperature regime, and in nitrate solution.
−Factor: seeds germinated in dark, at constant temperature equal to the mean of the alternating regime, and in water.
[b] The numbers given for percentage germination are mostly means taken over several chilling treatments, populations, seed types, etc., but not over treatments that involve variation in other factors listed in the table. In a few cases, the numbers are from selected treatments that demonstrate the effect.

mination (Wesson & Wareing, 1969b; Holm, 1972), the effect could not be attributed to improved oxygen availability alone. Holm (1972) further observed that imbibed *Abutilon theophrasti* and *Ipomoea purpurea* produced ethanol, acetone, and acetaldehyde when oxygen concentrations dropped below 6% and demonstrated that these compounds inhibited germination of seeds, even in normal air. He therefore proposed that moderate reduction in oxygen by respiration in the soil results in anaerobic seed metabolism, which produces volatile germination inhibitors. In the absence of air exchange, these enforce seed dormancy. Thus, tillage probably prompts germination of weed seeds both by venting volatile inhibitors from the surface soil and by moving deeply buried seeds to near-surface conditions where air exchange is improved.

Although ethylene and carbon dioxide concentrations are also commonly elevated in undisturbed soil, these compounds appear to play a small role in inhibiting seed germination. Ethylene affects germination of only a small proportion of weed species, and usually promotes, rather than inhibits, germination (Taylorson, 1979). Similarly, concentrations of carbon dioxide up to 5% tend to enhance, rather than inhibit, germination (Baskin & Baskin, 1987; Egley, 1995).

One of the most important cues promoting germination of seeds in the seed bank is light. In a classic study, Wesson and Wareing (1969a) collected soil at night, screened it in the dark, and then placed it in trays in a greenhouse in either the light or dark. Averaged over three experiments at different times of year, they found 12 times more dicot seedlings and 26 times more grass seedlings in the light treatment. Many subsequent studies have shown that germination of a great range of weed species is promoted by light (Taylorson, 1972; Stoller & Wax, 1974; Froud-Williams, 1985; Baskin & Baskin, 1986). Some species of weed seeds germinate in response to very small amounts of light. For example, conditional dormancy in *Datura ferox* and *Amaranthus retroflexus* can be broken by the equivalent of a few milliseconds of sunlight (Scopel, Ballaré & Sánchez, 1991; Gallagher & Cardina, 1998). Moreover, many species, like *Spergula arvensis* and *Stellaria media*, that lack light sensitivity when shed from the parent plant quickly develop it after incorporation into the soil (Wesson & Wareing, 1969b; Holm, 1972).

Because light-sensitive germination is controlled by the phytochrome system, light depleted in red wavelengths by passage through a plant canopy is inhibitory to germination of light-sensitive species (Górski, 1975). In fact, even some species with moderately high germination in the dark are severely inhibited by light that has passed through plant leaves (King, 1975; Silvertown, 1980). Thus, germination under established vegetation is held in check not only by the amount of light but also by its spectral composition.

Although the several factors discussed above promote germination individually, the effects are most pronounced when several factors combine. Vincent & Roberts (1977), Bostock (1978), Roberts & Benjamin (1979), and Kannangara & Field (1985) demonstrated that two- and three-way interactions among light, nitrate, and fluctuating temperature enhanced germination of 13 out of the 15 weed species they studied. Presumably, the several factors acting in concert provide a more certain signal that competition has been eliminated than any of the factors acting singly.

Germination in response to tillage is both a fact that must be dealt with in the design of agricultural systems and a tool for manipulation of weed populations. For example, shallow cultivation between crop rows is often preferable to deep cultivation. A shallow cultivation tends to eliminate the weeds that were prompted to germinate in response to seedbed preparation without cueing germination of many additional seeds. In contrast, a deep cultivation tends to bring up seeds that are then prompted to germinate by disturbance-related cues. Dynamics of the seed bank in response to tillage is discussed further in Chapter 4.

The germination response of weeds to soil disturbance can also be used to induce inappropriate germination. For example, species with broad seasonality of germination can be stimulated to establish at times that are unsuitable for survival to reproduction, thereby depleting the seed bank. A more common application is to use shallow cultivations with intervening rests before planting to flush out and kill many of the weeds that would otherwise establish with the crop. Use of this "false seedbed" method is analyzed in Chapter 4.

Not all species of weeds are sensitive to germination cues associated with soil disturbance. Most of these are relatively large seeded species (Table 2.4, below) that presumably have sufficient resources in the seedling stage to establish in the face of some competition from established vegetation. Many have hard, impermeable seed coats that prevent water uptake and germination, or other dormancy mechanisms that prevent germination until the seed coat is physically altered (Table 2.4). In the field, temperature extremes or desiccation typically break physical dormancy (Baskin & Baskin, 1998*a*, pp. 114–20). Response to these factors spreads germination over several years and, to some extent, also cues germination to appropriate times of the year. Some large-seeded weeds also have innate physiological dormancy mechanisms (Wareing & Foda, 1957). Thus, large-seeded weeds have mechanisms that match germination to appropriate environmental conditions, but only a few (e.g., *Solanum viarum* – Akanda, Mullahey & Shilling, 1996) sense the removal of competitors through a strong response to light, alternating temperature, or nitrate.

Survival of weed seeds in the soil

The seeds of most annual and stationary perennial weeds persist in the seed bank for at least a few years, and many remain viable for decades if conditions are favorable. Excavations from dated archaeological strata indicate that some agricultural weed species, including *Chenopodium album*, *Stellaria media*, and *Lamium purpureum*, probably remain viable for several hundred years (Odum, 1965), although movement of younger seeds into the strata by soil animals cannot be excluded with certainty. However, the cool, moist, dark, undisturbed environments of such sites are highly favorable to

Table 2.4. *Characteristics of some species with negligible response to germination cues associated with tillage*

Species	Seed size, mg	Innate dormancy[a]	Wounding/ hard seed[b]	Light[a]	Alternating temperature[a]	NO$_3$[a]	References
Abutilon theophrasti	4–10	0,(+)	H	0,+,(−)		0	LaCroix & Staniforth (1964), Holm (1972), Fawcett & Silfe (1978), Horowitz & Taylorson (1984, 1985), Warwick & Black (1988)
Avena ludoviciana		0,+	W	−			Thurston (1960), Froud-Williams (1985)
Bromus sterilis	8	0		0	0	0	Froud-Williams (1985), Thompson, Band & Hodgson (1993)
Convolvulus arvensis	10	0	H	0	(+)		Weaver & Riley (1982)
Xanthium strumarium	50–60	+	W	0	(+)		Wareing & Foda (1957), Kaul (1965), Weaver & Lechowicz (1983)

Notes:
[a] +, −, and 0 indicate that the factor promoted, inhibited or had no effect on germination in the studies cited. Parentheses indicate that the effect was weak. Blanks indicate that information was unavailable.
[b] H, hard seeds; W, wounding increases germination, but seeds are not hard in the usual sense.

Table 2.5. *Half-life and annual loss of seeds from soil for 20 weed species, computed from seed survival over a five-year period*[a]

	Loss per year (%)		Half-life (years)	
	Cultivated	Uncultivated	Cultivated	Uncultivated
Capsella bursa-pastoris	43	24	1.2	2.6
Chenopodium album	31	8	1.9	8.3
Euphorbia helioscopia	54	21	0.9	3.0
Fumaria officinalis	34	16	1.7	4.1
Matricaria matricarioides	33	28	1.8	2.1
Medicago lupulina	30	22	2.0	2.8
Papaver rhoeas	38	30	1.4	2.0
Poa annua	26	22	2.3	2.8
Polygonum aviculare	47	16	1.1	4.0
Polygonum convolvulus	50	25	1.0	2.4
Senecio vulgaris	[b]	45	[b]	1.2
Spergula arvensis	60	30	0.8	2.0
Stellaria media	54	32	0.9	1.8
Thlaspi arvense	50	10	1.0	6.8
Tripleurospermum maritimum ssp. *inodorum*	36	23	1.6	2.6
Urtica urens	37	17	1.5	3.7
Veronica hederifolia	[b]	13	[b]	5.1
Veronica persica	54	22	0.9	2.8
Vicia hirsuta	36	30	1.6	2.0
Viola arvensis	40	15	1.4	4.2

Notes:
[a] Seeds were mixed with the top 15 cm of soil and either left undisturbed or mixed four times per year.
[b] No seeds viable after five years.
Source: From Roberts & Feast (1972).

seed survival (Villiers, 1973), whereas seed longevity is much less in agricultural fields (Table 2.5). Often the logarithm of seed density plotted against time follows a straight line for weed seed banks (Roberts & Dawkins, 1967; Roberts & Feast, 1973), although in some cases the mortality rate is somewhat higher or lower during the first year (Figure 2.3). Thus, seed survival times are often better characterized by the half-life of the population rather than the maximum age achieved by the most persistent individual. Studies on weed seed survival lead to two general conclusions. First, a substantial portion of the seeds of even relatively persistent species disappears from the soil each year, and second, soil disturbance increases the rate of disappearance (Table 2.5) (Roberts & Feast, 1972; Lueschen & Andersen, 1980; Froud-Williams, Chancellor & Drennan, 1984; Warnes & Andersen, 1984; Barralis, Chadoeuf & Lonchamp, 1988).

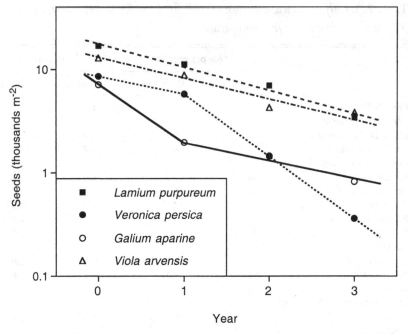

Figure 2.3 Decline in density of viable seeds through time in a field annually tilled and planted with winter wheat or winter oilseed rape. (After Wilson & Lawson, 1992.)

Sources of seed mortality

Factors affecting the rate of seed mortality in the soil include (i) the action of seed predators, including vertebrates, invertebrates, fungi, and bacteria, (ii) physiological aging and exhaustion of reserves through respiration, and (iii) germination at depths in the soil or times of year that are unsuitable for emergence. Strictly speaking, the latter involves the death of seedlings rather than seeds, but it is customarily treated as a source of seed mortality. Several studies have partitioned the sources of weed seed mortality, and of the three factors listed above, inappropriate germination often causes the greatest reduction in seed density (Roberts, 1972). For example, Schafer & Chilcote (1970) found that after burial for 60 days at 10 cm depth, 11% to 13% of *Lolium multiflorum* seeds were nonviable whereas 40% to 64% had died after germination. Zorner, Zimdahl & Schweizer (1984a, 1984b) and Gleichsner & Appleby (1989) found that *in situ* germination was the largest source of mortality for deeply buried *Avena fatua*, *Kochia scoparia*, and *Bromus rigidus* (*B. diandrus*) seeds, but that loss of viability increased as a cause of mortality with shallower placement. Wilson (1972) noted (i) that *A. fatua* lost dormancy more quickly at the soil surface and (ii) that the seeds on the soil surface rapidly lost weight whereas seeds that had been cultivated into the soil did not. He hypothesized that seeds on the soil surface physiologically initiated germination, but died prior to emergence of the radical. Many *Abutilon theophrasti* seeds on the soil surface imbibe and the seed coat breaks, but germination does not proceed further due to subsequent desiccation (C. L. Mohler, personal observation). In contrast, most *A. theophrasti* seeds buried 1 cm deep emerge successfully. Seed movement by tillage implements and seed survival at different depths in the soil profile are considered further in Chapter 4.

A problem with the above studies and observations is that the species investigated lack strong tillage-cued germination mechanisms and, with the exception of *K. scoparia*, all are relatively large-seeded. Whether the many small-seeded species that rely on environmental cues to inform them of proximity to the soil surface and lack of competition also suffer large mortality due to inappropriate germination remains to be determined. The technical problems of investigating causes of seed loss in small-seeded species with great longevity in the soil are substantial: few seeds are likely to lose viability or germinate inappropriately in any given time interval, and recovering tiny seedlings in the white thread stage is difficult.

Physiological aging of seeds involves loss of membrane integrity, deterioration of organelles, and accumulation of damage to DNA (Abdalla & Roberts, 1968; Villiers, 1973; Roberts, 1988). These aging processes proceed most rapidly when seeds are in warm conditions with seed moisture in the 8% to 15% range (Abdalla & Roberts, 1968; Villiers & Edgcumbe, 1975). In contrast, fully imbibed dormant seeds are metabolically active and apparently capable of repairing structural and genetic damage (Villiers & Edgcumbe, 1975; Elder & Osborne, 1993). These observations probably explain why mortality due to loss of viability increases toward the soil surface: conditions near the surface are warmer than deep in the soil profile and are periodically too dry to maintain seeds in a fully imbibed state. Eventually, even imbibed seeds die, presumably due to accumulation of lethal levels of damage to membranes and DNA (Osborne, 1980; Villiers, 1980). The extent to which depletion of food reserves is involved in the aging process is poorly studied.

Seed predators consume significant numbers of weed seeds in some agroecosytems. Prior to dispersal from the parent, predation is primarily by host-specific natural enemies. Pre-dispersal seed predators may occasionally consume a substantial proportion of the seeds produced (Forsyth & Watson, 1985), but particularly in annual crops, they may have difficulty in locating their host plants, as explained in the section "Survival after emergence" below. After seeds have dispersed from the parent, they are attacked by a range of generalist seed predators including birds, small mammals, earthworms, insects, and fungi (Wilson & Cussans, 1972; Grant, 1983; Brust & House, 1988; Fellows & Roeth, 1992). Seed predation, and the manipulation of agricultural systems to increase predation on weed seeds, are discussed further in Chapters 5 and 8.

What types of species survive better in soil?

Species vary greatly in their ability to survive in the soil seed bank, and certain broad patterns exist that predict which species survive well in the soil. Understanding trends in seed survival across species provides a means for targeting ecological management strategies at particular classes of weeds. It also provides a basis for guessing the likely seed persistence of unstudied species.

A first broad pattern across species is that many broadleaf weeds are able to survive in the seed bank for several decades (Chancellor, 1986), whereas seeds of only a few grass species survive in substantial numbers for more than 5 to 10 years (Dawson & Bruns, 1975; Froud-Williams, 1987, Baskin & Baskin, 1998b). Moreover, the seeds of some grass weeds do not persist longer than a single year (e.g., *Bromus diandrus* – Harradine, 1986; Gleichsner & Appleby, 1989; *Bromus sterilis* – Roberts, 1986), whereas few, if any, broadleaf agricultural weeds appear to have seed survival times as short as that. Both of the annual grasses tested by Conn & Deck (1995) (*Avena fatua* and *Hordeum jubatum*) were reduced to <1% viability by 3.7 years of burial and were completely gone after 9.7 years. In contrast, mean survival of the 13 annual broadleaf species they tested was 32% after 3.7 years, and all but one of the broadleaf species had at least a few surviving seeds after 9.7 years. Burnside *et al.* (1996) counted seedlings emerging from seed samples that had been left in the soil for 1 to 17 years. After 12 years of burial, the 14 annual broadleaf species averaged 10% germination, whereas germination of the 11 annual grasses averaged only 2%. Nevertheless, some economically important weedy annual grasses, notably *Echinochloa crus-galli*, *Bromus secalinus*, and several species of *Setaria*, retained moderate viability into their second decade of burial.

A second broad pattern among species is that annual and stationary perennial weeds tend to form persistent seed banks, whereas wandering perennials and woody weeds usually do not. For example, of the 22 terrestrial grass species tested by Roberts (1986), the weedy, short-lived *Avena fatua*, *A. sterilis* ssp. *ludoviciana*, and *Poa annua* had the greatest emergence 2 years after sowing, whereas many of the rhizomatous species were completely gone by that time. In the study of Burnside *et al.* (1996), seven stationary perennial broadleaf species (including biennials) averaged 37% germination after 8 years of burial and 25 annual broadleaf species averaged 15%, whereas nine wandering broadleaf perennials had an average germination of only 5%. In a study of the flora of northwestern Europe that emphasized arable weeds and grassland species, Thompson *et al.* (1998) showed greater persistence of monocarpic (single fruiting – in this context, annual and biennial) species relative to polycarpic (multiple fruiting) species.

A third pattern across species is that those with weak dormancy mechanisms often have shorter longevity than species with well-developed dormancy mechanisms (Bostock, 1978; Roberts, 1986). The greater persistence of species with dormancy is due to reduction of inappropriate germination rather than to greater retention of viability (Roberts, 1972). Experimental support for the importance of innate dormancy for seed survival comes from studies showing that seed lots of several weed species that were dormant when buried retained viability in the soil better than initially nondormant seed lots (Taylorson, 1970; Naylor, 1983; Zorner, Zimdahl & Schweizer, 1984*b*).

A fourth general pattern is that small, round-seeded species tend to persist in the seed bank longer than species with large or elongate seeds. Thompson, Band & Hodgson (1993) surveyed 97 species and found that few species with seeds larger than 2 mg or with variance in diaspore relative dimension greater than 0.2 persisted longer than 5 years in the soil. In the non-agricultural conditions in which seed persistence evolved, incorporation into the soil was probably more difficult for large or elongate seeds than for small, round ones. The latter can wash into cracks, or be ingested by earthworms more easily than the former. Since seed survival is lower on the soil surface than deep in the profile, species with a low probability of incorporation into the seed bank probably experienced little selection favoring mechanisms that allow long persistence. Large seeds are also more likely to be eaten by small mammals (Hulme, 1994), and possibly selection rarely favors mechanisms that allow long residency in the soil of large-seeded species for this reason. The exceptions to the general pattern occur primarily in the Leguminosae, Convolvulaceae, and Malvaceae, which include many species with large, hard seeds (Taylorson, 1987). Seeds of most of these species have toxins that may help defend them against mammalian seed predators (e.g., *Senna obtusifolia*, *Ipomoea purpurea*) (Kingsbury, 1964, p. 314; Friedman & Henika, 1991).

Seed persistence and weed management

Persistence of seeds in the soil has consequences for many aspects of weed management. By allowing a given generation of seeds to test the suitability of several growing seasons, a seed bank buffers annual species against a year in which little reproduction is possible (Cohen, 1966). This protects the weed against local extinction, but from the grower's point of view, it makes weeds with seed banks highly resistant to eradication. However, complete eradication is rarely necessary, and knowledge of seed longevity of a species allows some predictability regarding how long perfect control of the weed is required to reduce weed pressure by a given amount (Donald & Zimdahl, 1987). Similarly, rotation into a sod crop allows several years for mortality to reduce the seed bank (Thurston, 1966; Warnes & Andersen, 1984). The effect of rotation on seed banks is discussed further in Chapter 7.

Seed longevity of a species also has a large effect on its response to different tillage regimes, and, as explained in Chapter 4, may be an important factor contributing to the shift from broadleaf species to grasses with reduced tillage. Finally, accumulation of high densities of seeds in the soil allows dispersal in soil clinging to animals, vehicles, and tillage machinery, and this is probably an important route for dispersal of weeds between fields (see section "Dispersal of seeds and ramets" below).

Hazards of establishment

The period of establishment may be defined as the time between germination and the production of the first true leaf. This is the most poorly studied stage in the weed life cycle, except with respect to its sensitivity to herbicides. The few quantitative data available indicate that this phase of the life cycle represents a major bottleneck for some species (Boutin & Harper, 1991). Several mortality factors act on establishing weeds, including exhaustion of seed reserves, drought, seedling predation, disease, physical disturbance, and expression of morphological and genetic defects. Data on the effects of all these phenomena are scarce.

One of the most important factors is exhaustion of seed reserves during emergence. The probability of emergence for a newly germinated seedling is a function of depth of burial, the energy content of the seed, and the resistance of the soil. Although the soil in most tilled seedbeds is probably sufficiently loose to not greatly impede emergence, penetration of the shoot through compact soil requires more energy (Morton & Buchele, 1960), and this probably prevents some weed emergence in no-till systems (Mohler & Galford, 1997).

The seeds of most agricultural weeds weigh less than 2 mg and few exceed 10 mg (Table 2.6) (Stevens, 1932; Thompson, Band & Hodgson, 1993). Consequently, successful emergence requires that weed seeds germinate within a few centimeters of the soil surface (Chancellor, 1964; see also literature review and summary table in Mohler, 1993). In contrast to the many

Table 2.6. *Propagule weight of annual weeds and crops*

Weed species[a]	Propagule weight (mg)	Embryo plus endosperm weight (mg)	Crop species[b]	Propagule weight (mg)
Abutilon theophrasti	9.5	5.1	Maize	250
Ambrosia artemisiifolia	4.4	2.4	Soybean	220
Brassica kaber	2.2	1.8	Wheat	39
Chenopodium album	0.74	0.47	Oat	35
Amaranthus retroflexus	0.44	0.29	Rye	27

Notes:
[a] Weeds are the five most common annual weeds in agronomic crops in New York state (Bridges, 1992).
[b] Crops are the five annual agronomic crops with the greatest hectarage in New York state (New York Agricultural Statistics Service, 1994).
Source: Adapted from Mohler (1996).

small-seeded weeds that emerge best from a depth of 0.5 to 1.0 cm, most agronomic crops and many vegetable crops have much larger seeds (Table 2.6). Consequently, they are usually planted at 3 to 5 cm. As discussed in Chapter 4, this difference in emergence depth allows pre- and post-emergence cultivation in the crop row. In some systems, it also allows directed feeding of water and nutrients to the crop (Chapter 5). In addition, the difference in seed size between crops and weeds makes possible the use of crop residue and dust mulches (Chapter 5) and greatly facilitates the use of crop competition for weed management (Chapter 6). Conversely, the relatively small difference between the seed size of crop species and large-seeded weeds like *Xanthium strumarium*, *Avena fatua*, and *Ipomoea hederaceae* helps explain why these species are so difficult to control.

Herbivores and damping off fungi have their greatest impact on weed density during the establishment phase because very small plants have few resources for defense and recovery. Cover by residue and the crop canopy is a major factor regulating the effectiveness of naturally occurring generalist seedling predators (see Chapter 5). Frequently, the inundative release of biological weed control agents is most effective when the weeds are small (e.g., Pitelli, Charudattan & Devalerio, 1998).

The susceptibility of a weed to physical disturbance decreases as it grows. First, as the plant grows, stems and roots thicken and toughen with fiber. Consequently, impact with a hoe or cultivator tine is less likely to cause fatal breakage to a large old plant than to a small young one. Second, plants grow by repeated addition of metamers, units consisting of a leaf, the subtended bud(s), and an internode (White, 1979). Potentially, a weed can lose most of its shoot and still regrow into a full-sized plant, provided a single bud is left. Modular growth below ground similarly allows recovery from drastic damage to the root system. However, for most herbaceous dicot species, a seedling that is broken between the root and base of the cotyledons will not survive. At this stage the weed has only one shoot meristem, and its loss is fatal. Establishing monocot seedlings are somewhat less susceptible to damage than dicots because they lack the long hypocotyl between the root and shoot meristems, but they too may fail to recover following loss of a substantial portion of the cotyledon or primary root. Thus, very small weeds in the white thread and cotyledon stages are more easily controlled by mechanical means than are weeds that are more developed. Cultivation techniques specifically aimed at establishing weed seedlings are discussed in Chapter 4.

Surprisingly, the biology of mechanical weed management has been little studied. For example, conventional wisdom among farmers holds that rotary hoeing is most effective if the soil is not immediately wetted afterward by rain or irrigation, and some experimental evidence confirms this view (Lovely, Weber & Staniforth, 1958). However, the phenomenon has only been studied at the level of the field, and not at the level of the individual weed. To what extent is the elimination of weeds by a pre-emergence operation with a rotary hoe or tine weeder due to (i) direct damage, (ii) desiccation from loss of intimate contact with soil, or (iii) reburial of white thread seedlings that have already expended their seed reserves?

Vegetative growth and crop–weed competition

Once a cohort of weeds has established in a field, its success depends primarily on its survival, discussed in the following section, and its growth, discussed here. Two types of growth rate are relevant to understanding the growth potential of weeds. Absolute growth rate is the addition of biomass per unit time (g week^{-1}), whereas relative growth rate (RGR) is the biomass added per unit biomass per unit of time (g g^{-1} week^{-1}). In most species, RGR declines as the plant grows (Grime & Hunt, 1975; Spitters & Kramer, 1985; Ascencio & Lazo, 1997). This occurs because (i) a greater proportion of tissue is nonphotosynthetic in larger plants, (ii) maintenance respiration increases disproportionately with plant size, (iii) self-shading increases as plants grow, and (iv) larger leaves have less favorable source/sink relationships for photosynthesis (Chapin, Groves & Evans, 1989). Since RGR varies with plant size and with environmental conditions, the maximum RGR achieved by young plants in an optimal environment forms a useful basis for comparing species.

Agricultural weeds have the highest maximum RGR of any large category of plants. For example, in Grime & Hunt's (1975) analysis of growth rate of 132 British species, annuals had the highest maximum RGR of the several groups analyzed, and the agricultural annuals were mostly in the higher end of this class. Perennial agricultural weeds also have high RGR. For example, *Poa annua* and *Convolvulus arvensis* had the highest and third highest RGR measured. Grime & Hunt (1975) also compared occurrence of species in four RGR categories in 29 British habitats. Plants of manure piles had the greatest proportion of high RGR species, followed by those of enclosed pastures, arable land, and meadows. In short, productive agricultural habitats tend to favor plants with high RGR.

From a management perspective, the most important plants to compare with weeds are crops. Seibert & Pearce (1993) compared growth parameters of four weed and two crop species (Table 2.7). They found that RGR declined as seed size increased, such that *Xanthium strumarium*, an exceptionally large-seeded weed, behaved more like the crops. High RGR for the small-seeded weed species was primarily due to higher leaf area ratio (LAR, leaf area/plant weight) rather than higher net assimilation rate (NAR, change in plant weight/leaf area). That is, differences in growth rate due to seed size were attributable to morphology rather than physiology. The smaller-seeded species (weeds) put a greater proportion of plant mass into leaves (high LWR, leaf weight ratio) and had thinner leaves (high SLA, specific leaf area) than the large-seeded weed (*X. strumarium*) and the crops. The proportion of biomass invested in roots was lower in the weeds, but their root diameter was less so that total length of roots increased more quickly than in the crops. To some extent the particular patterns found by Seibert & Pearce (1993) probably depended on the choice of species. Chapin, Groves & Evans (1989) decreased this problem by comparing weed, domestic, and progenitor taxa in a single genus, *Hordeum*. They too found that seed size explained most of the variation in RGR, and again, the weeds had smaller seeds and higher RGR than the crops. The reason was that large seeds make large seedlings, and larger plants tend to have lower RGR regardless of whether the comparison is within a species or between species.

Because small-seeded weeds have a higher RGR than the larger-seeded crops, they tend to catch up in size eventually. As an extreme example, the initial 500-fold difference in the seed size of maize and redroot pigweed (Table 2.6) may be reduced to a two-fold difference in the size of the mature plants if each species is allowed to grow without competition (Mohler, 1996). Although the large initial size of most crop species gives them a lower RGR than many weeds, the larger size is still competitively advantageous. At emergence, the crop has a greater leaf area and a larger root system than the weed.

Table 2.7. Seed size and growth parameters for some weeds and crops

Species	Seed weight (mg)	RGR (g g^{-1} d^{-1})[a]	NAR (g dm^{-2} d^{-1})[b]	LAR (cm^2 g^{-1})[c]	SLA (cm^2 g^{-1})[d]	LWR (g g^{-1})[e]	RWR (g g^{-1})[f]	Root diameter (mm)[g]	RLI (cm cm^{-1} d^{-1})[h]
Amaranthus reroflexus	0.41	0.349	0.298	198	326	0.597	0.189	0.22	0.343
Chenopodium album	0.44	0.335	0.254	224	329	0.674	0.153	0.20	0.285
Abutilon theophrasti	7.8	0.244	0.145	190	326	0.583	0.214	0.46	0.274
Xanthium strumarium	38	0.187	0.224	136	237	0.560	0.217	0.35	0.227
Sunflower (domestic)	61	0.197	0.241	140	276	0.495	0.272	0.42	0.227
Soybean	158	0.155	0.176	132	242	0.539	0.241	0.64	0.201
Correlation with ln (seed weight)		−0.99**	−0.59	−0.94**	−0.86*	−0.84*	0.86*	0.86*	−0.93**

Notes:
[a] RGR: relative growth rate = g increase in plant weight g^{-1} plant weight d^{-1}.
[b] NAR: net assimilation rate = g increase in plant weight dm^{-2} leaf area d^{-1}.
[c] LAR: leaf area ratio = cm^2 leaf area g^{-1} plant weight.
[d] SLA: specific leaf area = cm^2 leaf area g^{-1} leaf weight.
[e] LWR: leaf weight ratio = g leaf weight g^{-1} plant weight.
[f] RWR: root weight ratio = g root weight g^{-1} plant weight.
[g] Average of root diameters at 0, 7, 14, 21, and 28 days after emergence.
[h] RLI: relative rate of root length increase from 0 to 28 days after emergence.
* Significant at $p < 0.05$ level.
** Significant at $p < 0.01$ level.
Source: From Seibert & Pearce (1993)

Therefore, the crop's absolute growth rate is initially greater, and usually remains greater for at least several weeks (Dunan & Zimdahl, 1991; Tanji, Zimdahl & Westra, 1997).

Use of the initial advantage conferred to the crop by relatively large size and high absolute growth rate is a key concept in ecological weed management (Mohler, 1996). A major strategy in most annual crops is to design the cropping system so that the initial size advantage still holds at the time the crop and weeds grow into physical contact. With few exceptions, both crops and weeds are adapted to open habitats, and both are intolerant of shade (Blackman & Black, 1959; Knake, 1972; Loomis & Connor, 1992, pp. 274–5; McLachlan et al., 1993; Bello, Owen & Hatterman-Valenti, 1995). Consequently, if the crop is in the superior position, it will suppress the growth of the weeds, whereas if the weeds grow above the crop canopy, then yield reduction is likely to be severe. Which outcome occurs depends on (i) the relative timing of emergence, (ii) the time course of height growth for the two species, and (iii) how rapidly the crop canopy closes. Factors that can be manipulated to affect the relative timing of emergence include planting date (Chapters 4 and 6), cultivation (Chapter 4), and mulch (Chapter 5). Factors that can be manipulated to affect the growth of crops relative to weeds include the planting date (Chapter 6), the use of allelopathic materials (Chapter 5), and the timing, type, and spatial distribution of fertilizer and irrigation water applied to the crop–weed community (Chapter 5). Finally, the speed with which the crop canopy closes can be increased through narrow row spacing, dense planting, intercropping, use of fast-growing cultivars, and choice of planting dates that optimize crop growth rate (Chapters 6 and 7).

The effectiveness of these tactics depends on the morphology of the weed species present in the field. For example, a tall, erect weed species is unlikely to be suppressed by a low-growing crop unless the emergence of the weed is substantially delayed relative to the crop. To some extent, weeds change shape in response to shade. For many broadleaf species, including *Polygonum arenastrum (P. aviculare), Cassia (Senna) obtusifolia, Abutilon theophrasti,* and *Xanthium strumarium,* branch length is reduced when the plant is shaded or crowded by competitors (Geber, 1989; Regnier & Stoller, 1989; Regnier & Harrison, 1993; Smith & Jordan, 1993). Shade or crowding also inhibits release of lateral buds in some species, particularly on the lower parts of the main shoot (*Abutilon theophrasti, Datura stramonium* but not *Xanthium strumarium* – Regnier & Stoller, 1989; *Amaranthus retroflexus* – McLachlan et al., 1993). These changes presumably channel resources into height growth rather than lateral spread and thereby improve competition for light. Several studies on *Amaranthus retroflexus* have shown that the vertical distribution of biomass and leaf area shifts upward as competition increases (Légère & Schreiber, 1989; McLachlan et al., 1993; Knezevic & Horak, 1998). Differences in branching of C. obtusifolia emerging on different dates and located at various distances from the soybean row allowed that species to maintain a nearly constant proportion of leaves above the crop, regardless of the timing of competition (Smith & Jordan, 1993). Since morphological flexibility allows some weed species to overtop even competitive crops like soybean, either the tactics discussed above need to create a substantial size differential between the crop and weed or else tall weeds will need to be dealt with by other means.

Plants compete for nutrients and water as well as for light. This is apparent from the many studies in which perennial cover crops growing beneath a main crop reduced yield less in a wet year or when irrigated, even though the cover crop was kept short (Chapter 7) (Mohler, 1995; Teasdale, 1998). Similarly, a short weed like *Taraxacum officinale* can reduce yield of a tall crop like sweet corn in a dry year (Mohler, 1991), even though shading by the crop reduces transpiration by the weed.

In annual cropping systems, the soil is unoccupied by roots at the beginning of the season, and usually nutrients and water are relatively abundant. Under such conditions, the outcome of below-ground competition between the crop and weed depends primarily on the rate at which the two species occupy the soil with roots, and their relative rates of uptake. Andrews & Newman (1970) showed that root density is critical in competition for nutrients. Because small-seeded species tend to have a higher rate of root elongation (Table 2.7) (Seibert & Pearce, 1993), weeds tend to rapidly occupy the soil volume to the detriment of the crop. Probably because weeds are adapted to exploit the brief pulse of nutrient availability that follows disturbance, they also usually have substantially higher macronutrient concentrations in the shoot than do the crops with which they compete (Vengris et al., 1953; Alkämper, 1976; Qasem, 1992). They thus sequester nutrients that would otherwise be available to the crop. Given that weeds have this double competitive advantage, fertilization often favors weeds more than crops (Chapter 5) (Vengris, Colby & Drake, 1955; Alkämper, 1976; Lawson & Wiseman, 1979; Dyck & Liebman, 1994). Consequently, directing water and nutrients toward the crop rather than the weeds is a critical component of weed management. This can be achieved by fertilizing and irrigating directly in the crop rooting zone, timing the application of fertilizer or manure to correspond to the needs of the crop, and using organic materials that inhibit root growth of weeds (Chapter 5).

Below-ground competition works differently in perennial systems where the soil is permanently occupied with roots (Grubb, 1994). In perennial communities, the competitive dominant is usually the species that can deplete the limiting resource to the lowest level (Tilman, 1982). Thus, in unfertilized grasslands, the superior competitor is usually the species that reduces the nutrient element in shortest supply (often nitrogen) to the lowest concentration when the several species are grown in monoculture (Tilman & Wedin, 1991a, 1991b). However, the ability to compete on low-nutrient soils is usually associated with a low ability to adjust growth rate in response to resource availability (Lambers & Poorter, 1992). Consequently, when grassland is fertilized, large, fast-growing species tend to overtop and shade out species adapted to low-nutrient conditions (Smith, Elston & Bunting, 1971). Since fast-growing species grown under favorable conditions tend to have higher concentrations of nitrogen (protein), minerals, and water, and lower concentrations of secondary compounds than slow-growing species (Lambers & Poorter, 1992), their palatability is high. That is, the potentially fast-growing species tend to be desirable forage species (i.e., the "crop"). Thus, fertilizing a weedy grassland tends to have the opposite effect on the competitive balance between crops and weeds of fertilizing a weedy annual crop. However, legumes complicate this picture somewhat. In low-nitrogen soils, nitrogen-fixing legumes tend to have an advantage relative to other species (Donald, 1961), despite relatively slow growth (Grime & Hunt, 1975). Where legumes have been abundant for a while, however, nitrogen accumulates, and taller-growing grasses tend to displace them (Turkington & Harper, 1979). Dung deposition also favors tall-growing grasses that displace legumes and contributes to a shifting mosaic of species in pasture communities (Lieth, 1960). Weed management in pastures is discussed further in Chapter 9.

Survival after emergence

Rates of natural mortality due to disease, herbivory, and drought are usually low for established weeds in annual crops. In the absence of post-emergence weed control, survival rates for annual weeds from the cotyledon stage to maturity usually lie between 25% and 75% (Table 2.8) (Chancellor & Peters, 1972; Naylor, 1972; Sagar & Mortimer, 1976; Weiss, 1981; Mack & Pyke, 1983; Lapointe *et al.*, 1984; Mohler & Callaway, 1992). Sometimes, however, rates of survival to maturity exceed 90% (Young, 1986) or approach 0% (Lindquist *et al.*, 1995) (Table 2.8). Given the high reproductive rates of annual weeds (see the following section), their reported survival rates seem surprisingly high, and probably indicate that most weed mortality occurs prior to emergence or is due to post-emergence control measures. Mortality rate usually decreases with increasing plant size or age (Weiss, 1981; Mohler & Callaway, 1992; Buhler & Owen, 1997).

Table 2.8. *Survival of seedlings emerging in maize in July from establishment to maturity (or for some A. theophrasti, death by hard frost)*

Species	Year	Till	No-till	Rye + Till	Rye + No-till
Amaranthus retroflexus	1986	0.27	0.44	0.25	0.46
	1987	0.33	0.55	0.20	0.55
	1991	0.78	0.82		
	1992	0.58	0.49		
Chenopodium album	1986	0	0	0.15	0
	1987	0.61	0.41	0.43	0.33
Portulaca oleracea	1986	0	0.04	0	0
	1987	–	0.31	–	–
Digitaria sanguinalis	1986	–	0.81	–	0.58
	1987	–	0.84	–	0.79
	1991	0.78	0.70		
	1992	0.60	0.16		
Abutilon theophrasti	1991	0.94	0.96		
	1992	0.63	0.48		

Notes:
No comparisons were significant except for till vs. no-till for *A. retroflexus* in 1986 and 1987, *C. album* in 1987, and *D. sanguinalis* in 1992.
Source: Data for 1986 and 1987 from Mohler & Callaway (1992); data for 1991 and 1992 from C. L. Mohler (unpublished).

Drought occasionally causes substantial mortality during the growth phase in some weed populations (Blackman & Templeman, 1938), but based on the several studies cited above, it does not appear to be a major limiting factor for most annual weeds. Usually drought will have greatest effect on population size during establishment rather than during growth and maturation. Once the weed has a well-established root system, a drought that kills many weeds is likely to severely damage the crop as well.

For introduced weed species, the low mortality rates may be partially due to escape from host-specific natural enemies. However, even in their native range most annual agricultural weeds probably escape serious attack because they represent unpredictable and ephemeral resources (Feeny, 1976). Crop rotation and year-to-year variation in the success of weed control practices create large fluctuations in the density of particular weed species. Moreover, weed populations tend to be patchy (e.g., Wilson & Brain, 1991; Cardina, Sparrow & McCoy, 1996), and because they are usually mixed in with a larger population of a more dominant plant species, namely the crop, they are probably hard for host-specific herbivores to locate (Root, 1973). In an exceptional case, monarch butterfly larvae defoliated 31% to 78% of young *Asclepias syriaca* in soybean in Minnesota (Yenish *et al.*, 1997). However, defoliation by monarch larvae was generally much less in maize due to lower weed seedling density, and in wheat because of a denser canopy during egg-laying by the adults. Thus, temporal variability, patchiness, and interference by the crop with search behavior of host-specific herbivores mitigate against effective control of weeds by these agents in annual crops. These obstacles could be overcome by mass application of host-specific herbivores. Mass application of herbivores for weed control has not been tried on a field scale, though Kremer & Spencer (1989a, 1989b) studied a seed-feeding scentless plant bug on *Abutilon theophrasti* for that purpose.

Many annual weed species, including *Amaranthus retroflexus* and *Chenopodium album*, are highly palatable to humans, indicating that they have poor physical and chemical defenses against generalist herbivores and pathogens (Feeny, 1977). However, management practices, especially tillage, tend to reduce populations of generalist enemies like mollusks (Hunter, 1967). Moreover, the growth rate of annual weeds is so high that once they are beyond the seedling stage they often increase in biomass faster than the herbivores can feed. Thus, once they have established, annual weeds usually escape control by generalist natural enemies as well as host-specific ones.

The generally small response of survival of annual weeds to variation in tillage and mulch (Table 2.8) indicates that manipulation of naturally occurring populations of herbivores and pathogens probably has limited potential for post-emergence weed management in annual cropping systems. This contrasts with the substantial management potential inherent in naturally occurring seed predators (see Chapter 8). Nevertheless, naturally occurring herbivores and pathogens may provide significant weed control in some systems, and these cases need to be identified, and programs developed that exploit this potential. For most annual weeds, the most effective approach to use of natural enemies for post-emergence control will usually be inundative release of pathogens (Chapter 8).

Survival rates of perennial weeds during the period from seedling to first reproduction are typically much lower than for annuals. For example, Mortimer (1976) established four perennials in small plots in grassland (type unspecified) that was (i) turned with a spade, (ii) killed with herbicides and the dead surface vegetation removed, or (iii) clipped to 7.5 cm and then left undisturbed. Proportion of plants surviving over an eight-month period ranged from 0.01 to 0.18 (Figure 2.4). *Poa annua* and *Plantago lanceolata* survival was several-fold lower in the undisturbed grassland than in the plots with inverted soil. In nearly every case, exclusion of invertebrates with insecticides and molluskicides increased survival, although many of the differences were too small to be significant individually (Figure 2.4). Other studies have found

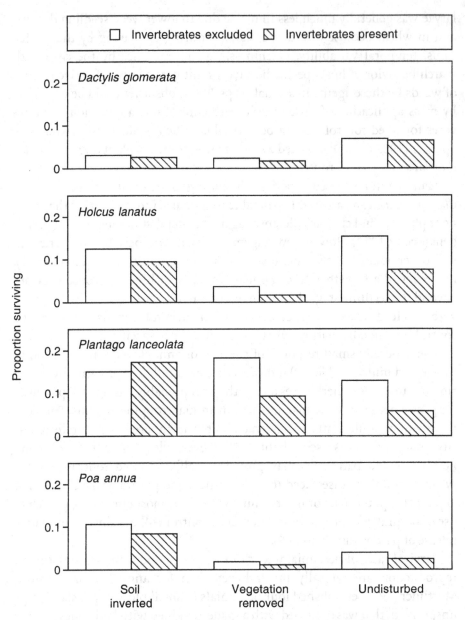

Figure 2.4 Survival of four perennial species from seedling to adult during an eight-month period (September to May) in three grassland disturbance conditions in northern Wales, UK. (After Mortimer, 1976.)

similarly low or even lower survival rates for perennial weeds in perennial vegetation (Forcella & Wood, 1986; McEvoy et al., 1993; Qi, Upadhyaya & Turkington, 1996). In contrast with annual weeds, perennial weeds in perennial vegetation like pastures and hay meadows remain in the same location for an extended period. This makes them easier for herbivores to find and allows the build-up of enemies to effective levels. Even more importantly, the dense, diverse crop provides shelter for the herbivores that attack young weeds, and the shady, moist conditions within the sward facilitate attack by pathogens. In addition, the weed suffers competitive pressure from the crop, and in pastures suffers from grazing and trampling by livestock. These factors can be manipulated to decrease weed survival in perennial systems (Chapters 8 and 9).

Life span and seed production

The potential postgermination life span of weeds in agricultural systems varies from a few months to decades. In most arable cropping systems, actual life span is rarely more than a few years due to periodic tillage. Some annuals are truly monocarpic: resources in vegetative tissues are remobilized to fill seeds and the plant senesces after seed set (e.g., *Chenopodium album*, *Setaria faberi*). However, many annuals shed seeds continuously through much of the growing season and for a substantial proportion of the weed's life span (e.g., *Galinsoga ciliata*, *Digitaria sanguinalis*).

Continuously fruiting annuals tend to dominate the weed flora of fall-sown cereals (Figure 2.5), perhaps because most sprawling species are continuously fruiting and a sprawling habit is well adapted for surviving winter conditions. The early seed production of continuously fruiting annuals like *Portulaca oleracea* and *Stellaria media* adapts them well to cropping systems in which disturbance occurs throughout the growing season; consequently, they are common weeds in gardens and vegetable farms. In contrast, the true monocarpic annuals are more sensitive to frequent weeding or cultivation, but because they do not expend resources on reproduction early in life, they are better able to grow tall and compete with large-statured crops. Consequently, monocarpic annuals tend to dominate the weed flora of tall crops such as maize, and crops such as oat that are rarely cultivated (Figure 2.5).

The seed production capacity of weeds varies greatly both within and between species. In most populations, a few individuals produce many seeds, whereas most individuals produce far fewer (Figure 2.6) (Salisbury, 1942; Mack & Pyke, 1983). This variation in seed production is largely the result of

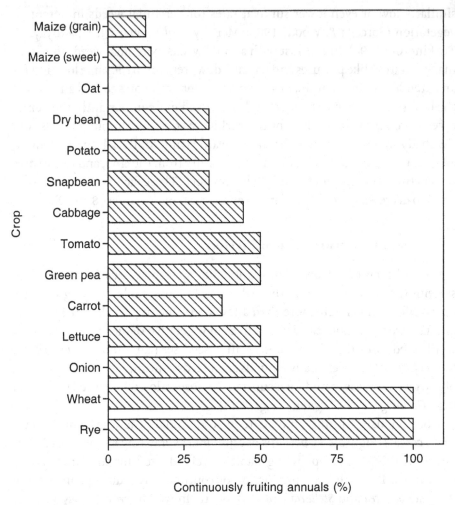

Figure 2.5 Continuous fruiting annuals as a percentage of the common annual weeds in 14 crops in New York state (data from Bridges, 1992). From top to bottom: spring grain crops (sweet corn grouped with maize grain), spring vegetable crops (roughly in decreasing order of competitive ability), winter grain crops.

exponential growth magnifying small differences in seed size, access to nutrients, proximity to crop plants, etc. The extreme skewness in the distribution of seed production over individuals in most weed populations indicates that hand, chemical, or mechanical killing of the largest weeds can reduce weed densities in subsequent crops even if all individuals are not destroyed. Mechanical methods for attacking the large weeds that emerge through crop canopies are discussed in Chapter 4.

Most annual weeds produce a few thousand seeds per individual when

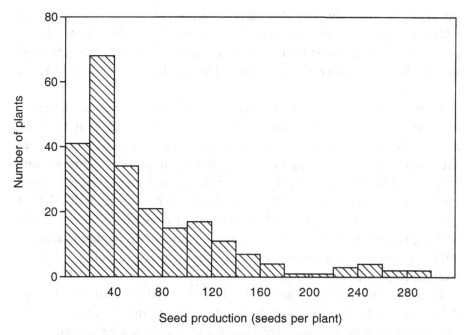

Figure 2.6 Distribution of estimated seed production by 231 individuals of *Amaranthus retroflexus* in no-till sweet corn plots. (C. L. Mohler & M. B. Callaway, unpublished data; see Mohler & Callaway, 1995.)

growing with minimal competition, though some produce 10 000 to 25 000 seeds per plant (Stevens, 1932; Salisbury, 1942), and a few like *Salsola iberica* and *Echinocloa crus-galli* may produce over 100 000 seeds per plant (Young, 1986; Norris, 1992). A few annuals (e.g., *Veronica hederifolia*) produce fewer than 100 seeds per individual (Salisbury, 1942; Boutin & Harper, 1991). Stationary perennial weeds show a similar range in seed production to annuals (Stevens, 1932; Salisbury, 1942), except that monocarpic perennials (biennials) tend to produce more seeds (Stevens, 1932), probably because the observed reproductive output is based on resources captured over more than one season of growth. Comparable data for wandering perennials are lacking, but given that they allot resources to vegetative spread, their seed production probably tends to be less on a per ramet basis. Some wandering perennials (e.g., biotypes of *Cynodon dactylon*) produce no viable seeds at all (Horowitz, 1972; Kigel & Koller, 1985).

Although most weed species potentially produce very many seeds per plant, the actual productivity in a crop is usually much less. C. L. Mohler & M. B. Callaway (unpublished data) found that *Amaranthus retroflexus* produced up to 253 000 seeds per plant, but that individuals emerging in unplanted plots in July as effects of an atrazine application dissipated averaged only 770 seeds per plant, probably due to a short growth period. Moreover, when growing with sweet corn, *A. retroflexus* averaged only 28 seeds per plant. Thus, cultural practices and competition from the crop act as important regulators of weed seed production (Zanin & Sattin, 1988; Mohler & Callaway, 1995; Blackshaw & Harker, 1997).

Several models have shown that including the effects of seed production on future crops lowers economic weed density thresholds by a factor of 3 to 8 relative to the effect of competition on the current crop alone (Cousens *et al.*, 1986; Doyle, Cousens & Moss, 1986; Bauer & Mortensen, 1992). Some authors have argued that the damage inflicted on future crops by seed production is so great that certain weeds should not be allowed to reproduce at all (*Abutilon theophrasti* – Zanin & Sattin, 1988; *Echinochloa crus-galli* – Norris, 1992). Although extreme efforts to prevent spread of new, localized populations are often justified, the economic utility of a zero tolerance policy for long-established populations remains to be demonstrated.

In any case, measures should be taken to reduce seed production. Depending on the phenology of the weed relative to the crop, a substantial proportion of potential seed production can sometimes be prevented by prompt post-harvest weed control measures (Young, 1986; Kegode, Forcella & Durgan, 1999). This is particularly true for cereals and early season vegetables where harvest of the crop releases the weeds from competition at a time in the season when temperature and day length allow rapid growth and maturation of previously suppressed weeds. For example, Webster, Cardina & Loux (1998) found that killing weeds in July or August following wheat harvest controlled 70% to 95% of various weed species in maize the following spring relative to control plots in which weeds were allowed to mature.

In some grain crops, a large portion of the weed seed produced passes through the combine. For example, Ballaré *et al.* (1987*a*) found that <2% of *Datura ferox* in soybean were shed prior to harvest, and that all three of the combines tested took up nearly all capsules. In such cases, if equipment were added to the combine to capture or destroy weed seeds rather than dispersing them with the chaff, substantial reductions in the annual addition of viable seeds to the seed bank could be achieved. Slagell Gossen *et al.* (1998) proposed attaching hammer mills or roller mills to grain combines to destroy weed seeds before they were returned to the field. They found that both types of mill killed a high percentage of *Bromus secalinus* seeds. In many crop–weed systems, however, the benefit of capturing or killing weed seeds in the combine would be small because most of the seeds disperse prior to harvest (Moss, 1983). Although seed collection and post-harvest weed control usually will not provide effective control by themselves, they can contribute substantially to integrated management of weed populations, especially if crop rotation provides some years in which the crop is removed early in the maturation period of the weed.

The capacity of wandering perennial species to produce vegetative propagules is also large. For example, single tubers of *Cyperus esculentus* planted in California and Zimbabwe grew into clones that in one year produced 6900 and 17 700 tubers, respectively (Tumbleson & Kommedahl, 1961; Lapham, 1985). Unlike seed production, which is necessarily preceded by a period of vegetative growth, vegetative reproduction in wandering perennials often begins early in life. Production of new tubers in *Cyperus esculentus* may begin as early as 3 weeks after tubers sprout (Bell *et al.*, 1962). Adventitious buds form on the roots of *Euphorbia esula* and *Cirsium arvense* within 1–2 and 6–8 weeks of seedling emergence, respectively (Selleck, 1958; Bakker, 1960). Consequently, the number of potential individuals produced is roughly proportional to the size of the plant, and tends to increase exponentially when interference is absent (Lapham, 1985). As a result, vigorous competition from a crop is important for reducing vegetative reproduction of wandering perennials (Håkansson, 1968; Håkansson & Wallgren, 1972). Strategies for mechanical control of wandering perennials are discussed extensively in Chapter 4.

Dispersal of seeds and ramets

The spatial scale of dispersal events

Various seed dispersal mechanisms work on greatly differing spatial scales (Figure 2.7). Dispersal of seeds by rain splash or explosive dehiscence of the fruit is generally not effective for more than a few meters, unless assisted by wind. At the other extreme, contaminated crop seed and other agricultural commodities have regularly transported weeds many hundreds of kilometers and have been major sources of intercontinental weed introductions.

Three categories of scale can be usefully distinguished for purposes of managing the spread of weeds: dispersal within fields, between fields, and between regions. Some natural dispersal processes operate primarily at the within-field scale (Figure 2.7). For a weed to spread long distances, other agencies must come into play. Wind, birds, mammals, and water may transport seeds between fields. Natural processes, however, rarely transport species between regions or continents, which is a large part of why regional floras were highly distinct prior to human commerce.

If a newly arrived weed persists within a field, its spread is likely because both natural processes and machinery will tend to disperse the species out

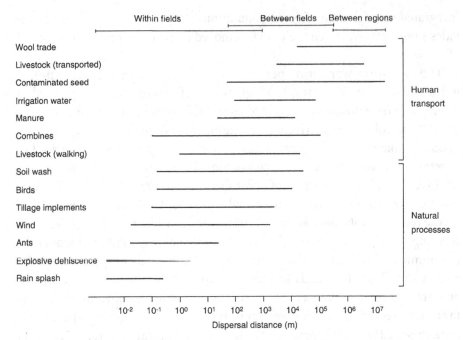

Figure 2.7 Estimated approximate range of dispersal distances for 14 processes that disperse weed seeds.

from its original location. Thus, vigilance and prompt eradication while the population is small are the best defenses against spread within a field. During eradication, care should be taken to avoid spreading the new weed around the field with tillage and cultivation implements.

Although wind, birds, and mammals occasionally move weeds between fields, most spread of weeds between fields is probably the result of human activity. Most of the discussion that follows focuses on dispersal of weeds between fields and regions, areas in which substantial management options exist.

Adaptations for seed dispersal

Weed species have a variety of adaptations for dispersal. Many members of the Compositae and other groups (e.g., *Asclepias* and *Epilobium* spp.) have hairs attached to the seeds that provide buoyancy to aid dispersal by wind. In other species, the plant breaks off at ground level (e.g., *Amaranthus albus*, *Sisymbrium altissimum*), or the inflorescence detaches as a unit (e.g., *Panicum capillare*) and rolls with the wind over the ground surface. Seeds of some species have hairs, spines, or hooks that adapt them to disperse in the fur of animals, and these are often equally effective at attaching to clothing (e.g.,

Cenchrus incertus, Arctium lappa). A few agricultural weeds, many of them woody, have fleshy fruits that entice birds to swallow the seeds (e.g., *Solanum* spp., *Toxicodendron radicans*). Since fruit-feeding birds usually lack the sort of alimentary tract required to digest seeds, these are mostly regurgitated or passed out with the feces, often at a considerable distance from the parent plant. Another small group of weed species have explosive dehiscence mechanisms that catapult seeds as much as a few meters from the parent (e.g., *Oxalis stricta* – Lovett Doust, MacKinnon & Lovett Doust, 1985). A very few weeds have oily bodies attached to the seeds (eliasomes) that entice ants to carry the seeds to their nests (e.g., *Fumaria officinalis*, *Euphorbia esula* – Pemberton & Irving, 1990). After the ants have bitten off the eliasome, the seeds are then discarded and may subsequently germinate. Seed-feeding ants also regularly disperse seeds accidentally during foraging. The effectiveness of these dispersal mechanisms are evaluated with further examples in Salisbury (1961, pp. 97–143) and Cousens & Mortimer (1995, pp. 55–85).

Although some agricultural weed species show obvious adaptations for dispersal, most do not. Of the 50 weeds of arable land discussed by Salisbury (1961), 76% lack any apparent adaptation for dispersal. Consequently, most weed seeds fall close to the parent plant. For example, Howard *et al.* (1991) found that *Bromus sterilis* and *B. interruptus* seeds shed in a winter wheat field fell in a normally distributed pattern about the parent plant with standard deviations of 31 cm and 19 cm, respectively.

Prior to dispersal by humans (see next section), species without obvious dispersal adaptations probably dispersed in soil washed along streams, in mud clinging to large animals, and in the guts of birds and mammals. The seeds of many weed species pass through the digestive tracts of grazing animals without damage (Kirk & Courtney, 1972; Takabayashi, Kubota & Abe, 1979; Blackshaw & Rode, 1991), and may be retained in the gut for several days (Burton, 1948; Özer, 1979), thereby allowing deposition at sites distant from their point of origin. Although most of the seeds ingested by seed-eating birds are probably destroyed, a few apparently pass through the digestive tract unharmed (Proctor, 1968; Aison, Johnson & Harger, 1984).

The frequency distribution of distance traveled by wind-dispersed seeds is typically very skewed (Smith & Kok, 1984; Feldman & Lewis, 1990). Consequently, most wind-dispersed seeds land within a few meters of the parent plant (Plummer & Keever, 1963; Michaux, 1989), but a few seeds may be caught in updrafts and occasionally travel far enough to reach nearby fields. Whereas only 10% of the arable weeds discussed by Salisbury (1961) are wind-dispersed, 28% of the species he lists as common in British upland grasslands have appendages that facilitate wind dispersal. Many of these wind-dispersed grassland weeds thrive on road margins, ditch banks, fence lines, and hedgerows where they are relatively free from trampling and grazing by animals and cutting by mowing machines. From there, they disperse into fields, especially during the establishment year of leys and during periods when pastures are rested from grazing. Consequently, preventing fruiting of these weeds in ruderal habitats adjacent to farm fields is an important part of their management.

Human dispersal of weeds: risks and potential reductions

Dispersal in contaminated seed

Weed contamination of crop seed has been a major source of introductions at all scales from continents to individual fields. Weed seeds are still regularly transported between countries in seed shipments (Tasrif et al., 1991; Huelma, Moody & Mew, 1996). Probably the only effective method for preventing this is inspection of large samples from every international shipment (Tasrif et al., 1991). This could be facilitated by computer visualization procedures that identify contaminated samples.

Improved seed-cleaning techniques and seed certification programs have greatly reduced the spread of weeds between farms in some regions, and have led to the near elimination of some weed species (Salisbury, 1961). However, even in developed countries, some growers still plant contaminated seed (Tonkin, 1982; Dewey, Thill & Foote, 1985; Dewey & Whitesides, 1990), and this is the norm in most developing countries. For example, Rao & Moody (1990) found an average of 3800 weed seeds per kg (17 species) after rice was processed by farmers in the Philippines, and 660 seeds per kg (15 species) after local commercial cleaning. Use of contaminated seed guarantees that the worst weeds will become ubiquitous throughout all the fields of a farm, and leads to spread between farms when seed is traded. In addition to mechanical cleaning, the propagation of weeds with crop seed can be greatly reduced by reserving one field or part of a field for production of next year's seed, and weeding this area intensively. This reflects the approach used to produce certified seed (Wellington, 1960), but may be more cost-effective than purchasing certified seed for many growers in both developed and developing countries.

Dispersal with manure, feed, and transported animals

Weed seeds may be moved to previously uninfested fields by application of manure. Mt. Pleasant & Schlather (1994) found a total of 13 grass and 35 broadleaf species in manure samples taken from 26 New York dairy farms. On most farms, the density of weed seeds in the manure was too low to significantly change seed density in the soil, but they concluded that manure did have a potential for spreading weeds. In particular, *Abutilon theophrasti* was probably introduced onto many New York farms during the last 30 years in contaminated feed, and then spread from field to field in manure once it had established. Weed density in manure may be greater in less-developed countries where animals are regularly grazed on weedy stubble after crop harvest. Dastgheib (1989) estimated that farmers in Iran were sowing nearly 10 million weed seeds per hectare per year with the sheep manure used to fertilize a wheat/paddy rice double-cropping system.

Mack (1981) documented the spread of *Bromus tectorum* in western North America. From a few initial introductions, probably in contaminated wheat seed, the weed first spread along rail lines and cattle trails, and then outward from these corridors to become a dominant species on much of the rangeland between the Cascade–Sierra Nevada and Rocky Mountains. Apparently, disposal of manure and bedding from cattle cars was particularly effective in creating secondary points of introduction. Since cattle are today often raised in one location, finished in another, and slaughtered at a third, the potential for spread of weeds during cattle transport remains large.

Several strategies can be used to reduce the risk of spreading weeds with manure. One is use of clean concentrates and fodder. Seeds cannot occur in the manure unless they are first present in the feed. Intensive pasture management can prevent weeds from going to seed (Chapter 9) (Sharrow & Mosher, 1982; Popay & Field, 1996). Mowing is also effective in this regard. At the very least, mill screenings should not be fed to animals unless they are first heat-treated to kill weed seeds. Ensiling is highly effective at killing seeds of most weed species (Zahnley & Fitch, 1941; Takabayashi, Kubota & Abe, 1979; Blackshaw & Rode, 1991). Consequently, ensiling forage that is contaminated with a problem weed may be preferable to direct feeding. However, a substantial percentage of *Abutilon theophrasti*, *Convolvulus arvensis*, and *Polygonum convolvulus* seeds can survive ensiling (Zahnley & Fitch, 1941; Blackshaw & Rode, 1991). High-temperature composting, anaerobic fermentation, or oven drying can greatly reduce the number of viable weed seeds in manure (Kirk & Courtney, 1972; Takabayashi, Kubota & Abe, 1979; Bloemhard et al., 1992; Šarapatka, Holub & Lhotská, 1993; Tompkins, Chaw & Abiola, 1998). However, the outside of a compost pile will not heat sufficiently to kill seeds, so simply piling the manure for a few weeks without turning may leave high densities of viable seeds (Cudney et al., 1992). Finally, manure from off the farm should be evaluated for weed seeds before transport.

Dispersal in raw wool and other bulk commodities

Weed species adapted to cling to animal fur may be transported thousands of kilometers in raw wool, and then dispersed with textile wastes. Several hundred species have apparently been introduced into Britain by this route (Dony, 1953; Salisbury, 1961), although many never became naturalized. Weeds probably also move in raw cotton. However, this remains to be documented, and the problem is likely smaller than with wool since cotton fields often have fewer weed species than sheep pastures.

The recent rapid shift of the textile industry from developed to developing countries is probably providing new opportunities for weed introductions in raw fiber. In addition to inspection and quarantine measures, introductions of weeds with textile raw materials can be curtailed by heat treatment or high-temperature composting of wastes prior to application to land.

Dispersal by machinery

Tillage machinery moves few seeds further than three meters within a field and moves most seeds only a meter or less (Howard et al., 1991; Rew, Froud-Williams & Boatman, 1996; Rew & Cussans, 1997; Mayer, Albrecht & Pfadenhauer, 1998). However, the few seeds that are carried long distances can form foci from which the weed can spread in future years. Roots, rhizomes, and tubers of wandering perennials can catch on the shanks of tine implements, particularly once the soil is loosened by primary tillage. Although most fragments do not move far, the few that do can spread an infestation over large areas in a single tillage operation (Schippers et al., 1993).

Few studies have quantitatively examined potential between-field movement of seeds in soil on tires and machinery. Schmidt (1989) observed over 3900 seedlings of 124 species emerging in the soil scraped from an automobile that had been used for field research in Germany. Mayer, Albrecht & Pfadenhauer (1998) found that seeds were moved between fields on tractor tires and a rotary tiller, but not by a plow, rotary harrow, or heavy cultivator. In their experiment, the equipment passed through 25 m of clean soil after encounter with the seeds.

Some idea of the potential for between-field dispersal in soil clinging to tillage machinery can be gathered from the density of seeds in soil. Many species commonly achieve densities of a few thousand seeds per m^2 (Jensen, 1969; Ball & Miller, 1989). Assuming a plowing depth of 20 cm, 1000 seeds per m^2 is about 1 seed per 200 g of soil. This represents a fairly small risk, except that (i) conservation tillage keeps seeds close to the surface where they are more likely to be picked up, (ii) field edges and headlands are typically plowed and cultivated last, and they are usually weedier than the rest of the field

(Marshall, 1989; Wilson & Aebischer, 1995), and (iii) tractor tires can pick up surface-lying seeds from along the field border as they leave. Consequently, seed movement in soil is probably the source of many new weed infestations, particularly of nearby fields. Nevertheless, the risk will usually be small until the weed becomes dense in the potential source field. Movement of vegetative propagules with soil on farm machinery is probably the major method of spread for some species that do not produce viable seeds (e.g., *Panicum repens* – Wilcut *et al.*, 1988).

Several studies on movement of weed seeds by combine harvesters indicate that most seeds are deposited within 10 m of the source but that some are dispersed as far as 50 m or more in the direction of travel (Ballaré *et al.*, 1987a; McCanny & Cavers, 1988; Howard *et al.*, 1991; Ghersa *et al.*, 1993; Rew, Froud-Williams & Boatman, 1996). Thus, combines can rapidly spread weeds throughout a field (Ballaré *et al.*, 1987b), with significant potential effects on crop yield (Maxwell & Ghersa, 1992).

The spread of weeds between fields by combines is probably also frequent, and prudence indicates that a combine should be cleaned before it is moved into a new field. McCanny & Cavers (1988) found that more seeds lodged in the central divider assembly of the maize header than elsewhere on the combine, and that this could be effectively cleaned by vacuuming. Data on the spread of weeds between fields by combines are badly needed. An interesting study could be made by cleaning trapped seeds out of combines each night as a custom combining operation works its way north through the midwestern USA during grain harvest. Comparison of the weed species removed from the combine after harvesting a field with the flora of the next field on the schedule would indicate the likelihood of long-distance spread of species by this route.

Dispersal in irrigation water

Seeds of most weed species can survive several months of immersion in fresh water (Comes, Bruns & Kelley, 1978), and most will float, particularly if chaff is retained on the seed, or if pieces of inflorescence fall onto the water. Consequently, many species of weed seeds disperse in irrigation water. Wilson (1980) found 77 species of weed seeds in samples of irrigation water in Nebraska, and Kelley & Bruns (1975) observed 77, 84, and 137 species in samples taken in eastern Washington in three years. In both studies, the density of seeds deposited was not sufficiently high to warrant concern, but the potential for introduction of new weed species to fields by irrigation water was substantial. Consequently, Kelley & Bruns (1975) recommended that irrigation water be screened to remove seeds. Both studies found that seed density in water increased as water traveled down canals with weedy banks. In contrast, Kelley & Bruns (1975) found no increase in seeds for water flowing in a canal whose banks were kept free of weeds by grazing, tillage, and burning. Thus, weed control on canal banks can reduce the dispersal of weeds into fields.

Weed dispersal as a management issue

The many studies cited in the preceding sections largely agree that the density of weed propagules dispersing into an area is usually insufficient to create substantial competitive pressure against crops. Instead, weed populations appear to reach competitively effective densities primarily through local population growth. The central problem that human-facilitated weed dispersal poses for management is therefore the prevention of new infestations, including both the arrival of new species onto farms and the multiplication of foci for local population growth within fields. Consequently, from the farmer's perspective, movement of weeds that are already widespread on the farm can largely be ignored, and efforts instead concentrated on preventing the spread of particular weeds that are both competitive and currently absent from all or much of the farm. From the weed scientist's perspective, the key issue with weed dispersal is prevention of the spread of economically damaging species through the landscape. Surprisingly little research directly addresses this problem. Effectively preventing the spread of weeds probably requires region-wide co-ordination of education and containment efforts analogous to infectious disease control activities of public health agencies. This is discussed further in Chapter 10.

Dispersal and spatial pattern

Weed populations usually have noticeably clumped spatial patterns. The most common pattern is for the frequency of occurrence of individuals to follow a negative binomial distribution (Zanin, Berti & Zuin, 1989; Wiles *et al.*, 1992; Mortensen, Johnson & Young, 1993; Mulugeta & Stoltenberg, 1997). For wandering perennial species, much of the clumped pattern is the result of vegetative growth. Variation in soil fertility, tilth, drainage, and the density and vigor of the crop causes variation in plant size of both perennial and annual weeds. In addition, variation in the size, burial depth, and genetic constitution among seeds, and in the time of emergence of seedlings, leads to great variation in the size of annual plants. The resulting clumped distribution of weed biomass in any given year leads to a clumped distribution of density the following year, since reproductive output is correlated with plant size (e.g., Mohler & Callaway, 1995), and most weed seeds only disperse a short distance. Even if a weed is initially distributed uniformly across a field, as might happen, for example, if it were introduced in contaminated seed, the high variance in reproductive output among individual plants (Figure 2.6) guarantees that the species will quickly become clumped. However, the initial distribution of a species within the field is usually far from uniform. Often the weed expands from an initial point of establishment, either as an expanding front, or with new inoculation points appearing elsewhere in the field by intermediate-distance dispersal from the original site (Cousens & Mortimer, 1995, pp. 217–42). Some weed populations may be entirely maintained by dispersal from an adjacent habitat that is more suitable for the plant's reproduction. In such cases, the species is likely to be more common along the edge of the field (Marshall, 1989; Wilson & Aebischer, 1995). However, even weeds that are well adapted to farm fields are often more abundant along field edges due to soil compaction, lower crop competition, and inefficient application of herbicides and cultivation (Wilson & Aebischer, 1995).

The persistent storage organs of perennial weeds and the persistent seed bank of many annual weeds insures that weed patches tend to remain in the same locations in successive seasons (Dieleman & Mortensen, 1999). Moreover, all the factors that disperse seeds within a field leave most seeds within a few meters (or less) of the parent plant. Consequently, once a clumped distribution of weeds is formed, it tends to persist. For example, Wilson & Brain (1991) found that *Alopecurus myosuroides* tended to occur in the same locations within a farm year after year. Patches tended to persist even through several years of sod. This is reasonable for a species with a persistent seed bank since dispersal forces are particularly weak after the seeds are in the soil, especially if the ground is not tilled. Thus, although variation in the soil conditions within a field doubtless contributes to the maintenance of some weed patches, the dynamics of reproduction seem sufficient to explain many of the clumped patterns observed.

Dispersal also creates patchiness at larger scales. For example, McCanny, Baugh & Cavers (1988) repeatedly surveyed wild *Panicum miliaceum* populations in two Ontario townships, and found it present in 10% to 20% of the tilled fields. They found that the number of infested fields did not change much, but that the species had a probability of local extinction of 17% to 48% depending on the year. As some populations went extinct, new ones formed by colonization of previously uninfested fields. Perennial weeds and species with persistent seed banks probably behave similarly, but on longer time scales.

Conclusions

The discussion in the preceding sections indicates that agricultural weeds generally share certain properties, including small seed size, high relative growth rate, low early absolute growth rate, intolerance to stress, and high reproductive capacity. They differ from crops in most of these respects, and these differences form the basis for a variety of weed management tactics.

Despite similarities among weeds, weed species differ with respect to longevity, ability to spread vegetatively, temporal pattern of seed production, relative seed size, ability of seeds to persist in the soil, and season of germination. Divergent life history characteristics allow different weed species to prosper in differing sorts of crop production systems and may require divergent management strategies for successful control.

The several life cycle stages of a weed provide separate opportunities for control. Constraining a weed population at several points in the life cycle by using multiple partial controls is the essence of integrated weed management and is the basic approach for meeting the objectives of weed management proposed in Chapter 1. Chapters 4 through 9 discuss methods for attacking weeds at various stages in their life cycle. Often, reduction in the number of individuals passing through a stage improves management options in succeeding stages. In some cases, a particular tactic may be quite impractical unless the population is constrained in other ways as well. Consequently, the potential effectiveness of a particular tactic may be much greater than is indicated by studies that treat the factor in isolation.

REFERENCES

Abdalla, F. H., & Roberts, E. H. (1968). Effects of temperature, moisture, and oxygen on the induction of chromosome damage in seeds of barley, broad beans, and peas during storage. *Annals of Botany*, New Series, **32**, 119–36.

Aison, S., Johnson, M. K., & Harger, T. R. (1984). Role of birds in dispersal of itchgrass (*Rottboellia exaltata* (L.) L.f.) seeds in the southeastern USA. *Protection Ecology*, **6**, 307–13.

Akanda, R. U., Mullahey, J. J., & Shilling, D. G. (1996). Environmental factors affecting germination of tropical soda apple (*Solanum viarum*). *Weed Science*, **44**, 570–4.

Alkämper, J. (1976). Influence of weed infestation on effect of fertilizer dressings. *Pflanzenshutz-Nachrichten Bayer*, **29**, 191–235.

Andrews, R. E., & Newman, E. I. (1970). Root density and competition for nutrients. *Oecologia Plantarum*, **5**, 319–34.

Ascencio, J., & Lazo, J. V. (1997). Growth evaluation during the vegetative phase of dicotyledonous weeds and under phosphorus deficiency. *Journal of Plant Nutrition*, **20**, 27–45.

Baker, H. G. (1965). Characteristics and modes of origin of weeds. In *The Genetics of*

Colonizing Species, ed. H. G. Baker, & G. L. Stebbins, pp. 147–68. New York: Academic Press.

Bakker, D. (1960). A comparative life-history study of *Cirsium arvense* (L.) Scop. and *Tussilago farfara* L., the most troublesome weeds in the newly reclaimed polders of the former Zuiderzee. In *The Biology of Weeds*, ed. J. L. Harper, pp. 205–22. Oxford, UK: Blackwell.

Ball, D. A., & Miller, S. D. (1989). A comparison of techniques for estimation of arable soil seedbanks and their relationship to weed flora. *Weed Research*, 29, 365–73.

Ballaré, C. L., Scopel, A. L., Ghersa, C. M., & Sánchez, R. A. (1987a). The demography of *Datura ferox* (L.) in soybean crops. *Weed Research*, 27, 91–102.

Ballaré, C. L., Scopel, A. L., Ghersa, C. M., & Sánchez, R. A. (1987b). The population ecology of *Datura ferox* in soybean crops: a simulation approach incorporating seed dispersal. *Agriculture, Ecosystems and Environment*, 19, 177–88.

Barralis, G., Chadoeuf, R., & Lonchamp, J. P. (1988). Longévité des semences de mauvaises herbes annuelles dans un sol cultivé. *Weed Research*, 28, 407–18.

Baskin, C. C., & Baskin, J. M. (1988). Germination ecophysiology of herbaceous plant species in a temperate region. *American Journal of Botany*, 75, 286–305.

Baskin, C. C., & Baskin, J. M. (1998a). *Seeds: Ecology, Biogeography, and Evolution of Dormancy and Germination*. San Diego, CA: Academic Press.

Baskin, C. C., & Baskin, J. M. (1998b). Ecology of seed dormancy and germination in grasses. In *Population Biology of Grasses*, ed. G. P. Cheplich, pp. 30–83. Cambridge, UK: Cambridge University Press.

Baskin, J. M., & Baskin, C. C. (1980). Ecophysiology of secondary dormancy in seeds of *Ambrosia artemisiifolia*. *Ecology*, 61, 475–80.

Baskin, J. M., & Baskin, C. C. (1983). Seasonal changes in the germination responses of buried seeds of *Arabidopsis thaliana* and ecological interpretation. *Botanical Gazette*, 144, 540–3.

Baskin, J. M., & Baskin, C. C. (1985). The annual dormancy cycle in buried weed seeds: a continuum. *BioScience*, 35, 492–8.

Baskin, J. M., & Baskin, C. C. (1986). Seasonal changes in the germination responses of buried witchgrass (*Panicum capillare*) seeds. *Weed Science*, 34, 22–4.

Baskin, J. M., & Baskin, C. C. (1987). Environmentally induced changes in the dormancy states of buried weed seeds. In *1987 British Crop Protection Conference – Weeds*, pp. 695–706. Croydon, UK: British Crop Protection Council.

Bauer, T. A., & Mortensen, D. A. (1992). A comparison of economic and economic optimum thresholds for two annual weeds in soybeans. *Weed Technology*, 6, 228–35.

Bell, R. S., Lachman, W. H., Rahn, E. M., & Sweet, R. D. (1962). *Life History Studies as related to Weed Control in the Northeast*. 1. Nutgrass, University of Rhode Island Agricultural Experiment Station Bulletin no. 364. Kingston, RI: University of Rhode Island Agricultural Experiment Station.

Bello, I. A., Owen, M. D. K., & Hatterman-Valenti, H. M. (1995). Effect of shade on velvetleaf (*Abutilon theophrasti*) growth, seed production, and dormancy. *Weed Technology*, 9, 452–5.

Benvenuti, S., & Macchia, M. (1995). Effect of hypoxia on buried weed seed germination. *Weed Research*, 35, 343–51.

Blackman, G. E., & Black, J. N. (1959). Physiological and ecological studies in the analysis of plant environment. XI. A further assessment of the influence of shading on the growth of different species in the vegetative phase. *Annals of Botany*, New Series, **23**, 51–63.

Blackman, G. E., & Templeman, W. G. (1938). The nature of the competition between cereal crops and annual weeds. *Journal of Agricultural Science*, **28**, 247–71.

Blackshaw, R. E., & Harker, K. N. (1997). Scentless chamomile (*Matricaria perforata*) growth, development, and seed production. *Weed Science*, **45**, 701–5.

Blackshaw, R. E., & Rode, L. M. (1991). Effect of ensiling and rumen digestion by cattle on weed seed viability. *Weed Science*, **39**, 104–8.

Bloemhard, C. M. J., Arts, M. W. M. F., Scheepens, P. C., & Elema, A. G. (1992). Thermal inactivation of weed seeds and tubers during drying of pig manure. *Netherlands Journal of Agricultural Science*, **40**, 11–19.

Bostock, S. J. (1978). Seed germination strategies of five perennial weeds. *Oecologia (Berlin)*, **36**, 113–26.

Boutin, C., & Harper, J. L. (1991). A comparative study of the population dynamics of five species of *Veronica* in natural habitats. *Journal of Ecology*, **79**, 199–221.

Brennan, T., Willemsen, R., Rudd, T., & Frenkel, C. (1978). Interaction of oxygen and ethylene in the release of ragweed seeds from dormancy. *Botanical Gazette*, **139**, 46–9.

Bridges, D. C. (ed.) (1992). *Crop Losses Due to Weeds in the United States – 1992*. Champaign, IL: Weed Science Society of America.

Brust, G. E., & House, G. J. (1988). Weed seed destruction by arthropods and rodents in low-input soybean agroecosystems. *American Journal of Alternative Agriculture*, **3**, 19–25.

Buchholtz, K. P., Grantz, R. L., Sweet, R. D., Timmons, F. L., & Vostral, H. J. (1967). Subcommittee on Standardization of Abbreviations, Terms and Definitions. *Weeds*, **15**, 388–9.

Buhler, D. D., & Owen, M. D. K. (1997). Emergence and survival of horseweed (*Conyza canadensis*). *Weed Science*, **45**, 98–101.

Burnside, O. C., Wilson, R. G., Weisberg, S., & Hubbard, K. G. (1996). Seed longevity of 41 weed species buried 17 years in eastern and western Nebraska. *Weed Science*, **44**, 74–86.

Burton, G. W. (1948). Recovery and viability of seeds of certain southern grasses and lespedeza passed through the bovine digestive tract. *Journal of Agricultural Research*, **76**, 95–103.

Cardina, J., Sparrow, D. H., & McCoy, E. L. (1996). Spatial relationships between seedbank and seedling populations of common lambsquarters (*Chenopodium album*) and annual grasses. *Weed Science*, **44**, 298–308.

Chancellor, R. J. (1964). The depth of weed seed germination in the field. In *Proceedings of the 7th British Weed Control Conference*, pp. 599–606. London: British Crop Protection Council.

Chancellor, R. J. (1986). Decline of arable weed seeds during 20 years in soil under grass and the periodicity of seedling emergence after cultivation. *Journal of Applied Ecology*, **23**, 631–7.

Chancellor, R. J., & Peters, N. C. B. (1972). Germination periodicity, plant survival and seed production in populations of *Avena fatua* L. growing in spring barley. In *Proceedings of the 11th British Weed Control Conference*, pp. 218–25. London: British Crop Protection Council.

Chapin, F. S. III, Groves, R. H., & Evans, L. T. (1989). Physiological determinants of growth rate in response to phosphorus supply in wild and cultivated *Hordeum* species. *Oecologia*, **79**, 96–105.

Chepil, W. S. (1946). Germination of weed seeds. I. Longevity, periodicity of germination, and vitality of seeds in cultivated soil. *Scientific Agriculture*, **26**, 307–46.

Cohen, D. (1966). Optimizing reproduction in a randomly varying environment. *Journal of Theoretical Biology*, **12**, 119–29.

Comes, R. D., Bruns, V. F., & Kelley, A. D. (1978). Longevity of certain weed and crop seeds in fresh water. *Weed Science*, **26**, 336–44.

Conn, J. S., & Deck, R. E. (1995). Seed viability and dormancy of 17 weed species after 9.7 years of burial in Alaska. *Weed Science*, **43**, 583–5.

Cousens, R., & Mortimer, M. (1995). *Dynamics of Weed Populations*. Cambridge, UK: Cambridge University Press.

Cousens, R., Doyle, C. J., Wilson, B. J., & Cussans, G. W. (1986). Modelling the economics of controlling *Avena fatua* in winter wheat. *Pesticide Science*, **17**, 1–12.

Cox, W. J., Zobel, R. W., van Es, H. M., & Otis, D. J. (1990). Tillage effects on some soil physical and corn physiological characteristics. *Agronomy Journal*, **82**, 806–12.

Cudney, D. W., Wright, S. D., Shultz, T. A., & Reints, J. S. (1992). Weed seed in dairy manure depends on collection site. *California Agriculture*, **46**(3), 31–2.

Dastgheib, F. (1989). Relative importance of crop seed, manure and irrigation water as sources of weed infestation. *Weed Research*, **29**, 113–16.

Dawson, J. H., & Bruns, V. F. (1975). Longevity of barnyardgrass, green foxtail and yellow foxtail seeds in soil. *Weed Science*, **23**, 437–40.

Dewey, S. A., & Whitesides, R. E. (1990). Weed seed analyses from four decades of Utah small grain drillbox surveys. *Proceedings of the Western Society of Weed Science*, **43**, 69–70.

Dewey, S. A., Thill, D. C., & Foote, P. W. (1985). *Weed Seed Contamination of Cereal Grain Seedlots: A Drillbox Survey*, Current Information Series no. 767. Moscow, ID: Cooperative Extension Service, Agricultural Experiment Station, University of Idaho.

Dieleman, J. A., & Mortensen, D. A. (1999). Characterizing the spatial pattern of *Abutilon theophrasti* seedling patches. *Weed Research*, **39**, 455–67.

Donald, C. M. (1961). Competition for light in crops and pastures. In *Mechanisms in Biological Competition, Symposium of the Society for Experimental Biology*, 15, pp. 283–313. New York: Academic Press.

Donald, W. W., & Zimdahl, R. L. (1987). Persistence, germinability, and distribution of jointed goatgrass (*Aegilops cylindrica*) seed in soil. *Weed Science*, **35**, 149–54.

Dony, J. G. (1953). Wool aliens in Bedfordshire. In *The Changing Flora of the British Isles*, ed. J. E. Lousley, pp. 160–3. Oxford, UK: Botanical Society of the British Isles.

Dou, Z., Fox, R. H., & Toth, J. D. (1995). Seasonal soil nitrate dynamics in corn as affected by tillage and nitrogen source. *Soil Science Society of America Journal*, **59**, 858–64.

Doyle, C. J., Cousens, R., & Moss, S. R. (1986). A model of the economics of controlling *Alopecurus myosuroides* Huds. in winter wheat. *Crop Protection*, **5**, 143–50.

Dunan, C. M., & Zimdahl, R. L. (1991). Competitive ability of wild oats (*Avena fatua*) and barley (*Hordeum vulgare*). *Weed Science*, **39**, 558–63.

Dyck, E., & Liebman, M. (1994). Soil fertility management as a factor in weed control: the effect of crimson clover residue, synthetic nitrogen fertilizer, and their interaction on emergence and early growth of lambsquarters and sweet corn. *Plant and Soil*, **167**, 227–37.

Edwards, M. M. (1969). Dormancy in seeds of charlock. IV. Interrelationships of growth, oxygen supply and concentration of inhibitor. *Journal of Experimental Botany*, **20**, 876–94.

Egley, G. H. (1995). Seed germination in soil: dormancy cycles. In *Seed Development and Germination*, ed. J. Kigel, & G. Galili, pp. 529–43. New York: Marcel Dekker.

Elder, R. H., & Osborne, D. J. (1993). Function of DNA synthesis and DNA repair in the survival of embryos during early germination and in dormancy. *Seed Science Research*, **3**, 43–53.

Fausey, J. C., & Renner, K. A. (1997). Germination, emergence, and growth of giant foxtail (*Setaria faberi*) and fall panicum (*Panicum dichotomiflorum*). *Weed Science*, **45**, 423–5.

Fawcett, R. S., & Silfe, F. W. (1978). Effects of field applications of nitrate on weed seed germination and dormancy. *Weed Science*, **26**, 594–6.

Feeny, P. (1976). Plant apparency and chemical defense. In *Recent Advances in Phytochemistry*, vol. 10, *Biochemical Interaction between Plants and Insects*, ed. J. W. Wallace, & R. L. Mansell, pp. 1–40. New York: Plenum Press.

Feeny, P. (1977). Defensive ecology of the Cruciferae. *Annals of the Missouri Botanical Garden*, **64**, 221–34.

Feldman, S. R., & Lewis, J. P. (1990). Output and dispersal of propagules of *Carduus acanthoides* L. *Weed Research*, **30**, 161–9.

Fellows, G. M., & Roeth, F. W. (1992). Factors influencing shattercane (*Sorghum bicolor*) seed survival. *Weed Science*, **40**, 434–40.

Fenner, M. (1978). Susceptibility to shade in seedlings of colonizing and closed turf species. *New Phytologist*, **81**, 739–44.

Forcella, F., & Wood, H. (1986). Demography and control of *Cirsium vulgare* (Savi) Ten. in relation to grazing. *Weed Research*, **26**, 199–206.

Forcella, F., Wilson, R. G., Dekker, J., Kremer, R. J., Cardina, J., Anderson, R. L., Alm, D., Renner, K. A., Harvey, R. G., Clay, S., & Buhler, D. D. (1997). Weed seed bank emergence across the corn belt. *Weed Science*, **45**, 67–76.

Forsyth, S. F., & Watson, A. K. (1985). Predispersal seed predation of Canada thistle. *Canadian Entomologist*, **117**, 1075–81.

Friedman, M., & Henika, P. R. (1991). Mutagenicity of toxic weed seeds in the Ames Test: jimson weed (*Datura stramonium*), velvetleaf (*Abutilon theophrasti*), morning glory (*Ipomoea* spp.), and sicklepod (*Cassia obtusifolia*). *Journal of Agricultural and Food Chemistry*, **39**, 494–501.

Froud-Williams, R. J. (1985). Dormancy and germination of arable grass-weeds. *Aspects of Applied Biology*, **9**, 9–18.

Froud-Williams, R. J. (1987). Survival and fate of weed seed populations: interaction with cultural practice. In *1987 British Crop Protection Conference – Weeds*, vol. 2, pp. 707–18. Croydon, UK: British Crop Protection Council.

Froud-Williams, R. J., Chancellor, R. J., & Drennan, D. S. H. (1984). The effects of seed burial and soil disturbance on emergence and survival of arable weeds in relation to minimal cultivation. *Journal of Applied Ecology*, **21**, 629–41.

Gallagher, R. S., & Cardina, J. (1998). Ecophysiological aspects of phytochrome-mediated germination in soil seed banks. *Aspects of Applied Biology*, **51**, 1–8.

Geber, M. A. (1989). Interplay of morphology and development on size inequality: a *Polygonum* greenhouse study. *Ecological Monographs*, **59**, 267–88.

Gebhardt, M. R., Daniel, T. C., Schweizer, E. E., & Allmaras, R. R. (1985). Conservation tillage. *Science*, **230**, 625–30.

Ghersa, C. M., Benech Arnold, R. L., & Martinez-Ghersa, M. A. (1992). The role of fluctuating temperatures in germination and establishment of *Sorghum halepense*. Regulation of germination at increasing depths. *Functional Ecology*, **6**, 460–8.

Ghersa, C. M., Martinez-Ghersa, M. A., Satorre, E. H., Van Esso, M. L., & Chichotky, G. (1993). Seed dispersal, distribution and recruitment of seedlings of *Sorghum halepense* (L.) Pers. *Weed Research*, **33**, 79–88.

Givnish, T. J. (1988). Adaptation to sun and shade: a whole-plant perspective. In *Ecology of Photosynthesis in Sun and Shade*, ed. J. R. Evans, S. von Caemmerer, & W. W. Adams III, pp. 63–92. East Melbourne, Australia: Commonwealth Scientific and Industrial Research Organization.

Gleichsner, J. A., & Appleby, A. P. (1989). Effect of depth and duration of seed burial on ripgut brome (*Bromus rigidus*). *Weed Science*, **37**, 68–72.

Górski, T. (1975). Germination of seeds in the shadow of plants. *Physiologia Plantarum*, **34**, 342–6.

Grant, J. D. (1983). The activities of earthworms and the fates of seeds. In *Earthworm Ecology*, ed. J. E. Satchell, pp. 107–22. London: Chapman & Hall.

Grime, J. P., & Hunt, R. (1975). Relative growth-rate: its range and adaptive significance in a local flora. *Journal of Ecology*, **63**, 393–422.

Grubb, P. J. (1994). Root competition in soils of different fertility: a paradox resolved? *Phytocoenologia*, **24**, 495–505.

Håkansson, S. (1968). Experiments with *Agropyron repens* (L.) Beauv. II. Production from rhizome pieces of different sizes and from seeds: various environmental conditions compared. *Lantbrukshögskolans Annaler*, **34**, 3–29.

Håkansson, S. (1983). Seasonal variation in the emergence of annual weeds: an introductory investigation in Sweden. *Weed Research*, **23**, 313–24.

Håkansson, S., & Wallgren, B. (1972). Experiments with *Sonchus arvensis* L. III. The development from reproductive roots cut into different lengths and planted at different depths, with and without competition from barley. *Swedish Journal of Agricultural Research*, **2**, 15–26.

Hald, A. B. (1999). The impact of changing the season in which cereals are sown on the diversity of the weed flora in rotational fields in Denmark. *Journal of Applied Ecology*, **36**, 24–32.

Harlan, J. R., & de Wet, J. M. J. (1965). Some thoughts about weeds. *Economic Botany*, **19**, 16–24.

Harradine, A. R. (1986). Seed longevity and seedling establishment of *Bromus diandrus* Roth. *Weed Research*, **26**, 173–80.

Henson, I. E. (1970). The effects of light, potassium nitrate and temperature on the germination of *Chenopodium album* L. *Weed Research*, **10**, 27–39.

Holm, R. E. (1972). Volatile metabolites controlling germination in buried weed seeds. *Plant Physiology*, **50**, 293–7.

Holm, L. G., Plucknett, D. L., Pancho, J. V., & Herberger, J. P. (1977). *The World's Worst Weeds*. Honolulu, HI: University Press of Hawaii.

Horowitz, M. (1972). Development of *Cynodon dactylon* (L.) Pers. *Weed Research*, **12**, 207–20.

Horowitz, M., & Taylorson, R. B. (1984). Hardseededness and germinability of velvetleaf (*Abutilon theophrasti*) as affected by temperature and moisture. *Weed Science*, **32**, 111–15.

Horowitz, M., & Taylorson, R. B. (1985). Behavior of hard and permeable seeds of *Abutilon theophrasti* Medic. (velvetleaf). *Weed Research*, **25**, 363–72.

Hosmani, M. M. (1995). *Striga (a Noxious Root Parasitic Weed)*, 2nd edn. Dharwad, India: S. M. Hosmani.

Howard, C. L., Mortimer, A. M., Gould, P., Putwain, P. D., Cousens, R., & Cussans, G. W. (1991). The dispersal of weeds: seed movement in arable agriculture. In *Brighton Crop Protection Conference – Weeds*, pp. 821–8. Farnham, UK: British Crop Protection Council.

Huelma, C. C., Moody, K., & Mew, T. W. (1996). Weed seeds in rice seed shipments: a case study. *International Journal of Pest Management*, **42**, 147–50.

Hulme, P. E. (1994). Post-dispersal seed predation in grassland: its magnitude and sources of variation. *Journal of Ecology*, **82**, 645 52.

Hunter, P. J. (1967). The effect of cultivations on slugs of arable land. *Plant Pathology*, **16**, 153–6.

Jensen, H. A. (1969). Content of buried seeds in arable soil in Denmark and its relation to the weed population. *Dansk Botanisk Arkiv*, **27**(2), 7–56.

Kannangara, H. W., & Field, R. J. (1985). Environmental and physiological factors affecting the fate of seeds of yarrow (*Achillea millefolium* L.) in arable land in New Zealand. *Weed Research*, **25**, 87–92.

Kaul, V. (1965). Physiological ecology of *Xanthium strumarium* Linn. II. Physiology of seeds in relation to its distribution. *Indian Botanical Society Journal*, **44**, 365–80.

Kelley, A. D., & Bruns, V. F. (1975). Dissemination of weed seeds by irrigation water. *Weed Science*, **23**, 486–93.

Kegode, G. O., Forcella, F., & Durgan, B. R. (1999). Limiting green and yellow foxtail (*Setaria viridis* and *S. glauca*) seed production following spring wheat (*Triticum aestivum*) harvest. *Weed Technology*, **13**, 43–7.

Kigel, J. (1994). Development and ecophysiology of Amaranths. In *Amaranth: Biology, Chemistry, and Technology*, ed. O. Paredes-López, pp. 39–73. Ann Arbor, MI: CRC Press.

Kigel, J., & Koller, D. (1985). Asexual reproduction of weeds. In *Weed Physiology*, vol. 1: *Reproduction and Ecophysiology*, ed. S. O. Duke, pp. 65–100. Boca Raton, FL: CRC Press.

King, T. J. (1975). Inhibition of seed germination under leaf canopies in *Arenaria serpyllifolia*, *Veronica arvensis* and *Cerastum* (sic) *holosteoides*. *New Phytologist*, **75**, 87–90.

Kingsbury, J. M. (1964). *Poisonous Plants of the United States and Canada*. Englewood Cliffs, NJ: Prentice-Hall.

Kirk, J., & Courtney, A. D. (1972). A study of the survival of wild oats (*Avena fatua*) seeds buried in farm yard manure and fed to bullocks. In *Proceedings of the 11th British Weed Control Conference*, pp. 226–33. London: British Crop Protection Council.

Knake, E. L. (1972). Effect of shade on giant foxtail. *Weed Science*, **20**, 588–92.

Knezevic, S. Z., & Horak, M. J. (1998). Influence of emergence time and density on redroot pigweed (*Amaranthus retroflexus*). *Weed Science*, **46**, 665–72.

Kremer, R. J., & Spencer, N. R. (1989a). Impact of a seed-feeding insect and microorganisms on velvetleaf (*Abutilon theophrasti*) seed viability. *Weed Science*, 37, 211–16.

Kremer, R. J., & Spencer, N. R. (1989b). Interaction of insects, fungi, and burial on velvetleaf (*Abutilon theophrasti*) seed viability. *Weed Technology*, 3, 322–8.

LaCroix, L. J., & Staniforth, D. W. (1964). Seed dormancy in velvetleaf. *Weeds*, 12, 171–4.

Lambers, H., & H. Poorter. (1992). Inherent variation in growth rate between higher plants: a search for physiological causes and ecological consequences. *Advances in Ecological Research*, 23, 187–261.

Lapham, J. (1985). Unrestricted growth, tuber formation and spread of *Cyperus esculentus* L. in Zimbabwe. *Weed Research*, 25, 323–9.

Lapointe, A.-M., Deschênes, J.-M., Gervai, P., & Lemieux, C. (1984). Biologie du chénopode blanc (*Chenopodium album*): influence du travail du sol sur la levée et de la densité du peuplement sur la croissance. *Canadian Journal of Botany*, 62, 2587–93.

Law, R., Bradshaw, A. D., & Putwain, P. D. (1977). Life-history variation in *Poa annua*. *Evolution*, 31, 233–46.

Lawson, H. M., & Wiseman, J. S. (1979). Competition between weeds and transplanted spring cabbage: effects of nitrogen top-dressing. *Horticultural Research*, 19, 25–34.

Lawson, H. M., Waister, P. D., & Stephens, R. J. (1974). Patterns of emergence of several important arable weed species. *British Crop Protection Conference Monographs*, 9, 121–35.

Légère, A., & Schreiber, M. M. (1989). Competition and canopy architecture as affected by soybean (*Glycine max*) row width and density of redroot pigweed (*Amaranthus retroflexus*). *Weed Science*, 37, 84–92.

Lieth, H. (1960). Patterns of change within grassland communities. In *The Biology of Weeds*, ed. J. L. Harper, pp. 27–39. Oxford, UK: Blackwell.

Lindquist, J. L., Maxwell, B. D., Buhler, D. D., & Gunsolus, J. L. (1995). Velvetleaf (*Abutilon theophrasti*) recruitment, survival, seed production, and interference in soybean (*Glycine max*). *Weed Science*, 43, 226–32.

Loomis, R. S., & Connor, D. J. (1992). *Crop Ecology*. Cambridge, UK: Cambridge University Press.

Lovely, W. G., Weber, C. R., & Staniforth, D. W. (1958). Effectiveness of the rotary hoe for weed control in soybeans. *Agronomy Journal*, 50, 621–5.

Lovett Doust, L., MacKinnon, A., & Lovett Doust, J. (1985). Biology of Canadian weeds. 71. *Oxalis stricta* L., *O. corniculata* L., *O. dillenii* Jacq. ssp. *dillenii* and *O. dillenii* Jacq. ssp. *filipes* (Small) Eiten. *Canadian Journal of Plant Science*, 65, 691–709.

Lueschen, W. E., & Andersen, R. N. (1980). Longevity of velvetleaf (*Abutilon theophrasti*) seeds in soil under agricultural practices. *Weed Science*, 28, 341–6.

Mack, R. N. (1981). Invasion of *Bromus tectorum* L. into western North America: an ecological chronicle. *Agro-Ecosystems*, 7, 145–65.

Mack, R. N., & Pyke, D. A. (1983). The demography of *Bromus tectorum*: variation in time and space. *Journal of Ecology*, 71, 69–93.

Marshall, E. J. P. (1989). Distribution patterns of plants associated with arable field edges. *Journal of Applied Ecology*, 26, 247–57.

Maxwell, B, D., & Ghersa, C. (1992). The influence of weed seed dispersion versus the effect of competition on crop yield. *Weed Technology*, 6, 196–204.

Mayer, F., Albrecht, H., & Pfadenhauer, J. (1998). The transport of seeds by soil-working implements. In *Weed Seedbanks: Determination, Dynamics and Manipulation*, ed. G. T. Champion, A. C. Grundy, N. E. Jones, E. J. P. Marshall & R. J. Froud-Williams, pp. 83–9. Wellesbourne, UK: Association of Applied Biologists.

McCanny, S. J., & Cavers, P. B. (1988). Spread of proso millet (*Panicum miliaceum* L.) in Ontario, Canada. II. Dispersal by combines. *Weed Research*, 28, 67–72.

McCanny, S. J., Baugh, M., & Cavers, P. B. (1988). Spread of proso millet (*Panicum miliaceum* L.) in Ontario, Canada. I. Rate of spread and crop susceptibility. *Weed Research*, 28, 59–65.

McEvoy, P. B., Rudd, N. T., Cox, C. S., & Huso, M. (1993). Disturbance, competition, and herbivory effects on ragwort *Senecio jacobaea* populations. *Ecological Monographs*, 63, 55–75.

McLachlan, S. M., Tollenaar, M., Swanton, C. J., & Weise, S. F. (1993). Effect of corn-induced shading on dry matter accumulation, distribution, and architecture of redroot pigweed (*Amaranthus retroflexus*). *Weed Science*, 41, 568–73.

McWilliams, E. L., Landers, R. Q., & Mahlstede, J. P. (1968). Variation in seed weight and germination in populations of *Amaranthus retroflexus* L. *Ecology*, 49, 290–6.

Michaux, B. (1989). Reproductive and vegetative biology of *Cirsium vulgare* (Savi) Ten. (Compositae: Cynareae). *New Zealand Journal of Botany*, 27, 401–14.

Mohler, C. L. (1991). Effects of tillage and mulch on weed biomass and sweet corn yield. *Weed Technology*, 5, 545–52.

Mohler, C. L. (1993). A model of the effects of tillage on emergence of weed seedlings. *Ecological Applications*, 3, 53–73.

Mohler, C. L. (1995). A living mulch (white clover) / dead mulch (compost) weed control system for winter squash. *Proceedings of the Northeastern Weed Science Society*, 49, 5–10.

Mohler, C. L. (1996). Ecological bases for the cultural control of annual weeds. *Journal of Production Agriculture*, 9, 468–74.

Mohler, C. L., & Callaway, M. B. (1992). Effects of tillage and mulch on the emergence and survival of weeds in sweet corn. *Journal of Applied Ecology*, 29, 21–34.

Mohler, C. L., & Callaway, M. B. (1995). Effects of tillage and mulch on weed seed production and seed banks in sweet corn. *Journal of Applied Ecology*, 32, 627–39.

Mohler, C. L., & Galford, A. E. (1997). Weed seedling emergence and seed survival: separating the effects of seed position and soil modification by tillage. *Weed Research*, 37, 147–55.

Mortensen, D. A., Johnson, G. A., & Young, L. J. (1993). Weed distribution in agricultural fields. In *Soil Specific Crop Management*, ed. P. C. Robert, R. H. Rust & W. F. Larson, pp. 113–24. Madison, WI: American Society of Agronomy.

Mortimer, A. M. (1976). Aspects of the seed population dynamics of *Dactylis glomerata* L., *Holcus lanatus* L., *Plantago lanceolata* L., and *Poa annua* L. In *Proceedings of the 1976 British Crop Protection Conference – Weeds*, pp. 687–94. Croydon, UK: British Crop Protection Council.

Morton, C. T., & Buchele, W. F. (1960). Emergence energy of plant seedlings. *Agricultural Engineering*, 41, 428–31, 453–5.

Moss, S. R. (1983). The production and shedding of *Alopecurus myosuroides* Huds. seeds in winter cereals crops. *Weed Research*, 23, 45–51.

Mt. Pleasant, J., & Schlather, K. J. (1994). Incidence of weed seed in cow (*Bos* sp.) manure and its importance as a weed source for cropland. *Weed Technology*, 8, 304–10.

Mulugeta, D., & Stoltenberg, D. E. (1997). Increased weed emergence and seed bank depletion by soil disturbance in a no-tillage system. *Weed Science*, **45**, 234–41.

Musselman, L. J., (ed.) (1987). *Parasitic Weeds in Agriculture*, vol. 1, Striga. Boca Raton, FL: CRC Press.

Naylor, R. E. L. (1972). Aspects of the population dynamics of the weed *Alopecurus myosuroides* Huds. in winter cereal crops. *Journal of Applied Ecology*, **9**, 127–39.

Naylor, J. M. (1983). Studies on the genetic control of some physiological processes in seeds. *Canadian Journal of Botany*, **61**, 3561–7.

New York Agricultural Statistics Service (1994). *New York Agricultural Statistics 1993-1994*. Albany, NY: New York State Department of Agriculture and Markets, Division of Statistics.

Nichols, M. L., & Reed, I. F. (1934). Soil dynamics. VI. Physical reactions of soil to moldboard surfaces. *Agricultural Engineering*, **15**, 187–90.

Norris, R. F. (1992). Case history for weed competition/population ecology: barnyardgrass (*Echinochloa crus-galli*) in sugarbeets (*Beta vulgaris*). *Weed Technology*, **6**, 220–7.

Odum, S. (1965). Germination of ancient seeds: floristic observations and experiments with archaeologically dated soil samples. *Dansk Botanisk Arkiv*, **24**, 1–70.

Osborne, D. J. (1980). Senescence in seeds. In *Senescence in Plants*, ed. K. V. Thimann, pp. 13–37. Boca Raton, FL: CRC Press.

Özer, Z. (1979). Über die Beeinflussung der Keimfähigkeit der Samen mancher Gründlandpflanzen beim Durchgang durch den Verdauungstrakt des Schafes und nach Mistgärung. *Weed Research*, **19**, 247–54.

Pemberton, R. W., & Irving, D. W. (1990). Elaiosomes on weed seeds and the potential for myrmecochory in naturalized plants. *Weed Science*, **38**, 615–19.

Philippi, T. (1993). Bet-hedging germination of desert annuals: beyond the first year. *American Naturalist*, **142**, 474–87.

Pieterse, A. H., Verkleij, J. A. C., & ter Borg, S. J. (eds) (1994). *Biology and Management of* Orobanche: *Proceedings of the 3rd International Workshop on* Orobanche *and Related* Striga *Research*. Amsterdam, The Netherlands: Royal Tropical Institute.

Pitelli, R. A., Charudattan, R., & Devalerio, J. T. (1998). Effect of *Alternaria cassiae*, *Pseudocercospora nigricans*, and soybean (*Glycine max*) planting density on the biological control of sicklepod (*Senna obtusifolia*). *Weed Technology*, **12**, 37–40.

Plummer, G. L., & Keever, C. (1963). Autumnal daylight weather and camphor-weed dispersal in the Georgia Piedmont region. *Botanical Gazette*, **124**, 283–9.

Pons, T. L. (1989). Breaking of seed dormancy by nitrate as a gap detection mechanism. *Annals of Botany*, **63**, 139–43.

Popay, A. I., & Roberts, E. H. (1970). Factors involved in the dormancy and germination of *Capsella bursa-pastoris* (L.) Medik. and *Senecio vulgaris* L. *Journal of Ecology*, **58**, 103–22.

Popay, I., & Field, R. (1996). Grazing animals as weed control agents. *Weed Technology*, **10**, 217–31.

Povilaitis, B. (1956). Dormancy studies with seeds of various weed species. *Proceedings of the International Seed Testing Association*, **21**, 88–111.

Proctor, V. W. (1968). Long-distance dispersal of seeds by retention in digestive tract of birds. *Science*, **160**, 321–2.

Qasem, J. R. (1992). Nutrient accumulation by weeds and their associated vegetable crops. *Journal of Horticultural Science*, **67**, 189–95.

Qi, M., Upadhyaya, M. K., & Turkington, R. (1996). Dynamics of seed bank and survivorship of meadow salsify (*Tragopogon pratensis*) populations. *Weed Science*, **44**, 100–8.

Rao, A. N., & Moody, K. (1990). Weed seed contamination in rice seed. *Seed Science and Technology*, **18**, 139–46.

Regnier, E. E., & Harrison, S. K. (1993). Compensatory responses of common cocklebur (*Xanthium strumarium*) and velvetleaf (*Abutilon theophrasti*) to partial shading. *Weed Science*, **41**, 541–7.

Regnier, E. E., & Stoller, E. W. (1989). The effects of soybean (*Glycine max*) interference on the canopy architecture of common cocklebur (*Xanthium strumarium*), jimsonweed (*Datura stramonium*), and velvetleaf (*Abutilon theophrasti*). *Weed Science*, **37**, 187–95.

Rew, L. J., & Cussans, G. W. (1997). Horizontal movement of seeds following tine and plough cultivation: implications of spatial dynamics of weed infestations. *Weed Research*, **37**, 247–56.

Rew, L. J., Froud-Williams, R. J., & Boatman, N. D. (1996). Dispersal of *Bromus sterilis* and *Anthriscus sylvestris* seed within arable field margins. *Agriculture, Ecosystems and Environment*, **59**, 107–14.

Roberts, E. H. (1972). Dormancy: a factor affecting seed survival in the soil. In *Viability of Seeds*, ed. E. H. Roberts, pp. 321–59. Syracuse, NY: Syracuse University Press.

Roberts, E. H. (1988). Seed aging: the genome and its expression. In *Senescence and Aging in Plants*, ed. L. D. Noodén & A. C. Leopold, pp. 465–98. San Diego, CA: Academic Press.

Roberts, H. A. (1984). Crop and weed emergence patterns in relation to time of cultivation and rainfall. *Annals of Applied Biology*, **105**, 263–75.

Roberts, H. A. (1986). Persistence of seeds of some grass species in cultivated soil. *Grass and Forage Science*, **41**, 273–6.

Roberts, H. A., & Benjamin, S. K. (1979). The interaction of light, nitrate and alternating temperature on the germination of *Chenopodium album*, *Capsella bursa-pastoris* and *Poa annua* before and after chilling. *Seed Science and Technology*, **7**, 379–92.

Roberts, H. A., & Dawkins, P. A. (1967). Effect of cultivation on the numbers of viable weed seeds in soil. *Weed Research*, **7**, 290–301.

Roberts, H. A., & Feast, P. M. (1972). Fate of seeds of some annual weeds in different depths of cultivated and undisturbed soil. *Weed Research*, **12**, 316–24.

Roberts, H. A., & Feast, P. M. (1973). Changes in the numbers of viable weed seeds in soil under different regimes. *Weed Research*, **13**, 298–303.

Roberts, H. A., & Lockett, P. M. (1975). Germination of buried and dry-stored seeds of *Stellaria media*. *Weed Research*, **15**, 199–204.

Roberts, H. A., & Neilson, J. E. (1980). Seed survival and periodicity of seedling emergence in some species of *Atriplex*, *Chenopodium*, *Polygonum* and *Rumex*. *Annals of Applied Biology*, **94**, 111–120.

Roberts, H. A., & Potter, M. E. (1980). Emergence patterns of weed seedlings in relation to cultivation and rainfall. *Weed Research*, **20**, 377–86.

Root, R. (1973). Organization of a plant–arthropod association in simple and diverse habitats: the fauna of collards (*Brassica oleracea*). *Ecological Monographs*, **42**, 95–124.

Sagar, G. R., & Mortimer, A. M. (1976). An approach to the study of the population dynamics of plants with special reference to weeds. *Applied Biology*, **1**, 1–43.

Salisbury, E. J. (1942). *The Reproductive Capacity of Plants*. London: Bell.

Salisbury, E. J. (1961). *Weeds and Aliens*. London: Collins.
Šarapatka, B., Holub, M., & Lhotská, M. (1993). The effect of farmyard manure anaerobic treatment on weed seed viability. *Biological Agriculture and Horticulture*, **10**, 1–8.
Sauer, J., & Struik, G. (1964). A possible ecological relation between soil disturbance, light-flash, and seed germination. *Ecology*, **45**, 884–6.
Schafer, D. E., & Chilcote, D. O. (1970). Factors influencing persistence and depletion in buried seed populations. II. The effects of soil temperature and moisture. *Crop Science*, **10**, 342–5.
Schippers, P., Ter Borg, S. J., Van Groenendael, J. M., & Habekotte, B. (1993). What makes *Cyperus esculentus* (yellow nutsedge) an invasive species? – A spatial model approach. In *Brighton Crop Protection Conference – Weeds*, pp. 495–504. Farnham, UK: British Crop Protection Council.
Schläpfer, B., & Ryser, P. (1996). Leaf and root turnover of three ecologically contrasting grass species in relation to their performance along a productivity gradient. *Oikos*, **75**, 398–406.
Schmidt, W. (1989). Plant dispersal by motor cars. *Vegetatio*, **80**, 147–52.
Scopel, A. L., Ballaré, C. L., & Sánchez, R. A. (1991). Induction of extreme light sensitivity in buried weed seeds and its role in the perception of soil cultivations. *Plant, Cell and Environment*, **14**, 501–8.
Seibert, A. C., & Pearce, R. B. (1993). Growth analysis of weed and crop species with reference to seed weight. *Weed Science*, **41**, 52–6.
Selleck, G. W. (1958). Life history of leafy spurge. In *Proceedings of the 15th Annual Meeting, North Central Weed Control Conference*, pp. 16–17.
Sharrow, S. H., & Mosher, W. D. (1982). Sheep as a biological control agent for tansy ragwort. *Journal of Range Management*, **35**, 480–2.
Silvertown, J. (1980). Leaf-canopy-induced seed dormancy in a grassland flora. *New Phytologist*, **85**, 109–18.
Slagell Gossen, R. R., Tyrl, R. J., Hauhouot, M., Peeper, T. F., Claypool, P. L., & Solie, J. B. (1998). Effects of mechanical damage on cheat (*Bromus secalinus*) caryopsis anatomy and germination. *Weed Science*, **46**, 249–57.
Smith, C. J., Elston, J., & Bunting, A. H. (1971). The effects of cutting and fertilizer treatments on the yield and botanical composition of chalk turf. *Journal of the British Grassland Society*, **26**, 213–19.
Smith, J. E., & Jordan, P. W. (1993). Sicklepod (*Cassia obtusifolia*) shoot structure as affected by soybean (*Glycine max*) interference. *Weed Science*, **41**, 75–81.
Smith, L. M., & Kok, L. T. (1984). Dispersal of musk thistle (*Carduus nutans*) seeds. *Weed Science*, **32**, 120–5.
Spitters, C. J. T., & Kramer, T. (1985). Changes in relative growth rate with plant ontogeny in spring wheat genotypes grown as isolated plants. *Euphytica*, **34**, 833–47.
Staver, C. (1991). The role of weeds in the productivity of Amazonian bush fallow agriculture. *Experimental Agriculture*, **27**, 287–304.
Steinbauer, G. P., & Grigsby, B. (1957a). Interaction of temperature, light, and moistening agent in the germination of weed seeds. *Weeds*, **3**, 175–82.
Steinbauer, G. P., & Grigsby, B. (1957b). Dormancy and germination characteristics of the seeds of four species of *Plantago*. *Proceedings of the Association of Official Seed Analysts of North America*, **47**, 158–64.

Stevens, O. A. (1932). The number and weight of seeds produced by weeds. *American Journal of Botany*, **19**, 784–94.

Stoller, E. W., & Wax, L. M. (1974). Dormancy changes and fate of some annual weed seeds in the soil. *Weed Science*, **22**, 151–5.

Takabayashi, M., Kubota, T., & Abe, H. (1979). Dissemination of weed seeds through cow feces. *Japan Agricultural Research Quarterly*, **13**, 204–7.

Tanji, A., Zimdahl, R. L., & Westra, P. (1997). The competitive ability of wheat (*Triticum aestivum*) compared to rigid ryegrass (*Lolium rigidum*) and cowcockle (*Vaccaria hispanica*). *Weed Science*, **45**, 481–7.

Tasrif, A., Sahid, I. B., Sastroutomo, S. S., & Latiff, A. (1991). Purity study of imported leguminous cover crops. *Plant Protection Quarterly*, **6**, 190–3.

Taylorson, R. B. (1970). Changes in dormancy and viability of weed seeds in soils. *Weed Science*, **18**, 265–9.

Taylorson, R. B. (1972). Phytochrome controlled changes in dormancy and germination of buried weed seeds. *Weed Science*, **20**, 417–22.

Taylorson, R. B. (1979). Response of weed seeds to ethylene and related hydrocarbons. *Weed Science*, **27**, 7–10.

Taylorson, R. B. (1987). Environmental and chemical manipulation of weed seed dormancy. *Reviews of Weed Science*, **3**, 135–54.

Teasdale, J. R. (1998). Cover crops, smother plants, and weed management. In *Integrated Weed and Soil Management*, ed. J. L. Hatfield, D. D. Buhler & B. A. Stewart, pp. 247–70. Chelsea, MI: Ann Arbor Press.

Thompson, K., Band, S. R., & Hodgson, J. G. (1993). Seed size and shape predict persistence in soil. *Functional Ecology*, **7**, 236–41.

Thompson, K., Bakker, J. P., Bekker, R. M., & Hodgson, J. G. (1998). Ecological correlates of seed persistence in soil in the north-west European flora. *Journal of Ecology*, **86**, 163–9.

Thurston, J. M. (1960). Dormancy in weed seeds. In *The Biology of Weeds*, British Ecological Society Symposium no. 1, ed. J. L. Harper, pp. 69–82. Oxford, UK: Blackwell.

Thurston, J. M. (1966). Survival of seeds of wild oats (*Avena fatua* L. and *Avena ludoviciana* Dur.) and charlock (*Sinapis arvensis* L.) in soil under leys. *Weed Research*, **6**, 67–80.

Tilman, D. (1982). *Resource Competition and Community Structure*. Princeton, NJ: Princeton University Press.

Tilman, D., & Wedin, D. (1991a). Plant traits and resource reduction for five grasses growing on a nitrogen gradient. *Ecology*, **72**, 685–700.

Tilman, D., & Wedin, D. (1991b). Dynamics of nitrogen competition between successional grasses. *Ecology*, **72**, 1038–49.

Tompkins, D. K., Chaw, D., & Abiola, A. T. (1998). Effect of windrow composting on weed seed germination and viability. *Compost Science and Utilization*, **6**, 30–4.

Tonkin, J. H. B. (1982). The presence of seed impurities in samples of cereal seed tested at the official seed testing station, Cambridge in the period of 1978–1981. *Aspects of Applied Biology*, **1**, 163–71.

Totterdell, S., & Roberts, E. H. (1980). Characteristics of alternating temperatures which stimulate loss of dormancy in seeds of *Rumex obtusifolius* L. and *Rumex crispus* L. *Plant, Cell and Environment*, **3**, 3–12.

Tumbleson, M. E., & Kommedahl, T. (1961). Reproductive potential of *Cyperus esculentus* by tubers. *Weeds*, **9**, 646–53.

Turkington, R., & Harper, J. L. (1979). The growth, distribution and neighbour relationships of *Trifolium repens* in a permanent pasture. I. Ordination, pattern and contact. *Journal of Ecology*, **67**, 201–18.

Vengris, J., Colby, W. G., & Drake, M. (1955). Plant nutrient competition between weeds and corn. *Agronomy Journal*, **47**, 213–16.

Vengris, J., Drake, M., Colby, W. G., & Bart, J. (1953). Chemical composition of weeds and accompanying crop plants. *Agronomy Journal*, **45**, 213–18.

Villiers, T. A (1973). Ageing and the longevity of seeds in field conditions. In *Seed Ecology, Proceedings of the 19th Easter School in Agricultural Sciences, University of Nottingham, 1972*, ed. W. Heydecker, pp. 265–88. University Park, PA: Pennsylvania State University Press.

Villiers, T. A. (1980). Ultrastructural changes in seed dormancy and senescence. In *Senescence in Plants*, ed. K. V. Thimann, pp. 39–66. Boca Raton, FL: CRC Press.

Villiers, T. A., & Edgcumbe, D. J. (1975). On the cause of seed deterioration in dry storage. *Seed Science and Technology*, **3**, 761–74.

Vincent, E. M., & Roberts, E. H. (1977). The interaction of light, nitrate and alternating temperature in promoting the germination of dormant seeds of common weed species. *Seed Science and Technology*, **5**, 659–70.

Wareing, P. F., & Foda, H. A. (1957). Growth inhibitors and dormancy in *Xanthium* seed. *Physiologia Plantarum*, **10**, 266–80.

Warnes, D. D., & Andersen, R. N. (1984). Decline of wild mustard (*Brassica kaber*) seeds in soil under various cultural and chemical practices. *Weed Science*, **32**, 214–17.

Warwick, S. I., & Black, L. D. (1988). The biology of Canadian weeds. 90. *Abutilon theophrasti*. *Canadian Journal of Plant Science*, **68**, 1069–85.

Weaver, S. E., & Lechowicz, M. J. (1983). The biology of Canadian weeds. 56. *Xanthium strumarium* L. *Canadian Journal of Plant Science*, **63**, 211–25.

Weaver, S. E., & Riley, W. R. (1982). The biology of Canadian weeds. 53. *Convolvulus arvensis* L. *Canadian Journal of Plant Science*, **62**, 461–72.

Weaver, S. E., & Thomas, A. G. (1986). Germination responses to temperature of atrazine-resistant and -susceptible biotypes of two pigweed (*Amaranthus*) species. *Weed Science*, **34**, 865–70.

Webster, T. M., Cardina, J., & Loux, M. M. (1998). The influence of weed management in wheat (*Triticum aestivum*) stubble on weed control in corn (*Zea mays*). *Weed Technology*, **12**, 522–6.

Weiss, P. W. (1981). Spatial distribution and dynamics of populations of the introduced annual *Emex australis* in south-eastern Australia. *Journal of Applied Ecology*, **18**, 849–64.

Wellington, P. S. (1960). Assessment and control of the dissemination of weeds by crop seeds. In *The Biology of Weeds*, ed. J. L. Harper, pp. 94–107. Oxford, UK: Blackwell.

Wesson, G., & Wareing, P. F. (1969*a*). The induction of light sensitivity in weed seeds by burial. *Journal of Experimental Botany*, **20**, 414–25.

Wesson, G., & Wareing, P. F. (1969*b*). The role of light in the germination of naturally occurring populations of buried seeds. *Journal of Experimental Botany*, **20**, 402–13.

White, J. (1979). The plant as a metapopulation. *Annual Review of Ecology and Systematics*, **10**, 109–45.

Wilcut, J. W., Dute, R. R., Truelove, B., & Davis, D. E. (1988). Factors limiting the distribution of cogongrass, *Imperata cylindrica* and torpedograss, *Panicum repens*. *Weed Science*, **36**, 577–82.

Wiles, L. J., Oliver, G. W., York, A. C., Gold, H. J., & Wilkerson, G. G. (1992). Spatial distribution of broadleaf weeds in North Carolina soybean (*Glycine max*) fields. *Weed Science*, **40**, 554–7.

Williams, J. T., & Harper, J. L. (1965). Seed polymorphism and germination. I. The influence of nitrates and low temperatures on the germination of *Chenopodium album*. *Weed Research*, **5**, 141–50.

Wilson, B. J. (1972). Studies of the fate of *Avena fatua* seeds on cereal stubble as influenced by autumn treatment. In *Proceedings of the 11th British Weed Control Conference*, pp. 242–7. London: British Crop Protection Council.

Wilson, B. J., & Brain, P. (1991). Long-term stability of distribution of *Alopecurus myosuroides* Huds. within cereal fields. *Weed Research*, **31**, 367–73.

Wilson, B. J., & Cussans, G. W. (1972). The effect of autumn cultivations on the emergence of *Avena fatua* seedlings. In *Proceedings of the 11th British Weed Control Conference*, pp. 234–41. London: British Crop Protection Council.

Wilson, B. J., & Lawson, H. M. (1992). Seedbank persistence and seedling emergence of seven weed species in autumn-sown crops following a single year's seeding. *Annals of Applied Biology*, **120**, 105–16.

Wilson, P. J., & Aebischer, N. J. (1995). The distribution of dicotyledonous arable weeds in relation to distance from the field edge. *Journal of Applied Ecology*, **32**, 295–310.

Wilson, R. G. Jr. (1980). Dissemination of weed seeds by surface irrigation water in western Nebraska. *Weed Science*, **28**, 87–92.

Yenish, J. P., Fry, T. A., Durgan, B. R., & Wyse, D. L. (1997). Establishment of common milkweed (*Asclepias syriaca*) in corn, soybean, and wheat. *Weed Science*, **45**, 44–53.

Young, F. L. (1986). Russian thistle (*Salsola iberica*) growth and development in wheat (*Triticum aestivum*). *Weed Science*, **34**, 901–5.

Zahnley, J. W., & Fitch, J. B. (1941). Effect of ensiling on the viability of weed seeds. *Journal of the American Society of Agronomy*, **33**, 816–22.

Zanin, G., & Sattin, M. (1988). Threshold level and seed production of velvetleaf (*Abutilon theophrasti* Medicus) in maize. *Weed Research*, **28**, 347–52.

Zanin, G., Berti, A., & Zuin, M. C. (1989). Estimation du stock semencier d'un sol labouré ou en semis direct. *Weed Research*, **29**, 407–17.

Zorner, P. S., Zimdahl, R. L., & Schweizer, E. E. (1984a). Sources of viable seed loss in buried dormant and non-dormant populations of wild oat (*Avena fatua* L.) seed in Colorado. *Weed Research*, **24**, 143–50.

Zorner, P. S., Zimdahl, R. L., & Schweizer, E. E. (1984b). Effect of depth and duration of seed burial on kochia (*Kochia scoparia*). *Weed Science*, **32**, 602–7.

CHARLES P. STAVER

3

Knowledge, science, and practice in ecological weed management: farmer–extensionist–scientist interactions

Introduction

Weed scientists usually cite pervasive crop yield losses due to weeds and substantial direct and indirect costs of weed control to justify research and extension budgets (see Chapter 1). Reductions in costs and yield losses should also be used to evaluate the progress of scientists in solving weed problems. Ultimately weed costs to agriculture are determined by how farmers and ranchers manage weeds, not by papers published or field days organized. In temperate and tropical regions, field crop farmers who use mechanization, cattle ranchers, dairy farmers, vegetable and fruit growers, and smallholders on hillsides all devote time and resources to weed management. What is the role of research and extension in enabling this wide diversity of farmers to manage their weeds better?

This chapter examines the implications of farmer–extensionist–scientist interactions for the development of improved weed management. The first sections review historically how humans have learned to manage weeds. The chapter then analyzes scientist, extensionist, and farmer perspectives on weeds. The final sections describe how farmers, extensionists, and scientists can collaborate to develop field- and farm-level weed management strategies better adapted to weed patchiness and uncertainty. Case studies from the USA and Central America illustrate possible working relations among scientists, extensionists, and farmers.

Three principles for making the on-farm management of weeds more efficient and cost-effective, less risky, and more environmentally sound figure prominently in the chapter:

1. *Farmers play a crucial role in the development of weed science.* They invent, adapt, and modify weed management techniques. To do this, they employ varied approaches including observation, logic, experimentation, extrapolation, and calculated risk-taking. They integrate information and recommendations from diverse sources, make decisions at scales of operation not generally addressed by the research and extension system, and form effective farmer-to-farmer communication networks.
2. *Programs to improve weed management by farmers should focus on strengthening farmer decision-making.* A process termed participatory learning for action illustrates an approach for strengthening farmer skills for goal-setting, experimentation, observation, record-keeping, and analysis, all key elements in decision-making. In this process, groups of farmers meet at critical moments before, during, and after the crop cycle to discuss current and alternative crop and weed management practices. Initially farmers analyze their personal and business goals, propose experiments, and suggest criteria for the evaluation of their decisions. During the crop cycle, farmers work as a group to improve their ecological reasoning through observation of weed composition and behavior across a spectrum of fields. They link observations with practices and evaluate the timeliness and effectiveness of each other's decisions. At the end of the crop cycle, they formulate improved weed management strategies based on their conclusions and propose a study plan for the next crop cycle.
3. *Weed science can benefit from a learning process that strengthens extensionists' and scientists' links with the temporal and spatial scale of farmer decision-making.* In a routine of regular interaction over several crop cycles, groups of farmers with research and extension cooperators can develop farmer- and researcher-initiated experimentation, field-scale monitoring, and analytical methods of crop and weed decision-making. This co-learning can contribute to the general effectiveness of the weed research and extension system by making it more responsive to the concerns of broader sectors of producers and society.

Knowledge and technology for weed management: an historical perspective

Humans have been managing weeds for over 10 000 years. The distinction between crops and weeds was one of the earliest human concepts distinguishing the beginning of agriculture (Rindos, 1984, pp. 137–43). Crops such as wheat and squash were among the first plants to be cultivated rather than simply gathered. Other species such as rye and maize were selected for deliberate planting and weeding somewhat later (Minc & Vandermeer, 1990).

Certain species, after undergoing initial domestication, subsequently lost favor as food sources and lapsed back into the weed complex or into the collected, but not cultivated, category. For example, *Setaria* in Mesoamerica (Minc & Vandermeer, 1990) and *Chenopodium* (Smith, 1992, pp. 103–32) in North America were domesticated early, but were abandoned as crops with the domestication of maize. Today species in these genera are important weeds.

Over thousands of years since the first planted fields and in diverse climatic zones, human society has continued to evolve techniques for crop production and weed control. This development has resulted in widening landscape disturbance and management, driven by increasing human population and changes in technology from stone, bronze, and iron through steel, petroleum, and computers.

Farmer and community learning has been central to the development of technology for crop production and weed control. Recent studies of farmer experimentation (e.g., Scoones & Thompson, 1994; Sumberg & Okali, 1995) indicate three important components in the farmer development of technology.

First, farmers cultivating the land year after year under diverse conditions of soils, crops, weeds, and weather accumulate a vast range of data on the effectiveness of their agricultural practices. In each field in each new planting season, farmers observe and adjust local crop production practices, although they are not conducting experiments. Although the observations in a single crop cycle only rarely lead to major changes in crop production or weed control techniques, farmers' gradual adjustments in cropping techniques over time have been the major force in the evolution of agricultural technology.

Second, farmers conduct tests or experiments in which they compare something new with their normal practices. In these experiments, farmers commonly test new physical inputs to crop production like crop species or varieties, cover crops, tools and equipment, or soil amendments. For such tests, farmers use the kitchen garden, a corner or strip of field close to their residence, or even an entire field. They may also test variations in management such as timing, spacing, or quantities of their usual inputs in a single or several seasons.

A third area in farmer technology development is the multiyear organization of the cropping system and the mix of activities that make up the farm operation. Whereas the testing of new physical inputs is primarily suited to planned experimentation, farmers use both planned testing and experiential learning under variable weather and market conditions to develop their cropping system organization and home consumption or income-generating strategies.

The rapid change in agriculture and weed control methods in Great Britain from the 17th to 19th century illustrates the central role played by farmers in the development of farming methods (Elliot *et al.*, 1977; Pretty, 1991, 1995, pp. 181–3). These centuries were characterized by increasing private tenancy of rural lands, urbanization, population growth, and industrialization. These factors influenced agricultural practices. During this period key technologies with weed control implications contributed to the intensification of crop production. Farmers developed crop rotations that included legumes like red clover and alternated cereals with "cleaning" crops such as turnip, potato, and forage beet. Cleaning crops planted in wide rows could be cultivated more easily during the crop cycle than the traditional small-grain cereals such as wheat and barley, which were categorized as "fouling" crops. Iron and steel parts for plows, harrows, and cultivators increased both the degree and precision of soil disturbance. In the 18th century, cultivation with horses improved labor efficiency in weed control. Improved seed cleaning equipment virtually eliminated *Agrostemma githago* and *Lolium temulentum*, weeds that had been extremely difficult to remove from small grains with prior methods (Elliot *et al.*, 1977). During this period new weeds continued to be introduced, as they had been in previous centuries. Weeds introduced in this period included *Cardaria draba*, *Veronica persica*, and *Galinsoga parviflora* (Godwin, 1960). Floristic composition also shifted in response to both changing weed control practices and other crop production factors. The effect of soil fertility on weed floristic composition, for example, was clearly shown in experiments at Rothamsted, England, started in 1843. In low fertility plots, species like *Equisetum arvensis* and *Aphanes arvensis* were found, whereas at intermediate and high fertility *Ranunculus arvensis* and *Stellaria media* were more frequent (Cousens & Mortimer, 1995, pp. 181–2). This suggests that the increasing use of animal manure, lime, ashes, bone meal, and green manure, as well as field drainage, also contributed to changing weed complexes.

The initial ideas for all these innovations originated with farmers and in local workshops (Pretty, 1991). Neighboring farmers and farmers from other regions observed the ideas in practice or learned of them through farmer clubs, books, and newspapers, all of which became increasingly common during the period. The number of books on agriculture, written primarily by farmers, increased from 2–10 in the last half of the 16th century to 150–400 by the beginning of the 19th century. Other farmers adapted the new ideas and reported their experiences and experiments at fairs, in printed material, and at farmer club meetings.

In the later half of the 19th century in the British Isles, Europe, and the USA, a new sector emerged in the development of agricultural knowledge. In

1843 Lawes and Gilbert, among the earliest proponents of the scientific method in agriculture, set up the first of their field experiments in Rothamsted (Johnston, 1994). Early agricultural scientists studied the effectiveness of mechanical weed control, rotations, and cover crops (Karlen et al., 1994). These concepts had been developed by farmers centuries earlier, and much of the early scientific agronomic knowledge was drawn from farmer practice.

The advent of the experiment station and the agricultural scientist brought about an important change in the development and spread of crop production technology. Whereas previously farmers were both the principal generators and users of technology, with the development of formal experimental science, a large proportion of technology generation moved off-farm. Thus, the generators and the users of technology separated into two different sectors (Busch & Lacy, 1983, pp. 5–36). Initially experiment stations maintained close links to the farm sector. However, with the emergence of disciplines within agricultural science, scientists distanced themselves from regular contact with farmers. They developed professional networks and journals to systematically document their work for their own use (Lockeretz & Anderson, 1993, pp. 26–7). In 1914 in the USA the separation in the generation and use of crop production technology was formally addressed through the establishment of the cooperative extension service under the Smith–Lever Act.

The development of herbicides, beginning in 1896 in a French vineyard with the chance discovery of the selective effects of copper sulfate on plants, further altered the relationship between weed technology generation and use. Advances in the laboratory sciences of chemistry and plant physiology led to the near simultaneous discovery in USA, England, and France of hormonal herbicides in the 1940s. This technological innovation was made with no input from farmers. Although farmers now have the choice of hundreds of different herbicides for a wide variety of crops, weed control has become a consumable, off-farm input in crop production. Herbicides must be purchased for each crop cycle. The separation between technology generation and use in the case of herbicides has been addressed by the public extension service, field sales representatives, and private crop consultants.

Farmers' intuitive understanding of weed ecology has formed part of crop production technology from the beginnings of agriculture. However, formal studies in weed ecology originated just in the past 50–75 years. Studies in weed ecology have only recently begun to affect the development of technologies for weed management. These studies have focused on the minimum weed-free period for different crops, the crop loss effects of different weed species, the dynamics of weed seed banks, and the physiology of vegetatively propagated weeds. In many cases, these concepts have been applied principally for managing herbicides. Weed ecology has also been useful in understanding why practices such as rotations, cover crops, and intercropping are effective in weed management, as other chapters in this book demonstrate.

Despite the expansion of the off-farm generation of knowledge about weed management in the past 50–75 years, farmers in temperate and tropical agriculture continue to experiment with machinery, crops, cropping systems, and farm organization for better weed control.

Much of the equipment for reducing herbicide applications, such as band applicators, wicks, and recirculating sprayers, or for combining spraying with other operations originated in farm workshops. During the 1970s and early 1980s more than 30% of the entries in the Ideas Competition at the Royal Norfolk Agricultural Show related to spraying (Sumberg & Okali, 1995, pp. 142–3).

In areas of the world where mechanization is less common, farmers face different weed problems and experiment with other methods. In Manya Krobo, a dry forest transition zone in Ghana, farmers in the past 60 years have gone from cocoa and long fallow cropping to bush fallow food cropping. More recently they have faced land and labor shortages, less reliable rainfall, and the spread of new herbaceous and woody weeds (Amanor, 1993). These include *Digitaria* and *Panicum* grasses, *Chromolaena odorata*, and *Leucaena* spp. They are currently experimenting with the conservation of tree seedlings and sprouts during weeding to reduce the invasion of savanna grasses, the use of cowpea and short-cycle cassava in different rotations with current crops to maintain productivity, and selective fallow management to promote native tree and shrub species.

Contrasting perspectives of farmers, extensionists, and scientists on weeds

Over the past 50–100 years human society has been formulating and using knowledge about weeds from different perspectives. These perspectives represent the views and experience of individuals and groups who have common or similar experience with weeds and recognize similar rules for processing information. In a broader context, these perspectives have been referred to as knowledge communities (Marglin, 1990; Hess, 1995, pp. 2–4). Three communities with the strongest interests in weed knowledge are farmers, extensionists, and scientists, including scientists in both industry and public institutions. Other important sectors include policy makers, regulators, consumers, and environmentalists (Campbell, 1994).

People in each knowledge community have interests in weeds for their own reasons and have their own implicit standards for what constitutes knowledge (Table 3.1). First, each community has accepted methods for generating or defining what counts as knowledge. For example, farmers value what they or their neighbors have tried more than what has been shown on a distant experiment station. Second, each community has accepted procedures for communicating knowledge. Researchers, for example, give more credence to articles on replicated experiments with significant statistics than to verbal descriptions of weed problems. Third, the spatial scale and time period for knowledge formulation and application varies among knowledge communities. Figure 3.1 illustrates the separation in time and scale of themes of interest to researchers and farmers. Scientists are interested in principles, recommendations, or products for wide application. Farmers need weed knowledge for local and particular use.

Each of the three knowledge communities can be typified by how they handle the uncertainty that characterizes crop production (Table 3.1). Researchers use formal analysis and replication under controlled conditions, usually in the laboratory or in small plots (Figure 3.1). They block, average, and eliminate outliers and failed experiments, working on a time-scale defined by administrative procedures such as tenure reviews, thesis deadlines, and grant evaluations. Extensionists work more locally than scientists and closer to crop production time. They build their weed extension programs from research results, practical publications, on-farm trials, and contacts with farmers. To take into account local conditions, extensionists develop more specific recommendations than researchers. During abnormal crop cycles, they respond with troubleshooting and special updates for their clients. Farmers are time and location specific in their application of weed knowledge to a single field in a given year within the context of the whole farm and possible off-farm activities. They make decisions about weed management based on partial and uncertain data. When they plan the crop cycle, they use accumulated experience and specific past information about the field, but cannot be sure what the new crop cycle will bring. Once the crop cycle begins, they modify their decisions based on weather, input availability and prices, and expected crop value. As a result, farmer methods for handling uncertainty include best-bet practices, contingency planning, adaptive response, and loss-cutting.

Although farmers, extensionists, and researchers can easily be distinguished

Table 3.1. How farmers, extensionists, and scientists differ in their generation and communication of knowledge about weed management

	Farmers	Extensionists	Scientists
What methods are acceptable for new knowledge generation?	· practice of crop production · experimentation · observations of correlations and patterns · verification of others' experience with personal experience	· review of applied research and practical publications · annual training and recommendation updates · on-farm trials and demonstrations · farm visits	· replicated experiments and statistical analysis · computer modeling · writing articles and books · product patenting
What channels are acceptable for knowledge communication?	· exchange of experience with other farmers · media oriented to crop production · visits, field days, training · consultation with local farmer experts and extension	· practical and trade publications · professional meetings · network with other extensionists · product promotion literature · planning workshops	· network with scientists of similar interests · journals · scientific meetings · advisory and expert meetings · product promotion literature
What is the spatial scale for weed management?	· field by field · logistics of whole farm · community	· target groups · influential farmers · troubleshooting on request	· general principles and mechanisms · recommendations for zones of similar conditions
What is the time horizon for weed management?	· farm planning · crop cycle planning · day to day	· annual work plan by crop cycle to reach clients · troubleshooting on request	· rhythm of funding · advance to publishable papers · promotion and review
How are variability, heterogeneity, and change in crop production managed?	· contingency planning · adaptive response · best bet · cut loses	· recommendations for zones of similar conditions · recommendation updates · troubleshooting	· averaging and blocking · elimination of outliers and failed experiments · definition of zones of similar conditions · decision models

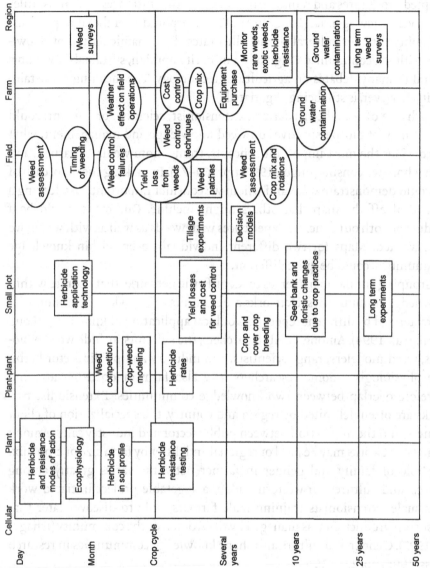

Figure 3.1 Location in time and space of scientist research topics (squares) and farmer interests (circles). (After Firbank, 1991.)

in any country based on their contrasting perspectives (Table 3.1), no knowledge community is rigid and unchanging. In practice, each community is a loose network of interacting individuals. These individuals face institutional rules or community traditions. In some interactions, individuals follow accepted procedures and reinforce the rules. In other instances, they resist rules and develop new agendas (Long, 1992). The incorporation of the concept of sustainability into agricultural science illustrates the dynamic nature of knowledge within a community. Originally a minority opinion, sustainability is now central to many debates on agricultural technology. The meaning of sustainability, however, is still being negotiated.

Each one of us, whether farmer, extensionist, scientist, or student, could draw a map of our perspectives on weed knowledge and management, what we consider the most important concepts, from whom we have learned, and with whom we consult (Engel, 1997, pp. 160–73). In making the map, each of us would demonstrate what and whom we consider important, also leaving a great deal off the map that others might include. Our maps would be a product of both our concrete experiences with weeds as well as with whom we have worked. Maps for two different individuals, even within knowledge communities, could be quite different.

Grouping similar maps serves to identify the networks that operate within knowledge communities. Networks are subunits of knowledge communities characterized by different repertoires or local application of knowledge (Long & Villareal, 1994). Among weed researchers, the subunits include weed biologists, weed modelers, range scientists, and industry and public sector herbicide physiologists. Some researchers may also farm or do extension, and therefore overlap between two knowledge communities. Extensionist networks are often delimited by region and country, the specialization of client farmers, and the distinction between public sector and industry sales. Among farmers, networks may be local or regional and differ by farm size or crop mix. The role of family and gender in farmer networks varies greatly among regions and cultures. In western Sudan, a vegetable project initiated work with male extensionists training male farmers, only to discover later that most crop production was managed by the women of the community (Ishag et al., 1997). Gender and culture also shape knowledge communities in research (Hess, 1995, pp. 27–32).

This description of knowledge communities is pertinent to improving weed management. To reduce crop losses to weeds and the costs of their control, the three communities must have effective and productive linkages. Formal linkages through systems of research and extension have used different modalities such as technology transfer, training and visit, and more recently participation (Roling, 1988, pp. 36–62). Interfaces between knowledge communities, however, are just as often characterized by gaps, discontinuities, and differences, precisely because they represent the point of contact between communities with contrasting objectives and procedures for generating knowledge (Long & Villareal, 1994). These discontinuities may be significant when indigenous farm communities interface with government extension services staffed with urban-born technicians who have limited field experience, not an uncommon situation in Africa, Asia, and Latin America. Cultural and class prejudices, mutual lack of respect, and divergent interests in such cases may produce reinforcing negative images and limit productive interaction. Even when farmers, extensionists, and scientists have a common culture and similar preferences in crop technology, each sector uses different portions of the total pool of weed management knowledge. For example, as shown in the case study from Iowa in this chapter, the three communities often disagree about which knowledge is more relevant and about which themes need further attention.

Fortunately, just as knowledge communities change internally based on the social interactions among their groups and individuals, the way they interface also changes. To reduce crop losses to weeds and the costs of weed control, effective work at the interface between farmers, extension, and research is crucial (Engel, 1997, pp. 21–44). Both the nature of weeds and the demands of decision-making in crop production indicate directions for more effective interactions among farmers, extensionists, and scientists.

Weed patchiness and uncertainty: the challenge to improving weed management

Weeds in a crop field are distributed irregularly, with patches of high density as well as patches with few weeds (Cardina, Johnson & Sparrow, 1997). Spatially these patches may be relatively stable from year to year, a product of localized seed rain, a relatively immobile seed bank, the clonal spread of vegetatively propagated weeds, and the patchiness of the soil environment (Colbach, Forcella & Johnson, 2000).

Although many weed species may be present in a field, only a limited number are important for crop management (Johnson *et al.*, 1995). Each of these species has a defined life cycle with a relatively defined phenology around which weed and crop management practices are usually organized.

This patchiness of weeds presents difficulties for monitoring and recording weed abundance and composition at the field and farm scale. Weeds reduce crop losses at the scale of individual crop plants. Control practices, while usually applied over the entire field, act against individual weeds. In addition, fields across the agricultural landscape are different in their weed patchiness and composition, due to founder effects, differing patterns of cropping and weed control practices, soil and drainage, and location in the landscape. For example, Johnson et al. (1995) found that a particular weed species will not necessarily have the same degree of clumping in different fields. How to improve farmer planning and decision-making through simpler and more accurate scouting of weed patches, estimation of damage, and extrapolation of weed dynamics is a major challenge to farmers, extensionists, and weed scientists.

In addition to being patchy, the presence of weeds in crop fields and across the agricultural landscape is uncertain. For a specific field in a specific season, when weeds will germinate, how fast they will grow in relation to the crop, how much seed they will produce, and how effective crop growth and weed control practices will be are difficult to predict (Ghersa & Holt, 1995).

First, the individuals of a weed species have a wider range of response to weather conditions than the individuals of a crop planted in the same field. Although each weed species has a relatively defined life cycle, individual weeds within a species show a range of responses to moisture and temperature cues (Chapter 9) (Dekker, 1997).

Second, unpredictable variations in weather during and between seasons affect weed germination and growth, the relative development of the weeds and the crop, and the effectiveness of weed control measures. In a single location in Minnesota, for example, variable weather conditions from 1991 to 1994 included lingering snowpacks, a late cold snap, an exceptionally wet spring, an exceptionally dry spring, and midseason droughts (Forcella et al., 1996). These affected date of soil preparation and planting, date of weed germination, and herbicide effectiveness. Thus, although weed patches may show some stability across years, actual weed density, weed phenology in relation to the crop, and weed seed production may be much more variable than weed patch location and thus harder to predict.

Third, over several cropping seasons, nondirectional random shifts in weed composition due to weather fluctuations and semipredictable directional shifts in weed composition due to cropping patterns occur simultaneously. This interaction contributes an additional dimension of uncertainty to weed management. This may be further complicated by the occasional invasion of new weed species.

Lastly, farmers manage weeds according to different criteria and constraints depending on the year. Small farmers are routinely affected by family illnesses, economic crises, and low crop prices. Large-scale farmers often suffer from machinery breakdown, labor problems, and unexpected changes in the cost of inputs. All these factors can affect the nature and timeliness of weed control measures.

Thus, on the one hand, weeds are sufficiently predictable that farmers can use routine control practices to produce crops without being overrun by weeds. On the other hand, weeds are rarely eliminated altogether, due to the localized mismatch between routine control practices and the uncertainty of weed patchiness. Farmer decision-making in weed management aims to minimize this mismatch for more efficient and less risky crop production. How have advances by researchers and extensionists taken into account weed patchiness and uncertainty for the wide diversity of the world's farmers?

Since the early 1900s, the routine use of uniformly applied agrichemical inputs on the better croplands has produced impressive increases in yields and labor productivity, first in temperate and later in tropical agriculture. Through multiyear replicated experiments, scientists conducted input–output research to identify the best broadly applied levels and combinations of different inputs, each of which has a specific, short-term purpose. Extensionists and later private crop consultants promoted the use of improved varieties, chemical fertilizers, and pesticides. This simple production model based on the efficient assemblage of purchased inputs into an end product resembles an industrial process (Levins, 1986). In the USA maize production quadrupled from 1940 to 1990 with fewer farmers and less land in production (Hossner & Dibb, 1995). In more recent years, in China rice and wheat yields have doubled and quadrupled, respectively (Hossner & Dibb, 1995).

Recently, science and society have begun to realize that a crop field is not a factory, but rather part of a living and responsive system. Herbicide resistance in weeds, floristic shifts to harder-to-control weeds, ground and surface water pollution, and human health effects are now routinely recognized as part of the risks and external costs of using herbicides and other agrichemicals (Chapter 1).

In many tropical countries, the standardized, high-input approach to increased crop yields has not fit productively with the complex landscapes, incipient infrastructure, and the diverse human cultures and cropping systems of smallholders (Pretty, 1995, pp. 31–3). As a result, input use has generally been irregular and crop yield responses modest and inconsistent. In the countries of Central America, *Phaseolus* bean yields have not increased consistently during the past 30 years (FAOSTAT, 1999). In Nicaragua, for example, since 1965 bean yields have fluctuated from 0.5 to 0.9 Mg ha^{-1}, but the long-term yield has increased only slightly (unpublished Nicaraguan Central Bank files, 1998). Similarly, coffee yields have fluctuated from 0.3 to 0.8 Mg ha^{-1}.

The Nicaraguan Coffee Growers' Union (UNICAFE, 1998) found that coffee yields could double with improved management and on better sites could quadruple with higher inputs and better management. During recent years, pesticide poisonings and the expenditure of foreign currency for herbicide imports have also fluctuated, although the fluctuations have not been correlated with crop yields (Beck, 1997).

These efforts in temperate and tropical regions to improve methods of crop production, including weed control, point to an important and still unfolding lesson. Progress has been possible, but not without costs and failures. Further progress will depend on the ability of scientists, extensionists, and farmers to work within increasingly complex expectations. At one time the objective was simply higher crop yields. Current goals include higher crop yields, protection of human health and the environment, improvement of soil and water quality, and greater market competitiveness. These factors are represented on the horizontal axis in Figure 3.2 as the increasing ecological, social, and economic complexity affecting crop production. In the case of weed management, this complexity is a product of several factors: changes in the larger social and economic context of agriculture, increasing understanding of weed ecology and crop production, and a need to ameliorate past negative results, such as herbicide resistance, and minimize them in the future.

In confronting this complexity, weed management has progressed from initially simple input–output research (e.g., trials on herbicide rates and cover crop species) through site-specific recommendations (e.g., herbicide rates by soil types) and problem-solving research (e.g., limiting competition between crops and cover crops) to predictive models and decision aids (e.g., Kropff & van Laar, 1993; Forcella et al., 1996). On one hand, these changes can be interpreted as the successive fine tuning of recommendations for broadscale, uniform weed management. This possibility for the vertical axis in Figure 3.2 originates from the perspective that the natural world can be increasingly predicted and controlled. However, these advances can also be seen as first steps in the development of adaptive management (Holling, 1978, pp. 1–21; Roling & Wagemakers, 1998). In conventional modern agriculture, most research aims to provide farmers with general technologies and recommendations suited for average situations. In contrast, the aim of adaptive management is a progressive increase in farmers' ability to develop and adapt a range of technologies for a local fit under variable and uncertain situations. Adaptive management assumes that decisions on weed control are based on less than perfect information, that control measures are not completely effective, and that each crop season provides additional data for the farmer on the development of more effective weed management. This is an appropriate response to the

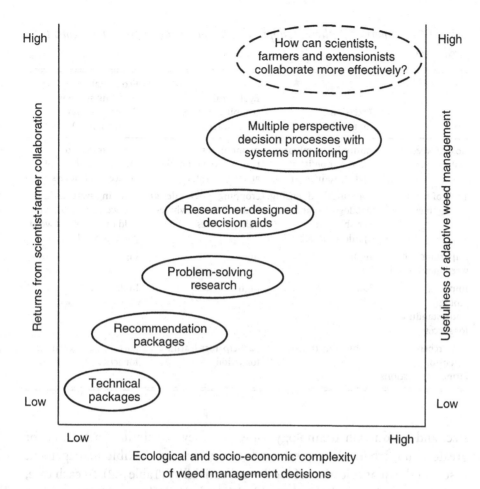

Figure 3.2 Relation of the complexity of weed management decisions to the need for adaptive management approaches and to returns from farmer–extensionist–scientist collaboration. (After Coutts, 1994, p.7.)

increasing ecological, social, and economic complexities that confront programs for the improvement of on-farm weed management.

Three approaches to farmer management of weed patchiness and uncertainty

Three contrasting approaches for practical adaptive weed management can be identified: precision agriculture, mechanized ecological cropping systems, and ecological cropping systems with hand tools and manual labor and occasional animal power for soil preparation. As the brief descriptions below demonstrate, these three approaches diverge in level of input use, field

Table 3.2. *Comparison of alternate approaches for improving on-farm management of weed variability*

	Precision monoculture	Ecological cropping systems (mechanized)	Ecological cropping systems (hand tools and animal power)
Goal of weed management	minimize/eliminate weed competition and reproduction	maintain easy-to-manage weed complex at acceptable levels	maintain easy-to-manage weed complex at acceptable levels
Tools of weed management	computerized sensing, datalogging, and variable-rate application technology	cropping systems design and well-timed whole-field practices	cropping systems design and well-timed whole-field practices and weed patch control
Importance of weed monitoring	high	high	high
Importance of ecological understanding for success	low	high	high
Researcher–extensionist–farmer interactions	technology transfer	participatory learning for action	participatory learning for action

size, and production technology. However, they are similar in aiming for greater land, labor, and capital productivity through flexible management based on the quantification of local weed variability (Table 3.2). In each case, farmers are trying to make better decisions by tailoring practices to weed patchiness rather than using routine uniform practices. Decisions about what practices to use are based on field-to-field and within-field monitoring for timely matching of practices to weed composition and patches.

Precision agriculture employs computerized spatial information for crop management (Lass & Callihan, 1993; Roberts, Rust & Larson, 1995; National Research Council, 1997, pp. 26–43). In this approach, real time yield monitoring on the harvester is connected with satellite-linked global positioning and geographic information systems to produce a detailed yield map that can be overlain on detailed soil maps. Variable-rate seeders and fertilizer applicators make possible the within-field fine-tuning of seed and fertilizer rates according to soil production potential. The more precise use of inputs may be directed to reduced environmental pollution from fertilizers or to further increase yields (Blackmore et al., 1995). Realization of either goal is subject to weather unpredictability. For example, Jaynes & Colvin (1997) showed that yields from specific localities in a field were below average one season and above average in others. This complicates the fine-tuning of input levels, since weather may be difficult to predict.

Weed observations can also be incorporated into the data system to locate problem areas for additional observation from one crop to the next. Variable-rate applicators permit fine-tuning herbicide applications to soil type or patches of particular weed species. Sprayer prototypes controlled by weed sensors have also been developed to apply post-emergent herbicides only where weeds are present (Thompson, Stafford & Miller, 1991). Patch spraying has been calculated to save from 9% to 97% in herbicide use, compared to field-wide application for the control of the perennial weed *Elymus (Elytrigia) repens* in cereal grains in England (Rew *et al.*, 1996). Little saving occurred when weed patches were extensive, a wide buffer was sprayed at the patch edges, and the areas below threshold were sprayed with a lower herbicide dose. High savings resulted when weed patches were few and concentrated, no buffer at the patch edges was sprayed, and the areas below threshold were not sprayed.

The second approach, also in mechanized agriculture, builds on the multiple interactions among diverse living organisms and the physical environment in a crop sequence to minimize the impact of weed variability and uncertainty. Weeds are managed by manipulating a diversity of factors that negatively affect weed population dynamics and favor the crop over the weed in crop–weed interactions. Non-noxious weed complexes are maintained at below-threshold levels through crop rotation, cover crops, timely tillage and cultivation, crop residue management, choice of crop varieties, and other tactics. The ecological approach focuses on the design of multiple-year cropping systems that suppress weeds rather than directed, short-term control of weed patches. This approach is well illustrated by the study of potato-based rotations under three contrasting weed management treatments (conventional, reduced input, and mechanical) and two soil managements (Gallandt *et al.*, 1998) described in Chapter 5. In the southern Brazilian state of Santa Catarina, thousands of farmers routinely use a diversity of green manures either intercropped with main crops or as covers during fallow periods to prevent soil erosion, suppress weeds, reduce weeding costs, and build soil tilth (Bunch, 1993). Green manure management is mechanized with animal-drawn implements, which flatten standing cover crops, conserving the cut biomass on the soil surface, and clear a narrow furrow for planting. Since 1987 the combination of green manures, animal manures, and soil and moisture conservation has produced yield increases of over 65% for maize and soybeans. Labor costs for weeding and plowing have also declined. As both the potato-based system and the minimum-till green manure system show, the mechanized ecological approach is cumulative over seasons and buffered against weed patchiness and uncertainty by crop vigor, low weed levels, and the use of a diversity of practices.

A third approach applies the ecological cropping systems practices mentioned above like green manures, intercropping, and rotations, but uses hand tools, sometimes animal power, and only occasionally, if at all, tractor-drawn implements. This represents an extension of traditional smallholder agricultural systems, which are well suited to the localized management of weed patchiness. The close interaction between mental and manual labor and the speed at which work takes place permits continuous interpretation of field and crop conditions and simultaneous adjustment of how each practice is carried out. Farmers hand-weeding a field can observe local variation in both crop stand and weed severity and simultaneously customize their management plan to patchiness. Roguing primarily noxious weeds before they produce seed, planting cover crops in large gaps in the crop stand, and selectively applying mulch in potentially severe weed patches are examples of mosaic weed management in response to weed patchiness.

Crop production based on the deliberate and opportunistic adaptation to site heterogeneity still characterizes some indigenous agriculture (Richards, 1985; Salick, 1989) and home gardens, but has diminished among many small farmers who frequently use uniform, field-wide practices in spite of high within-field variability. Their ability to innovate has been overloaded in many rural communities by rapid social and economic displacement, shrinking farm size, and accumulated land degradation (Blaikie, 1987, pp. 117–37). In addition, local perspectives on how to innovate in crop production and weed control have been sidetracked by input-linked credit programs and promotion activities of the commercial input sector (van der Ploeg, 1993).

Bean production in Central America in a slash-mulch short-fallow rotation illustrates the management of vegetation heterogeneity for improved cropping. In this system bean seed is thrown into standing one- to three-year fallow vegetation, which is then slashed as mulch to promote bean germination and control weeds. No further weeding is used. This system is low cost, has low labor requirements, and is soil conserving, but has also been criticized as low yielding (Thurston *et al.*, 1994). Farmers using this system readily identified good (e.g., *Ageratum conyzoides*, *Melinis minutiflora*, *Melanthera aspera*) and undesirable (e.g., *Rottboellia cochinchinensis*, *Pteridium aquilinum*) fallow species (G. Melendes, unpublished data). A shift to tillage and fertilizer and pesticide inputs increased cropping frequency and yields, but cost more for inputs, required more labor, and was not feasible on sloping lands. Upgrading fallows by planting patches of *Tithonia diversifolia* or vining legumes like *Canavalia ensiformis* decreased less desirable vegetation, increased fallow biomass and nutrient content of the mulch, and also increased bean yields up to 50% (G. Melendes, unpublished data). Upgraded fallows of mixed vegetation had higher bean yields with less pest damage than pure improved fallows, suggesting that farmers should introduce certain higher-biomass-producing species in patches without eliminating selected resident fallow species.

Adaptive management and farmer–extensionist–scientist interactions

Adaptive weed management following any of the three approaches can contribute to lower weed losses and costs. How might farmers, extensionists, and scientists interact to make this happen?

Precision agriculture is likely to follow the technology transfer diffusion model that has been employed for the promotion of other purchased inputs in crop production (Table 3.2). Off-farm scientists develop the sensing devices for commercial farm machinery, the links for satellite communication, software for field data interpretation, and genetically engineered crop varieties. Select groups of retailers and innovative farmers then pilot-test the products. Custom applicators, crop consultants, and data processing services, with support from public extension in regions with large demand, will make the technology available through contracts. Their focus is likely to be on larger producers who contract large quantities of inputs (Nowak, 1997). Several questions are pending (Hewitt & Smith, 1996; National Research Council, 1997). Will the components of precision agriculture be available to the majority of producers? Who will have access to the data banks of crop yield response by season and soil type? Will the technology be used for yield maximization for a few farmers or to provide the information base for less risky, more environmentally sound agriculture practiced by a majority of farmers who use mechanization? Hamilton (1995, pp. 146–51) concluded that making computerized data analysis/decision aids more hands-on and transparent would improve farmer decision-making. Alessi (1996) proposed that a wider and more diverse group of farmers should participate in pilot-testing of precision agriculture technology. In this way the technology would serve broader community interests.

The interaction among farmers, extensionists, and scientists appropriate for the development of the ecological management of weed patchiness and uncertainty can be termed *participatory learning for action* (Table 3.2) (after Hamiliton, 1995, p. 14). This process draws on field and landscape perspectives of weed control (Firbank, 1993; Cousens & Mortimer, 1995, pp. 217–42,

Cardina, Johnson & Sparrow, 1997) and on principles of farmer learning and decision-making. Other terminology that refers to similar procedures includes farmer participatory research (Okali, Sumberg & Farrington, 1994), participatory technology development (Haverkort, van der Kamp & Waters-Bayer, 1991), indigenous knowledge and technology development (Brokensha, Warren & Werner, 1980), participatory research and development (Chambers, 1995), and participatory action research (Whyte, 1991). This or related processes are being used in farmer networks in USA, Europe, Australia, Africa, Asia, and Latin America (Scoones & Thompson, 1994; Thrupp, 1996; Veldhuizen et al., 1997).

Why the term participatory learning for action? First, the focus is on all the *actions* in weed management by farmers. Actions include not just the field practices employed, but information acquisition and use by farmers and their planning and decision-making processes. Second, *learning* (rather than research) suggests a broad approach to inquiry based on experiments, field observation, group analysis of data from monitoring routine practice, and any other tools that improve weed management. Third, the process should be *participatory*. That is, farmers, extensionists, and scientists should all contribute to and learn from the process.

Participatory learning for ecological weed management: a proposal

The development of ecological weed management depends on the collective ability of farmers, extensionists, and scientists to convert local weed information into an improved understanding of weed ecology. Special attention must be taken to develop principles of weed ecology that are applicable in improved farmer planning and decision-making. The following proposal offers a starting point for a "learning to learn" process on farmers' management of weed patchiness and uncertainty. The process should be open-ended and evolving. Initial learning and experiences among farmers, extensionists, and scientists become the basis for planning future steps. Four themes are discussed here: the role of groups in participatory learning, farmers' decision-making as an organizing principle, methods for observing and understanding weeds, and farmer communication with other farmers.

Why participatory learning involves farmer groups

The process for developing agricultural technology and information follows a series of steps. Problems and opportunities are identified either through formal analysis or by intuition. Next, those participating in technology generation prioritize problems and opportunities and convert them into options or treatments. Third, options are tested in experiments and studies. Formal data and informal observations are collected and evaluated. Finally, participants review the results and plan further steps for the generation of technology and information. To understand who participates in agricultural technology generation, those who make decisions and influence outcomes at each phase must be identified (Nelson, 1994).

Biggs (1989) proposed four categories of researcher–farmer cooperation in technology generation, depending on who makes the decisions in each step. In a contractual relationship, researchers make all the decisions, but conduct the research on-farm, primarily to gain access to a wider range of soil and climatic conditions than is generally available on research stations. In a consultative relationship, researchers consult farmers about their problems and their views, but prioritize, design, and implement the research and interpret the results themselves. In a collaborative relationship, researchers and farmers work together to define problems and possible options for testing, share responsibilities for plot implementation and evaluations, and plan together further actions. In a collegial relationship, researchers support the farmer or farmers' group in implementing their own technology development efforts. Figure 3.3, a matrix of the steps in technology generation, contrasts two of these cases, scientist-run on-farm trials and farmers' experiments.

Participatory learning for action, also shown in Figure 3.3, is collaborative and collegial in its decision-making. In this approach, scientists and extension agents do not act as experts, even though they possess a wealth of information about weeds. Neither are they the only ones who know about weeds nor those solely responsible for solutions to weed problems. In the context of participatory learning, scientists and extensionists take on two new roles. First, they are facilitators of group inquiry and decision-making. This means they promote the development of learning within the group, insure that decision-making is participatory, and keep the process focused on a time frame and spatial scale relevant to farmers. Second, instead of acting as lecturers, researchers and extension staff employ their technical resources and analytical skills to promote group inquiry. Diagnostic field tours to visualize weed problems, reconstruction of recent weed trends in different fields to evaluate decision-making, and field exercises that reveal cause and effect can promote group analysis and plans for action. Scientists and extensionists provide information in the form of suggestions rather than recommendations. These suggestions may come from scientists' personal experience, the results of experiments, and technical literature.

Farmer–extensionist–researcher learning can be more effective with a

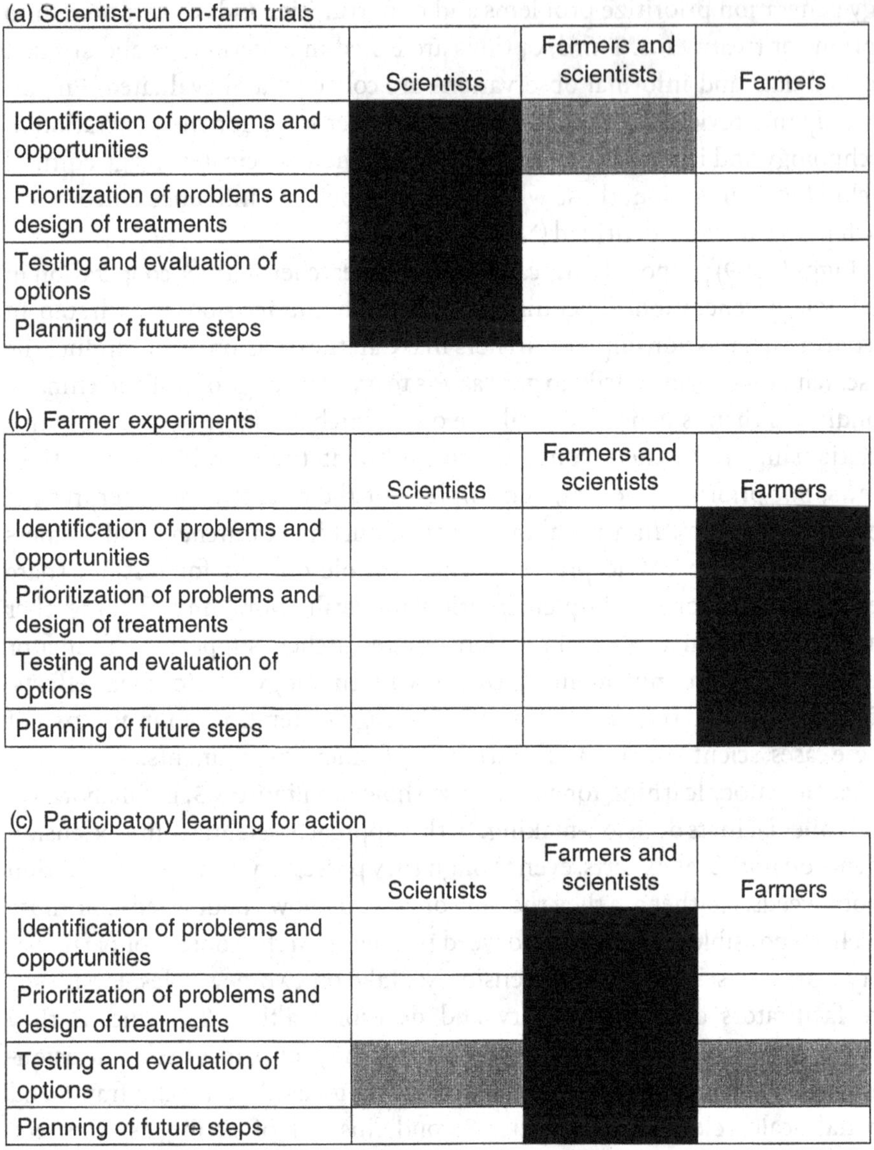

(a) Scientist-run on-farm trials

	Scientists	Farmers and scientists	Farmers
Identification of problems and opportunities	■	▨	
Prioritization of problems and design of treatments	■		
Testing and evaluation of options	■		
Planning of future steps	■		

(b) Farmer experiments

	Scientists	Farmers and scientists	Farmers
Identification of problems and opportunities			■
Prioritization of problems and design of treatments			■
Testing and evaluation of options			■
Planning of future steps			■

(c) Participatory learning for action

	Scientists	Farmers and scientists	Farmers
Identification of problems and opportunities		■	
Prioritization of problems and design of treatments		■	
Testing and evaluation of options	▨	■	▨
Planning of future steps		■	

Figure 3.3 Research on-farm: who makes the decisions in each step of the experimental process. Blocks shaded in lighter tones represent possible complementary activities. (After Okali *et al.*, 1994, pp. 95–6.)

group focus (Bryant & White, 1984, pp. 14–32; Pretty, 1995, pp. 147–9). The debate, dialogue, and exchange of ideas in a group generates motivation and promotes creativity. This process also insures that everyone has a more equal access to all relevant information. Data collection and group discussions among 10 to 25 farmers better represent the spectrum of field conditions and farmer experiences in a community or region. This spectrum is especially important for developing management strategies for weed patchiness and uncertainty. Additionally, in group meetings farmers outnumber researchers and extensionists, helping to promote a focus on field problems and practical solutions and the use of uncomplicated language. Similarly, the participation of several extension staff and more than one researcher, possibly from different disciplines, serves to broaden the opinions and perspectives.

Farmers' decision-making as an organizing principle

Farmers facing a new planting season base their decisions on different types of information: experience accumulated from previous seasons, their neighbors' experiences, specific data about each of their fields, expectations about prices, resources, and weather, and technical recommendations from public extension and commercial promotion (see Table 3.1). This information, together with each farmer's goals, provides the basis for a tentative plan. This is often a minor variation on a routine developed over time in response to the local conditions (Aubry, Papy & Papillon, 1998). As the expected planting period nears, farmers modify their decisions based on additional observations and information about the season and the status of other fields. Once the field is planted, crop management plans are adjusted. As the season unfolds, farmers modify their expectations and decisions frequently. This routine of iterative decision-making with uncertain and incomplete information is dictated by a seasonal schedule.

A group of farmers, extensionists, and researchers meeting regularly to analyze and improve farmer weed management should logically follow a routine that parallels the seasonal schedule of decision-making based on crop phenology (Figure 3.4). This sequence can be adapted to commercial vegetable growing, cash grain production, perennial crops, or management-intensive grazing. At each meeting, the group discusses what information is available for decision-making, how it was generated, what information farmers use to make decisions, what options they are considering, what information they would like to have, and how they will determine whether they have made good decisions. Generally a good decision is one that made the best use of the available information rather than one that simply produced fortunate outcomes. This approach has been used with tens of thousands of Indonesian farmers to improve integrated pest management in rice (Roling & van der Fliert, 1994), and has been likened to the case study approach used in many business schools for management education (Useem, Setti & Pincus, 1992).

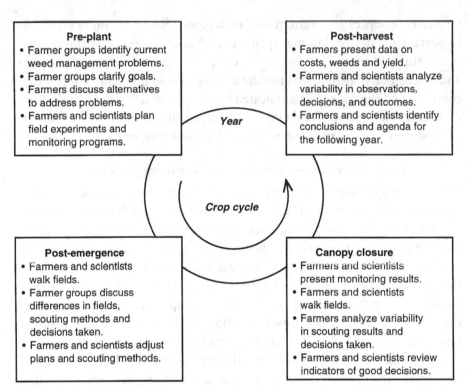

Figure 3.4 An illustrative yearly cycle for scientist collaboration with farmer groups. The objectives are to observe field-scale weed presence and to analyze farmer decision-making in the use of weed control practices.

Methods to measure and study weeds

To link their different perspectives on weeds, farmers, extensionists, and researchers need complementary methods to observe and quantify weed species composition, abundance, and distribution. Few would disagree that better quantification of weeds should lead to better management. Cousens & Mortimer (1995, pp. 291–2) suggested that observations should have enough precision to estimate current weed levels and predict trends. Similarly, Firbank (1993) proposed that weed monitoring should capture differences in order of magnitude. Hamilton (1995, pp. 101–4), in his study of adult learning methods with Australian grain farmers, concluded that observation methods should place less emphasis on high accuracy. He suggested that observation should facilitate group analysis of different situations by establishing relative magnitudes. With this type of observation, farmers more readily extrapolated results to other fields or farms.

Research on methods for field level scouting in weed science has concentrated on the decision to apply post-emergent herbicides (Marshall, 1988;

Total weed cover (%)	< 5 %	5-25 %	25-50 %	>50 %
	IIIII IIIII IIIII	IIIII IIIII IIIII IIIII IIIII IIIII II	III	
Weeds present by growth stage	Vegetative growth		Flowers and seeds	
Major weed # 1 *Digitaria*	IIIII IIIII IIIII IIIII IIIII IIIII IIIII IIIII IIIII II			
Major weed # 2 *Bidens*	IIIII IIIII IIIII IIIII IIII			
Major weed # 3 *Cyperus*	IIIII IIIII IIIII IIIII IIIII IIIII IIIII IIIII IIII			
Other broadleaf annuals	IIIII IIIII IIIII IIIII IIIII IIIII IIIII III			
Other broadleaf perennials	III			
Other grasses	I			

Figure 3.5 Data sheet used by farmers to record weeds in annual crops in Nicaragua. Farmers observe both total cover and the presence of major types of weeds by phenological stage in 50 circular quadrats. This example was taken in a maize field 25 days after planting and before the first weeding.

Berti *et al.*, 1992; Forcella, 1993; Gold, Bay & Wilkerson, 1996; Johnson *et al.*, 1996). The primary objective is the one-time determination of whether the mean weed population density in a field is below or above a threshold that triggers application of a post-emergent herbicide.

In the context of farmer decision-making and a weed working group routine, additional reasons can be identified for documenting weed abundance, weed floristic composition, and weed patterns at the field and landscape levels. These include analysis of the timeliness of practices in farmer fields and researcher experiments, the evaluation of field-scale trials, and the comparison of weed dynamics among experiments and fields, and across years. For example, the format in Figure 3.5 was designed for use with smallholder maize and bean producers in Nicaragua. In a 15–30 minute walk through their fields (0.5–2.0 ha), farmers determine total weed cover and the presence and reproductive status of different weeds in 50 circular quadrats 25–30 cm in diameter. In a later group discussion, farmers compare problem weeds in different fields, total weed cover and crop stage, variability within each field, and the likelihood that the floristic composition will change based on current weed control practices. This method does not generate a spatial map, but it does provide information to analyze decisions on field-wide weed control and particular practices directed at specific weeds or patches.

The development of simple methods for on-farm use that combine accuracy and time efficiency is a major practical challenge to weed ecologists and weed extension specialists. Finding workable sampling and data recording methods for use in group discussion and for comparisons in time and space is one of the major initial areas of collaboration among farmers, extensionists, and scientists when developing new programs in participatory learning for action. A few basic guidelines are available from previous studies. A larger number of small units offers more precision than a smaller number of large units (Lemieux, Cloutier & Leroux, 1992). Greater sampling intensity is needed for accurate assessments of species that are less common (Marshall, 1988). Spatial distributions of weeds cannot be estimated with arithmetic interpolation from quadrat counts (Marshall, 1988). Transects can be used efficiently for sampling cover in large land areas (Morrison *et al.*, 1993). Which is the best method? When should farmers sample? How frequently should they sample? Answers to these questions depend on the producers' interest, the types of weeds, the type of crop, field size, and specific concerns of the group. Midwestern USA maize and soybean growers and Central American maize and bean farmers grow similar crops, but would have very different discussions about weed variability and uncertainty, and would propose different observation methods.

Equipped with shared methods for weed measurement, farmers, extensionists, and researchers can develop site-specific and group-specific learning approaches as illustrated in the case studies of this chapter. These may be derived from individual or group initiatives, and vary in their degrees of collaboration as shown in Figure 3.6. The farmer group context keeps both scientist and farmer activities focused on farmer management of variability and uncertainty for fewer weeds and higher yields.

Farmers, extensionists, and scientists each have different potential rewards from participatory learning for action. Table 3.3 indicates how scientists who are worried about funding and publications, extension staff needing to cover their district with limited budget and time, and farmers who are concerned with crop prices and too much or too little rain might benefit from a working group routine based on participatory learning for action. Each plays an ample role in the advance of weed management; each has expectations to meet and procedures to follow in their own knowledge communities; and each has opportunities for creative working relationships with other sectors.

Farmer communication with other farmers

In Europe and the USA and especially in Latin America, Africa, and Asia, there are many, many more farmers than extensionists, and many more extension agents than scientists. With these proportions, how can participa-

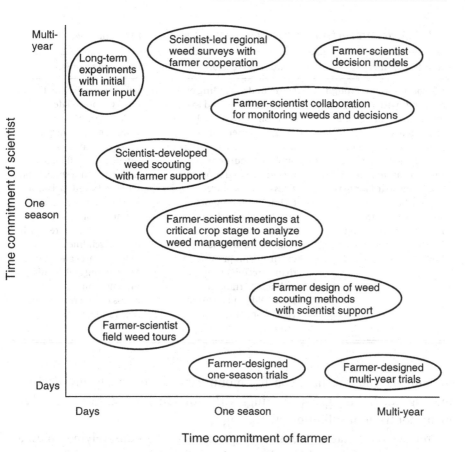

Figure 3.6 Relative time commitments of farmers and scientists in different, but complementary, learning approaches for improved weed management. (After Van Huis & Meerman, 1997.)

tory learning for action, which proposes that scientists work directly with farmer groups, contribute to improved decision-making in weed management among millions of farm households?

The possibility for widespread impact of participatory learning for action resides in a three-stage process that begins in pilot areas and expands outward through organized extension programs and informal farmer and rural household communication networks.

The first stage, described in the three previous sections, focuses on creating a nucleus of methods, results, and experienced individuals in pilot groups. This stage may appear costly in time, although Nelson (1994) found that standard demonstration plots were only slightly lower cost than farmer experimentation groups. The slightly greater cost pays off because the experienced

Table 3.3. *Rewards to farmers, extensionists, and scientists for working together*

Rewards to farmers	Rewards to extension	Rewards to scientists
· new perspectives through exchange with other farmers and with scientists and extensionists · structured analysis of information and procedures for decision-making; a sounding board for new approaches · source of ideas for short-term problem solving · better understanding of how to manage weeds	· better understanding of how farmers observe and make decisions as basis for better design of extension programs · in-depth understanding of on-farm conditions, including weed problems; assessment of current technologies for more effective feedback to researchers · farmers as partners in extension programs rather than as recipients of technology transfer · pilot fields and farms for visits from other farmer groups	· new perspectives on farmer observation and decision-making criteria for weed management to improve research strategies · intellectual challenge of understanding spatial and temporal weed variability at field and landscape levels · definition of new research directions integrated with other disciplines · practical cases and examples for teaching and training presentations · access to data from many fields and farms

individuals and the methods for monitoring weeds, analyzing decision-making, and linking group meetings with individual actions form critical elements for the multiplication of learning.

In the second stage, extensionists promote new groups, relying on established farmer groups. Although scientists are not present, the principles of participatory learning continue: farmer experimentation, field observation, group analysis of plans and decisions, and discussion of new weed management methods. In addition, extensionists strengthen farmer-to-farmer exchanges among pilot groups and new groups with surrounding farmers. Other household members and diverse non-farm sectors of rural communities may also be incorporated into these exchanges. Much needs to be learned about how ideas spread in rural social networks; insights on this subject would contribute to more effective facilitation of farmer-to-farmer exchanges (Box, 1989; Engel, 1997; Selener, Chenier & Zelaya, 1997).

The configuration of the third phase represents a wider spread of the participatory learning approach in informal rural communication networks and a potential for partnerships between farmer networks and formal research organizations. For example, in the Netherlands, horticultural study groups begun by growers to compensate weak research programs now make up a national federation that is developing links with government research programs (Oerlemans, Proost & Rauwhorst, 1997). In Colombia, farmer experimentation groups have formed their own umbrella organization to find funding and provide technical advice (Ashby *et al.*, 1999). What are the essential components for a vitalized co-learning network that encompasses research, public and private extension, and formal and informal farmer to farmer communication? The answer will depend on the experiences generated in such diverse areas as California, Iowa, Holland, Peru, Cuba, Vietnam, Philippines, Senegal, and many other countries (Thrupp, 1996; Veldhuizen *et al.*, 1997).

Farmers, extensionists, and scientists learning together: four examples

The case studies described in the following pages represent a spectrum of farmer–extensionist–scientist collaboration to improve weed management and crop production. The first case describes a study controlled primarily by scientists, whereas the last case describes a farmer-to-farmer approach. The cases complement one another in illustrating the four concepts of participatory learning described previously. Each case is innovative in some dimension. However, in all the cases widespread improvement of farmer capacity for ecological weed management remains a challenge.

California tomato cropping systems: farmers advise a research station study

In the late 1980s a group of scientists representing ten different disciplines at the University of California at Davis established a 12-year replicated comparison of conventional, low-input, and organic tomato-based cropping systems [see Temple *et al.* (1994*a*, 1994*b*) for details of treatments and working procedures]. Tomato is one of the most economically important crops in the state. The group decided to work on the experiment station to insure rigor in the operation of the experiment and in data collection. However, they were also interested in using best farmer practices in each system to achieve profitability rather than comparing predetermined fixed treatments. They recruited an advisory committee of two conventional and two organic tomato growers and two county extension advisors. Farmers, extension advisors, and scientists have been meeting every two weeks to analyze the status of the experiment and to plan upcoming crop management activities.

This California case is innovative in its multidisciplinary focus, in the treatment flexibility based on the use of best farmer practices, and in the incorporation of farmers as advisors. A comparison with the concepts of participatory learning for action suggests that learning was done primarily on the scientists' terms. Scientists formed the majority of the group; all data were generated on the research station in the 8-ha main experiment and a 3.2-ha satellite plot; data were taken for credibility with scientific peers. Scientists gained insight into grower decision-making, particularly the difficulty of learning to use cover crops and new farm machinery, the difficulties of growing new crops for the first time, and the uncertainty created by weather and prices.

However, this approach had few mechanisms for comparisons with a greater diversity of on-farm conditions or involvement of larger groups of growers in co-learning. Yield averages for the county were the reference for comparison of experimental crop yields, but other local farm data were not available, for example, on weed abundance or floristics or grower weed control methods. Farmer advisors recommended the use of transplants instead of direct seeding in the organic and low-input tomato for easier weed control, and a longer growing period for green manures (Lanini et al., 1994). Were scientists adopting practices already used by most organic growers? By the third year, tomato yields were similar in the three systems, although weed biomass was significantly higher in the organic and low-input systems. The collection of data by growers on weeds, soils, and pests in their fields could have provided useful reference points for the experiment as well as a basis for broader farmer–extensionist–scientist discussions on variability among fields and farms.

Iowa grain cropping: farmers design and run replicated trials

During the midwestern USA farm crisis of the 1980s, a group of Iowa farmers organized the Practical Farmers of Iowa (PFI) (Harp, 1996). They felt that university research and extension programs were unresponsive to farmers' economic and environmental problems. Organized in five chapters across the state, they test alternative management practices such as lower nitrogen fertilizer rates and ridge tillage without herbicides. Scientists from Iowa State University were recruited to collaborate on experimental design and data analysis. At an annual winter planning meeting in each chapter, farmers and scientists meet to discuss research ideas and to draw up experimental procedures. Individual farmers identify problems that interest them. The on-farm research is conducted in replicated trials, usually with two treatments and six replicates, on each farm. Plots are the length of a field and a single or double planter width (Thompson & Thompson, 1990). Farmers plant the trials and collect data with support from researchers and students. PFI has attracted over 10 000 people to summer field days and farm tours in eight years. Farmers and researchers also present results at other extension events.

A recent evaluation identified specific strengths in the PFI approach (Harp, 1996). Scientists and farmers developed a common language based on the mutual understanding of each others' constraints and opportunities. Scientists learned more about farmers' research needs, while farmers had a channel to influence the university research agenda. For university research and extension programs, the PFI network guaranteed accelerated diffusion of results. Participating farmers from PFI developed leadership skills by organizing a farmer-managed research program to identify lower-cost cropping practices with reduced environmental impact.

Several difficulties were also identified (Harp, 1996). University staff cited problems with colleagues and job tenure from on-farm work. The trials from individual farms provided only site-specific results that were difficult to publish in scientific journals. The adopted procedure of standardized treatments for multifarm trials conflicted with PFI philosophy that prioritized individual farmer decisions about treatments. Farmers also found that opportunity costs of data collection and trial management were high because these activities made little immediate contribution to farm profits.

From 1987 to 1994, PFI farmers conducted 394 trials, including 78 on weed management. Fifty-one trials on maize and soybean demonstrated that ridge tillage without herbicides suffered no yield reductions and had lower production costs (Harp, 1996). These results and others concerning fertilizer reductions provided assurance for farmers contemplating input reduction. However, a more diverse participatory learning process that included field monitoring, group analysis of farmer planning and decision-making, and reviews of weed patterns could have promoted more extensive farmer–scientist collaboration. This broader range of co-learning activities might also have allowed the active participation of farmers beyond those who were motivated to run replicated trials. Improved management of the spatial and temporal variability of weeds requires that more farmers document weed numbers and distribution over multiple seasons and then use these data to discuss their criteria for decisionmaking.

Ground cover in coffee: farmers observe weeds and propose management alternatives

In Central America coffee is often produced on sloped land, either under shade with low inputs or in open sun with pesticides and high levels of fertilizers (Rice & Ward, 1996). Yields vary from 200 to 2700 kg of green beans ha^{-1}, depending on soils, climates, farm size, and farmer resources. The diversity of growing conditions and the layout of coffee fields create spatial and temporal variability in weeds, insect pests, and diseases. Improving yields and lowering production costs requires farmer management of heterogeneous conditions.

In the early 1990s, integrated pest management (IPM) specialists from the Center for Teaching and Research in Tropical Agriculture (CATIE) began a program in Nicaragua to improve pest management in coffee (Staver *et al.*, 1995; Staver, 1998). In parallel with field studies of pest dynamics and replicated trials with new management practices, the team began work with groups of smallholder coffee growers. The team quickly found that growers had extensive practical experience, but had an incomplete understanding of pests and the reasons for their variability. Farmers were also uncertain about when to use specific practices and how to evaluate whether one practice was better than another. These difficulties limited their efforts to test and adapt alternative management options.

To improve grower capacity to evaluate practices and make decisions based on observation and ecological reasoning, the CATIE team has been using participatory training procedures. Farmer groups meet every six to eight weeks during the yearly crop cycle to discuss what they know about pests, how they decide to manage them, and to fill in gaps in their understanding with practical exercises. A nearby coffee field is used as a laboratory to observe pests and their variability and to learn scouting methods. Between sessions farmers analyze their own fields and bring results to the next meeting. Data variability among farmers, fields, and seasons is used to promote discussion on why pest populations fluctuate, the effectiveness of current practices, and the relationship between pest levels and control decisions. In this process farmers begin to develop their own ecological logic of how to manage pests better and become eager to test alternative practices in a group plot or in their own fields.

In the case of weeds, the participatory training takes into account farmer knowledge of weed types. In a quick walk through a nearby coffee field, farmers choose a weed they consider very damaging to coffee, one not so damaging, and one that does not affect coffee plants. With these weeds in hand, the group discusses the different species one by one, how they grow and reproduce, and what type of damage they do to coffee plants. Individual weeds are grouped into categories by growth habit. The group also analyzes other ground covers in the coffee field such as leaf litter, discusses the consequences of bare soil, and identifies the weeds and other ground covers that protect the soil without competing much with coffee. To measure the cover of each weed type in the field, the group uses a simple transect method, and analyzes how the current control practices favor certain weeds and reduce others. Finally, the group discusses the best combinations of ground cover and identifies practices to implement in their comparison plot. Progress is evaluated in follow-up sessions every two months. Have growers sampled weeds? Are new practices being implemented? How effective are they? How much do they cost?

The framework for training based on weed types, sampling methods, and discussions of practices was developed by scientists, but has only advanced by incorporating coffee farmers' opinions and ideas. Scientists learn key words used by farmers, and farmers adopt scientists' vocabulary to refer to new ecological concepts. Standard weed scouting procedures have permitted comparisons between different fields and groups. Farmer groups have proposed that different weed species be used as ground cover depending on location. The approach has been successful in generating discussions about weed types and the importance of ground cover for soil conservation. Many farmers have tried selective weeding instead of total weeding. However, farmer groups have established few experimental plots to test alternative weed management practices. These farmers of low-yielding coffee may be stating that weed studies are not very important until they improve other problems, such as low coffee plant density and vigor, that have a greater impact on yields and profitability than do weeds.

Cover crops in Central America: farmers show other farmers

Velvetbeans are aggressive annual vining legumes of the genus *Mucuna*. They originated in Asia and were introduced into Florida via the Caribbean in the 1870s for use as a cover crop in citrus groves. [See Buckles (1993, 1994) for more details on the rise and fall of *Mucuna* in the USA and Central America.] From 1900 to 1920 nearly 1 500 000 ha were planted with *Mucuna* spp. as green manure and animal feed in the southern USA. Numerous studies and extension bulletins were also completed on velvetbean in this period. By the 1940s velvetbean had disappeared in the southern USA due to the availability of cheap nitrogen fertilizers and the spread of soybean.

In the 1920s banana companies introduced velvetbean into Central America. They promoted its use in association with maize that was cultivated by banana workers on company lands. After maize harvest, mules used for banana transport grazed these fields.

When mules were replaced with tractors, the use of velvetbean declined on banana plantations, but spread to peasant fields. From the late 1930s onwards, the use of velvetbean for weed control and as a green manure has spread among communities of Guatemala, Belize, southern Mexico, and Honduras. In Atlantida, Honduras, from the 1970s to the early 1990s the number of maize producers using velvetbean increased from 10% to over 60%, largely based on farmer-to-farmer communication (Buckles, 1994). The different regions where velvetbean is used vary greatly in length of dry season and rainfall, factors that farmers have taken into account when adapting their systems for velvetbean use (CIDICCO, 1995).

Many nongovernmental development organizations (NGOs) throughout Central America over the past 10 years have begun to promote the use of leguminous annual cover crops with small farmers. Velvetbean and other cover crops are viewed not only as valuable for improving soil fertility and reducing weed control costs, but also as a way to motivate farmers to experiment with solutions to their problems with their own resources (Bunch, 1982). Farmers who have learned to manage velvetbean through experiments on their own farms play a key role in the farmer-to-farmer approach. These farmers, who are known as promoters, often begin by asking a group of farmers why they think their yields have declined. Promoters speak of similar problems in their home region and describe the use of leguminous cover crops. The group may visit other farmers already working with cover crops. Promoters offer small amounts of seed for multiplication and testing. As farmers observe results and harvest seed, they encourage other farmers to try cover crops. Regional and national farmers' meetings are often organized to promote the exchange of results (e.g., Buckles & Arteaga, 1993; Lopez, 1993).

Does the dramatic spread of velvetbean in recent years represent improved farmer capacity for managing crop production and weed control or the fortuitous, but temporary, solution to a special combination of production problems? Will the farmer experimentation and farmer-to-farmer communication promoted by NGOs lead to improved farmer capacity or simply more efficient technology transfer? How will farmers who learned to use velvetbean from other farmers for weed control in maize respond to changing maize prices or new pest problems? A collapse in maize prices in the Atlantida region of Honduras would jeopardize the velvetbean–maize rotation, since velvetbean does not intercrop easily with annual crops other than maize. Recently, severe infestations of *Rottboellia cochinchinensis* have been reported in velvetbean fallows in Honduras, leading to land abandonment (Triomphe, 1996). These examples suggest that while farmer-to-farmer networks can be low cost and effective, there is a role for strategic, on-going links between farmer networks and research and extension systems. Scientists and extension staff may lack immediate solutions to problems like maize prices or the invasion of *R. cochinchinensis*. Nevertheless, a process of co-learning for improved weed management can achieve several objectives: scientists focused more clearly on integrated approaches to field problems; extensionists directed toward increased farmer capacity for decision-making rather than on technology transfer; and farmers communicating ecological knowledge rather than technological novelties with other farmers.

A concluding note

The four case studies demonstrate that there are diverse starting points for strengthening participatory ecological approaches to farmer weed management. These case studies and examples are drawn from both developed and developing countries. In the USA and Europe, a reduced farm population occupies the best land and is supported by a substantial information infrastructure. In Asia, Latin America, and Africa, a large rural population farms much marginal land with only minimal support from an erratically funded infrastructure for education and information. For improved weed management as part of a more efficient, productive, and resource-conserving agriculture, the two worlds can both be well served by strengthening decision-making capabilities among farmers. This can be accomplished by more effectively linking extensionists, scientists, and farmers.

REFERENCES

Alessi, S. (1996). Commentary: the idea of precision agriculture. *Consortium News*, 9, 5–11. Madison, WI: Consortium for Sustainable Agriculture Research and Extension.

Amanor, K. (1993). Farmer experimentation and changing fallow ecology in the Krobo District of Ghana. In *Cultivating Knowledge*, ed. W. de Boef, K. Amanor & K. Wellard, pp. 35–43. London: Intermediate Technology Publications.

Ashby, J., Braun, A., Brekelbaum, T., Gracia, T., Guerrero, M., Quiros, C., & Roa, J. (1999). *Investing in Farmer Researchers: Experience in Latin America*. Cali, Colombia: International Center for Tropical Agriculture (CIAT).

Aubry, C., Papy, F., & Capillon, A. (1998). Modeling decision-making processes for annual crop management. *Agricultural Systems*, 56, 45–65.

Beck, I. (1997). Uso y riesgo de plaguicidas en Nicaragua (1985–1995). *Memoria del Primer Encuentro Nacional de la Red de Acción sobre Plaguicidas de América Latina-Nicaragua*, August 1997, pp. 10–17.

Berti, A., Zanin, G., Baldoni, G., Grignoni, C., Mazzoncini, M., Montemucco, P., Tei, F., Vazzana, C. & Viggiani, P. (1992). Frequency distribution of weed counts and applicability of a sequential method to integrated weed management. *Weed Research*, 32, 39–44.

Biggs, S. (1989). *Resource-Poor Farmer Participation in Research: A Synthesis of Experiences from Nine Agricultural Research Systems*, On-Farm Client-Oriented Research Comparative Study Paper no. 3. The Hague, Netherlands: International Service for National Agricultural Research.

Blackmore, B., Wheeler, P., Morris, J., Morris, R., & Jones, R. (1995). The role of precision farming in sustainable agriculture. In *Site-Specific Management for Agricultural Systems*, ed. P. Robert, R. Rust & W. Larson, pp. 777–93. Madison, WI: American Society of Agronomy.

Blaikie, P. (1987). *The Political Economy of Soil Erosion*. New York: Wiley.

Box, L. (1989). Knowledge, networks and cultivators: cassava in the Dominican Republic. In *Encounters at the Interface, a Perspective on Social Discontinuities in Rural Development*,

Wageningen Sociological Studies no. 27, ed. N. Long, pp. 165–83. Wageningen, Netherlands: Agricultural University.

Brokensha, D., Warren, D., & Werner, O. (1980). *Indigenous Knowledge Systems and Development*. Washington, DC: University Press of America.

Bryant, C., & White, L. (1984). *Managing Rural Development with Small Farmer Participation*. West Hartford, CT: Kumarian Press.

Buckles, D. (ed.) (1993). *Gorras y sombreros: caminos hacia la colaboracion entre tecnicos y campesinos*. Mexico, DF: International Maize and Wheat Improvement Center (CIMMYT).

Buckles, D. (1994). *El frijol terciopelo – una planta nueva con historia*. Mexico, DF: International Maize and Wheat Improvement Center (CIMMYT).

Buckles, D., & Arteaga, L. (1993). Extension campesino a campesino de los abonos verdes en la Sierra Madre, Veracruz, Mexico. In *Gorras y sombreros: caminos hacia la colaboracion entre tecnicos y campesinos*, ed. D. Buckles, pp. 51–62. Mexico, DF: International Maize and Wheat Improvement Center (CIMMYT).

Bunch, R. (1982). *Two Ears of Corn: A Guide to People-Centered Agricultural Improvement*. Oklahoma City, OK: World Neighbors.

Bunch, R. (1993). *EPAGRI's Work in the State of Santa Catarina, Brazil: Major New Possibilities for Resource-Poor Farmers*. Teguicigalpa, Honduras: Asociación de Consejeros para una Agricultura Sostenible, Ecológica y Humana (COSECHA).

Busch, L., & Lacy, W. (1983). *Science, Agriculture, and the Politics of Research*. Boulder, CO: Westview Press.

Campbell, A. (1994). Community first: landcare in Australia. In *Beyond Farmer First*, ed. I. Scoones & J. Thompson, pp. 252–7. London: Intermediate Technology Publications.

Cardina, J., Johnson, G., & Sparrow, D. (1997). The nature and consequence of weed spatial distribution. *Weed Science*, **45**, 364–73.

Chambers, R. (1995). Paradigm shifts and the practice of participatory research and development. In *Power and Participatory Development*, ed. N. Nelson & S. Wright, pp. 30–42. London: Intermediate Technology Publications.

CIDICCO (1995). *Management Practices for Working with Velvetbean*, Noticias sobre cultivos de cobertura no. 5. Teguicigalpa, Honduras: Centro Internacional de Documentación e Información sobre Cultivos de Cobertura (CIDICCO).

Colbach, N., Forcella, F., & Johnson, G. (2000). Spatial and temporal stability of weed populations over five years. *Weed Science*, **48**, 366–77.

Cousens, R., & Mortimer, M. (1995). *Dynamics of Weed Populations*. Cambridge, UK: Cambridge University Press.

Coutts, J. (1994). *Process, Paper Policy and Practice. A Case Study of a Formal Extension Policy in Queensland, Australia, 1987–1994*. Wageningen, Netherlands: Agricultural University. Published doctoral dissertation.

Dekker, J. (1997). Weed diversity and weed management. *Weed Science*, **45**, 357–63.

Elliot, J., Holmes, J., Lockhart, J., & MacKay, D. (1977). The evolution of methods of weed control in British agriculture. In *Weed Control Handbook*, vol. 1, ed. J. Fryer & R. Makepeace, pp. 11–23. Oxford, UK: Blackwell.

Engel, P. (1997). *The Social Organization of Innovation: A Focus on Stakeholder Interaction*. Amsterdam, Netherlands: Royal Tropical Institute.

FAOSTAT (1999). *FAOSTAT Statistics Database Agriculture*. Internet: http://apps.fao.org/cgi-bin/nph-db.pl?subset=agriculture.

Firbank, L. (1991). Interactions between weeds and crops. In *The Ecology of Temperate Cereal Fields*, ed. L. Firbank, N. Carter, J. Darbyshire & G. Potts, pp. 209–31. Boston, MA: Blackwell.

Firbank, L. (1993). The implication of scale on the ecology and management of weeds. In *Landscape Ecology and Agroecosystems*, ed. R. Bunce, L. Ryszkowski & M. Paoletti, pp. 91–104. Boca Raton, FL: Lewis Publishers.

Forcella, F. (1993). Value of managing within-field variability. In *Soil Specific Crop Management*, ed. P. Robert, R. Rust & W. Larson, pp. 125–32. Madison, WI: American Society of Agronomy – Crop Science Society of America – Soil Science Society of America.

Forcella, F., King, R., Swinton, S., Buhler, D., & Gunsolus, J. (1996). Multi-year validation of a decision aid for integrated weed management in row crops. *Weed Science*, **44**, 650–61.

Gallandt, E., Liebman, M., Corson, S., Porter, G., & Ullrich, S. (1998). Effects of pest and soil management systems on weed dynamics in potato. *Weed Science*, **46**, 238–48.

Ghersa, C., & Holt, J. (1995). Using phenology prediction in weed management: a review. *Weed Research*, **35**, 461–70.

Godwin, H. (1960). The history of weeds in Britain. In *The Biology of Weeds: A Symposium of the British Ecological Society*, ed. J. L. Harper, pp. 1–10. Oxford, UK: Blackwell.

Gold, H., Bay, J., & Wilkerson, G. (1996). Scouting for weeds, based on the negative binomial distribution. *Weed Science*, **44**, 504–10.

Hamilton, N. (1995). *Learning to Learn with Farmers: A Case Study of an Adult Learning Extension Project Conducted in Queensland, Australia 1990–1995*. Wageningen, Netherlands: Agricultural University. Published doctoral dissertation.

Harp, A. (1996). Iowa, USA: an effective partnership between the Practical Farmers of Iowa and Iowa State University. In *New Partnerships for Sustainable Agriculture*, ed. A. Thrupp, pp. 127–36. Washington, DC: World Resources Institute.

Haverkort, B., van der Kamp, J., & Waters-Bayer, A. (eds) (1991). *Joining Farmers' Experiments: Experiences in Participatory Technology Development*. London: Intermediate Technology Publications.

Hess, D. (1995). *Science and Technology in a Multicultural World*. New York: Columbia University Press.

Hewitt, T., & Smith, K. (1996). Social and economic issues related to precision farming. *Consortium News*, **9**, 5–9. Madison, WI: Consortium for Sustainable Agriculture Research and Extension.

Holling, C. (ed.) (1978). *Adaptive Environmental Assessment and Management*. New York: John Wiley.

Hossner, L., & Dibb, D. (1995). Reassessing the role of agrochemicals in developing country agriculture. In *Agriculture and Environment: Bridging Food Production and Environmental Protection in Developing Countries*, ed. A. Juo & R. Freed, pp. 17–32. American Society of Agronomy Special Publication no. 60. Madison, WI: American Society of Agronomy.

Ishag, S., Al Fakie, O., Adam, M., Adam, Y., Bremer, K., & Mogge, M. (1997). Extension through farmer experimentation in Sudan. In *Farmers' Research in Practice*, ed. L.

Veldhuizen, A. Waters-Bayer, R. Ramirez, D. Johnson & J. Thompson, pp. 89–108. London: Intermediate Technology Publications.

Jaynes, D., & Colvin, T. (1997). Spatiotemporal variability of corn and soybean yield. *Agronomy Journal*, **89**, 30–7.

Johnson, G., Mortensen, D., Young, L., & Martin, A. (1995). The stability of weed seedling population models and parameters in eastern Nebraska corn (*Zea mays*) and soybean (*Glycine max*) fields. *Weed Science*, **43**, 604–11.

Johnson, G., Mortensen, D., Young, L., & Martin, A. (1996). Parametric sequential sampling based on multistage estimation of the negative binomial parameter k. *Weed Science*, **44**, 555–9.

Johnston, A. (1994). The Rothamsted classical experiments. In *Long-Term Experiments in Agricultural and Ecological Sciences*, ed. R. Leigh & A. Johnston, pp. 9–38. Wallingford, UK: CAB International.

Karlen, D., Varvel, G., Bullock, D., & Cruse, R. (1994). Crop rotations for the 21st century. *Advances in Agronomy*, **53**, 1–45.

Kropff, M., & van Laar, H., eds. (1993). *Modeling Crop–Weed Interactions*. Wallingford, UK: CAB International.

Lanini, T., Zalom, F., Marois, J., & Ferris, H. (1994). Researchers find short-term insect problems, long-term weed problems. *California Agriculture*, **48**, 27–33.

Lass, L., & Callihan, R. (1993). GPS and GIS for weed surveys and management. *Weed Technology*, **7**, 249–54.

Lemieux, C., Cloutier, D., & Leroux, G. (1992). Sampling quackgrass (*Elytrigia repens*) populations. *Weed Science*, **40**, 534–41.

Levins, R. (1986). Perspectives on IPM: from an industrial to an ecological model. In *Ecological Theory and Integrated Pest Management*, ed. M. Kogan, pp. 1–18. New York: John Wiley.

Lockeretz, W., & Anderson, M. (1993). *Agricultural Research Alternatives*. Lincoln, NE: University of Nebraska Press.

Long, N. (1992). Introduction. In *Battlefields of Knowledge*, ed. N. Long & A. Long, pp. 3–15. New York: Chapman & Hall.

Long, N., & Villareal, M. (1994). The interweaving of knowledge and power in development interfaces. In *Beyond Farmer First*, ed. I. Scoones & J. Thompson, pp. 41–55. London: Intermediate Technology Publications.

Lopez, M. (1993). El primer encuentro nacional de experimentadores campesinos, Matagalpa, Nicaragua. In *Gorras y sombreros: caminos hacia la colaboracion entre tecnicos y campesinos*, ed. D. Buckles, pp. 65–8. Mexico, DF: International Maize and Wheat Improvement Center (CIMMYT).

Marglin, S. (1990). Losing touch: the cultural conditions of worker accommodation and resistance. In *Dominating Knowledge*, ed. F. Marglin & S. Marglin, pp. 217–82. New York: Oxford University Press.

Marshall, E. (1988). Field-scale estimates of grass weed populations in arable land. *Weed Research*, **28**, 191–8.

Minc, L., & Vandermeer, J. (1990). The origin and spread of agriculture. In *Agroecology*, ed. R. Carroll, J. Vandermeer & P. Rosset, pp. 65–111. New York: McGraw-Hill.

Morrison, J., Huang, C., Lightle, D., & Daughtry, C. (1993). Residue measurement techniques. *Journal of Soil and Water Conservation*, **48**, 479–83.

National Research Council (1997). *Precision Agriculture in the 21st Century*. Committee on Assessing Crop Yield: Site-Specific Farming, Information Systems, and Research Opportunities/Board of Agriculture. Washington, DC: National Academy Press.

Nelson, K. (1994). Participation, empowerment, and farmer evaluations: a comparative analysis of IPM technology generation in Nicaragua. *Agriculture and Human Values*, **11**, 109–25.

Nowak, P. (1997). A sociological analysis of site specific management. In *The State of Site Specific Management for Agriculture*, eds. F. Pierce & E. Sadler, pp. 397–422. Madison, WI: American Society of Agronomy.

Oerlemans, N., Proost, J., & Rauwhorst, J. (1997). Farmers' study groups in the Netherlands. In *Farmers' Research in Practice*, ed. L. Veldhuizen, A. Waters-Bayer, R. Ramirez, D. Johnson & J. Thompson, pp. 263–77. London: Intermediate Technology Publications.

Okali, C., Sumberg, J., & Farrington, J. (1994). *Farmer Participatory Research*. London: Intermediate Technology Publications.

Pretty, J. (1991). Farmers' extension practice and technology application: agricultural revolution in 17–19th century Britain. *Agriculture and Human Values*, **8**, 132–48.

Pretty, J. (1995). *Regenerating Agriculture*. Washington, DC: John Henry Press.

Rew, L., Cussans, G., Mugglestone, M., & Miller, P. (1996). A technique for mapping the spatial distribution of *Elymus repens*, with estimates of the potential reduction in herbicide use from patch spraying. *Weed Research*, **36**, 283–92.

Rice, R., & Ward, J. (1996). *Coffee, Conservation, and Commerce in the Western Hemisphere*. Washington, DC: Natural Resources Defense Fund/Smithsonian Migratory Bird Center.

Richards, P. (1985). *Indigenous Agricultural Revolution: Ecology and Food Production in West Africa*. Boulder, CO: Westview Press.

Rindos, D. (1984). *The Origins of Agriculture: An Evolutionary Perspective*. New York: Academic Press.

Roberts, P., Rust, R., & Larson, W. (1995). *Site-Specific Management for Agricultural Systems*. Madison, WI: American Society of Agronomy.

Roling, N. (1988). *Extension Science: Information Systems in Agricultural Development*. Cambridge, UK: Cambridge University Press.

Roling, N., & van der Fliert, E. (1994). Transforming extension for sustainable agriculture: the case for integrated pest management in rice in Indonesia. *Agriculture and Human Values*, **11**, 96–108.

Roling, N., & Wagemakers, A. (1998). A new practice: facilitating sustainable agriculture. In *Facilitating Sustainable Agriculture*, ed. N. Roling & A. Wagemakers, pp. 3–22. Cambridge, UK: Cambridge University Press.

Salick, J. (1989). Ecological basis of Amuesha agriculture, Peruvian Upper Amazon. *Advances in Economic Botany*, **7**, 189–206.

Scoones, I., & Thompson, J. (eds) (1994). *Beyond Farmer First*. London: Intermediate Technology Publications.

Selener, D., Chenier, J., & Zelaya, R. (1997). *Farmer-to-Farmer Extension: Lessons from the Field*. Quito, Ecuador: International Institute of Rural Reconstruction.

Smith, B. (1992). *Rivers of Change: Essays on Early Agriculture in Eastern North America*. Washington, DC: Smithsonian Institution Press.

Staver, C. (1998). Managing ground cover heterogeneity in coffee (*Coffea arabica* L.) under managed shade: from replicated plots to farmer practice. In *Agroforestry in Sustainable Agricultural Systems*, ed. L. Buck, J. Lassoie & E. Fernandes, pp. 67–96. Boca Raton, FL: CRC Press.

Staver, C., Aguilar, A., Aguilar, V., & Somarriba, S. (1995). Selective weeding: ground cover and soil conservation in coffee in Nicaragua. *Institute for Low External Input Agriculture (ILEIA) Newsletter*, **11**, 22–3.

Sumberg, J., & Okali, C. (1995). *Farmers' Experiments: Creating Local Knowledge*. London: Lynne Rienner Publishers.

Temple, S., Friedman, D., Somasco, O., Ferris, H., Scow, K., & Klonsky, K. (1994a). An interdisciplinary, experiment station-based participatory comparison of alternative crop management systems for California's Sacramento Valley. *American Journal of Alternative Agriculture*, **9**, 64–71.

Temple, S., Somasco, O., Kirk, M., & Friedman, D. (1994b). Conventional, low-input and organic agriculture compared. *California Agriculture*, **48**, 14–19.

Thompson, J., Stafford, J., & Miller, P. (1991). Potential of automatic weed detection and selective herbicide application. *Crop Protection*, **10**, 254–9.

Thompson, R., & Thompson, S. (1990). The on-farm research program of Practical Farmers of Iowa. *American Journal of Alternative Agriculture*, **5**, 163–7.

Thrupp, L. (ed.) (1996). *New Partnerships for Sustainable Agriculture*. Washington, DC: World Resources Institute.

Thurston, D., Smith, M., Abawi, G., & Kearl, S. (eds.) (1994). *Tapado Slash/Mulch: How Farmers Use It and What Researchers Know about It*. Ithaca, NY: Cornell International Institute for Food, Agriculture and Development/Cornell University.

Triomphe, B. (1996). Seasonal nitrogen dynamics and long-term changes in soil properties under the mucuna-maize cropping system on the hillsides of northern Honduras. PhD dissertation, Cornell University, Ithaca, NY.

UNICAFE (1998). *Evaluación Intermedia del Proyecto: Fortalecimiento de los Servicios de Transferencia de Tecnología y Gestión Empresarial a los Pequeños y Medianos Caficultores*. Managua, Nicaragua: Unión Nicaraguense de Cafetaleros (UNICAFE).

Useem, M., Setti, L., & Pincus, J. (1992). The science of Javanese management: organizational alignment in an Indonesian development programme. *Public Administration and Development*, **12**, 447–71.

van der Ploeg, J. (1993). Potatoes and knowledge. In *An Anthropological Critique of Development: The Growth of Ignorance*, ed. M. Hobart, pp. 209–27. New York: Routledge.

Van Huis, A., & Meerman, F. (1997). Can we make IPM work for resource-poor farmers in sub-Saharan Africa? *International Journal of Pest Management*, **43**, 313–20.

Veldhuizen, L., Waters-Bayer, A., Ramirez, R., Johnson, D., & Thompson, J. (eds) (1997). *Farmers' Research in Practice*. London: Intermediate Technology Publications.

Whyte, W. (1991). *Participatory Action Research*. Newbury Park, CA: Sage Publications.

CHARLES L. MOHLER

4

Mechanical management of weeds

Introduction

Physical removal of weeds by soil disturbance prior to planting, and by hoeing and hand-weeding during crop growth are undoubtedly the oldest forms of agricultural weed management. Farmers and agricultural equipment manufacturers continue to develop this ancient tradition of mechanical weed control through the refinement of hand tools and the invention of new tillage and weeding machinery. The purpose of this chapter is to explore the ways in which tillage before crop planting and mechanical weed control methods after planting interact with the ecology of weeds, and to use that understanding to suggest strategies for weed management.

Tillage and cultivation affect weeds in three distinct ways. First, they uproot, dismember, and bury growing weeds and dormant perennating organs. Second, they change the soil environment in ways that can promote germination and establishment of weeds or, less commonly, inhibit germination and establishment. Third, they move weed seeds vertically and horizontally, and this affects the probability that seedlings emerge, survive, and compete with the crop. The second of these effects was discussed in Chapter 2. The first and third are addressed in this chapter.

Each of the tools used for tillage and cultivation disturbs the soil in a unique way. In particular, tools vary with respect to their working depth and the degree to which they invert the soil column, break up soil aggregates, and shake weed roots free from the soil. A general principle underlying this chapter is that *the impact of tillage or cultivation on a species depends on the interaction between the nature of the soil disturbance and the life history characteristics of the weed.* The size, position, and physiology of shoots and underground organs have a large influence on the weed's ability to survive a particular type of disturbance. Moreover, the size, longevity, and germination characteristics of seeds largely determine how they respond to redistribution in the soil column by farm machinery.

A second general principle is that *the timing of tillage or cultivation determines how effective the operation is for weed management*. Timing is critical in several respects. First, obtaining a desired action on the soil and weeds requires proper timing relative to season and weather. Second, a given weed species will be more susceptible to a certain type of disturbance at some stages in its development than at others. Finally, the stage of crop development affects the degree and type of disturbance that the crop can tolerate.

A third principle is that *mechanical weed management is most effective when multiple operations are performed in a planned sequence*. Disturbance can be used to manage weeds at several points in the crop cycle. Tillage prior to planting can bury extant vegetation and disrupt roots and rhizomes. Shallow cultivation prior to emergence and close to young crop plants can kill small weeds before they establish. Shallow cultivation is largely ineffective, however, unless the soil has first been prepared by proper tillage. Deeper cultivation between rows can dig out weeds and throw soil into the crop row to bury young weeds. By the time the crop is large enough to stand the impact of soil thrown around the stems, however, many weeds will be too big to bury unless early germinating weeds are suppressed, for example, by over-the-row cultivation. Finally, tillage between harvest and the next crop can be used to suppress perennials and flush seeds from the soil. Thus, tillage and cultivation for weed management require conscious planning of the sequence of soil disturbances throughout the crop cycle.

Tillage: pros and cons

Tillage prior to planting a crop can be used to meet a variety of objectives, including weed control, seedbed preparation, and residue management (Buckingham & Pauli, 1993, p. 2). From a weed management perspective, tillage re-initiates ecological succession, allowing dominance by early successional annual crops rather than the perennial species that naturally come to dominate undisturbed vegetation.

Tillage has been criticized as a cause of erosion and destroyer of soil tilth. Indeed, when applied without soil conservation measures or used in inappropriate soil and weather conditions, some types of tillage can promote erosion or loss of soil structure (Dickey *et al.*, 1984; Andraski, Mueller & Daniel, 1985; Gebhardt *et al.*, 1985; Langdale *et al.*, 1994). When properly used, however, tillage can enhance water infiltration (Unger & Cassel, 1991), facilitate management of soil fertility (Randall, 1984), and help warm cold soils (Johnson & Lowery, 1985; Cox *et al.*, 1990; Coolman & Hoyt, 1993). It can also increase the proportion of crop seeds that produce established plants (Carter & Barnett, 1987; Griffith et al., 1988; Cox et al., 1992) and improve root growth through better aeration, reduced bulk density, and lower soil resistance to penetration (Bauder, Randall & Swann, 1981; Cox et al., 1990). All these effects potentially improve crop productivity.

Moreover, although reduced tillage practices are often advisable, soil can usually be conserved effectively without complete elimination of tillage if other conservation practices are used. These include cover crops, soil building crop rotations, contour plowing and planting, and sod berms and waterways. Integrated use of such practices can improve soil properties and greatly reduce erosion relative to conventional tillage cropping systems (Cacek, 1984; Reganold, Elliott & Unger, 1987). Jackson (1988) compared two adjacent Ohio farms on an erosion-prone soil. One was in no-till management; the other was regularly tilled but had been treated with soil building rotations, cover crops, manure, and reduced compaction practices for 70 years. The tilled farm showed no indication of erosion and had lower bulk density and higher infiltration rates and soil organic matter than the no-till farm (Jackson, 1988). Most erosion attributed to tillage results from the exposure of soil to wind and rain that occurs when surface organic matter is buried, rather than from soil disturbance *per se*. Chapter 5 addresses ways for maintaining surface organic matter on tilled land.

Mechanical management of perennial weeds

Effects of tillage practices on established weeds

Different tillage implements move the soil in different ways and therefore have substantially different effects on weed populations (Table 4.1). Moldboard plows invert the soil (Nichols & Reed, 1934), and consequently tend to bury growing weeds with relatively little dismemberment. Chisel plows and field cultivators invert the soil to a lesser extent than moldboard plows, but still tend to bury weeds. The primary action of a chisel plow is to create a wake of soil rolling back from the tool. Plants in the immediate track of the blade are likely to be uprooted, as will small weeds that are affected by the lateral cracking of the surface soil. However, damage to well-rooted plants more than a few centimeters from the blade is more likely to occur by burial than by uprooting. In contrast, sweep plows and field cultivators tend to heave up the soil vertically as it passes over the sweep, as well as throw it laterally away from the shank. The blade severs the roots of large weeds, uproots smaller weeds, and buries both with soil. The degree to which weeds and residue are mixed into the soil depends on the angle of the blade relative to

Table 4.1. *Effectiveness of tillage implements for uprooting, dismemberment, and burial of weeds*

Implement	Uprooting	Dismemberment	Burial
Moldboard plow	Good	Poor	Good
Chisel plow	Moderate	Poor	Moderate
Field cultivator	Moderate to good[a]	Moderate	Moderate
Sweep plow	Poor	Moderate[b]	Poor
Disks	Moderate	Good	Moderate
Rotary tiller	Moderate	Good	Moderate

Notes:
[a] Depending on depth of operation relative to weed roots.
[b] Especially good at severing shoots from roots, but poor at fragmenting plants.

the soil surface: the shallowly angled sweeps of a sweep plow tend to sever roots and loosen the soil above with little mixing relative to the more steeply angled and closer spaced blades of most field cultivators. Finally, disks and rotary tillers chop up weeds and crop residue, and mix them into the soil profile to the depth of penetration. Disks tend to cut the weeds whereas rotary tillers tear, but the effect on the plants is often similar. Because both moldboard and chisel plows leave large clods and intact weeds, operations with these implements are generally followed by use of disks, harrows, or some other implement to further chop and mix the surface soil prior to planting. Consequently, tillage usually subjects weeds to a variety of destructive actions.

Depending on the implement, tillage thus chops, uproots, or buries established weeds. Usually tillage occurs early enough in the life cycle of annual plants to be fatal to essentially all individuals, regardless of the method employed, though young grasses sometimes survive noninversion tillage (Moss, 1985a; Cavers & Kane, 1990). Although many studies have noted an increase in perennial weed species with reduced tillage (Jones, 1966; Pollard & Cussans, 1976; Froud-Williams, Drennan & Chancellor, 1983; Koskinen & McWhorter, 1986; Conn, 1987; Buhler *et al.*, 1994), surprisingly little research has related the mechanical action of implements to damage inflicted on the weeds. For example, it would be useful to know how different implements affect the size distribution of *Elytrigia repens* rhizome fragments, the vertical depth distribution of *Cyperus* spp. tubers, or the percentage of damaged buds on *Taraxacum officinale* taproots.

Consideration of differences in growth habit among perennial weed species allows some prediction as to the type of tillage most likely to be effective (Table 4.2). Here the basic types of weeds discussed in Chapter 2 are

Table 4.2. Susceptibility of perennating weeds with different growth habits to uprooting, dismemberment, and burial by tillage implements

Growth form	Uproot	Sever root and shoot	Fragment storage organ	Bury	Examples
Wandering perennials					
rhizomes, etc., below tillage depth	Very low	Moderate	Very low	Moderate	*Convolvulus arvensis, Asclepias syriaca*
rhizomes, etc., above tillage depth	Low	Moderate	Moderate/propagate	Moderate	*Elytrigia repens, Sonchus arvensis, Cynodon dactylon*
with bulb, tuber etc.	Low	Moderate	Very low	Moderate	*Cyperus rotundus, Arrhenatherum elatius* var. *bulbosum*
Stationary perennials					
with taproot, bulb, etc.	Low	Low	Moderate/propagate	High	*Taraxacum officinale, Rumex crispus, Allium vineale*
with fibrous root	Moderate	High	High	High	*Plantago major*

further subdivided according to the scheme of Håkansson (1982). The effect of tillage on a wandering perennial depends on whether the implement reaches the storage roots or rhizomes, or merely severs the vertical branches. Similarly, the presence or absence of a taproot, bulb, or similar storage organ influences the response of nonwandering species to tillage. Annuals with a phenology that is out of phase with the tillage operation behave as perennials. For example, an autumn-germinating annual presents well-established, overwintering plants to a spring tillage operation and thus responds as would a perennial.

Perennials like *Asclepias syriaca* and *Convolvulus arvensis* in which a large proportion of the perennating roots or rhizomes lie below the plow layer are susceptible to tillage primarily through exhaustion of reserves. Because the rhizome lies deep in the soil, a long, usually large-diameter vertical shoot is required to get the growing point to the soil surface. Consequently, replacement of shoots that are removed by tillage requires a substantial investment of energy. However, because the dormant buds on the deep rhizomes are immune to all except extraordinarily deep tillage, substantial control of a population by tillage alone may require fallowing the field in order to perform several operations (see below).

Although weeds with shallow perennating roots and rhizomes can be uprooted and chopped by tillage implements, this often has only a short-term benefit since rhizome fragments develop new shoots and new roots grow to supply exposed pieces of shoot. Moreover, separation of buds onto a multitude of small root or rhizome fragments can actually increase the productivity of some weeds by releasing buds from apical dominance (Håkansson, 1968b; Håkansson & Wallgren, 1972a, 1976; Bourdôt, Field & White, 1982). In addition, if the implement drags fragments around the field, tillage may effectively disperse the weed, as well as propagate it. Nevertheless, as detailed below, shallowly wandering perennials are susceptible to a variety of mechanical measures.

Some perennials spread by rhizomes but perennate by means of a bulb or tuber. These species may pose particular problems, because the tubers tend to survive tillage intact and the substantial food storage allows emergence after deep burial. *Cyperus rotundus*, which has been labeled the world's worst weed species (Holm *et al.*, 1977), is of this type. However, this and similar species may be attacked by tillage that destroys the shoot shortly after emergence, or that exposes the tubers to desiccation (see subsequent sections).

The effect of tillage on nonwandering perennial weeds depends on whether the species has a taproot or equivalent storage organ. Uprooting species with a taproot is relatively ineffective (Table 4.2) unless the root can be brought to the surface to desiccate during dry weather. Simply removing the leaves is ineffective because they quickly regrow from the well-supplied root crown. Burial may be an effective control strategy if the storage organ is severely fragmented.

Most nonwandering species without taproots (e.g., *Plantago major*) are easily controlled by tillage (Table 4.2). In this regard, *Poa annua*, which commonly perennates (Law, Bradshaw & Putwain, 1977), is an exception that proves the rule. This species is notoriously hard to kill with shallow tillage methods because soil clings to the dense root mat (Bates, 1948). However, when clumps are successfully broken up, the fragments often die quickly. Since most nonwandering weeds without taproots have a rosette growth form with small storage reserves and limited capacity for stem elongation, burial is often highly effective. Consequently, these species are rarely problem weeds in tilled systems.

Various strategies exist for mechanical management of perennial weeds. Which approach is most effective against a particular species depends on the growth and stress response characteristics of the weed. Effectiveness of all these strategies is improved by understanding the growth cycle of the weed.

How the timing of tillage affects the growth of perennial weeds

The effect of tillage on a perennial weed varies depending on the phenology of the species relative to the timing of tillage. In general, *a perennial weed in a seasonal climate is most vulnerable to damage shortly after reserves in the perennating organs have been converted to new shoots.*

In a series of studies in southern Sweden, Håkansson & Wallgren (Håkansson, 1963, 1967, 1969a; Håkansson & Wallgren, 1972b, 1976) found that although *Allium vineale*, *Sonchus arvensis*, and *Agropyron (Elytrigia) repens* perennate by different means, all were most susceptible to damage by burial at the point when their perennating organs reached minimum mass. For *A. repens* this occurred when three to four leaves had formed on the new shoots – just prior to initiation of tillers and new rhizomes (Håkansson 1967; Håkansson & Wallgren 1976). In Sweden the three-to-four leaf stage was reached in late May for undisturbed plants, but in warmer climates overwintering leaves may resupply rhizomes before the new growth reaches this size. Majek, Erickson & Duke (1984) similarly found that *A. repens* in New York was most susceptible to tillage in early May, just before formation of new rhizomes.

For *Allium vineale*, maximum susceptibility to tillage due to depletion of stored reserves occurred primarily in the spring following autumn germination of bulbs and was difficult to induce by cultural practices. In contrast, provided *Agropyron repens* and *S. arvensis* were metabolically active, fragmentation induced the plants to divert resources from storage organs to shoots. A second tillage operation could then kill the fragments at their most susceptible stage.

Exhaustion of storage reserves by fallow cultivation

All types of perennial weeds can be reduced in abundance by repeated tillage or cultivation. Complete eradication of severe infestations of some perennial weeds may require many operations over two or more growing seasons (Bakke *et al.*, 1944; Timmons & Bruns, 1951; Hodgson, 1970). Such intensive tillage should be avoided for reasons of soil conservation, but these studies provide insight into the effects of tillage on perennials. Bakke *et al.* (1944) found that during the first year of fallow cultivation, most of the weight lost from roots of *Convolvulus arvensis* was due to depletion of carbohydrates; large numbers of roots did not die until the second year of fallowing.

The optimal period between successive cultivations depends on the depth of cultivation (Timmons & Bruns, 1951; Håkansson, 1969*b*) and season of the year. However, many perennials decline most rapidly when cultivation is repeated at two to four week intervals (Table 4.3). If the interval is longer than optimal, new growth has an opportunity to replenish perennating organs, and eradication will be delayed. If the interval between cultivations is shorter than optimal, the maximum amount of stored carbohydrates will not yet have been converted to shoot growth prior to each operation, and more operations will be required for eradication. The interval between cultivations can be prolonged once the weeds begin to weaken.

Often full eradication is not required, and in many situations significant reduction in weed pressure can be accomplished with one or two extra operations during the normal fallow season. In a New Zealand study, two rotary cultivations in early spring were as effective as glyphosate for controlling *Achillia millefolium* in spring barley (Bourdôt & Butler, 1985). In Florida, two diskings reduced *Imperata cylindrica* rhizome density more than any herbicide tested, although only the combination of herbicides and disking gave good control (Willard *et al.*, 1996). In southern England, tine cultivation following barley harvest was more effective than dalapon plus aminotriazole in controlling a mixture of *Agropyron (Elytrigia) repens* and *Agrostis gigantea* (Hughes & Roebuck, 1970). Other studies have similarly shown the effectiveness of fallow season tillage for management of perennial weeds (Fail, 1956; Lym & Messersmith, 1993).

Exhaustion of storage reserves by chopping and burying storage organs

Shallowly wandering perennials can be managed by a strategy in which the perennating roots or rhizomes are cut into small fragments by

Table 4.3. *Number of operations, interval between operations, and time required to eradicate some perennial weeds by fallow cultivation directed primarily toward exhaustion of perennating organs*

Species	Experiments[b]	Minimum operations[a]			Minimum time to eradicate			Reference
		Operations	Interval[c] (days)	Eradication time (years)	Operations	Interval[c] (days)	Eradication time (years)	
Convolvulus arvensis	8	18	20	2.2	26	12	1.8	Timmons & Bruns (1951)
Centaurea repens	2	22	25	2.5	24	18	2.0	Timmons & Bruns (1951)
Lepidium draba	2	14	28	1.8	18	18	1.6	Timmons & Bruns (1951)
Apocynum cannabinum	1	17	28	3	17	28	3	Timmons & Bruns (1951)
Sorghum halepense	3	7	28	1	7	28	1	Timmons & Bruns (1951)
Cyperus rotundus	1	16	~21	2				Smith & Mayton (1938)
Cirsium arvense	1				10	18	2	Hodgson (1970)
Agropyron repens	4				4	25	0.5	Fail (1956)
Agrostis spp.	3				3	19	0.2	Fail (1956)
Agropyron repens[d]	7				4	25	<1	Håkansson (1969b)

Notes:
[a] A strategy that minimizes the number of cultivations often requires longer to achieve complete eradication.
[b] Sample size for means shown in the other columns.
[c] Interval between cultivations during the first year. Usually the interval was increased in subsequent years as the weeds weakened.
[d] Plants buried to controlled depths.

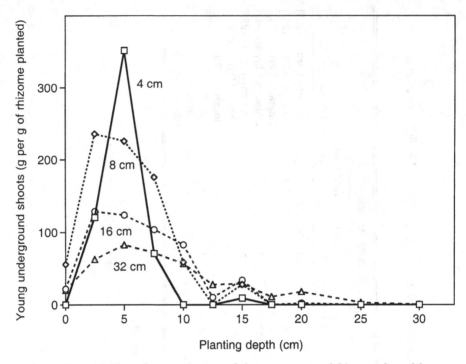

Figure 4.1 Mass of new underground shoots per gram of rhizome planted for *Agropyron (Elytrigia) repens* rhizomes of several lengths planted at various depths. (Drawn from data in Håkansson, 1968b.)

shallow tillage, and these are then buried by deep plowing. Although field-scale trials of this procedure have not been published, the parameters of the method have been worked out for *Agropyron (Elytrigia) repens* (Vengris, 1962; Håkansson, 1968b, 1971), *Sonchus arvensis* (Håkansson & Wallgren, 1972b), *Holcus mollis* and *Agrostis gigantea* (Håkansson & Wallgren, 1976), *Achillia millefolium* (Bourdôt, 1984; Field & Jayaweera, 1985), and *Mentha arvensis* (Ivany, 1997).

The procedure is illustrated by data on *Agropyron repens*. Håkansson (1968b, 1971) planted rhizomes of 4, 8, 16, and 32 cm length in October at 11 depths down to 30 cm. Rhizomes established poorly at the soil surface, and only the longest pieces produced a substantial weight of new rhizomes by the following autumn (Figure 4.1). Mechanically moving most rhizomes completely to the surface would be difficult, however, so this finding is not very useful for management. At 2.5 to 5 cm depth, the rhizomes established well, and the shorter pieces produced more shoots and new rhizomes per gram of rhizome planted because a greater percentage of buds sprouted (Figure 4.1). In contrast, however, productivity declined with deeper planting, and few shoots from the 4-cm fragments reached the surface from depths greater than 10 cm. This led to death of the original rhizomes, without replacement by new rhizomes (Figure 4.1). Similar results have been obtained for several different years, genetic materials, soils, and planting dates (Håkansson 1968a, 1968b, 1971; Håkansson & Wallgren, 1976), and for rhizomes that were allowed to sprout prior to burial (Vengris, 1962). Since most *A. repens* rhizomes occur in the top 10 to 15 cm of soil (Håkansson, 1968b), breaking them into small pieces by rotary tillage or disking is feasible. Subsequent burial will be most complete if plowing inverts the soil as fully as possible. Fortunately, the depth distribution of new rhizomes developed from fragments that managed to survive burial was little affected by the planting depth (Håkansson, 1969c).

In further work, Håkansson (1968a, 1971) found that a crop of white mustard caused a greater percentage decrease in the biomass of *A. repens* originating from small, deeply planted rhizomes relative to large or shallowly planted rhizomes. This happened because shoots from the deeply planted pieces emerged slowly, thereby giving the crop an opportunity to grow up and shade the weed, and the small reserves in the shorter fragments led to weaker growth. The synergistic interaction between crop competition and tillage in this study illustrates the importance of integrating ecological management procedures – simply chopping and burying the rhizomes would not be expected to decrease *A. repens* density in a fallow or a weakly competitive crop. The chop-and-bury plus crop competition approach to the management of shallowly spreading perennials is particularly notable because it results from understanding of weed ecology rather than from the development of new technology.

Exposure of perennating organs to desiccation and freezing

Desiccation of roots and rhizomes by summer fallowing was a traditional method of controlling wandering perennials in Europe. The soil was plowed and allowed to dry into clods. These were stirred occasionally with a plow or heavy cultivator to completely desiccate roots and rhizomes (Bates, 1948; Travers, 1950). Foster (1989) indicated that the same procedure can be used to manage *Rumex obtusifolius* and *R. crispus*.

In Botswana, Phillips (1993) found that an extra moldboard plowing substantially reduced subsequent growth of *Cynodon dactylon* and increased sorghum grain yield, especially if it occurred during the dry season when tillage promoted desiccation of rhizomes. Similarly, in Nicaragua Vargas *et al.* (1990) found that plowing dry soil at the end of a four-month dry season caused *Cyperus rotundus* tubers to die of desiccation. This greatly reduced shoot density of the weed in the subsequent crop (Figure 4.2).

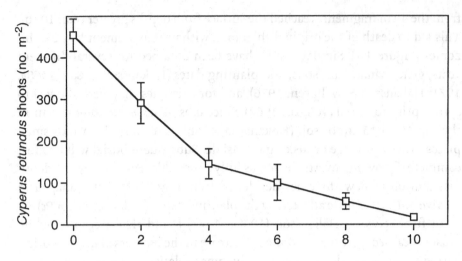

Figure 4.2 Density of shoots of *Cyperus rotundus* 14 days after irrigation in response to the period of desiccation between dry plowing and irrigation. All treatments were first irrigated on the same day. (Drawn from data in Vargas *et al.*, 1990.)

In regions with cold winters, some perennial weeds can be killed by freezing damage to the perennating organs. Schimming & Messersmith (1988) studied the temperatures required to kill overwintering perennial buds of four species. *Cirsium arvense* suffered 90% mortality at $-12\,°C$, but the other species were more cold tolerant. Since minimum soil temperature increases with depth, effective exposure of roots and rhizomes to lethal temperatures requires working these organs to the surface. Development of winter hardiness in the autumn is an energy-consuming activity, and thus forcing resprouting by autumn cultivation may increase sensitivity to freezing (Schimming & Messersmith, 1988).

Physical removal of perennating organs from the field

Severe infestations of species with perennating organs near the surface may be substantially reduced in density by removal of roots and rhizomes from the field. Typically, the soil is plowed, roots or rhizomes are worked to the surface with a spring tooth harrow and then pulled to the edge of the field with a rake or harrow (Travers, 1950). Rhizomes should be broken as little as possible during primary tillage, since longer pieces are more easily sorted to the surface (Kouwenhoven & Terpstra, 1979). With current technology, removal of roots and rhizomes is only practical on small fields; on large fields, much of this material falls through the tines and remains in the field. In principle, machines to strain out and collect storage roots and rhizomes should be possible.

Effects of tillage on weed seedling density

Effect of the timing of tillage on weed seedling density

The timing of tillage and seedbed preparation has large effects on the density of weeds in the subsequent crop. As explained in Chapter 2, most weed species germinate during specific periods of the year. If the tillage and seedbed preparation occur prior to the bulk of that period, most individuals will emerge after planting and potentially compete with the crop. In contrast, if planting is delayed until after most of the weeds have emerged then most can be eliminated by tillage. Three critical questions are addressed here. (1) What are the general rules regarding planting date for various phenological categories of crops and weeds? (2) How does one determine the optimum seedbed preparation/planting date? (3) How should soil disturbance be manipulated to best exploit the seasonality of weed emergence?

Regarding question (1), the weeds that are most problematical for an annual crop are those that germinate around the time the crop is planted. Those that come up much earlier in the year will be eliminated during seedbed preparation, and those that emerge only much later in the season will not be competitive against the crop. The remaining species will be of two types: (i) those that germinate opportunistically in response to disturbance over a broad range of soil climatic conditions and (ii) seasonal species whose peak emergence occurs after the earliest practical planting date for the crop. A shift in planting date is not a viable management strategy for weeds of the first sort (Ghafar & Watson, 1983). The latter group of seasonal germinating species is thus the focus of this discussion. With regard to these, two strategies are possible.

First, if the crop is capable of emerging earlier in the season than most of the weed species in the seed bank, then planting the crop as early as possible may be a viable strategy. By the time the weeds are ready to germinate, the crop will have a competitive advantage due to its larger size, and if the crop casts sufficient shade, some weed germination and emergence may be prevented or at least further delayed. Temperate zone spring cereal grains have this sort of phenological relationship to their weeds, and although delayed planting may eliminate some spring germinating weeds (Spandl, Durgan & Forcella, 1998), it is often counterproductive (Deschenes & Dubuc, 1981; Légère, 1997). The competitive advantages of early planting of spring cereals are discussed further in Chapter 6.

A second strategy is to delay seedbed preparation and planting until after many of the weeds have germinated and are susceptible to destruction by soil disturbance. This approach appears to be most successful for spring-planted row crops, particularly maize and soybean, and for autumn-planted grains. Delaying planting from 25 April until 15 May in Wisconsin reduced in-row weed density in rotary hoed maize by an average of 55% (Mulder & Doll, 1994). However, late planting also reduced maize yield in one of two years. Similarly, late-planted soybean often has lower weed densities and lower percentage yield loss to weeds relative to early-planted soybean, but may suffer some reduction in yield due to shortened growing season (Weaver, 1986; Gunsolus, 1990; Buhler & Gunsolus, 1996). A substantial percentage of fall-germinating weeds, including *Alopecurus myosuroides*, *Bromus secalinus*, *Avena ludoviciana*, *Veronica persica*, and *Lamium purpureum*, can be eliminated by delaying the planting of winter wheat, but late planting often reduces yields (Moss, 1985a; Koscelny et al., 1991; Christensen, Rasmussen & Olesen, 1994; Singh et al., 1995; Cosser et al., 1997).

Delayed planting can also be useful for weed management in some spring-seeded vegetable crops (Wellbank & Witts, 1962), but will generally not be useful in summer-seeded vegetables. The latter are typically infested with weeds like *Galinsoga ciliata* and *Portulaca oleracea* that germinate over a wide range of dates in response to tillage. Note, however, that this type of weed is highly susceptible to stale and false seedbed techniques (see below).

In all the studies discussed above, the seasonal cycle was dominated by changes in the temperature regime. Whether similar planting date strategies exist for some crop–weed combinations in the wet–dry seasonal tropics remains to be determined.

Forcella and others have addressed question (2) regarding determination of the optimum planting date through a series of models that predict the percentage emergence of individual weed species based on soil temperature and moisture data (Forcella, 1993, 1998; Harvey & Forcella, 1993; King & Oliver, 1994; Wilen, Holt & McCloskey, 1996). The models predict relative weed pressure at successive potential planting dates as a proportion of the density that would occur without seedbed preparation. The time to plant occurs when the expected weed density falls below a threshold the grower believes he/she can control with the weed management tools available, or, alternatively, when the expected yield loss due to further delay in planting exceeds the expected yield loss due to weeds. A problem with these models is that they predict the percentage of ultimate emergence that has occurred by a given date rather than the actual density. The model of Reese & Forcella (1997) is presently paramaterized for 15 weed species common in the central USA.

Forcella, Eradat-Oskoui & Wagner (1993) used a related approach to determine the optimal seeding date for maize and soybeans in the northern midwest of the USA. They computed two functions of yield versus planting date. One expressed weed-free yield as a function of growing degree-days, precipitation, and crop maturity group. The second expressed the relation of yield to weed density after planting for various planting dates. When graphed, the intersection of the two curves indicated the planting date corresponding to the maximum crop yield at a given weed density. The optimum planting date depended greatly on both weed species and density. The optimum date was early when the weed was an early-emerging species (e.g., *Chenopodium album*), and later when the weed emerged later in the season (*Setaria* spp.). In all cases examined, increase in weed density delayed the optimal seeding date. In an extreme case, increase in *Setaria* density in soybean from 40 to 200 m^{-2} shifted the optimal planting date from 28 April to 7 June.

Thus, both empirical studies and models indicate that delayed planting to eliminate weeds from spring row crops and fall cereals generally reduces yields relative to the ideal situation of early planting in weed-free conditions. However, *when weed control is imperfect, yield usually increases with delayed planting, and the optimum delay relative to weed-free conditions increases with the density of germinable seeds.*

Question (3) above asked what sort of tillage systems work best with a delayed seeding approach to weed management. Note in this regard that the models of Forcella and his colleagues were developed in the context of a tillage regime with fall plowing and superficial seedbed preparation. If soil is moldboard plowed shortly before planting, then depletion of the surface seed bank by delayed tillage is largely irrelevant; seeds previously in enforced dormancy will be brought to the surface, and many of these will subsequently germinate and infest the crop. Thus, *delayed planting works best when deep tillage is avoided or occurs well in advance of a shallow seedbed preparation.*

Type of tillage affects weed seedling density

As explained in Chapter 2, tillage modifies the soil environment in ways that promote the germination of weed seeds. How this promotion of germination by tillage translates into weed seedling density is less than transparent, however, and the large literature on the effects of tillage on seedling density is highly contradictory. For example, some studies have found more weeds in tilled plots (Roberts & Feast, 1972), whereas others have found more without tillage (Moss, 1985b; Cardina, Regnier & Harrison, 1991; Mohler & Callaway, 1992; Stahl *et al.*, 1999). Moreover, many studies have found the effects of tillage to vary among species (Chancellor 1964a; Pollard & Cussans,

1976; Pollard *et al.*, 1982; Froud-Williams, Drennan & Chancellor, 1983; Buhler & Daniel, 1988; Buhler & Oplinger, 1990), or among sites (Wilson & Cussans, 1972; Froud-Williams, Drennan & Chancellor, 1983; Buhler & Mester, 1991), or among years of an experiment (Wilson, 1981, 1985; Roman, Murphy & Swanton, 1999). Several authors have reviewed the extensive and often confusing literature on the effects of primary tillage on weed populations (Cussans, 1975, 1976; Froud-Williams, Chancellor & Drennan, 1981; Froud-Williams, 1987; Mohler, 1993; Buhler, 1995).

Much of the difficulty in understanding the effects of tillage on weed density can be resolved by recognizing that the vertical distribution of seeds in the seed bank is a critical factor affecting seed survival, germination, and emergence (Mohler, 1993). If most seeds are near the surface and plowing buries them deeply, then tillage will reduce seedling density. In contrast, if most seeds are deeply buried, then plowing may increase seedling density by bringing seeds to the surface. Although this is obvious, remarkably few studies of the effects of tillage on weed populations give any information indicating the distribution of the seed bank prior to initiation of the experiment. A few exceptions include Roberts (1963), Moss (1985*b*), and Van Esso, Ghersa & Soriano (1986).

The effect of tillage on the density of weeds in the subsequent crop is a function of (i) how tillage redistributes seeds in the soil profile, (ii) the capacity of weed species to emerge from various depths in the soil, and (iii) how depth in the soil affects survival of weed seeds. Each of these issues is discussed before considering their combined effect on weed density.

Redistribution of weed seeds by tillage

In the absence of tillage, seeds infiltrate into an agricultural soil via cracks, the activities of soil fauna, and frost action. This infiltration is slow. For example, Moss (1985*b*) found 92% of *Alopecurus myosuroides* seeds in the top 2.5 cm of soil 10 months after sowing. After 34 months, 78% were still in the top 2.5 cm. Other studies show similar results (Weaver & Cavers 1979; Van Esso, Ghersa & Soriano, 1986).

Tillage implements move seeds vertically to different extents. Most studies of seed movement have observed the distribution of seeds or beads that were sown on the soil surface and then tilled in with one or more tillage operations (Pawlowski & Malicki, 1968; Moss, 1988; Staricka *et al.*, 1990; Yenish *et al.*, 1996). Others have compared the distribution of natural seed populations before and after tillage (Russel & Mehta, 1938; Roberts, 1963; Wicks & Somerhalder, 1971; Yenish, Doll & Buhler, 1992). Hulburt & Menzel's (1953) measurements of how surface-applied ^{32}P was mixed by disking and rototilling may indicate the way surface seeds are moved by these operations.

Typically, a single moldboard plowing of surface seeds results in a skewed, bell-shaped distribution of seed density with depth (Figure 4.3). The direction and degree of skewing has varied among studies (Mohler, 1993), perhaps due to variation in depth of operation and degree of soil turning by the plows. A variety of factors affect the degree of inversion. These include design of the moldboard and coulters, speed of operation, amount of roots and crop residue present, and soil structure, texture, and moisture.

In contrast with moldboard plowing, a single operation with other implements produces a monotonic decline in seed density (Figure 4.3). A second operation (usually the following season) makes the distribution more uniform with all implements (Figure 4.3), but seeds still tend to be concentrated near the surface with tine and disk type implements.

Using colored beads placed at different depths, Cousens & Moss (1990) and Grundy, Mead & Burston (1999) developed data on movement to and from multiple layers of the soil and then used these data to model seed distributions after multiple tillage events. The distributions predicted by Cousens & Moss (1990) for a single cohort of seeds moved repeatedly by a single type of implement were similar to seed distributions after multiple tillage events observed in empirical studies (Figure 4.3) (Röttele & Koch, 1981; Van Esso, Ghersa & Soriano, 1986). Cousens & Moss (1990) also considered the situation where weeds shed seeds onto the surface each year. The stable distributions predicted under the assumptions of the model after many cycles were similar to those for a single cohort after one tillage operation (Figure 4.4). Published seed distributions after many years of moldboard plowing with seed input (Roberts & Stokes, 1965; Fay & Olson, 1978) are more uniform than predicted by Cousens & Moss (1990). In real agricultural fields, the distribution of seeds is rarely stable because years of high seed input usually occur sporadically among years of successful control and little seed input. Following one or more high-input years, the distribution would look something like those in Figure 4.3 for one tillage operation. After one or more years of good weed control, the distribution would become more uniform.

Tillage and the emergence of weed seedlings

Germination of weed seeds is more likely near the soil surface because seeds are more likely to experience light, fluctuating temperatures, and other factors that commonly promote germination (see Chapter 2). Moreover, if a seed does germinate, emergence from the soil is more likely if the seed is near

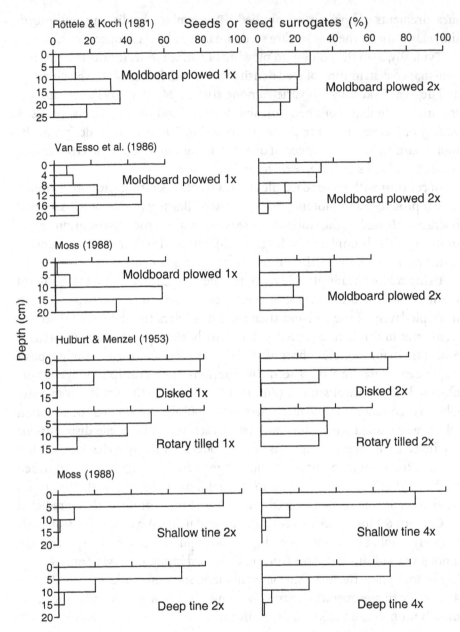

Figure 4.3 Vertical distribution of seeds and soil markers following one or two operations with various tillage implements.

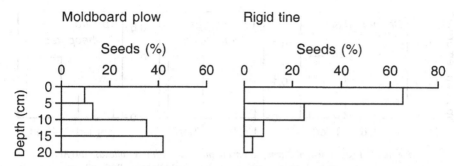

Figure 4.4 Projected stable vertical seed distributions for a weed reproducing according to the population model of Doyle, Cousens & Moss (1986) and subjected to yearly moldboard plow or rigid tine tillage. (Redrawn from Cousens & Moss, 1990.)

the surface because the likelihood of exhaustion of nutrient reserves prior to emergence is less (Vleeshouwers, 1997) and the period of exposure of the upwardly growing shoot to hazards in the soil is shorter.

Chancellor (1964b) determined the depth from which seedlings of 19 species emerged in field conditions, and Mohler (1993) reviewed 21 studies on seedling emergence from weed seeds placed at various depths. These and more recent studies (MacDonald, Brecke & Shilling, 1992; Horak & Sweat, 1994; Cussans et al., 1996; Yenish et al., 1996; Prostko et al., 1997; Fausey & Renner, 1997; Grundy & Mead, 1998) show that most individuals of most weed species arise from the top 2–4 cm of soil. However, for many species some individuals emerge from deeper soil layers, and a small percentage of some large-seeded species emerge from 10 cm or more (Stoller & Wax, 1973; Horak & Sweat, 1994; Cussans et al., 1996). About half of the species examined showed a monotonic decline in emergence with decreasing depth; for the remainder, shallow burial increased emergence (Mohler, 1993), probably by improving water uptake. Clearly, although deep burial of weed seeds generally prevents seedling emergence, shallow incorporation into the soil affects various species differently.

In addition to indirectly affecting emergence via seed distribution, tillage changes soil properties that affect emergence. Cussans et al. (1996) showed that weed species varied in their emergence response to clod size. More importantly, emergence through compact (untilled) soil is more difficult than through loose soil (Figure 4.5) (Morton & Buchele, 1960; Mohler & Galford, 1997; Vleeshouwers, 1997).

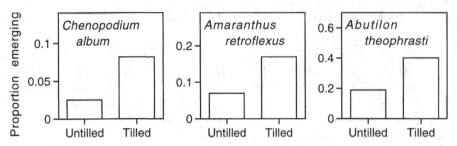

Figure 4.5 Proportion of *Chenopodium album*, *Amaranthus retroflexus*, and *Abutilon theophrasti* emerging after planting into either tilled or untilled soil with an apparatus that produced minimal soil disturbance. See Mohler & Galford (1997).

Tillage and seed survival

Within a given season, the density of seedlings is the integral over all depths of the product of the number of seeds at a given depth and the probability of a seed at that depth producing an emerged seedling (Mohler, 1993). However, to understand the effects of tillage over several seasons, the survival of seeds in the soil must be considered as well.

Roberts & Dawkins (1967) performed an early experiment relating seed survival to tillage. They turned the soil at three-month intervals, at six-month intervals, or left it undisturbed. No seed production was allowed in the plots. Annual sampling of the seed bank indicated that the rate of decline in number of seeds present was relatively constant over years and increased with frequency of soil disturbance (Figure 4.6a). Other studies corroborate this finding (Roberts, 1962; Roberts & Feast, 1973a, 1973b). In the study of Roberts & Dawkins (1967), the decrease in viable seeds with tillage was largely accounted for by an increase in the number of emerged seedlings (Figure 4.6b), but the generality of this result is unknown.

Tillage affects seed survival in three ways (Figure 4.7). First, the tillage operation itself may stimulate germination, for example, by exposing the seeds to a light flash or by scarifying them, and if germination then occurs at a depth that does not allow emergence or at a time of year or in weather conditions that do not allow establishment, then germination will lead to death. Second, changes in soil conditions due to tillage may also stimulate germination in conditions unsuitable for establishment. Finally, action of seed predators, pathogens, and damaging physical influences generally decreases with greater depth in the soil, and, as discussed above, tillage redistributes seeds in the soil profile. For simplicity, germination under conditions that do not allow establishment is referred to below as inappropriate germination and treated as a type of seed mortality, although technically, death occurs in the seedling rather than seed stage.

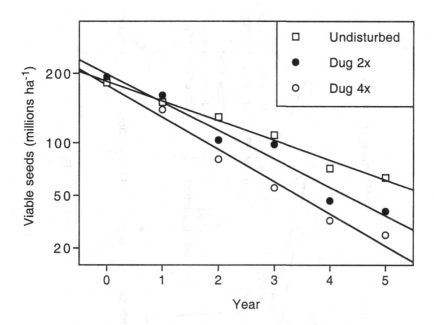

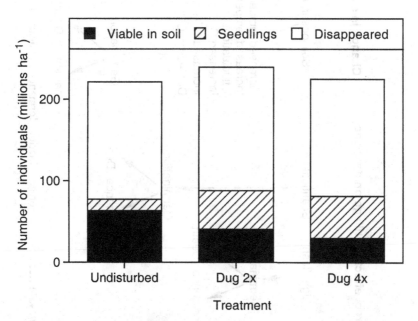

Figure 4.6 (a) Seed survival of mixed species of weed seeds in soil that was either undisturbed, or stirred by digging two times or four times per year. (b) Fates of seeds in the same experiment. (Drawn from data in Roberts & Dawkins, 1967.)

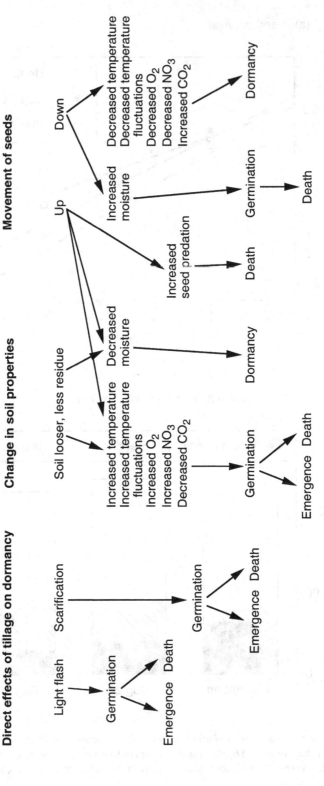

Figure 4.7 Effects of tillage on germination, dormancy, and death of weed seeds.

Tillage can directly induce inappropriate germination by scarification of hard seeds and by providing a light flash. However, inappropriate germination due to scarification of hard-seeded species like *Abutilon theophrasti* and *Convolvulus arvensis* during tillage is probably a small source of mortality since (i) usually only a fraction of seeds in the seed bank are hard, (ii) most of these probably escape scarification during any given tillage operation, and (iii) some of the germinating seeds emerge as seedlings. If a seed of a light-sensitive species receives a brief flash of light prior to reburial beyond the depth from which emergence can occur, the seed may be stimulated to germinate inappropriately. This is more likely with tillage methods like moldboard plowing or rototilling that move large amounts of soil vertically. The effect probably kills a larger proportion of small-seeded than large-seeded species, since most small-seeded species must be close to the surface for successful emergence, and most large-seeded species lack a light requirement (see Chapter 2). The role of both scarification and light flash in tillage-induced seed mortality requires experimental investigation.

Tillage probably also promotes the inappropriate germination of weed seeds by the same mechanisms that it stimulates appropriate germination, namely, by increasing temperatures and the amplitude of temperature fluctuations, modifying the soil atmosphere, and changing the chemistry of the soil solution, particularly nitrate concentration. However, stimulation of inappropriate germination by tillage may be less than stimulation of appropriate germination, since the magnitude of all of these effects declines with increasing depth, and seeds that germinate close to the surface often produce emerged seedlings. The magnitude of inappropriate germination due to changes in the soil environment induced by tillage needs to be examined.

Seed mortality often varies with depth in the soil (reviewed by Mohler, 1993), and as tillage moves seeds, it exposes them to changes in mortality risk. However, in most studies to date, the seeds were confined in packets or containers that excluded most seed predators. Even more important, seedling emergence could not be observed. Since germination is generally greater for seeds closer to the surface due to light, warmer soil, and fluctuating temperatures (see Chapter 2), the greater apparent mortality often observed near the surface may have been due to death of individuals that would have emerged had they not been restrained. For a few studies (Stoller & Wax, 1973; Dawson & Bruns, 1975; Froud-Williams, Chancellor & Drennan, 1983; Moss, 1985b), Mohler (1993) computed seed mortality by subtracting the number of emergents from the number of seeds that disappeared. These data indicated that survival of seeds not producing seedlings decreased with depth as often as it increased (Mohler, 1993). However, several additional studies on *Avena fatua* and *Helianthus annuus* have shown increased survival when seeds were incorporated into the soil (Banting, 1966; Wilson, 1972; Wilson & Cussans, 1972, 1975; Robinson, 1978), and recent work on several broadleaf species (Mohler, 1999) showed consistent increase in seed survival with depth. In general, biological activity increases toward the soil surface, and action of seed predators and pathogens should be greatest there. In addition, desiccation and exhaustion of dormant seeds is more likely near the surface than in the cooler, moister soil below. Nevertheless, the pattern of seed mortality with depth apparently varies among species, and probably with soil and cultural conditions as well.

Plowing and other tillage that raises deeply buried seeds will bring many sufficiently near the surface to prompt germination. However, only some of these successfully emerge, and the tillage operation has essentially killed those that fail. This effect is highly dependent on the germination biology of the species present. Moreover, it cannot be studied simply by placing seeds at different depths, since seeds in the soil are likely to have different dormancy status than seeds from the laboratory shelf or refrigerator.

In summary, tillage decreases the weed seed bank. Frequent tillage decreases the seed bank more rapidly than infrequent tillage. Tillage that produces more vertical displacement of seeds probably creates more true seed mortality, though shallow tillage may decrease the seed bank more rapidly through seedling emergence. Except for a few studies that have distinguished death by germination or disappearance from loss of viability (Sanchez del Arco, Torner & Fernandez Quintanilla, 1995), few data are available on the causes of seed mortality in field conditions.

Synthesis of the effects of tillage on seeds and seedlings

Mohler (1993) used an analytical model to explore the multiple effects of tillage on emergence. The model assumed that emergence declined exponentially with depth of the seed and that seed survival increased with depth to an asymptote. The model consisted of equations that predicted for various tillage regimes the proportion of an initial seed bank that emerged as seedlings in each successive year, assuming no additional seed input. Depth of tillage and species properties including dormancy, ability to emerge through soil, and near-surface seed survival were varied to determine conditions under which one tillage regime resulted in fewer weed seedlings than another.

Emergence was greater from no-till than from plow or rotary tillage during the initiation year for most realistic parameter values (Mohler, 1993). Assuming no innate seed dormancy, in the next year tillage had more seedlings than no-till, unless the species only emerged when near the surface and near-surface seed survival was good. Seed dormancy extended the range of seed survival and emergence ability conditions over which no-till produced more seedlings than till. In later years, tilled regimes generally had more seedlings than no-till regardless of dormancy.

Systematic examination of 15 field studies in which seed return to the soil was prevented, tillage occurred once per year, and some indication was given as to how the seed bank was vertically distributed at the beginning of the experiment (Bibbey, 1935; Chancellor, 1964*a*; Wilson & Cussans, 1972, 1975; Wilson, 1978, 1981, 1985; Lueschen & Andersen, 1980; Froud-Williams, 1983; Froud-Williams, Chancellor & Drennan, 1984; Schweizer & Zimdahl, 1984; Moss, 1985*b*, 1987; Buhler & Daniel, 1988; Egley & Williams, 1990) showed that empirical results generally paralleled model predictions (Mohler, 1993). For example, Wilson (1981) found that the density of *Avena fatua* emerging from a surface sowing of seeds decreased in the order no-till, tine tillage, plow tillage during the initial year of treatment, but that the following year plow tillage had a greater density than no-till or tine tillage.

This analysis leads to several suggestions for management of seed banks with tillage.

- *The high density of weeds often observed during the first year of reduced tillage is frequently a transitory phenomena.* Provided seed shed can be prevented that year, a great reduction in weed density is likely in subsequent years.
- *Change from a plow tillage regime to a reduced tillage regime is likely to be more successful if it is preceded by at least one year of good weed control.* Good weed control before transition to reduced tillage will help insure that the surface seed bank is relatively depleted by emergence rather than enriched by seed shed.
- *If many long-viable seeds have been mixed into the soil by past tillage, the best strategy may be frequent, consistently shallow tillage to deplete the surface seed bank.* If this approach is taken, prevention of seed shed is important.
- *Finally, if weed control fails and many seeds are shed onto an otherwise relatively clean soil, the best strategy may be to plow as deeply as possible, and then use shallow tillage in subsequent years to prevent returning the seeds to the surface.* This will be a particularly valuable tactic for managing species with short to moderate seed longevity.

Thus, a flexible tillage strategy that takes into account the seed longevity and probable distribution of critical weed species in the soil is likely to facilitate other means of weed management. Although analysis of seed distribution is beyond the capacity of most growers, an understanding of the way in which tillage implements move seeds, coupled with a history of seed shed in a field, should provide sufficient qualitative information for choosing the most appropriate tillage method for the situation.

Many authors have noted an increase in density of annual grass weeds with reduced tillage (Pollard & Cussans, 1981; Froud-Williams, Drennan & Chancellor, 1983; Wrucke & Arnold, 1985; Cardina, Regnier & Harrison, 1991; Teasdale, Beste & Potts, 1991; Swanton et al., 1999). For example, S. R. Moss (personal communication) summarized 13 comparisons from several studies on *Alopecurus myosuroides* (Pollard et al., 1982; Wilson, Moss & Wright, 1989; Clarke & Moss, 1991) and found that plowing reduced density by an average of 63% (range 17% to 98%) relative to shallow tine tillage. This shift toward grasses with reduced tillage has been attributed to less effective grass herbicides (Wrucke & Arnold, 1985), emergence response to burial depth of seeds (Buhler, 1995), or to peculiarities of the cropping system or the particular weed species present (Swanton, Clements & Derksen, 1993).

Although these phenomena may play a role, a major factor is the shorter seed longevity of most grass weeds relative to many annual broadleafs (see Chapter 2). Species with relatively short-lived seeds require regular seed input to maintain high densities. When shed seeds are plowed under each year, most perish before they return to the surface by subsequent tillage. For a single cohort, the model of Mohler (1993) predicts higher density with plow tillage in the second and subsequent years because burial enforces dormancy and protects some seeds from near-surface causes of mortality. However, for species with short-lived seeds, the number of emergents from these older cohorts is low regardless of tillage regime. For species with short-lived seeds, the most recently shed cohort of seeds largely determines weed density, and this cohort will produce fewer emerged individuals if it is plowed under. In contrast, for many broadleaf species with long-lived seed banks, the most recently shed cohort may be a small fraction of the total seeds in the soil, in which case plowing may increase weed density.

Population dynamics of annual weeds in ridge tillage

The effect of ridge tillage on weeds is distinctive from other types of tillage in that seeds do not move randomly relative to the crop row. In ridge tillage (sometimes referred to as till-plant) a crop is planted on ridges formed the previous growing season (Figure 4.8). During planting, the surface of the ridge is scraped into the inter-row valleys. Seeds that were shed onto the ridge the previous season are thus moved to the valleys where seedlings can be destroyed by inter-row cultivation. Finally, ridges are rebuilt for the next season by hilling up around the crop stems during the final cultivation.

Wicks & Somerhalder (1971) sampled soil for weed seeds at various dis-

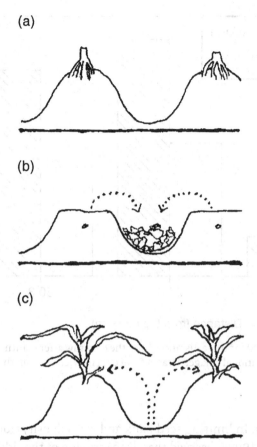

Figure 4.8 The changing microtopography of a ridge-tilled field. After harvest, the ridges are intact (a). At planting, crop residues and soil at the tops of ridges are thrown into the inter-rows (b). At the final cultivation, soil and decomposed residue are hilled back onto the ridges (c). (Redrawn from Forcella & Lindstrom, 1988a.)

tances from the crop row in ridge tilled and conventionally plowed maize plots. Before planting, seed density in the surface 7.6 cm was higher in ridge till, especially in the crop row. After planting, seed density in the crop row was lower and seed density in the inter-row area was higher in ridge till than in conventional tillage (Figure 4.9). Similarly, Forcella & Lindstrom (1988a) demonstrated that large numbers of seeds were moved away from the crop during planting. However, despite the many individuals that germinated in the inter-row and were killed during cultivation, the ridging operation also returned many seeds back to the ridge. Germination after the last cultivation led to weed populations that, though not competitive with the crop, were sufficient to maintain the seed bank. Sensitivity analysis of a ridge tillage weed population model indicated that movement of a large percentage of seeds from the

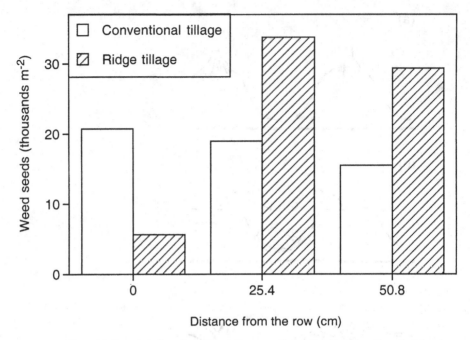

Figure 4.9 Density of seeds at three distances from the crop row after planting in ridge tillage and conventional tillage. (Drawn from data in Wicks & Somerhalder, 1971.)

ridge to the furrow is critical in limiting seed bank and weed density (Jordan, 1993). Accordingly, Buhler (1998) observed that deep scraping of the ridges at planting reduced seedling densities of several species relative to shallow scraping.

In Forcella & Lindstrom's (1988a, 1988b) study, overall weed seed density was similar in ridge and conventionally tilled treatments that were rotated between maize and soybean, but in continuous maize, seed densities were higher with ridge till. The difference may have been due to greater soil cracking in continuous maize, allowing seeds to fall below the depth to which the ridge was scraped at planting. The higher seed density resulted in greater seedling density in ridge till in one of two years. These results show how interactions among agronomic practices can affect weed management.

Comparison of ridge tillage plus rotary hoeing with conventional tillage plus herbicides in 51 on-farm trials in Iowa showed no yield difference, only slightly higher average weed densities, and substantial cost savings with ridge tillage (Exner, Thompson & Thompson, 1996). The long experience of many farmers with this tillage/weed management system has resulted in several recommendations (Thompson & Thompson, 1984; Cramer et al., 1991, pp. 25–9).

- *Approximately 5 cm of soil should be scraped from the ridge at planting.* This is sufficient to remove residue and most weed seeds shed the previous season while retaining the ridge bases intact. Maintenance of the ridge bases facilitates cultivator guidance and fosters development of organic matter and macropores in the crop row.
- *The ridge should be wide (~30 cm) and relatively flat topped so that the planter cleans soil from a substantial strip on either side of the row.* Otherwise, weeds will emerge from surface seeds near the row where they will be difficult to control.
- A winter cover crop on the ridges slows emergence and growth of weeds in the spring. Since small weeds are more susceptible to the scraping action of the planter, *a cover crop helps insure that the planter will destroy any weeds that establish prior to planting.*
- *Ridge-building should occur early enough in the crop growth cycle and tools kept far enough from the row to avoid root pruning the crop.* Root pruning is always an issue in inter-row cultivation, but it is more likely in ridge tillage systems because the tools works deeper relative to the base of crop plants.

As with all reduced tillage methods, populations of perennial weeds sometimes increase with ridge tillage (Clements *et al.*, 1996). Another significant problem is that the method is suited only to crops planted in widely spaced rows. In particular, rotation with sod crops requires destruction of the ridges (Cox *et al.*, 1992). Despite these problems, ridge tillage is an important soil conservation and weed management system, particularly in regions where maize, sorghum, and soybean are the predominant crops.

Stale seedbed and false seedbed

Since tillage promotes germination of many weed species (see Chapter 2), tillage followed by destruction of weed seedlings with minimal further soil disturbance often leads to lower weed density in the crop. This is referred to as the stale seedbed method of planting. The technique is especially useful for providing reduced competition early in the development of small-seeded or slowly establishing crops like onion and carrot. Much recent work has explored application of the technique to soybean, and in this context it is largely used to avoid tillage during late spring on clay soils (Heatherly *et al.*, 1993; Lanie *et al.*, 1993, 1994; Oliver *et al.*, 1993). Usually, removal of the weeds is accomplished with a herbicide, but flaming can also be used. Balsari, Berruto & Ferrero (1994) found that a single flaming four days after irrigation and one day before transplanting lettuce seedlings reduced weed densities by 62% and produced a net income similar to chemical treatment with propyzamide (pronamide).

The false seedbed procedure works in a manner similar to the stale seedbed. In this technique, preparation of a seedbed is followed by one or more superficial cultivations at about one-week intervals prior to planting the crop. This reduces the pool of germinable seeds in the surface soil and can reduce weed density in the crop. Usually the cultivation is kept shallow so that few additional seeds are brought within emergence distance of the surface. Because firming the soil of a prepared seedbed promotes weed emergence (Roberts & Hewson, 1971), the soil should be rolled after all cultivations except the final one before planting.

Although the false seedbed procedure is widely used by organic growers (Wookey, 1985; Stopes & Millington, 1991), it has received little scientific study. Johnson & Mullinix (1995, 1998) found that two shallow passes with a rotary tiller equaled two applications of glyphosate for reducing weed pressure on subsequent peanut crops, and two cultivations before planting were better than glyphosate for weed control in cucumber. In a study on rapeseed in Alberta, Darwent & Smith (1985) compared delayed seeding and preplanting removal of *Avena fatua* with cultivation or nonresidual herbicides to use of trifluralin and early seeding. Although weed density and biomass were statistically lower with trifluralin, control with delayed planting and cultivation was sufficient to give consistently good yields. In contrast, Robinson & Dunham (1956) found no advantage to cultivation prior to seedbed preparation in the production of soybean in Minnesota. Apparently, weed seeds were protected from germination in clods worked up by sweep cultivation of firm soil.

The false seedbed technique has several limitations. First, it can only be effective if the soil is warm and moist enough to allow germination of weed seeds. For example, Baumann & Slembrouck (1994) spring-tine harrowed two and three weeks after plowing a grass sod, then one week later prepared a final seedbed and planted carrots. They found no difference in weed density between this treatment and one that was plowed and planted on the same schedule but without the preplant harrowing, probably because the soil was dry and not conducive to seed germination prior to harrowing. A second limitation of the procedure is that the soil is kept bare and loose for an additional period, and this may promote erosion. A third limitation is that yield may be lost if planting is delayed by the preplanting cultivation. A fourth problem is that although cultivations prior to planting may greatly reduce one set of species, if planting is delayed, a new set of species may become physiologically ready to germinate. In this case, weed composition may shift without a change in total abundance. The first and second of these problems should be least in irrigated agriculture, since in these systems soil moisture is controlled by the grower and the land is flat. The third and forth problems are likely to be least in systems where the growing season is long relative to the crop's developmental requirements, and therefore the grower has flexibility with regard to planting date. Despite its limitations, the false seedbed technique has broad applicability.

Basic principles of mechanical weeding

This and succeeding sections discuss methods and implements for physically removing weeds from crops. Most of these implements act by cutting or uprooting the weeds with tools that disturb the soil. These implements are commonly referred to as *cultivators*. In addition to this large class of implements, *thermal and electric weeders* damage weed tissues by a discharge of heat, cold, or electricity. The most common of these are the various types of *flame weeders*. Other implements include *weed pullers* and *mowers*. All these implements may be classified according to where they work relative to the crop row. *Inter-row* cultivators remove weeds from the area between crop rows. In contrast, *in-row* weeders specifically attack weeds in the crop row. *Near-row* cultivators and weeders may or may not affect weeds in the inter-row, but are able to harm weeds closer to the crop row than is commonly the case for most inter-row cultivators. Finally, some machines act similarly on both the in-row and inter-row areas, and these are referred to here as *full-field* machines. Full-field cultivators are usually used prior to or just after crop emergence. The most difficult weeds to remove with cultivators and other types of weeders are those that establish close to crop plants. Consequently, much of the discussion will focus on implements that are effective against weeds in and near the crop row.

Although this and succeeding sections focus on machine-powered implements, most of the principles governing cultivation apply equally to hand and animal-powered tools. Also, many of the implements discussed in the following sections, including inter-row sweep and shovel cultivators, rolling cultivators, basket weeders, and weeding harrows, have hand and animal-powered analogs (Intermediate Technology Publications, 1985, pp. 12–55; Alström, 1990, pp. 98–131). Moreover, most in-row and near-row weeding tools are simple, low-draught machines that could easily be mounted on an animal-pulled toolbar. The low speed and fine position control possible with an animal-drawn implement is ideal for these tools, and they could potentially reduce some of the most arduous labor in the smallholder cropping cycle. Even small increases in mechanization, such as a shift from hand hoeing to use of a push weeder, greatly decrease weeding time (Tewari, Datta & Murthy, 1993), thereby improving the timeliness of weeding and crop yields.

Mechanical weeding is guided by several simple principles.

1. *Row-oriented cultivators should work the same number of rows as the planter, or a simple fraction of this number.* Otherwise, imperfect spacing between adjacent planter passes will lead to improper placement of tools relative to some rows, with consequent damage to the crop and poor control of weeds in the inter-row.
2. *The action of the cultivator must be appropriate for the growth stages of the weeds and crop.* The timing and number of cultivations required depend on the growth rate of the target species and the size range over which it is susceptible to the implement. Based on many years of farming experience, Bender (1994, pp. 35–7) suggested that staggered planting of crops facilitates timely cultivation by reducing bottlenecks due to weather.

    The degree to which precise timing is critical depends on how closely the implement works to the crop row. In-row weeders and full-field implements cannot dig deeply without damaging the crop. For these machines, operations must be timed to catch the weeds after they have germinated but before they become well rooted, and delaying cultivation may allow many to escape (VanGessel et al., 1998; Fogelberg & Dock Gustavsson, 1999). Implements that work close to, but not in, the row have a larger window within which the work can be performed, but still require careful attention to timing. In contrast, timing is less critical with most inter-row cultivators. For example, Mt. Pleasant & Burt (1994) found that timing of cultivation with a shovel cultivator had little effect on either weed biomass or maize yield.
3. *Creation and maintenance of a size differential between the crop and the weeds facilitates effective mechanical weed control.* Most sophisticated mechanical weed management programs begin with a stale seedbed or pre-emergence cultivation to delay emergence of weeds relative to emergence of the crop. Full-field, in-row, and near-row weeding can then increase in depth and degree of soil movement as the crop grows larger. For many row crops (e.g., maize, sorghum, potato), once the crop is well established, soil can be thrown into the row to cover small weeds. However, because of the high growth rate of most agricultural weeds, this will only be effective if the first cohorts that germinate following crop planting have been killed previously.
4. *The effectiveness of cultivation decreases as weed density increases.* This occurs for several reasons. First, some proportion of weeds in the crop row will escape even a well-planned and carefully executed cultivation program. Dieleman, Mortensen & Martin (1999) found that the proportion escaping was constant over a wide range of densities. Thus, if the density is high, the escapes may cause yield reduction (Buhler, Gunsolus & Ralston,

1992; Schweizer, Westra & Lybecker, 1994; Eberlein *et al.*, 1997). In contrast, if the initial density is moderate, the few that survive will cause little competitive stress in robust agronomic crops and can be cost-effectively hand rogued out of high-value crops. Second, soil clings better to a dense mass of roots than to individual strands of root. Consequently, rerooting is a bigger problem when weeds are dense. Finally, some implements do not penetrate well when roots bind the soil together and the soil surface is lubricated by green plant tissue. If perennial weeds are abundant in the field or the seed bank is thought to be high, steps should be taken prior to planting to reduce weed density, and the field should be rotated into a crop that will tolerate vigorous cultivation and hilling of soil about the stems.

5. *Effective cultivation requires good tilth, careful seedbed preparation, and adequate soil drainage.* Tilth is critical. Good tilth facilitates stripping soil from weed roots. It also reduces the probability of knocking over crop plants with clods when soil is thrown into the row. Moreover, shallowly working tools are relatively ineffective in cloddy soil because (i) seedlings of some species emerge from greater depth in cloddy soil (Cussans *et al.*, 1996) and the tools cannot reach them without harming the crop, (ii) when clods are moved, seedlings emerge that otherwise could not reach the soil surface, and (iii) seedlings in clods may successfully establish after cultivation if rain or irrigation subsequently allows the clods to merge into the soil matrix (Mohler, Frisch & Mt. Pleasant, 1997). All these factors argue for practices that improve soil structure, including cover crops, manuring, rotation with sod crops, and controlling wheel traffic. They also argue for delaying tillage until soil moisture conditions are appropriate, even if this entails a delay in planting.

   Even in soil with good structure, clods will form if the seedbed preparation is inadequate to eliminate them. For many large-seeded crops, a coarse seedbed is not detrimental to establishment and may be beneficial in reducing erosion (Burwell & Larson, 1969). However, for the reasons mentioned above, it may be disadvantageous during cultivation. For shallowly working implements, a level seedbed facilitates depth control. For some, it is mandatory.

   Because timeliness is critical to the success of most in-row, near-row, and full-field cultivation, adequate soil drainage may make the difference between successful weed management and substantial crop loss. When storm events are following in close succession with short rain-free periods between, adequate tile drainage may allow cultivation on fields where it would otherwise be impossible.

6. *Cultivation (and tillage) in the dark stimulates germination of fewer weed seeds*

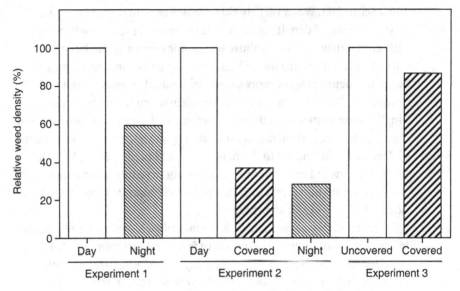

Figure 4.10 Density of weeds in treatments tilled in the dark relative to treatments tilled in light. (Drawn from data in Ascard, 1994a.)

than cultivation in daylight. As explained in Chapter 2, light stimulates seed germination in many weed species. During cultivation or tillage, seeds may be exposed to a brief flash of light and then buried again. Consequently, tillage and cultivation at night, or with implements that are covered with light-excluding canopies, often results in lower weed densities (Hartmann & Nezadal, 1990; Ascard, 1994a; Scopel, Ballaré & Radosevich, 1994; Botto *et al.*, 1998; Gallagher & Cardina, 1998). Buhler (1997) showed that dark tillage reduced densities of several small-seeded broadleaf weeds, but that it did not reduce densities of annual grasses or large-seeded broadleafs.

Even for generally light-sensitive species, some seeds do not require light for germination, and others will end up near enough to the surface to satisfy their light requirement regardless of how or when the operation was performed. Consequently, dark cultivation only reduces, but does not eliminate, weed emergence. Variation in species composition, dormancy state of light-sensitive species, distribution of seeds in the soil column and degree of soil mixing probably all contribute to variation in the results of dark tillage experiments (Figure 4.10). With regard to soil mixing, Jensen (1995) demonstrated that density of *Chenopodium album* emerging after 0 to 16 harrowings in daylight increased monotonically over the full range of soil disturbance. In contrast, maximum emergence was reached with four harrowings when operations were performed in the dark. Presumably, when tillage was performed in daylight, additional seeds were exposed to a light flash with each operation.

The optimum strategy for using light-sensitive weed germination may be to perform primary tillage in the light, wait for emergence, and then prepare and plant the final seedbed in the dark (Hartmann & Nezadal, 1990; Ascard, 1994a; Melander, 1998). So far, no studies have reported on the effectiveness of dark cultivation after planting. A possible strategy for post-planting cultivation would be to perform shallow full-field cultivations in the dark to minimize weeds in the crop row. Early inter-row cultivation could then be done in the light to help clean out the seed bank, since usually 100% of young weeds in the inter-row can be killed by subsequent operations. The final cultivation would then be done in the dark or with light-shielded equipment to minimize further emergence.

7. *Attentive timing relative to changing weather and soil conditions can improve the effectiveness of cultivation*. Rotary hoes are ineffective when the ground is too wet (Lovely, Weber & Staniforth, 1958). Flame weeders work best when leaf surfaces are dry (Parish, 1990). Most cultivators are more effective during hot dry weather since, under these conditions, uprooted weeds desiccate quickly without rerooting (Terpstra & Kouwenhoven, 1981). Thus, planning cultivation with the weather forecast in mind frequently improves results.

Machinery for mechanical weeding

A cultivator consists of a frame and one or more types of tooling that engage with the soil and weeds. Most commonly, cultivators are belly mounted under the tractor or carried on a three-point hitch. Front mounting is also possible for some implements. Belly- or front-mounted implements are easier to guide and less prone to damage the crop because the driver can see the position of at least one set of tools relative to the row and because small changes in the tractor's direction do not cause large shifts in the implement's position. However, wider, longer (to accommodate more tools), and higher-clearance implements generally require a rear mount. Automatic guidance systems and human-steered implements are discussed below.

A variety of tools are available for physical weed control. The amount and type of information available on these implements vary greatly. For most equipment, comparative data are meager, and some devices have received no scientific study at all. Consequently, much of the information discussed in the following sections is based on the personal experience of the author, or derived from discussion with farmers and colleagues that have used the implements.

Inter-row cultivators: shovels and sweeps, rolling cultivators, and rotary tillers

Tools designed to work between crop rows can dig moderately deeply (typically 5 to 10 cm) without harming the crop. Consequently, complete destruction of even large weeds is often possible during inter-row cultivation. However, unless precautions are taken, vigorous soil movement can bury young crop plants, and deep digging can damage the roots of larger crops. To avoid crop damage, these tools usually work only 50% to 70% of the soil surface, leaving weeds in the crop row unharmed. However, Melander & Hartvig (1997) have shown that safe cultivation within 2.5 cm of the row is possible if the crop is small and protected with shields and a good cultivator guidance system. At later stages of crop growth, sweeps may be used to hill-up soil in the crop row to bury small weed seedlings.

Sweeps and shovels are the most commonly used cultivation tools. They are simple and durable. They vary greatly in width, shape, and pitch (Figure 4.11). Generally, soil movement away from the shank increases with width and pitch, and decreases as the angle between the leading edges increases. "Goosefoot" style shovels (Figure 4.11c) move less soil than standard shovels and sweeps (Figure 4.11b, e). S-shaped (Danish) shanks vibrate more, which brings weeds to the surface and helps shake soil loose from weed roots, but they are less robust than C-shaped shanks.

Multiple shanks provide flexibility. For example, Mohler, Frisch & Mt. Pleasant (1997) used 2.5 cm spikes (Figure 4.11d, top) nearest the row when maize was young to reduce soil movement toward the row, but changed to 10 cm sweeps to throw soil into the row at the last cultivation. However, minimum-tillage cultivators designed to operate in high crop residue usually have one shank with a single broad sweep per inter-row. This design presents less metal at the ground level to snag debris. Also, these cultivators have a coulter in front of each shank to cut residue so it can flow past the shank. In minimum-tillage machines, hilling up around the crop is accomplished with wing attachments that increase the lateral displacement of soil, or with disk hillers (see below).

The usual tooling on a rolling cultivator consists of gangs of wheels with stout, curved tines that cut and dig out weeds as they roll over the ground. Alternatively, disk gangs can be used for cultivation in high residue. In either case, each gang is mounted on a separate tube, with two gangs per inter-row. Adjusting the angle relative to the direction of travel controls aggressiveness

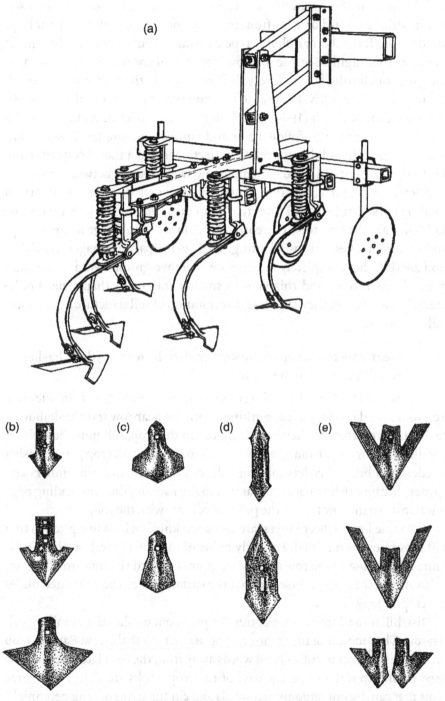

Figure 4.11 (a) One gang of a parallel gang sweep cultivator equipped with disk hillers. (b) Cultivating shovels. (c) Goosefoot shovels. (d) Reversible-point shovels. (e) Sweeps.

and the amount of soil movement. Depending on the setting of the gangs, soil flow is strictly toward or away from the row. Rolling cultivators dig out large weeds less effectively than shovel-type cultivators, but they work the surface soil more thoroughly. Because soil flow is strictly in one direction, and because the gangs can be tilted to work very shallowly next to the crop row, rolling cultivators can safely cultivate closer to the crop than can shovel cultivators. Mt. Pleasant, Burt & Frisch (1994) found higher weed cover in maize following rolling cultivation than following shovel cultivation. Mohler, Frisch & Mt. Pleasant (1997) found that the relative effectiveness of a shovel cultivator and a rolling cultivator equipped with inter-row sweeps varied between years.

Rotary tillers for cultivation consist of gangs of power-take-off driven rotating curved or L-shaped tines that chop up weeds and mix them into the soil. They are the best tools currently available for strip tillage in cover crops and living mulches prior to planting. Their principal advantages in cultivation are that they completely incorporate all above-ground weed tissues, and probably chop roots and rhizomes to smaller fragments than other implements. Nevertheless, they can cause deterioration of soil structure by excessive pulverization.

Near-row tools: vegetable knives, disk hillers, spyders, basket weeders, and brush weeders

Several tools have been invented to cultivate 5 to 12 cm from the crop. To avoid root damage, typical working depth for near-row tools is shallow: 2 to 5 cm. Furthermore, when working close to the crop, soil must flow either parallel with, or away from, the row to avoid burying small crop plants. Basket weeders and brush weeders can work closer to the row than disk hillers and spyders because their rotation is parallel to the row and thus the leading edge is no closer to the row than is the point of contact with the soil.

Vegetable knives (beet knives) are L-shaped knives with a low pitch so that soil movement is minimal. For early cultivations, the vertical portion of the knife is run close to the row and the tip points toward the inter-row. At later cultivations, the tip is reversed so that the surface soil can be cultivated under the crop canopy.

Disk hillers and spyders are optional equipment on shovel and rolling cultivators. They mount in the front-most position, next to the row. Early in crop growth they are set to cut soil and weeds away from the row; later they may be used to hill up soil around the base of the crop. Disks are sharp, aggressive tools that can dig out large annual weeds and cut the stems of rank perennials. They also perform well in heavy crop residue. Spyders are star-shaped wheels (Figure 4.12) that dig rather than cut the soil. They are smaller diameter (32

Figure 4.12 Spyders. (Redrawn from Schweizer, Westra & Lybecker, 1994.)

cm) than most disk hillers, and this allows them to work closer to some crops. Also, when cutting soil away from the row, they leave a loose soil layer next to the row rather than a smooth shoulder, and this probably reduces soil drying rather than encouraging it.

Basket weeders consist of two sets of rotating wire cages. The forward cages are ground driven; the rear cages are driven by a chain connected to the forward cages, and turn twice as fast. Penetration is shallow, 2 to 5 cm, but the soil is thoroughly worked. Consequently, few small weeds escape substantial damage even if they are not completely uprooted. The implement is unsuited to stony ground because rocks bend the baskets out of shape and can become caught between adjacent wires.

Two types of brush weeders are currently in use. One consists of power-take-off driven polypropylene brushes on a horizontal shaft; these work parallel to the crop row (Figure 4.13). They uproot small weeds, and shear off larger ones (Pedersen, 1990). The soil flow is primarily parallel to the row, which, in conjunction with narrow tunnel shields (6 to 20 cm wide), allows cultivation very close to small crop plants. Another type of cultivator is required once the crop plants grow too large to move easily through the shields. A second type uses pairs of unshielded brushes on rotating vertical shafts. Melander (1997, 1998) used this type to control weeds within 1.5 cm of onion rows without damaging the crop. Fogelberg & Dock Gustavsson (1998, 1999) showed that a vertical-axis brush weeder could be used for in-row weeding of young carrots because the weeds were more prone to uprooting than carrots when both were in the two-to-four leaf stage. Brush weeders resist clogging with large weeds and debris (Geier & Vogtmann, 1987), and

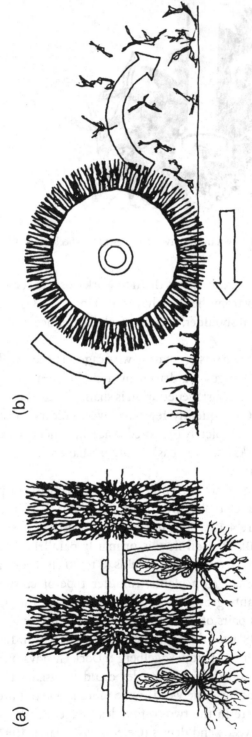

Figure 4.13 Brush hoe. (a) View from rear, showing tunnel shields. (b) Side view. (Redrawn from Pedersen, 1990.)

may work in wet soils (Weber, 1994). An additional advantage is that, like basket weeders, they leave a loose, uniform soil surface that is not conducive to weed germination. Horizontal-axis brush weeders require a flat seedbed for consistent depth of operation (Geier & Vogtmann, 1987).

Full-field and in-row cultivation tools: weeding harrows, rotary hoes, rubber-finger weeders, spinners, torsion weeders, and spring hoes

The ecology behind all cultivators that attack weeds in the crop row is essentially the same: disturb the shallow surface layer of soil above the rooting depth of the crop, thereby killing very small weeds without uprooting crop plants. All in-row and full-field cultivating tools are primarily successful against small-seeded weeds that must germinate near the soil surface to emerge, and in large-seeded crops that can be planted relatively deeply. Conversely, these tools are ineffective against large-seeded and perennial weeds that emerge from below the depth of operation, and they cannot be used in small-seeded crops unless the crop is transplanted. Deeper planting allows more aggressive weeding.

Weeds are fully susceptible to in-row and full-field tools only when in the white thread and early cotyledon phases of development. Most weeds larger than this will escape unless the tools are used so aggressively that the crop is damaged. Consequently, successful in-row cultivation requires repeated removal of weeds while they are still very small. Such repeated working of the soil might seem contrary to good soil conservation. However, the operations are so superficial that damage to soil structure by the implements themselves is probably negligible, although repeated wheel traffic can contribute to compaction.

Full-field implements affect the crop row and inter-row areas equally. They are primarily useful in close-planted crops like cereals, and for control of weeds in the crop row of wider spaced crops like maize, sorghum, and soybean when these will be cultivated with an inter-row machine later in the season. Full-field cultivators are of two types: rotary hoes and weeding harrows.

A rotary hoe consists of two ranks of wheels, each bearing 16 spoon-like projections. The wheels are attached to the tool bar by spring-loaded arms to allow movement over obstacles. The ground-driven wheels typically penetrate to a depth of 2–4 cm and flick up soil and small weeds as they turn. To disturb the soil effectively, the machine must operate at high speed (11–21 km h^{-1}). This allows rapid weeding of large areas.

Lovely, Weber & Staniforth (1958) found that whereas three timely rotary hoeings reduced subsequent weed dry weight by 72 % in solid seeded soybean,

Table 4.4. *Weed control and soybean yield with timely and untimely rotary hoeing and various soil conditions*

Treatment	Hoe passes	Weed dry weight		Soybean yield	
		kg ha^{-1}	% control	kg ha^{-1}	% of max
Timely + hand weed	2	0	100	2620	100
Timely	3	530	72	2280	87
Untimely	2	1160	38	1950	74
Timely, wet after	3	1250	33	2150	81
Untimely, wet after	2	1390	25	1880	71
Timely, wet before	3	1250	33	2420	92
Untimely, wet before	2	1660	11	1880	72
Weedy check	–	1860	0	1610	61

Source: Adapted from Lovely, Weber & Staniforth (1958).

two untimely hoeings reduced it only 38% (Table 4.4). Similarly, Mulder & Doll (1993) found that three rotary hoeings controlled weeds better than two hoeings, and Mohler, Frisch & Mt. Pleasant (1997) found one to two hoeings inadequate in some years. However, VanGessel *et al.* (1995*a*) found one well-timed rotary hoeing as good as two. Mohler & Frisch (1997) found a rotary hoe about as effective as a spring tine weeding harrow for weed management in oat.

Wet soil conditions reduce the effectiveness of rotary hoeing (Table 4.4). Wet ground limits soil movement by the implement, and rainfall or irrigation soon after rotary hoeing probably reduces the percentage of weeds that die by desiccation.

Harrows vary greatly in design, but all consist of a frame with many downward pointing, small diameter tines. Although chain harrows and spike-tooth harrows are still successfully used for weed control, spring-tine harrows are rapidly replacing these more traditional designs. The tines on spring-tine harrows are either spring steel wires (typically 4–7 mm diameter), or else rigid metal fingers attached to the frame with a spring. Their popularity as weeding tools comes from a greater ability to adjust down pressure and hence aggressiveness, coupled with the ability of the tines to spring over or around well-rooted crop plants. Nevertheless, comparison trials have not demonstrated superiority of the spring-tine design over older types of harrows (Rasmussen, 1992*a*; Wilson, Wright & Butler, 1993).

In cereals, harrows are most commonly used pre-emergence or at emergence, and again when the crop has two to three leaves. Cultivation between these stages will usually bury too many crop plants. Harrows are also some-

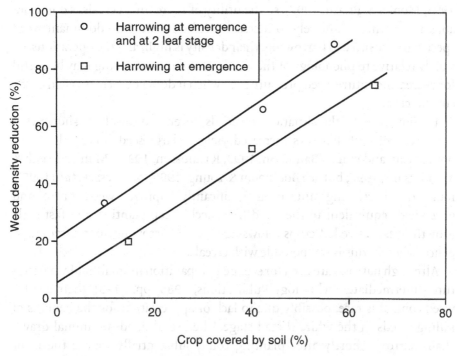

Figure 4.14 Weed control as a function of the percentage of narrow row field pea covered by soil during harrowing. (Redrawn from Rasmussen, 1992b.)

times use to comb sprawling and vining weeds like *Stellaria media* and *Convolvulus arvensis* out of cereals shortly before stem elongation of the crop (Kress, 1993; Wilson, Wright & Butler, 1993). In this application, the tines do not penetrate the soil. With large-seeded row crops and transplants, opportunities for harrowing are more continuous, although beans are sensitive at the crook stage (VanGessel et al., 1995a).

Several studies have shown a positive, often linear, relation between percentage weed control and degree of crop covering (Figure 4.14) (Rasmussen, 1990, 1991, 1992b, 1993; Rydberg, 1994). Rasmussen (1991) used this relation to develop a model for predicting yield response to harrowing and showed that maximum yield generally occurs at substantially less than full weed control. In his model system with spring barley, maximum yield was obtained in the range of 40% to 75% weed control for *Brassica napus* sown as a weed, but at 0% weed control (no harrowing) for *Phacelia tanacetifolia*. Several empirical studies in cereals have also found no consistent increase in yield with harrowing relative to the weedy check (Stiefel & Popay, 1990; Peruzzi et al., 1993; Steinmann & Gerowitt, 1993; Rydberg, 1994; Rasmussen & Svenningsen,

1995; Mohler & Frisch, 1997), so the utility of harrowing cereals is questionable. Harrowing is unlikely to increase yield unless the predominant weed species are sensitive to harrowing, their density is high, and the operations are timely relative to phenology of the weeds. However, harrowing may be useful for decreasing future weed pressure even when it does not increase yield of the current crop.

In contrast with the situation in cereals, several studies have shown that weed control with harrows increased yield of large-seeded crops like pea, broad bean, and maize (Baumann, 1992; Rasmussen, 1992b; Mohler, Frisch & Mt. Pleasant, 1997; but see Boerboom & Young, 1995). VanGessel et al. (1995a) found that harrowing pinto bean significantly improved weed control and gave yields equivalent to the weed-free check. Deep planting and fast early growth of large-seeded crops allows aggressive harrowing soon after emergence, whereas this is not possible with cereals.

Although harrows are used for seedbed preparation in smallholder agriculture (Intermediate Technology Publications, 1985, pp. 21–3), their use for weed control is rare, possibly due to lack of appreciation for the benefits of killing weeds in the white thread stage. The use of hand- or animal-drawn chain harrows shortly after planting could substantially reduce the labor requirements for later hand-weeding in some systems.

Several tools using different soil-moving principles are available for in-row weed control in row crops. These are all precision tools that must be carefully set for depth, distance from the row, and in some cases, angle, to achieve good weed control without damaging crop plants. They are best used in a front-mount or belly-mounted position, or with a guidance system. A smooth, flat seedbed improves the consistency of weed control.

Torsion weeders are spring steel rods that reach within a few centimeters of the crop row and travel 2 to 3 cm below the soil surface (Figure 4.15). The compressive action of the springs causes the soil in the row to boil up, thereby disturbing weeds in the row that are not yet well rooted. Larger weeds next to the row may be sheared off as well. Spring hoes work in a similar manner, but are more robust and aggressive (Figure 4.15). Both types of tools are usually mounted on inter-row cultivators forward of the shovels or rolling gangs. Several studies have shown improved weed control and crop yield by cultivators equipped with these tools in combination with spinners and spyders (see below) relative to standard shovel cultivators (Figure 4.16) (Schweizer, Westra & Lybecker, 1994; VanGessel et al., 1995b, 1998; Mohler, Frisch & Mt. Pleasant, 1997).

Stiff, heavy-duty spring hoes are available for work in orchards and vineyards. These scrape the soil surface free of weeds in and near the row. A castor at the tip allows the tool to bend past trunks without scraping the bark.

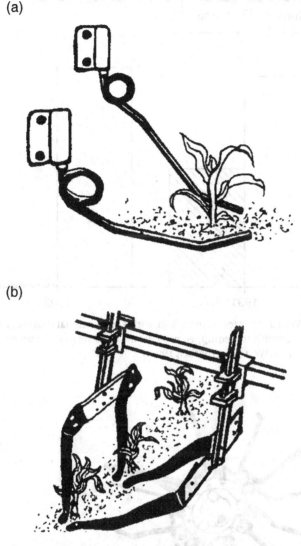

Figure 4.15 (a) Torsion weeders and (b) spring hoes. (Redrawn from Schweizer, Westra & Lybecker, 1994.)

Spinners are ground-driven, in-row weeders that consist of a basket-like arrangement of spring steel wires that scratch laterally across the crop row (Figure 4.17). They are normally used in pairs to increase the proportion of the row area that is worked. Usually the depth is set such that the deepest penetration is a little above the planting depth and squarely in the row. Alternatively, the tool can be set so that the deepest penetration occurs next to the row, with the tool scratching across the row with an ascending or descending stroke. A direct strike by the tip of a wire can cut off a crop plant, but usually mortality

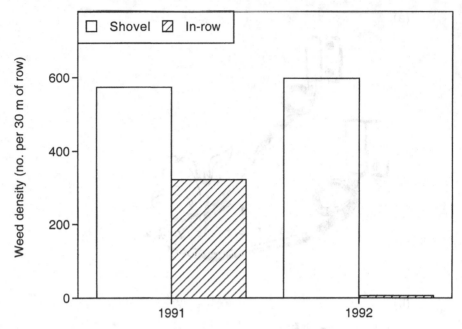

Figure 4.16 Weed density following tillage with a conventional shovel cultivator or with a shovel cultivator equipped with in-row weeding tools. (Drawn from data in Schweizer, Westra & Lybecker, 1994.)

Figure 4.17 Spinners. (Redrawn from Schweizer, Westra & Lybecker, 1994.)

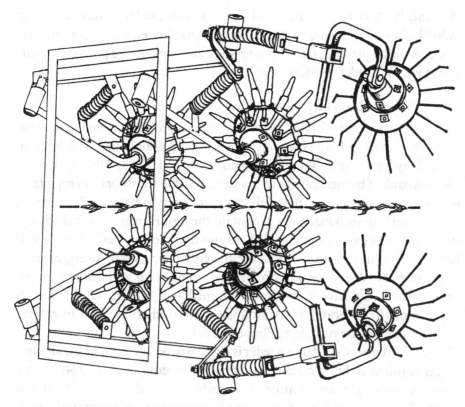

Figure 4.18 Rubber-finger weeder.

from properly set tools will be no more than a few percent, and is often negligible. Cultivation late on a sunny day when crop stems are less turgid reduces mortality. Several studies have examined spinners in conjunction with other in-row and near-row tools (Schweizer, Westra & Lybecker, 1994; VanGessel *et al.*, 1995b; Mohler, Frisch & Mt. Pleasant, 1997), but so far, controlled investigations of these tools have been minimal.

Rubber-finger weeders consist of two pairs of ground-driven wheels equipped with rubber fingers that stir the surface soil in the row, but bend around well-rooted crop plants. These are followed by wire baskets that aggressively stir the area adjacent to the row (Figure 4.18). The rubber fingers are stiff, so weeding in the row requires that crop stems be tough and well rooted. The implement is commonly used to weed nursery stock.

Rotary hoes, weeding harrows, and some in-row tools commonly reduce crop density by several percent (Rasmussen, 1991, 1992a; Buhler, Gunsolus & Ralston, 1992; Mulder & Doll, 1993; Rydberg, 1994; Mohler, Frisch & Mt.

Pleasant, 1997). Generally, this stand loss appears as randomly spaced missing individuals, rather than as blighted row sections. Consequently, growers may find that a high planting rate improves yield and competitive pressure on surviving weeds (Bender, 1994, p. 39).

Thermal and electric weeders

Flame weeders briefly expose weeds to a propane or butane flame at 800–1000°C (Ascard, 1995a), which disrupts cell membranes and leads to rapid dehydration (Ellwanger, Bingham & Chapell, 1973; Ellwanger et al., 1973). A bank of burners can flame a wide area to kill weeds before crop planting or crop emergence, or to defoliate plants prior to harvest (Tawczynski, 1990). Shielding such machines to contain the heat increases their efficiency. Irrigation a few days before planting ensures that the first flush of weeds will have emerged in time to flame before the crop is up. Alternatively, burners directed toward the row can control in-row weeds in crops like maize, onion, and cabbage that have a protected terminal bud (Figure 4.19) (Geier & Vogtmann, 1988; Ascard, 1990; Holmøy & Netland, 1994), and in cotton, which has a corky stem (Seifert & Snipes, 1996).

The effectiveness of a flame weeder is best described in terms of the amount of gas required to kill a certain percentage of weeds (Ascard, 1994b). For a given machine, gas consumption is usually regulated by varying ground speed. The amount of gas required for 95% control varies substantially with weed species and size. For example, Ascard (1994b) required 1.5- to 2-fold more propane per hectare to control *Sinapis alba* in the two-to-four leaf stage than to control it in the none-to-two leaf stage. Many common broadleaf species such as *Chenopodium album*, *Stellaria media*, and *Senecio vulgaris* can be well controlled by gas doses of less than 50 kg ha^{-1} when young (Ascard, 1995b). In contrast, grasses and broadleaf species with protected buds (e.g., *Matricaria inodora*, *Phleum pratense*, and *Poa annua*) are relatively resistant to flaming and can be controlled only with large gas doses, or not at all (Rahkonen & Vanhala, 1993; Ascard, 1995b).

Other thermal weed control approaches have been tested, including infrared irradiation and freezing with liquid nitrogen or carbon dioxide snow, but these appear to be less efficient than treatment with an open flame (Parish, 1989; Fergedal, 1993). Microwave heating of soil greatly decreases weed emergence, but appears impractical at field scales (Barker & Craker, 1991). Concentration of solar radiation with a Fresnel lens is energy-efficient and effective at killing young weeds (Johnson et al., 1989), but implements based on this technology are likely to be bulky and slow. Similarly, hot-water weeders are in use, but the large volume of water that must be heated (9000 to

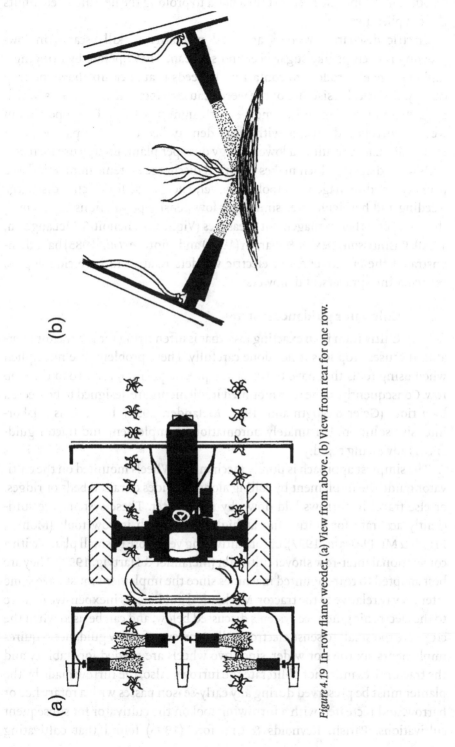

Figure 4.19 In-row flame weeder. (a) View from above. (b) View from rear for one row.

11 000 L ha^{-1}) and the need to trail a hood to prolong the heating effect limits their application.

Electric discharge weeders are used primarily to kill escapes in low-growing row crops like sugar beet and soybean. They operate by bringing a high-voltage electrode into contact with weeds that stick up above the crop canopy. Electrical resistance of the weeds causes vaporization of fluids, which disrupts tissues (Vigneault, Benoit & McLaughlin, 1990). The proportion of weeds controlled decreases with weed density because more pathways for energy discharge result in a lower energy dose per plant. Energy use increases with weed density, which makes electrical discharge systems impractical as a primary weed management tool. However, energy use for electric discharge weeding and herbicides was similar for low-density populations (e.g., 5 m^{-2}) that escaped other management measures (Vigneault, Benoit & McLaughlin, 1990). Rasmusson, Dexter & Warren (1979) and Diprose *et al.* (1985) have demonstrated the effectiveness of electric weeders relative to herbicide wipers, recirculating sprayers, and mowers.

Cultivator guidance systems

Cultivation is an exacting task that is often tiring for the tractor operator. It causes crop loss if not done carefully. These problems are multiplied when using tools that have to travel at a precise position relative to the crop row. Consequently, some rear-mounted implements are designed to be steered by a rider (Geier & Vogtmann, 1987; Melander, 1997), but this is a labor-intensive solution. Fortunately, automation of implement and tractor guidance is advancing rapidly.

The simplest approach is purely mechanical. Wheels mounted on the cultivator guide the implement by rolling along the sides of raised beds or ridges, or else travel in furrows laid down by the planter. These systems are sufficiently accurate for cultivating at high speeds with in-row tools (Mohler, Frisch & Mt. Pleasant, 1997), or for cultivating very close to small plants with a conventional inter-row shovel cultivator (Melander & Hartvig, 1997). They are best adapted to rear-mounted machines since the implement must have some lateral sway relative to the tractor. Mechanical guidance is inexpensive relative to the electronic guidance systems discussed below, and can be used when the crops are too small to sense electronically. However, furrow guidance requires implements six rows or wider, since two wheels are needed for stability and the tractor tires must not obliterate the furrows. Also, the furrow made by the planter must be preserved during any early-season passes with a rotary hoe or harrow, and recreated with a furrowing tool on the cultivator for subsequent cultivations. Parish, Reynolds & Crawford (1995) found that cultivating cotton using a mechanical guidance system substantially reduced costs by allowing a narrower band of herbicides over the crop row.

Electronic guidance systems usually have wands that sense the presence of the crop row. The simplest merely alert the operator when the implement strays to one side, but most models control a steering device. Various steering devices correct the cultivator's position by (i) shifting the lift arms of the three-point hitch, (ii) shifting the cultivator laterally relative to the hitch, (iii) rotating the cultivator slightly on the hitch, (iv) turning it slightly using disk coulters, or (v) turning the tractor steering wheel (Cramer, 1988; Bowman, 1991, 1997, pp. 30–3). The last approach results in a longer delay between error and correction on rear-mounted machines, but it allows the driver to watch for jamming and other problems. It is also the only approach that is well adapted to belly- or front-mounted machines, although in principle, the second approach could work if the implement was attached to the tractor by a laterally sliding carriage.

Generally, crops like maize and sorghum that are flexible when young cannot be reliably sensed until 12–15 cm tall; beans can be detected at 7–10 cm. Some systems can guide electronically off a planter-made furrow when the crop is too small to be sensed. Tian, Slaughter & Norris (1997) recently developed software for detecting rows of tomato seedlings by computer interpretation of a video image. When incorporated into a guidance system, this should allow guided cultivation of very young crops. Van Zuydam, Sonneveld & Naber (1995) used a laser beacon to provide precision tractor guidance for several types of field operation, including cultivation.

Rapid progress in the computer industry opens possibilities for further development of automatically guided cultivators. These could potentially operate under the control of artificial intelligence software, distinguish weeds from crops using a variety of electronic sensors and image processing devices (Zhang & Chaisattapagon, 1995), and selectively destroy weeds with a flexible array of tools. Development of such machines does not require new technological breakthroughs, but may be unnecessarily sophisticated for most cropping systems.

Systems of mechanical weed management

Effective mechanical weed control typically requires several machines. These need to be appropriate for the type of crop, timing of crop development, tillage practices, and type of weed problem. That is, various mechanical weed management practices need to be integrated into a program that in turn is integrated with other ecological management methods.

A mechanical weed management program commonly used by organic maize and soybean growers in the midwestern USA consists of two to three rotary hoeings followed by two cultivations with sweeps or shovels. It is well adapted to both ridge tilled and flat tilled fields (Thompson & Thompson, 1984; Gunsolus, 1990; Mulder & Doll, 1993). In this system the rotary hoe reduces weed density and delays establishment of the weeds relative to the crop. At the first inter-row cultivation, the crop is usually protected from burial by shields. As the crop develops, large amounts of soil are thrown around the plant bases to bury weeds. The machines used are simple, robust, and pulled at high speeds, allowing rapid cultivation of large fields. Although the weed control is not as complete as that achieved by more sophisticated devices, this poses few problems in highly competitive crops.

Mechanical weed control programs without herbicides in high-value vegetable crops vary greatly in detail, but many share common elements (Grubinger & Else, 1996; Bowman, 1997, pp. 67–86). A false or stale seedbed procedure is often used to reduce initial weed density. Harrowing of large-seeded crops or flaming of small-seeded crops then further reduces weed density prior to crop emergence. After the crop is up, the emphasis is often on frequent cultivation close to the crop row using a basket weeder, brush weeder, or vegetable knives. Inter-rows may be cultivated with shovels after the crop is large. The value of the crop usually makes hand roguing of weeds in the crop row economically viable. Consequently, few weeds set seed, and this facilitates weed management and minimizes the cost of hand-weeding in subsequent years.

Each grower needs to find the right mix of implements to meet the particular situations presented by the soils, climate, and crops grown on the farm. Multiple implements usually are required to meet the diversity of regularly encountered weeding tasks; additional machines may be useful in unusual circumstances.

Comparison of chemical and mechanical weed management

Only a few studies have compared current herbicide programs with modern cultivation programs that include full-field and in-row implements. VanGessel *et al.* (1995*b*) examined various combinations of rotary hoeing, alachlor, in-row cultivation, and a post-emergence herbicide chosen by a decision aid. One rotary hoeing plus in-row cultivation resulted in weed control and yields similar to alachlor plus post-emergence herbicide. Schweizer, Westra & Lybecker (1994) similarly concluded that in-row cultivation could successfully control weeds in maize. Mohler, Frisch & Mt. Pleasant (1997) found that weed control and maize yields with rotary hoeing plus in-row weeding were equivalent to an herbicide control treatment in two years, but lower in the third due to untimely rotary hoeing. Inter-row cultivation plus row flaming provided yields and weed control equivalent to appropriate herbicide programs in maize, set onion, and cabbage, but not in direct-seeded onion (Geier & Vogtmann, 1988; Ascard, 1990; Netland, Balvoll & Holmøy, 1994).

Most recent studies of mechanical weed control have focused on cultivation in conjunction with banded or reduced rates of herbicides. These integrated systems have generally performed well (Kouwenhoven, Wevers & Post, 1991; Eadie et al., 1992; Hartzler et al., 1993; Mulder & Doll, 1993; Buhler et al., 1994; Mt. Pleasant, Burt & Frisch, 1994; Parks et al., 1995; Mulugeta & Stoltenberg, 1997), with some exceptions (Snipes & Mueller, 1992; Buhler, Gunsolus & Ralston, 1993). Bridgemohan & Brathwaite (1989) found inter-row cultivation better than any herbicide for control of *Rottboellia cochinchinensis* in maize.

Often, cultivation has improved weed control or decreased the impact of weeds on yield even when herbicides were applied at full rates (Glaz, Ulloa & Parrado, 1989; Steckel, DeFelice & Sims, 1990; Shaw, Newsom & Smith, 1991; Buhler et al., 1994; Mt. Pleasant, Burt & Frisch, 1994; Steckel & DeFelice, 1995; Newsom & Shaw, 1996). Even when weeds have been effectively controlled by chemicals or hand-weeding, cultivation often increases yield, presumably due to better management of soil moisture or improved root growth (Prihar & Van Doren, 1967; Russel, Fehr & Mitchell, 1971; Hauser, Cecil & Dowler, 1973; Whitaker, Heinemann & Wischmeier, 1973; Johnson, 1985; Snipes & Mueller, 1992; Snipes et al., 1992).

Contrary to frequently voiced concerns that higher labor expenses make mechanical weeding uncompetitive with herbicides, analyses show that costs for the two approaches are often similar (Mulder & Doll, 1993; Schweizer, Westra & Lybecker, 1994; Mohler, Frisch & Mt. Pleasant, 1997). Moreover, integrated systems using cultivation with reduced rate or banded herbicides often provide equivalent yield at lower cost than chemical control alone (Bicki, Wax & Sipp, 1991; Mulder & Doll, 1993). Although replacement of a substantial proportion of herbicide use with cultivation is often economically profitable on a per hectare basis, cultivation is difficult on large, specialized farms because labor and machinery are often insufficient for timely operations. Forces driving such inefficiencies are discussed in Chapter 11.

Surprisingly, cultivation often requires less energy than typical herbicide programs. This is because most herbicides require large amounts of energy for feed stocks, process energy, packaging, and transportation (Pimentel,

1980). For example, based on data in Green (1987) and Clements *et al.* (1995), two rotary hoeings plus two inter-row cultivations in maize requires 585 MJ ha^{-1} of fuel, whereas a typical herbicide program of 2.2 kg ha^{-1} of metolachlor plus 0.56 kg ha^{-1} atrazine uses 714 MJ ha^{-1}. A single application of glyphosate at 1.68 kg ha^{-1} uses the equivalent of 763 MJ ha^{-1}. Flame weeding is considerably more energy intensive: a 50 kg ha^{-1} propane flame weeding uses 2700 MJ ha^{-1} (Ascard, 1995a). In contrast, herbicides that are applied at low rates probably require relatively little energy. Considering whole crop production systems, herbicides range from a significant proportion of the total energy input for Iowa corn (14%, Pimentel & Burgess, 1980) and Ohio soybean (13%, Scott & Krummel, 1980) to a minor input for Florida cabbage (1.6%, How, 1980) and a trivial input for Kansas wheat (0.15%, Briggle, 1980). These values would probably change little if mechanical management were used instead. When weed control represents a substantial energy input, however, some energy savings are possible with integrated systems (Clements *et al.*, 1995).

Directions for future research

The improvement of mechanical weed management requires a broad research agenda, including biological, agronomic, and engineering studies.

Much work is needed on the mechanisms whereby tillage affects perennial weed populations. How do the size distributions of root and rhizome fragments compare following tillage with different implements, and how does this affect a fragment's survival probability and subsequent rate of growth? Do particular species survive better when buried with the shoot attached or with the shoot severed from the roots? How do these factors interact with soil temperature and moisture conditions in determining the degree of control by tillage?

Knowledge is equally sparse on the mechanisms whereby tillage affects seed banks. In particular, it is not even clear whether most depletion of seed banks by tillage is due primarily to additional seedling emergence, as implied by Figure 4.6b, or whether the tillage-related mortality factors listed in Figure 4.7 play an important role for some weed populations. Ultimately, the emergence models discussed in the section "Effect of the timing of tillage on weed seedling density" need to be extended to predict not just when emergence will occur, but how many seedlings will emerge, and how this relates to the tillage regime.

The destruction of weed seedlings during cultivation also requires mechanistic investigation. When using tools that disturb only a shallow surface layer, which species must be killed in the white thread stage, which remain susceptible after emergence, and how does this relate to development of the weeds' root systems? Which species can recover from shallow burial? How does the rate of kill for particular species relate to soil moisture status at the time of disturbance? To what extent are seedlings of particular species killed by burial vs. dismemberment vs. desiccation, and does the cause of death vary with the implement? Fogelberg & Dock Gustavsson (1999) showed that a vertical-axis brush weeder killed more weeds by uprooting than by burial, though both mechanisms contributed significantly to weed mortality. Similar quantitative data are unavailable for most implements.

Much highly applied work is also needed to provide a scientific basis for cultivation recommendations. Many implements have not been studied systematically for performance under different speeds, angles, depths, or position relative to the row for even a single crop, but such data are needed for a variety of crops, each in a range of phenological stages. Quantitative European work on weeding harrows, flame weeders, and brush weeders provides models for the type of research that is needed (Rasmussen, 1991; Ascard, 1994b; Melander, 1997). Similarly, anecdotal observations and arguments from first principles have led to recommendations here and elsewhere regarding the role of tilth, soil texture, and clod size on the performance of cultivation implements, but scientific documentation to support these recommendations is largely lacking. This lack of information is analogous to making herbicides available to growers without supplying information on crop tolerance, appropriate rate, or timing of application.

A more difficult subject in need of investigation is root pruning during cultivation. Few data are available on how close or how deep various implements can be used without damage to crop root systems. Obviously, this depends on crop species and size, and probably also depends on growing conditions and soil properties. Further work is needed that builds on Russel, Fehr & Mitchell's (1971) study of soybean response to root damage by cultivators under various environmental conditions.

Regarding the engineering of mechanical weeders, the critical challenge at present is not to create implements based on new principles, but rather to make the present equipment more usable. In particular, most tools must be set for depth and distance from the row, and often angle as well. Provision of cranks, rulers, and angle gauges on supporting shanks and brackets could greatly simplify these adjustments (Mattsson, Nylander & Ascard, 1990). Ideally, a cultivator should be tuned to the crop size and soil conditions each time it is used, and this is essential for precision implements working in or close to the row. However, the nut-and-bolt work required to adjust most machines is frustrating and potentially interferes with timely use of the implement.

Guidance systems that allow controlled cultivation of small plants are also sorely needed. An important application of these will be for cultivation of crops planted in close rows (Olsen, 1995).

Finally, further research is needed on the integration of mechanical weed management with other ecological and chemical weed management tactics. The complexity of the potential interactions indicates that this work will need to extend beyond multifactorial empirical studies. A promising approach may be to link crop–weed competition models, which typically take weed density as an external forcing function, with weed population models, which focus on weed density, but generally take a simplified approach to competition. Despite the long history of tillage and cultivation, much biological and agronomic research is still needed to achieve the full potential of this ancient approach to weed management.

REFERENCES

Alström, S. (1990). *Fundamentals of Weed Management in Hot Climate Peasant Agriculture*. Crop Production Science no. 11. Uppsala, Sweden: Swedish University of Agricultural Sciences.

Andraski, B. J., Mueller, D. H., & Daniel, T. C. (1985). Effects of tillage and rainfall simulation date on water and soil losses. *Soil Science Society of America Journal*, **49**, 1512–7.

Ascard, J. (1990). Thermal weed control in onions. *Veröffentlichungen der Budesanstalt für Agrarbiologie Linz/Donau*, **20**, 175–88.

Ascard, J. (1994a). Soil cultivation in darkness reduced weed emergence. *Acta Horticulturae*, **372**, 167–77.

Ascard, J. (1994b). Dose–response models for flame weeding in relation to plant size and density. *Weed Research*, **34**, 377–85.

Ascard, J. (1995a). Thermal weed control by flaming: biological and technical aspects. *Institutionen för Lantbruksteknik (Swedish University of Agricultural Sciences)*, Alnarp, Report, **200**, 1–61.

Ascard, J. (1995b). Effects of flame weeding on weed species at different developmental stages. *Weed Research*, **35**, 397–411.

Bakke, A. I., Gaessler, W. G., Pultz, L. M., & Salmon, S. C. (1944). Relation of cultivation to depletion of root reserves in European bindweed at different soil horizons. *Journal of Agricultural Research (Washington)*, **69**, 137–47.

Balsari, P., Berruto, R., & Ferrero, A. (1994). Flame weed control in lettuce crop. *Acta Horticulturae*, **372**, 213–22.

Banting, J. D. (1966). Studies on the persistence of *Avena fatua*. *Canadian Journal of Plant Science*, **46**, 129–40.

Barker, A. V., & Craker, L. E. (1991). Inhibition of weed seed germination by microwaves. *Agronomy Journal*, **83**, 302–5.

Bates, G. H. (1948). *Weed Control*. London: E. & F. N. Spon.
Bauder, J. W., Randall, G. W. & Swann, J. B. (1981). Effect of four continuous tillage systems on mechanical impedance of a clay loam soil. *Soil Science Society of America Journal*, **45**, 802–6.
Baumann, D. T. (1992). Mechanical weed control with spring tine harrows (weed harrows) in row crops. In *9th International Symposium on the Biology of Weeds*, 16–18 September 1992, Dijon, France, pp. 123–8. Paris: European Weed Research Society.
Baumann, D. T. & Slembrouck, I. (1994). Mechanical and integrated weed control systems in row crops. *Acta Horticulturae*, **372**, 245–52.
Bender, J. (1994). *Future Harvest: Pesticide Free Farming*. Lincoln, NE: University of Nebraska Press.
Bibbey, R. O. (1935). The influence of environment upon the germination of weed seeds. *Scientific Agriculture*, **16**, 141–50.
Bicki, T. J., Wax, L. M. & Sipp, S. K. (1991). Evaluation of reduced herbicide application strategies for weed control in coarse-textured soils. *Journal of Production Agriculture*, **4**, 516–19.
Boerboom, C. M. & Young, F. L. (1995). Effect of postplant tillage and crop density on broadleaf weed control in dry pea (*Pisum sativum*) and lentil (*Lens culinaris*). *Weed Technology*, **9**, 99–106.
Botto, J. F., Scopel, A. L., Ballaré, C. L. & Sánchez, R. A. (1998). The effect of light during and after soil cultivation with different tillage implements on weed seedling emergence. *Weed Science*, **46**, 351–7.
Bourdôt, G. W. (1984). Regeneration of yarrow (*Achillea millefolium* L.) rhizome fragments of different length from various depths in the soil. *Weed Research*, **24**, 421–9.
Bourdôt, G. W. & Butler, J. H. B. (1985). Control of *Achillea millefolium* L. (yarrow) by rotary cultivation and glyphosate. *Weed Research*, **25**, 251–8.
Bourdôt, G. W., Field, R. J. & White, J. G. H. (1982). Yarrow: numbers of buds in the soil and their activity on rhizome fragments of varying length. *New Zealand Journal of Experimental Agriculture*, **10**, 63–7.
Bowman, G. (1991). Killer cultivators: guidance systems worthy of NASA make weeding faster, easier. *The New Farm*, **13**(3), 26–35.
Bowman, G. (ed.) (1997). *Steel in the Field: A Farmer's Guide to Weed Management Tools*. Beltsville, MD: Sustainable Agriculture Network.
Bridgemohan, P. & Brathwaite, R. A. I. (1989). Weed management strategies for the control of *Rottboellia cochinchinensis* in maize in Trinidad. *Weed Research*, **29**, 433–40.
Briggle, L. W. (1980). Introduction to energy use in wheat production. In *Handbook of Energy Utilization in Agriculture*, ed. D. Pimentel, pp. 109–16. Boca Raton, FL: CRC Press.
Buckingham, F. & Pauli, A. W. (1993). *Tillage*, 3rd edn. Moline, IL: Deere & Co.
Buhler, D. D. (1995). Influence of tillage systems on weed population dynamics and management in corn and soybean in the central USA. *Crop Science*, **35**, 1247–58.
Buhler, D. D. (1997). Effects of tillage and light environment on emergence of 13 annual weeds. *Weed Technology*, **11**, 496–501.
Buhler, D. D. (1998). Effect of ridge truncation on weed populations and control in ridge-tillage corn (*Zea mays*). *Weed Science*, **46**, 225–30.
Buhler, D. D., & Daniel, T. C. (1988). Influence of tillage systems on giant foxtail, *Setaria*

faberi, and velvetleaf, *Abutilon theophrasti*, density and control in corn, *Zea mays*. *Weed Science*, **36**, 642–7.

Buhler, D. D., & Gunsolus, J. L. (1996). Effect of date of preplant tillage and planting on weed populations and mechanical weed control in soybean (*Glycine max*). *Weed Science*, **44**, 373–9.

Buhler, D. D., & Mester, T. C. (1991). Effect of tillage systems on the emergence depth of giant (*Setaria faberi*) and green foxtail (*Setaria viridis*). *Weed Science*, **39**, 200–3.

Buhler, D. D., & Oplinger, E. S. (1990). Influence of tillage systems on annual weed density and control in solid-seeded soybean (*Glycine max*). *Weed Science*, **38**, 158–65.

Buhler, D. D., Gunsolus, J. L., & Ralston, D. F. (1992). Integrated weed management techniques to reduce herbicide inputs in soybean. *Agronomy Journal*, **84**, 973–8.

Buhler, D. D., Gunsolus, J. L., & Ralston, D. F. (1993). Common cocklebur (*Xanthium strumarium*) control in soybean (*Glycine max*) with reduced bentazon rates and cultivation. *Weed Science*, **41**, 447–53.

Buhler, D. D., Doll, J. D., Proost, R. T., & Visocky, M. R. (1994). Interrow cultivation to reduce herbicide use in corn following alfalfa without tillage. *Agronomy Journal*, **86**, 66–72.

Burwell, R. E., & Larson, W. E. (1969). Infiltration as influenced by tillage-induced random roughness and pore space. *Soil Science Society of America Proceedings*, **33**, 449–52.

Cacek, T. (1984). Organic farming: the other conservation farming system. *Journal of Soil and Water Conservation*, **39**, 357–60.

Cardina, J., Regnier, E., & Harrison, K. (1991). Long-term tillage effects on seed banks in three Ohio soils. *Weed Science*, **39**, 186–94.

Carter, P. R., & Barnett, K. H. (1987). Corn-hybrid performance under conventional and no-tillage systems after thinning. *Agronomy Journal*, **79**, 919–26.

Cavers, P. B., & Kane, M. (1990). Responses of proso millet (*Panicum miliaceum*) seedlings to mechanical damage and/or drought treatments. *Weed Technology*, **4**, 425–32.

Chancellor, R. J. (1964a). Emergence of weed seedlings in the field and the effects of different frequencies of cultivation. In *Proceedings of the 7th British Weed Control Conference*, pp. 599–606. London: British Crop Protection Council.

Chancellor, J. J. (1964b). The depth of weed seed germination in the field. In *Proceedings of the 7th British Weed Control Conference*, pp. 607–13. London: British Crop Protection Council.

Christensen, S., Rasmussen, G., & Olesen, J. E. (1994). Differential weed suppression and weed control in winter wheat. *Aspects of Applied Biology*, **40**, 335–42.

Clarke, J. H., & Moss, S. R. (1991). The occurrence of herbicide resistant *Alopecurus myosuroides* (blackgrass) in the United Kingdom and strategies for its control. In *Brighton Crop Protection Conference: Weeds*, pp. 1041–8. Farnham, UK: British Crop Protection Council.

Clements, D. R., Weise, S. F., Brown, R., Stonehouse, D. P., Hume, D. J., & Swanton, C. J. (1995). Energy analysis of tillage and herbicide inputs in alternative weed management systems. *Agriculture, Ecosystems and Environment*, **52**, 119–28.

Clements, D. R., Benoit, D. L., Murphy, S. D., & Swanton, C. J. (1996). Tillage effects on weed seed return and seedbank composition. *Weed Science*, **44**, 314–22.

Conn, J. S. (1987). Effects of tillage and cropping sequence on Alaskan weed vegetation: studies on land under cultivation for eleven years. *Soil and Tillage Research*, **9**, 265–74.

Coolman, R. M., & Hoyt, G. D. (1993). The effects of reduced tillage on the soil environment. *HortTechnology*, **3**, 143–5.

Cosser, N. D., Gooding, M. J., Thompson, A. J., & Froud-Williams, R. J. (1997). Competitive ability and tolerance of organically grown wheat cultivars to natural weed infestations. *Annals of Applied Biology*, **130**, 523–35.

Cousens, R., & Moss, S. R. (1990). A model of the effects of cultivations on the vertical distribution of weed seeds within the soil. *Weed Research*, **30**, 61–70.

Cox, W. J., Zobel, R. W., van Es, H. M., & Otis, D. J. (1990). Tillage effects on some soil physical and corn physiological characteristics. *Agronomy Journal*, **82**, 806–12.

Cox, W. J., Otis, D. J., van Es, H. M., Gaffney, F. B., Snyder, D. P., Reynolds, K. R., & van der Grinten, M. (1992). Feasibility of no-tillage and ridge tillage systems in the Northeastern USA. *Journal of Production Agriculture*, **5**, 111–17.

Cramer, C. (1988). 5 'better ideas' for low-cost weed control. *The New Farm*, **10**(3), 16–19.

Cramer, C., Bowman, G., Brusko, M., Cicero, K., Hofstetter, B., & Shirley, C. (1991). *Controlling Weeds with Fewer Chemicals*. Emmaus, PA.: Rodale Institute.

Cussans, G. W. (1975). Weed control in reduced cultivation and direct drilling systems. *Outlook for Agriculture*, **8**, 240–2.

Cussans, G. W. (1976). The influence of changing husbandry on weeds and weed control in arable crops. In *Proceedings of the 1976 British Crop Protection Conference: Weeds*, pp. 1001–8. Croydon, UK: British Crop Protection Council.

Cussans, G. W., Raudonius, S., Brain, P., & Cumberworth, S. (1996). Effects of depth of seed burial and soil aggregate size on seedling emergence of *Alopecurus myosuroides*, *Galium aparine*, *Stellaria media* and wheat. *Weed Research*, **36**, 133–41.

Darwent, A. L., & Smith, J. H. (1985). Delayed seeding for wild oat control in rapeseed in northwest Alberta. *Canadian Journal of Plant Science*, **65**, 1101–6.

Dawson, J. H., & Bruns, V. F. (1975). Longevity of barnyard grass, green foxtail, and yellow foxtail seeds in soil. *Weed Science*, **23**, 437–40.

Deschenes, J.-M., & Dubuc, J.-P. (1981). Effets de l'humidité du sol, ses dates de semis et des mauvaises herbes sur le rendement des céréales. *Canadian Journal of Plant Science*, **61**, 851–7.

Dickey, E. C., Shelton, D. P., Jasa, P. J., & Peterson, T. R. (1984). Tillage, residue and erosion on moderately sloping soils. *Transactions of the American Society of Agricultural Engineers*, **27**, 1093–9.

Dieleman, J. A., Mortensen, D. A., & Martin, A. R. (1999). Influence of velvetleaf (*Abutilon theophrasti*) and common sunflower (*Helianthus annuus*) density variation on weed management outcomes. *Weed Science*, **47**, 81–9.

Diprose, M. F., Fletcher, R., Longden, P. C., & Champion, M. J. (1985). Use of electricity to control bolters in sugar beet (*Beta vulgaris* L.): a comparison of the electrothermal with chemical and mechanical cutting methods. *Weed Research*, **25**, 53–60.

Doyle, C. J., Cousens, R., & Moss, S. R. (1986). A model of the economics of controlling *Alopecurus myosuroides* Huds. in winter wheat. *Crop Protection*, **5**, 143–50.

Eadie, A. G., Swanton, C. J., Shaw, J. E., & Anderson, G. W. (1992). Banded herbicide applications and cultivation in a modified no-till corn (*Zea mays*) system. *Weed Technology*, **6**, 535–42.

Eberlein, C. V., Patterson, P. E., Guttieri, M. J., & Stark, J. C. (1997). Efficacy and economics of cultivation for weed control in potato (*Solanum tuberosum*). *Weed Technology*, **11**, 257–64.

Egley, G. H., & Williams, R. D. (1990). Decline of weed seeds and seedling emergence over five years as affected by soil disturbances. *Weed Science*, **38**, 504–10.

Ellwanger, T. C. Jr., Bingham, S. W., & Chapell, W. E. (1973). Physiological effects of ultra-high temperatures on corn. *Weed Science*, **21**, 296–9.

Ellwanger, T. C. Jr., Bingham, S. W., Chapell, W. E., & Tolin, S. A. (1973). Cytological effects of ultra-high temperatures on corn. *Weed Science*, **21**, 299–303.

Exner, D. N., Thompson, R. L., & Thompson, S. N. (1996). Practical experience and on-farm research with weed management in an Iowa ridge tillage-based system. *Journal of Production Agriculture*, **9**, 496–500.

Fail, H. (1956). The effect of rotary cultivation on the rhizomatous weeds. *Journal of Agricultural Engineering Research*, **1**, 68–80.

Fausey, J. C., & Renner, K. A. (1997). Germination, emergence, and growth of giant foxtail (*Setaria faberi*) and fall panicum (*Panicum dichotomiflorum*). *Weed Science*, **45**, 423–5.

Fay, P. K., & Olson, W. A. (1978). Technique for separating weed seed from soil. *Weed Science*, **26**, 530–3.

Fergedal, S. (1993). Weed control by freezing with liquid nitrogen and carbon dioxide snow: a comparison between flaming and freezing. In *Non-Chemical Weed Control, Communications of the 4th International Conference of the International Federation of Organic Agricultural Movements*, Dijon, France, ed. J. M. Thomas, pp. 153–6. [No city]: International Federation of Organic Agricultural Movements.

Field, R. J., & Jayaweera, C. S. (1985). Regeneration of yarrow (*Achillea millefolium* L.) rhizomes as influenced by rhizome age, fragmentation and depth of soil burial. *Plant Protection Quarterly*, **1**, 71–3.

Fogelberg, F., & Dock Gustavsson, A.-M. (1998). Resistance against uprooting in carrots (*Daucus carota*) and annual weeds: a basis for selective mechanical weed control. *Weed Research*, **38**, 183–90.

Fogelberg, F., & Dock Gustavsson, A.-M. (1999). Mechanical damage to annual weeds and carrots by in-row brush weeding. *Weed Research*, **39**, 469–79.

Forcella, F. (1993). Seedling emergence model for velvetleaf. *Agronomy Journal*, **85**, 929–33.

Forcella, F. (1998). Real-time assessment of seed dormancy and seedling growth for weed management. *Seed Science Research*, **8**, 201–9.

Forcella, F., Eradat-Oskoui, K., & Wagner, S. W. (1993). Application of weed seedbank ecology to low-input crop management. *Ecological Applications*, **3**, 74–83.

Forcella, F., & Lindstrom, M. J. (1988a). Movement and germination of weed seeds in ridge-till crop production systems. *Weed Science*, **36**, 56–9.

Forcella, F., & Lindstrom, M. J. (1988b). Weed seed populations in ridge and conventional tillage. *Weed Science*, **36**, 500–3.

Foster, L. (1989). The biology and non-chemical control of dock species *Rumex obtusifolius* and *R. crispus*. *Biological Agriculture and Horticulture*, **6**, 11–25.

Froud-Williams, R. J. (1983). The influence of straw disposal and cultivation regime on the population dynamics of *Bromus sterilis*. *Annals of Applied Biology*, **103**, 139–48.

Froud-Williams, R. J. (1987). Survival and fate of weed seed populations: interactions with cultural practice. In *Proceedings of the 1987 British Crop Protection Conference: Weeds*, pp. 707–18. Croydon, UK: British Crop Protection Council.

Froud-Williams, R. J., Chancellor, R. J., & Drennan, D. S. H. (1981). Potential changes in weed floras associated with reduced-cultivation systems for cereal production in temperate regions. *Weed Research*, **21**, 99–109.

Froud-Williams, R. J., Chancellor, R. J., & Drennan, D. S. H. (1983). Influence of cultivation regime upon buried weed seeds in arable cropping systems. *Journal of Applied Ecology*, **20**, 199–208.

Froud-Williams, R. J., Chancellor, R. J., & Drennan, D. S. H. (1984). The effects of seed burial and soil disturbance on emergence and survival of arable weeds in relation to minimal cultivation. *Journal of Applied Ecology*, **21**, 629–41.

Froud-Williams, R. J., Drennan, D. S. H., & Chancellor, R. J. (1983). Influence of cultivation regime on weed floras of arable cropping systems. *Journal of Applied Ecology*, **20**, 187–97.

Gallagher, R. S., & Cardina, J. (1998). The effect of light environment during tillage on the recruitment of various summer annuals. *Weed Science*, **46**, 214–16.

Gebhardt, M. R., Daniel, T. C., Schweizer, E. E., & Allmaras, R. R. (1985). Conservation tillage. *Science*, **230**, 625–30.

Geier, V., & Vogtmann, H. (1987). The multiple row brush hoe – a new tool for mechanical weed control. *Bulletin for Organic Agriculture*, **1**, 4–6.

Geier, V., & Vogtmann, H. (1988). Weed control without herbicides in corn crops. In *Global Perspectives on Agroecology and Sustainable Agricultural Systems, Proceedings of the 6th International Conference of the International Federation of Organic Agricultural Movements*, vol. 2, eds. P. Allen & D. van Dusen, pp. 483–9. Santa Cruz, CA: Agroecology Program, University of California at Santa Cruz.

Ghafar, Z., & Watson, A. K. (1983). Effect of corn (*Zea mays*) seeding date on the growth of yellow nutsedge (*Cyperus esculentus*). *Weed Science*, **31**, 572–5.

Glaz, B., Ulloa, M. F., & Parrado, R. (1989). Cultivation, cultivar, and crop age effects on sugarcane. *Agronomy Journal*, **81**, 163–7.

Green, M. B. (1987). Energy in pesticide manufacture, distribution and use. In *Energy in Plant Nutrition and Pest Control, Energy in World Agriculture*, vol. 2, ed. Z. R. Helsel, pp. 165–77. Amsterdam, Netherlands: Elsevier.

Griffith, D. R., Kladivko, E. J., Mannering, J. V., West, T. D., & Parsons, S. D. (1988). Long-term tillage and rotation effects on corn growth and yield on high and low organic matter, poorly drained soils. *Agronomy Journal*, **80**, 599–605.

Grubinger, V., & Else, M. J. (1996). *Vegetable Farmers and their Weed-Control Machines*. Video, 76 min. Burlington, VT: University of Vermont Extension System.

Grundy, A. C., & Mead, A. (1998). Modelling the effect of seed depth on weed seedling emergence. In *Weed Seedbanks: Determination, Dynamics and Manipulation*, ed. G. T. Chapman, A. C. Grundy, N. E. Jones, E. J. P. Marshall & R. J. Froud-Williams, pp. 75–82. Wellesbourne, UK: Association of Applied Biologists.

Grundy, A. C., Mead, A., & Burston, S. (1999). Modelling the effect of cultivation on seed movement with application to the prediction of weed seedling emergence. *Journal of Applied Ecology*, **36**, 663–78.

Gunsolus, J. L. (1990). Mechanical and cultural weed control in corn and soybeans. *American Journal of Alternative Agriculture*, **5**, 114–19.

Håkansson, S. (1963). *Allium vineale* L. as a weed, with special reference to the conditions in south-eastern Sweden. *Växtodling (Plant Husbandry)*, **19**, 1–208.

Håkansson, S. (1967). Experiments with *Agropyron repens* (L.) Beauv. I. Development and growth, and the response to burial at different developmental stages. *Lantbrukshögskolans Annaler*, **33**, 823–74.

Håkansson, S. (1968a). Experiments with *Agropyron repens* (L.) Beauv. II. Production from rhizome pieces of different sizes and from seeds. Various environmental conditions compared. *Lantbrukshögskolans Annaler*, **34**, 3–29.

Håkansson, S. (1968*b*). Experiments with *Agropyron repens* (L.) Beauv. III. Production of aerial and underground shoots after planting rhizome pieces of different lengths at varying depths. *Lantbrukshögskolans Annaler*, **34**, 31–51.

Håkansson, S. (1969*a*). Experiments with *Sonchus arvensis* L. I. Development and growth, and the response to burial and defoliation in different developmental stages. *Lantbrukshögskolans Annaler*, **35**, 999–1030.

Håkansson, S. (1969*b*). Experiments with *Agropyron repens* (L.) Beauv. IV. Response to burial and defoliation repeated with different intervals. *Lantbrukshögskolans Annaler*, **35**, 61–78.

Håkansson, S. (1969*c*). Experiments with *Agropyron repens* (L.) Beauv. VI. Rhizome orientation and life length of broken rhizomes in the soil, and reproductive capacity of different underground shoot parts. *Lantbrukshögskolans Annaler*, **35**, 869–94.

Håkansson, S. (1971). Experiments with *Agropyron repens* (L.) Beauv. X. Individual and combined effects of division and burial of the rhizomes and competition from a crop. *Swedish Journal of Agricultural Research*, **1**, 239–46.

Håkansson, S. (1982). Multiplication, growth and persistence of perennial weeds. In *Biology and Ecology of Weeds*, ed. W. Holzner & N. Numata, pp. 123–35. The Hague, Netherlands: Dr. W. Junk.

Håkansson, S., & Wallgren, B. (1972*a*). Experiments with *Sonchus arvensis* L. II. Reproduction, plant development and response to mechanical disturbance. *Swedish Journal of Agricultural Research*, **2**, 3–14.

Håkansson, S., & Wallgren, B. (1972*b*). Experiments with *Sonchus arvensis* L. III. The development from reproductive roots cut into different lengths and planted at different depths, with and without competition from barley. *Swedish Journal of Agricultural Research*, **2**, 15–26.

Håkansson, S., & Wallgren, B. (1976). *Agropyron repens* (L.) Beauv., *Holcus mollis* L. and *Agrostis gigantea* Roth as weeds – some properties. *Swedish Journal of Agricultural Research*, **6**, 109–20.

Hartmann, K. M., & Nezadal, W. (1990). Photocontrol of weeds without herbicides. *Naturwissenschaften*, **77**, 158–63.

Hartzler, R. G., van Kooten, B. D., Stoltenberg, D. E., Hall, E. M., & Fawcett, R. S. (1993). On-farm evaluation of mechanical and chemical weed management practices in corn (*Zea mays*). *Weed Technology*, **7**, 1001–4.

Harvey, S. J., & Forcella, F. (1993). Vernal seedling emergence model for common lambsquarters (*Chenopodium album*). *Weed Science*, **41**, 309–16.

Hauser, E. W., Cecil, S. R., & Dowler, C. C. (1973). Systems of weed control for peanuts. *Weed Science*, **21**, 176–80.

Hauser, E. W., Dowler, C. C., Jellum, M. D., & Cecil, S. R. (1974). Effects of herbicide–crop rotation on nutsedge, annual weeds, and crops. *Weed Science*, **22**, 172–6.

Heatherly, L. G., Wesley, R. A., Elmore, C. D., & Spurlock, S. R. (1993). Net returns from stale seedbed plantings of soybean (*Glycine max*) on clay soil. *Weed Technology*, **7**, 972–80.

Hodgson, J. M. (1970). The response of Canada thistle ecotypes to 2,4-D, amitrole, and intensive cultivation. *Weed Science*, **18**, 253–5.

Holm, L. G., Plucknett, D. L., Pancho, J. V., & Herberger, J. P. (1977). *The World's Worst Weeds*. Honolulu, HI: University Press of Hawaii.

Holmøy, R., & Netland, J. (1994). Band spraying, selective flame weeding and hoeing in late white cabbage. Part I. *Acta Horticulturae*, 372, 223–34.

Horak, M. J., & Sweat, J. K. (1994). Germination, emergence, and seedling establishment of buffalo gourd (*Cucurbita foetidissima*). *Weed Science*, 42, 358–63.

How, R. B. (1980). Cabbage. In *Handbook of Energy Utilization in Agriculture*, ed. D. Pimentel, pp. 181–4. Boca Raton, FL: CRC Press.

Hulburt, W. C., & Menzel, R. G. (1953). Soil mixing characteristics of tillage implements. *Agricultural Engineering*, 34, 702–8.

Hughes, R. G., & Roebuck, J. F. (1970). Comparison of systems of perennial grass weed control in spring barley. In *Proceedings of the 10th British Weed Control Conference*, pp. 105–10. London, UK: British Crop Protection Council.

Intermediate Technology Publications. (1985). *Tools for Agriculture: A Buyer's Guide to Appropriate Equipment*, 3rd edn. London: Intermediate Technology Publications.

Ivany, J. A. (1997). Effect of rhizome depth in soil on emergence and growth of field mint (*Mentha arvensis*). *Weed Technology*, 11, 149–51.

Jackson, M. (1988). Amish agriculture and no-till: the hazards of applying the USLE to unusual farms. *Journal of Soil and Water Conservation*, 43, 483–6.

Jensen, P. K. (1995). Effect of light environment during soil disturbance on germination and emergence pattern of weeds. *Annals of Applied Biology*, 127, 561–71.

Johnson, D. W., Krall, J. M., Delaney, R. H., & Pochop, L. O. (1989). Response of monocot and dicot weed species to Fresnel lens concentrated solar radiation. *Weed Science*, 37, 797–801.

Johnson, M. D., & Lowery, B. (1985). Effect of three conservation tillage practices on soil temperature and thermal properties. *Soil Science Society of America Journal*, 49, 1547–52.

Johnson, R. R. (1985). A new look at cultivation. *Crops and Soils*, 37, 13–16.

Johnson, W. C. III, & Mullinix, B. G. Jr. (1995). Weed management in peanut using stale seedbed techniques. *Weed Science*, 43, 293–7.

Johnson, W. C. III, & Mullinix, B. G. Jr. (1998). Stale seedbed weed control in cucumber. *Weed Science*, 46, 698–702.

Jones, R. (1966). Effect of seed-bed preparation on the weed flora of spring barley. In *Proceedings of the 8th British Weed Control Conference*, pp. 227–8. London: British Crop Protection Council.

Jordan, N. (1993). Simulation analysis of weed population dynamics in ridge-tilled fields. *Weed Science*, 41, 468–74.

King, C. A., & Oliver, L. R. (1994). A model for predicting large crabgrass (*Digitaria sanguinalis*) emergence as influenced by temperature and water potential. *Weed Science*, 42, 561–7.

Koscelny, J. A., Peeper, T. E., Solie, J. B., & Solomon, S. G. Jr. (1991). Seeding date, seeding rate, and row spacing affect wheat (*Triticum aestivum*) and cheat (*Bromus secalinus*). *Weed Technology*, 5, 707–12.

Koskinen, W. C., & McWhorter, C. G. (1986). Weed control in conservation tillage. *Journal of Soil and Water Conservation*, 41, 365–70.

Kouwenhoven, J. K., & Terpstra, R. (1979). Sorting action of tines and tine-like tools in the field. *Journal of Agricultural Engineering Research*, 24, 95–113.

Kouwenhoven, J. K., Wevers, J. D. A., & Post, B. J. (1991). Possibilities of mechanical post-emergence weed control in sugar beet. *Soil and Tillage Research*, **21**, 85–95.

Kress, W. (1993). *Successful Weed Control with Flexible and Hoeing Harrows*. Gesellschaft für Boden, Technik, Qualität Bundesverband für Ökologie in Land- und Gartenbau, Merkblatt 6.1.

Langdale, G. W., Alberts, E. E., Bruce, R. R., Edwards, W. M., & McGregor, K. C. (1994). Concepts of residue management: infiltration, runoff, and erosion. In *Crops Residue Management*, ed. J. L. Hatfield & B. A. Stewart, pp. 109–24. Boca Raton, FL: CRC Press.

Lanie, A. J., Griffin, J. L., Reynolds, D. B., & Vidrine, P. R. (1993). Influence of residual herbicides on rate of paraquat and glyphosate in stale seedbed soybean (*Glycine max*). *Weed Technology*, **7**, 960–5.

Lanie, A. J., Griffin, J. L., Vidrine, P. R., & Reynolds, D. B. (1994). Weed control with non-selective herbicides in soybean (*Glycine max*) stale seedbed culture. *Weed Technology*, **8**, 159–64.

Law, R., Bradshaw, A. D., & Putwain, P. D. (1977). Life-history variation in *Poa annua*. *Evolution*, **31**, 233–46.

Légère, A. (1997). Cereal planting dates as a tool in the management of *Galeopsis tetrahit* and associated weed species in spring barley and oat. *Crop Protection*, **16**, 117–25.

Lovely, W. G., Weber, C. R., & Staniforth, D. W. (1958). Effectiveness of the rotary hoe for weed control in soybeans. *Agronomy Journal*, **50**, 621–5.

Lueschen, W. E., & Andersen, R. N. (1980). Longevity of velvetleaf (*Abutilon theophrasti*) seeds in soil under agricultural practices. *Weed Science*, **28**, 341–6.

Lym, R. G., & Messersmith, C. G. (1993). Fall cultivation and fertilization to reduce winterhardiness of leafy spurge (*Euphorbia esula*). *Weed Science*, **41**, 441–6.

MacDonald, G. E., Brecke, B. J., & Shilling, D. G. (1992). Factors affecting germination of dogfennel (*Eupatorium capillifolium*) and yankeeweed (*Eupatorium compositifolium*). *Weed Science*, **40**, 424–8.

Majek, B. A., Erickson, C., & Duke, W. B. (1984). Tillage effects and environmental influences on quackgrass (*Agropyron repens*) rhizome growth. *Weed Science*, **32**, 376–81.

Mattsson, B., Nylander, C., & Ascard, J. (1990). Comparison of seven inter-row weeders. *Veröffentlichungen der Bundesanstalt für Agrarbiologie, Linz/Donau*, **20**, 91–107.

Melander, B. (1997). Optimization of the adjustment of a vertical axis rotary brush weeder for intra-row weed control in row crops. *Journal of Agricultural Engineering Research*, **68**, 39–50.

Melander, B. (1998). Interactions between soil cultivation in darkness, flaming and brush weeding when used for in-row weed control in vegetables. *Biological Agriculture and Horticulture*, **16**, 1–14.

Melander, B., & Hartvig, P. (1997). Yield responses of weed-free seeded onions [*Allium cepa* (L.)] to hoeing close to the row. *Crop Protection*, **16**, 687–91.

Mohler, C. L. (1993). A model of the effects of tillage on emergence of weed seedlings. *Ecological Applications*, **3**, 53–73.

Mohler, C. L. (1999). Effects of planting depth and simulated crop residue on seed survival of three annual broadleaf weeds. *Weed Science Society of America Abstracts*, **39**, 27–8.

Mohler, C. L., & Callaway, M. B. (1992). Effects of tillage and mulch on the emergence and survival of weeds in sweet corn. *Journal of Applied Ecology*, **29**, 21–34.

Mohler, C. L., & Frisch, J. C. (1997). Mechanical weed control in oats with a rotary hoe and tine weeder. *Proceedings of the Northeastern Weed Science Society*, **51**, 2–6.

Mohler, C. L., & Galford, A. E. (1997). Weed seedling emergence and seed survival: separating the effects of seed position and soil modification by tillage. *Weed Research*, **37**, 147–55.

Mohler, C. L., Frisch, J. C., & Mt. Pleasant, J. (1997). Evaluation of mechanical weed management programs for corn (*Zea mays*). *Weed Technology*, **11**, 123–31.

Morton, C. T., & Buchele, W. F. (1960). Emergence energy of plant seedlings. *Agricultural Engineering*, **41**, 428–31, 453–5.

Moss, S. R. (1985*a*). The effect of drilling date, pre-drilling cultivations and herbicides on *Alopecurus myosuroides* (black-grass) populations in winter cereals. *Aspects of Applied Biology*, **9**, 31–9.

Moss, S. R. (1985*b*). The survival of *Alopecurus myosuroides* Huds. seeds in soil. *Weed Research*, **25**, 201–11.

Moss, S. R. (1987). Influence of tillage, straw disposal system and seed return on the population dynamics of *Alopecurus myosuroides* Huds. in winter wheat. *Weed Research*, **27**, 313–20.

Moss, S. R. (1988). Influence of cultivation on the vertical distribution of weed seeds in the soil. In *8ième Colloque International sur la Biologie, L'Ecologie et la Systematique des Mauvaises Herbes*, pp. 71–80. Dijon, France: European Weed Research Society.

Mt. Pleasant, J., & Burt, R. F. (1994). Time of cultivation in corn: effects on weed levels and grain yields. *Proceedings of the Northeastern Weed Science Society*, **48**, 66.

Mt. Pleasant, J., Burt, R. F., & Frisch, J. C. (1994). Integrating mechanical and chemical weed management in corn (*Zea mays*). *Weed Technology*, **8**, 217–23.

Mulder, T. A., & Doll, J. D. (1993). Integrating reduced herbicide use with mechanical weeding in corn (*Zea mays*). *Weed Technology*, **7**, 382–9.

Mulder, T. A., & Doll, J. D. (1994). Reduced input corn weed control: the effects of planting date, early season weed control, and row-crop cultivator selection. *Journal of Production Agriculture*, **7**, 256–60.

Mulugeta, D., & Stoltenberg, D. E. (1997). Weed and seedbank management with integrated methods as influenced by tillage. *Weed Science*, **45**, 706–15.

Netland, J., Balvoll, G., & Holmøy, R. (1994). Band spraying, selective flame weeding and hoeing in late white cabbage. Part II. *Acta Horticulturae*, **372**, 235–43.

Newsom, L. J., & Shaw, D. R. (1996). Cultivation enhances weed control in soybean (*Glycine max*) with AC 263,222. *Weed Technology*, **10**, 502–7.

Nichols, M. L., & Reed, I. F. (1934). Soil dynamics. VI. Physical reactions of soil to moldboard surfaces. *Agricultural Engineering*, **15**, 187–90.

Oliver, L. R., Klingaman, T. E., McClelland, M., & Bozsa, R. C. (1993). Herbicide systems in stale seedbed soybean (*Glycine max*) production. *Weed Technology*, **7**, 816–23.

Olsen, H. J. (1995). Determination of row position in small-grain crops by analysis of video images. *Computers and Electronics in Agriculture*, **12**, 147–62.

Parish, S. (1987). Weed control ideas from Europe visit. *New Farmer and Grower*, **16**, 8–12.

Parish, S. (1989). Weed control: testing the effects of infrared radiation. *Agricultural Engineer*, **44**(3), 53–5.

Parish, S. (1990). A procedure for assessing flame treatments under controlled conditions. *Veröffentlichungen der Bundesanstalt für Agrarbiologie, Linz/Donau*, **20**, 189–96.

Parish, R. L., Reynolds, D. B., & Crawford, S. H. (1995). Precision-guided cultivation techniques to reduce herbicide inputs in cotton. *Applied Engineering in Agriculture*, **11**, 349–53.

Parks, R. J., Curran, W. S., Roth, G. W., Hartwig, N. L., & Calvin, D. D. (1995). Common lambsquarters (*Chenopodium album*) control in corn (*Zea mays*) with postemergence herbicides and cultivation. *Weed Technology*, **9**, 728–35.

Pawlowski, F., & Malicki, L. (1968). Effect of different methods of ploughing on vertical distribution of weed seeds in soil formed of loess. *Annales Universitatis Mariae Curie-Sklodowska Lublin-Polonia*, **23**, 161–74.

Pedersen, B. T. (1990). Test of the multiple row brush hoe. *Veröffentlichungen der Bundesanstalt für Agrarbiologie, Linz/Donau*, **20**, 109–25.

Peruzzi, A., Silvestri, N., Gini, N. & Coli, A. (1993). Weed control of winter cereals by means of weeding harrows: first experimental results. *Agricoltura Mediterranea*, **123**, 236–42.

Phillips, M. C. (1993). Use of tillage to control *Cynodon dactylon* under small-scale farming conditions. *Crop Protection*, **12**, 267–72.

Pimentel, D. (1980). Energy inputs for the production, formulation, packaging, and transport of various pesticides. In *Handbook of Energy Utilization in Agriculture*, ed. D. Pimentel, pp. 45–8. Boca Raton, FL: CRC Press.

Pimentel, D., & Burgess, M. (1980). Energy inputs in corn production. In *Handbook of Energy Utilization in Agriculture*, ed. D. Pimentel, pp. 67–84. Boca Raton, FL: CRC Press.

Pollard, F., & Cussans, G. W. (1976). The influence of tillage on the weed flora of four sites sown to successive crops of spring barley. In *Proceedings of the 1976 British Crop Protection Conference: Weeds*, pp. 1019–28. Croydon, UK: British Crop Protection Council.

Pollard, F., & Cussans, G. W. (1981). The influence of tillage on the weed flora in a succession of winter cereal crops on a sandy loam soil. *Weed Research*, **21**, 185–90.

Pollard, F., Moss, S. R., Cussans, G. W., & Froud-Williams, R. J. (1982). The influence of tillage on the weed flora in a succession of winter wheat crops on a clay loam soil and a silt loam soil. *Weed Research*, **22**, 129–36.

Prihar, S. S., & Van Doren, D. M. Jr. (1967). Mode of response of weed-free corn to post-planting cultivation. *Agronomy Journal*, **59**, 513–16.

Prostko, E. P., Wu, H.-I., Chandler, J. M., & Senseman, S. A. (1997). Modeling weed emergence as influenced by burial depth using the Fermi–Dirac distribution function. *Weed Science*, **45**, 242–8.

Rahkonen, J., & Vanhala, P. (1993). Response of a mixed weed stand to flaming and use of temperature measurements in predicting weed control efficiency. In *Non-Chemical Weed Control, Communications of the 4th International Conference of the International Federation of Organic Agricultural Movements*, Dijon, France, ed. J. M. Thomas, pp. 173–81.

Randall, G. W. (1984). Role of crop nutrition technology in meeting future needs. In *Future Agricultural Technology and Resource Conservation*, ed. B. C. English, J. A. Maetzold, B. R. Holding & E. O. Heady, pp. 354–67. Ames, IA: Iowa State University Press.

Rasmussen, J. (1990). Selectivity – an important parameter on establishing the optimum harrowing technique for weed control in growing cereals. *Proceedings of the European Weed Research Society Symposium 1990, Integrated Weed Management in Cereals*, pp. 197–204. Paris: European Weed Research Society.

Rasmussen, J. (1991). A model for prediction of yield response in weed harrowing. *Weed Research*, **31**, 401–8.

Rasmussen, J. (1992a). Testing harrows for mechanical control of annual weeds in agricultural crops. *Weed Research*, **32**, 267–74.

Rasmussen, J. (1992b). Experimental approaches to mechanical weed control in field peas. In *Proceedings of the 9th International Symposium on the Biology of Weeds*, 16–18 September 1992, Dijon, France, pp. 129–38. Paris: European Weed Research Society.

Rasmussen, J. (1993). The influence of harrowing used for post-emergence weed control on the interference between crop and weeds. In *Proceedings of the European Weed Research Society Symposium, Quantitative Approaches in Weed and Herbicide Research and their Practical Application*, Braunschweig, 1993, pp. 209–17. Paris: European Weed Research Society.

Rasmussen, J., & Svenningsen, T. (1995). Selective weed harrowing in cereals. *Biological Agriculture and Horticulture*, **12**, 29–46.

Rasmusson, D. D., Dexter, A. G., & Warren, H. III. (1979). The use of electricity to control weeds. *Proceedings of the North Central Weed Control Conference*, **34**, 66.

Reese, C., & Forcella, F. (1997). *WeedCast User's Manual in MS Word 2.0 for Windows*. Internet: http://www.infolink.morris.mn.US/~lwink/products/weedcast.htm.

Reganold, J. P., Elliott, L. F., & Unger, Y. L. (1987). Long-term effects of organic and conventional farming on soil erosion. *Nature*, **330**, 370–2.

Roberts, H. A. (1962). Studies on the weeds of vegetable crops. II. Effect of six years of cropping on the weed seeds in the soil. *Journal of Ecology*, **50**, 803–13.

Roberts, H. A. (1963). Studies on the weeds of vegetable crops. III. Effect of different primary cultivations on the weed seeds in the soil. *Journal of Ecology*, **51**, 83–95.

Roberts, H. A., & Dawkins, P. A. (1967). Effect of cultivation on the numbers of viable weed seeds in soil. *Weed Research*, **7**, 290–301.

Roberts, H. A., & Feast, P. M. (1972). Fate of seeds of some annual weeds in different depths of cultivated and undisturbed soil. *Weed Research*, **12**, 316–24.

Roberts, H. A., & Feast, P. M. (1973a). Emergence and longevity of seeds of annual weeds in cultivated and undisturbed soil. *Journal of Applied Ecology*, **10**, 133–43.

Roberts, H. A., & Feast, P. M. (1973b). Changes in the numbers of viable weed seeds in soil under different regimes. *Weed Research*, **13**, 298–303.

Roberts, H. A., & Hewson, R. T. (1971). Herbicide performance and soil surface conditions. *Weed Research*, **11**, 69–73.

Roberts, H. A., & Stokes, F. G. (1965). Studies on the weeds of vegetable crops. V. Final observations on an experiment with different primary cultivations. *Journal of Applied Ecology*, **2**, 307–15.

Robinson, R. G. (1978). Control by tillage and persistence of volunteer sunflower and annual weeds. *Agronomy Journal*, **70**, 1053–6.

Robinson, R. G., & Dunham, R. S. (1956). Pre-planting tillage for weed control in soybeans. *Agronomy Journal*, **48**, 493–5.

Roman, E. S., Murphy, S. D., & Swanton, C. J. (1999). Effect of tillage and *Zea mays* on *Chenopodium album* seedling emergence and density. *Weed Science*, **47**, 551–6.

Röttele, M., & Koch, W. (1981). Verteilung von Unkrautsamen im Boden und Konsequenzen für die Bestimmung der Samendichte. *Zeitschrift für Pflanzenkrankheiten und Pflanzenschutz*, Supplement, **9**, 383–91.

Russel, E. W., & Mehta, N. P. (1938). Studies in soil cultivation. VIII. The influence of the seed bed on crop growth. *Journal of Agricultural Science*, **28**, 272–98.

Russel, W. J., Fehr, W. R., & Mitchell, R. L. (1971). Effects of row cultivation on growth and yield of soybeans. *Agronomy Journal*, **63**, 772–4.

Rydberg, T. (1994). Weed harrowing: the influence of driving speed and driving direction on degree of soil covering and the growth of weed and crop plants. *Biological Agriculture and Horticulture*, **10**, 197–205.

Sanchez del Arco, M. J., Torner, C., & Fernandez Quintanilla, C. (1995). Seed dynamics in populations of *Avena sterilis* ssp. *ludoviciana*. *Weed Research*, **35**, 477–87.

Schimming, W. K., & Messersmith, C. G. (1988). Freezing resistance of overwintering buds of four perennial weeds. *Weed Science*, **36**, 568–73.

Schweizer, E. E., & Zimdahl, R. L. (1984). Weed seed decline in irrigated soil after six years of continuous corn (*Zea mays*) and herbicides. *Weed Science*, **32**, 76–83.

Schweizer, E. E., Westra, P., & Lybecker, D. W. (1994). Controlling weeds in corn (*Zea mays*) rows with an in-row cultivator versus decisions made by a computer model. *Weed Science*, **42**, 593–600.

Scopel, A. L., Ballaré, C. L., & Radosevich, S. R. (1994). Photostimulation of seed germination during soil tillage. *New Phytologist*, **126**, 145–52.

Scott, W. O., & Krummel, J. (1980). Energy used in producing soybeans. In *Handbook of Energy Utilization in Agriculture*, ed. D. Pimentel, pp. 117–21. Boca Raton, FL: CRC Press.

Seifert, S., & Snipes, C. E. (1996). Influence of flame cultivation on mortality of cotton (*Gossypium hirsutum*) pests and beneficial insects. *Weed Technology*, **10**, 544–9.

Shaw, D. R., Newsom, L. J., & Smith, C. A. (1991). Influence of cultivation timing on chemical control of sicklepod (*Cassia obtusifolia*) in soybean (*Glycine max*). *Weed Science*, **39**, 67–72.

Singh, S., Malik, R. K., Panwar, R. S., & Balyan, R. S. (1995). Influence of sowing time on winter wild oat (*Avena ludoviciana*) control in wheat (*Triticum aestivum*) with isoproturon. *Weed Science*, **43**, 370–4.

Smith, E. V., & Mayton, E. L. (1938). Nut grass eradication studies. II. The eradication of nut grass, *Cyperus rotundus* L., by certain tillage treatments. *Journal of the American Society of Agronomy*, **30**, 18–21.

Snipes, C. E., & Mueller, T. C. (1992). Cotton (*Gossypium hirsutum*) yield response to mechanical and chemical weed control systems. *Weed Science*, **40**, 249–54.

Snipes, C. E., Colvin, D. L., Patterson, M. G., & Crawford, S. H. (1992). Cotton (*Gossypium hirsutum*) yield response to cultivation timing and frequency. *Weed Technology*, **6**, 31–5.

Spandl, E., Durgan, B. R., & Forcella, F. (1998). Tillage and planting date influence foxtail (*Setaria* spp.) emergence in continuous spring wheat (*Triticum aestivum*). *Weed Technology*, **12**, 223–9.

Stahl, L. A. B., Johnson, G. A., Wyse, D. L., Buhler, D. D., & Gunsolus, J. L. (1999). Effect of tillage on timing of *Setaria* spp. emergence and growth. *Weed Science*, **47**, 563–70.

Staricka, J. A., Burford, P. M., Allmaras, R. R., & Nelson, W. W. (1990). Tracing the vertical distribution of simulated shattered seeds as related to tillage. *Agronomy Journal*, **82**, 1131–4.

Steckel, G. J., & DeFelice, M. S. (1995). Reducing johnsongrass (*Sorghum halepense*) interference in corn (*Zea mays*) with herbicides and cultivation. *Weed Technology*, 9, 53–7.

Steckel, L. E., DeFelice, M. S., & Sims, B. D. (1990). Integrating reduced rates of postemergence herbicides and cultivation for broadleaf weed control in soybeans (*Glycine max*). *Weed Science*, 38, 541–5.

Steinmann, H.-H., & Gerowitt, B. (1993). Mechanical control of *Galium aparine* in winter wheat. In *Non-Chemical Weed Control, Communications of the 4th International Conference of the International Federation of Organic Agricultural Movements*, Dijon, France, ed. J. M. Thomas, pp. 273–7.

Stiefel, W., & Popay, A. I. (1990). Weed control in organic arable crops. *Proceedings of the 43rd New Zealand Weed and Pest Control Conference*, pp. 138–41.

Stoller, E. W., & Wax, L. M. (1973). Periodicity of germination and emergence of some annual weeds. *Weed Science*, 21, 574–80.

Stopes, C., & Millington, S. (1991). Weed control in organic farming systems. In *Brighton Crop Protection Conference: Weeds*, pp. 185–92. Farnham, UK: British Crop Protection Council.

Swanton, C. J., Clements, D. R., & Derksen, D. A. (1993). Weed succession under conservation tillage: a hierarchical framework for research and management. *Weed Technology*, 7, 286–97.

Swanton, C. J., Shrestha, A., Roy, R. C., Ball-Coelho, B. R., & Knezevic, S. Z. (1999). Effect of tillage systems, N, and cover crop on the composition of weed flora. *Weed Science*, 47, 454–61.

Tawczynski, C. (1990). Fire your pests. *The New Farm*, 12(2), 10–13.

Teasdale, J. R., Beste, C. E., & Potts, W. E. (1991). Response of weeds to tillage and cover crop residue. *Weed Science*, 39, 195–9.

Terpstra, R., & Kouwenhoven, J. K. (1981). Inter-row and intra-row weed control with a hoe-ridger. *Journal of Agricultural Engineering Research*, 26, 127–34.

Tewari, V. K., Datta, R. K., & Murthy, A. S. R. (1993). Field performance of weeding blades of a manually operated push-pull weeder. *Journal of Agricultural Engineering Research*, 55, 129–41.

Thompson, R., & Thompson, S. (1984). Forget foxtail. *The New Farm*, 6(5), 18–20.

Tian, L., Slaughter, D. C., & Norris, R. F. (1997). Outdoor field machine vision identification of tomato seedlings for automated weed control. *Transactions of the American Society of Agricultural Engineers*, 40, 1761–8.

Timmons, F. L., & Bruns, V. F. (1951). Frequency and depth of shoot-cutting in eradication of certain creeping perennial weeds. *Agronomy Journal*, 43, 371–5.

Travers, S. J. (1950). Weed control by good husbandry. *Journal of the Ministry of Agriculture*, 57, 264–70.

Unger, P. W., & Cassel, D. K. (1991). Tillage implement disturbance effects on soil properties related to soil and water conservation: a literature review. *Soil and Tillage Research*, 19, 363–82.

Van Esso, M. L., Ghersa, C. M., & Soriano, A. (1986). Cultivation effects on the dynamics of a Johnson grass seed population in the soil profile. *Soil and Tillage Research*, 6, 325–35.

VanGessel, M. J., Wiles, L. J., Schweizer, E. E., & Westra, P. (1995a). Weed control efficacy and pinto bean (*Phaseolus vulgaris*) tolerance to early season mechanical weeding. *Weed Technology*, **9**, 531–4.

VanGessel, M. J., Schweizer, E. E., Lybecker, D. W., & Westra, P. (1995b). Compatibility and efficiency of in-row cultivation for weed management in corn (*Zea mays*). *Weed Technology*, **9**, 754–60.

VanGessel, M. J., Schweizer, E. E., Wilson, R. G., Wiles, L. J., & Westra, P. (1998). Impact of timing and frequency of in-row cultivation for weed control in dry bean (*Phaseolus vulgaris*). *Weed Technology*, **12**, 548–53.

van Zuydam, R. P., Sonneveld, C., & Naber, H. (1995). Weed control in sugar beet by precision guided implements. *Crop Protection*, **14**, 335–340.

Vargas, M., Sarria, M., Staver, C., Dinarte, S., Aker, C., & Soto, E. (1990). Dry soil tillage for purple nutsedge control in Western Nicaragua: effects of tillage method and exposure time. In *Memorias 3er Congreso Internacional MIP (Manejo Integrado de Plagas)*, pp. 252–4. Managua, Nicaragua: CATIE.

Vengris, J. (1962). The effect of rhizome length and depth of planting on the mechanical and chemical control of quackgrass. *Weeds*, **10**, 71–4.

Vigneault, C., Benoit, D. L., & McLaughlin, N. B. (1990). Energy aspects of weed electrocution. *Reviews of Weed Science*, **5**, 15–26.

Vleeshouwers, L. M. (1997). Modelling the effect of temperature, soil penetration resistance, burial depth and seed weight on pre-emergence growth of weeds. *Annals of Botany*, **79**, 553–63.

Weaver, S. E. (1986). Factors affecting threshold levels and seed production of jimsonweed (*Datura stramonium* L.) in soyabeans (*Glycine max* (L.) Merr.). *Weed Research*, **26**, 215–23.

Weaver, S. E., & Cavers, P. B. (1979). Dynamics of seed populations of *Rumex crispus* and *Rumex obtusifolius* (Polygonaceae) in disturbed and undisturbed soil. *Journal of Applied Ecology*, **16**, 909–17.

Weber, H. (1994). Mechanical weed control with a row brush hoe. *Acta Horticulturae*, **372**, 253–60.

Wellbank, P. J., & Witts, K. J. (1962). Effect of sowing date on competition of weeds with a crop. *Rothamsted Experimental Station, Report for 1962*, 95.

Whitaker, F. D., Heinemann, H. G., & Wischmeier, W. H. (1973). Chemical weed controls affect runoff, erosion, and corn yields. *Journal of Soil and Water Conservation*, **28**, 174–6.

Wicks, G. A., & Somerhalder, B. R. (1971). Effect of seedbed preparation for corn on distribution of weed seed. *Weed Science*, **19**, 666–8.

Wilen, C. A., Holt, J. S., & McCloskey, W. B. (1996). Predicting yellow nutsedge (*Cyperus esculentus*) emergence using degree-day models. *Weed Science*, **44**, 821–9.

Willard, T. R, Shilling, D. G., Gaffney, J. F., & Currey, W. L. (1996). Mechanical and chemical control of cogongrass (*Imperata cylindrica*). *Weed Technology*, **10**, 722–6.

Wilson, B. J. (1972). Studies of the fate of *Avena fatua* seeds on cereal stubble, as influenced by autumn treatment. In *Proceedings of the 11th British Weed Control Conference*, pp. 242–7. London, UK: British Crop Protection Council.

Wilson, B. J. (1978). The long term decline of a population of *Avena fatua* L. with different cultivations associated with spring barley cropping. *Weed Research*, **18**, 25–31.

Wilson, B. J. (1981). The influence of reduced cultivations and direct drilling on the long-term decline of a population of *Avena fatua* L. in spring barley. *Weed Research*, **21**, 23–8.

Wilson, B. J. (1985). Effect of seed age and cultivation on seedling emergence and seed decline of *Avena fatua* L. in winter barley. *Weed Research*, **25**, 213–19.

Wilson, B. J., & Cussans, G. W. (1972). The effect of autumn cultivations on the emergence of *Avena fatua* seedlings. In *Proceedings of the 11th British Weed Control Conference*, pp. 234–41. London, UK: British Crop Protection Council.

Wilson, B. J., & Cussans, G. W. (1975). A study of the population dynamics of *Avena fatua* L. as influenced by straw burning, seed shedding and cultivations. *Weed Research*, **15**, 249–58.

Wilson, B. J., Moss, S. R., & Wright, K. J. (1989). Long term studies of weed populations in winter wheat as affected by straw disposal, tillage and herbicide use. In *Brighton Crop Protection Conference: Weeds*, pp. 131–6. Farnham, UK: British Crop Protection Council.

Wilson, B. J., Wright, K. J., & Butler, R. C. (1993). The effect of different frequencies of harrowing in the autumn or spring on winter wheat, and on the control of *Stellaria media* (L.). vill., *Galium aparine* L. and *Brassica napus* L. *Weed Research*, **333**, 501–6.

Wookey, C. B. (1985). Weed control practice on an organic farm. In *Proceedings of the 1985 British Crop Protection Conference: Weeds*, pp. 577–82. Croydon, UK: British Crop Protection Council.

Wrucke, M. A., & Arnold, W. E. (1985). Weed species distribution as influenced by tillage and herbicides. *Weed Science*, **33**, 853–6.

Yenish, J. P., Doll, J. D., & Buhler, D. D. (1992). Effects of tillage on vertical distribution and viability of weed seed in soil. *Weed Science*, **40**, 429–33.

Yenish, J. P., Fry, T. A., Durgan, B. R., & Wyse, D. L. (1996). Tillage effects on seed distribution and common milkweed (*Asclepias syriaca*) establishment. *Weed Science*, **44**, 815–20.

Zhang, N., & Chaisattapagon, C. (1995). Effective criteria for weed identification in wheat fields using machine vision. *Transactions of the American Society of Agricultural Engineers*, **38**, 965–74.

MATT LIEBMAN AND CHARLES L. MOHLER

5

Weeds and the soil environment

Introduction

One of the distinguishing characteristics of terrestrial plants is that they spend a significant portion of their lives unable to travel farther than they can grow. As a consequence of the sessile, fixed root habit, the resource environment in which plants grow and reproduce is a very local phenomenon and interactions among neighboring plants are common (Harper, 1977, p. 4). The sessile habit makes it possible to suppress weeds through manipulations of soil conditions.

Given the similarity of most terrestrial plant species in their requirements for sunlight, water, and nutrients, it is not surprising that weeds compete with crops for resources and reduce crop yields. Conversely, crop plants exert a large competitive effect on associated weeds (see Chapter 6). A key insight from ecology, however, is that outcomes of competitive interactions between plants are highly dependent on environmental conditions, especially soil-related factors. As Harper (1977, p. 369) noted, "there is a very extensive literature in which it is demonstrated repeatedly that the balance between a pair of species in mixture is changed by the addition of a particular nutrient, alteration of the pH, change in the level of the water table, application of water stress or of shading."

Ecological studies have also revealed that plant abundance and distribution are affected by the availability of appropriate sites for germination and establishment (Grubb, 1977). Because plant species differ in their recruitment responses to soil conditions (Harper, 1977, pp. 111–47), manipulations of soil chemical, physical, and biological characteristics can lead to higher or lower densities of particular species even before competitive interactions begin.

To date, relatively little research has focused on weed management through manipulations of soil conditions other than herbicide application.

Thus a major task for the development of ecological weed management is to identify soil management strategies that predictably reduce weed growth and reproduction, while enhancing crop performance. Although this work is just beginning, sufficient data exist to suggest four general principles for managing weeds through manipulations of soil temperature, moisture, nutrient, and residue conditions.

1. *Before and after seasons of crop production, weed seeds and perennating structures can be killed by altering soil moisture and temperature regimes.* Attacking weeds when crops are not present allows the use of non-selective techniques, such as solar heating and flooding, that are potentially lethal to both crops and weeds. Destruction of weed propagules during "off-seasons" reduces the density of weeds present when crops are sown.
2. *During seasons of crop production, resources can be made differentially accessible to crops and inaccessible to weeds.* Both the location and timing of resource availability may be regulated by management activities. Emphasis should be placed on providing resources to the crop where and when it is best able to use them, and on depriving weeds of resources during critical periods of growth and development.
3. *When resources are accessible to both crops and weeds, differential responses between species to soil conditions can be exploited to stress weeds and enhance crop performance.* Differences can exist between crops and weeds in their germination and growth responses to soil thermal regimes, nutrient sources, residue amendments, and temporal patterns of moisture and nutrient availability. When these differences exist, the goal of soil management is to create conditions that favor crops and place weeds at a disadvantage. Plants given an initial growth advantage over their neighbors by manipulation of soil conditions can become superior competitors for light for the remainder of the growing season.
4. *Shifts in weed community composition occur in response to manipulation of soil conditions.* Because soil management can have strong effects on plant performance, and because species are affected unequally by changes in soil conditions, shifts in weed community composition toward tolerant taxa are likely to occur unless multiple weed management tactics are used. Soil management practices should be components of multitactic weed management strategies that subject weeds to diverse types of stress.

In the following sections of this chapter, we examine applications of these principles in a wide range of agricultural ecosystems. This survey is not meant to be a complete catalog of management practices. Rather, our intent is to stimulate consideration of the many possibilities that soil management offers to weed management and to suggest potentially fruitful lines of research.

Temperature management

Seed germination and plant growth occur only within a range of temperatures, and often are retarded by temperature reductions within the lower end of this range. On a field scale, the most practical means of reducing soil temperature involves the retention of crop residue on the soil surface. The consequences for weeds and crops of using residue to lower soil temperature are discussed later in this chapter.

Elevation of soil temperature can be used to kill weeds, and may be particularly useful for attacking seeds and other structures that resist control by other means. A common example in temperate regions is post-harvest burning of cereal straw and stubble, which can raise soil surface temperature above 200 °C and kill or reduce the germination of weed seeds there and within plant debris (Young, Ogg & Dotray, 1990; Giovanninni et al., 1993). In tropical slash-and-burn systems, burning vegetation can raise temperatures above 200 °C at the soil surface and above 100 °C at 1 cm depth, which lowers total seed density, though it favors the emergence of heat-stimulated species (Ewel et al., 1981; Uhl et al., 1981). These effects not withstanding, the resulting air pollution, degradation of soil organic matter, and potential for creating wildfires argue against broadscale burning as a weed management practice.

Solarization

Soil solarization (also called solar heating, plastic mulching, or soil tarping) is an approach for thermal weed suppression that is suited to regions with seasonally high temperatures and intense sunlight. Under such conditions, soil is covered for several weeks with a polyethylene sheet that traps solar energy and raises soil temperature substantially above ambient levels (Stapleton & DeVay, 1986; Katan, 1987; Bell, Elmore & Durazo, 1988). Because crops as well as weeds are susceptible to heat stress, solarization is performed before crops are sown. Irrigation is generally used to wet the soil before or shortly after polyethylene tarps are laid because (i) weed seeds and perennating structures are physiologically less tolerant of high temperatures under moist, rather than dry conditions; (ii) moist soil conducts heat more effectively than dry soil; and (iii) adequate moisture promotes biological activity in the soil that may increase weed mortality (Horowitz, Regev & Herzlinger, 1983; Katan, 1987; Elmore, Roncoroni & Giraud, 1993). To maximize heating, polyethylene is laid close to the soil surface and sealed with soil at the edges (Stapleton & DeVay, 1986). Transparent mulch is superior to black mulch for raising soil temperature and reducing weed populations (Horowitz, Regev & Herzlinger, 1983; Standifer, Wilson & Porche-Sorbet, 1984).

Solarization with transparent polyethylene sheets has multiple effects on the soil environment. Typically, increases are observed in soil ethylene and carbon dioxide concentrations, nutrient availability (nitrogen, phosphorus, calcium, and magnesium), and moisture content of upper soil layers (Egley, 1983; Horowitz, Regev & Herzlinger, 1983; Rubin & Benjamin, 1984; Stapleton & DeVay, 1986). The increase in daily maximum temperatures in surface soil layers is illustrated by data from a field experiment conducted in central India by Kumar et al. (1993). Solarization with transparent polyethylene tarps (100 μm thickness) increased mean maximum soil temperatures at 5, 10, and 15 cm depth by 9, 7, and 7 °C, respectively. Over a 32-day period when mean maximum air temperature was 39 °C, maximum soil temperature at 5 cm depth exceeded 50 °C on 32 days, exceeded 55 °C on 23 days, and exceeded 60 °C on 7 days; soil temperature at 5 cm depth in the untarped control treatment never exceeded 50 °C. In the southern USA (Mississippi), Egley (1983) noted that maximum soil temperature at 1.3 cm depth under clear polyethylene sheets reached 65–69 °C for 3–4 hours, compared with 43–50 °C in uncovered soil; at 5.1 cm depth the increase in maximum temperature due to tarping was about 10 °C. Horowitz, Regev & Herzlinger (1983) and Rubin & Benjamin (1984) observed similar tarping effects on maximum soil temperature in field experiments in Israel.

Solarization is thought to increase weed seed mortality through direct thermal damage to cell structure and metabolism, toxic effects of gases produced within soil by decomposing organic matter and metabolizing seeds, and microbial attack on seeds and perennating structures weakened by elevated temperature (Horowitz, Regev & Herzlinger, 1983; Rubin & Benjamin, 1984).

Field experiments conducted in the southern and western USA, Israel, India, and other locations have demonstrated that soil solarization can substantially reduce viable seed densities of many weed species and weed emergence in subsequent crops (Horowitz, Regev & Herzlinger, 1983; Standifer, Wilson & Porche-Sorbet, 1984; Bell, Elmore & Durazo, 1988; Kumar et al., 1993). Several important points emerge from such studies.

First, mortality of weed seeds due to solarization is usually greater close to the soil surface than deeper in the soil profile. For example, in a solarization experiment conducted by Standifer, Wilson & Porche-Sorbet (1984), reductions in numbers of germinable *Poa annua* seeds were largely restricted to the upper 6 cm of soil. Reductions in annual *Cyperus* species occurred mostly in the upper 4 cm of soil, though negative effects on *Echinochloa crus-galli* seeds were apparent to 15 cm depth. Greater weed seed mortality near the soil surface and lower mortality with depth reflect the distribution of heat in the soil profile created by solarization (Rubin & Benjamin, 1984).

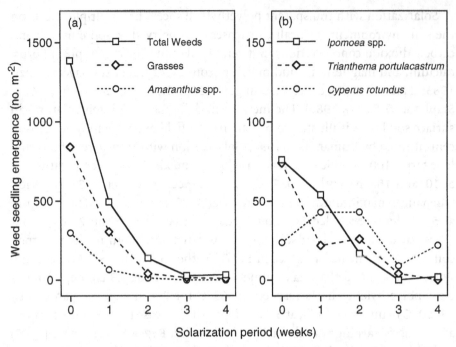

Figure 5.1 Effects of soil solarization on weed seedling emergence from natural seed populations in a field experiment conducted in Mississippi, USA. Plots were covered with transparent polyethylene tarps for 1 to 4 weeks; an untarped control treatment (0 weeks of solarization) was included in the experiment. (After Egley, 1983.)

Second, weed seedling emergence decreases as the solarization period increases in duration. Egley (1983) observed that relative to an unsolarized treatment, emergence of annual grass species was reduced by 64%, 95%, 98%, and 99% when soil was covered with polyethylene for 1, 2, 3, and 4 weeks, respectively (Figure 5.1a). Emergence of other annual species – *Amaranthus* spp., *Ipomoea* spp., and *Trianthema portulacastrum* – was also reduced strongly as the length of solarization increased (Figure 5.1a, b). However, none of the solarization treatments significantly affected emergence of the perennial sedge *Cyperus rotundus* (Figure 5.1b).

The latter result illustrates a third aspect of solarization: weed species can differ greatly in their susceptibility to the technique. Annual weeds tend to be more susceptible to solar heating than perennial species (Horowitz, Regev & Herzlinger, 1983; Kumar *et al.*, 1993), probably because perennials can arise from roots and rhizomes that are buried too deeply to be affected by surface heating. Similarly, large-seeded annual species that can emerge from greater soil depths generally suffer less than small-seeded annual species that can emerge successfully only near the soil surface (Rubin & Benjamin, 1984). Thus, shifts in weed species composition should be expected following solarization, and the need for additional management tactics should be anticipated.

In addition to weed suppression, advantages of soil solarization include suppression of soil-borne crop pathogens, such as *Verticillium dahliae* and *Fusarium oxysporum*, and increases in crop yield (Katan, 1987; Bell, Elmore & Durazo, 1988; Kumar *et al.*, 1993; Stapleton & DeVay, 1995). Disadvantages include the reliance on a relatively expensive input (polyethylene mulch); a lack of fully effective control of many perennial and some annual weed species; the need to remove and dispose of large amounts of plastic; and the need to take a field out of production, albeit temporarily (Bell, Elmore & Durazo, 1988). Because the cost of solarization is high ($750 to $1500 ha^{-1} – Stapleton & DeVay, 1995), the technique is most appropriate for high-value crops that would otherwise have large labor requirements and significant expenses for weed control. The development of new thin-layer mulch materials that are temporarily effective for sealing the soil surface, but which are photodegradable or biodegradable, has been suggested as a means of lowering the costs of solarization and mulch disposal (Stapleton & DeVay, 1995). Also needed is research to identify how solarization can be combined with other management practices for better control of a broader spectrum of weed species. For example, Elmore, Roncoroni & Giraud (1993) found that control of the perennial grass *Cynodon dactylon* by solarization was improved by tillage prior to applying polyethylene tarps. As noted later in this chapter (see section "Residue effects on weed and crop performance"), residues of certain crop species suppress weeds chemically as the residues decompose. Incorporation of cover crop residues into soil before solarization can increase control of soil-borne plant diseases (Gamliel & Stapleton, 1997) and this combination of practices might also increase weed suppression.

Water management

Drainage and irrigation are widely used to improve soil moisture conditions for crop production. Conversely, water can be added or withheld to prevent weed germination, or to stress or kill weeds. Weed suppression through moisture management can be achieved through non-selective methods, used when crops are absent, and selective methods, used when weeds are growing in mixture with crops. The reactions of individual weed species and weed communities to moisture manipulations are determined by the timing, location, and magnitude of alterations of soil moisture conditions.

Flooding fields when crops are not present

Flooding can severely reduce the germination, growth, and survival of weed species unadapted to anaerobic conditions and can be used in certain agroecosystems to suppress perennial weeds before or after periods of crop production. McWhorter (1972) reported that flooding soil for 14 days when water temperature was $\geq 20\,°C$ prevented establishment of the perennial grass *Sorghum halapense* from rhizomes. Because flooding is practical on thousands of hectares of level land in the lower Mississippi River valley that are infested with *S. halapense*, McWhorter (1972) suggested that excellent control might be obtained by flooding for two weeks during summer months, when water temperatures are high; substantial control might also be obtained by flooding for four weeks in March or April, which would permit crop production during the summer of the same year.

The combination of flooding and subsequent ice-encasement, which can kill plants because of the accumulation of CO_2 and other toxic metabolic products, may also be used to manage certain weed species. Ransom & Oelke (1983) reported that fall flooding and subsequent freezing killed all corms of *Alisma triviale*, an important broadleaf perennial weed infesting cultivated fields of wild rice (*Zizania palustris*) in northern Minnesota. Flooding alone failed to suppress corm viability of *A. triviale*; exposure to freezing temperatures and ice-encasement were also required.

Water management in rice production systems

Water management is a key factor influencing the density, productivity, and species composition of weeds infesting rice, the staple food for about one-half of the world's population. Rice can be grown under a wide range of environmental conditions (Purseglove, 1985, pp. 161–99), but most rice is grown with flood irrigation for at least a portion of the crop cycle. Because rice is tolerant of flooding but many weed species are not, differential responses between rice and associated weeds to moisture conditions comprise an important component of weed management for the crop. Planting may consist of broadcasting or drilling seeds into dry soil followed by irrigation (dry-seeding); broadcasting seeds into flooded fields (water-seeding); or transplanting young seedlings into wet or flooded soil. In addition to raising soil levees or bunds around rice fields to retain water, water retention may be improved by sealing wet soil (puddling) with a harrow or animal treading before planting the crop.

The potential weed flora of rice fields includes terrestrial, semiaquatic, and aquatic species of monocots and dicots. The actual abundance of different weed species is strongly affected by water management, field preparation methods, and rice seeding practices (Moody, 1991). For example, Sarkar & Moody (1983) reported that weed biomass was 72–106 kg ha^{-1} when rice was planted in puddled fields, but 1519–1582 kg ha^{-1} when it was planted in dry fields. The relative abundance of different weed taxa was also affected by dry versus puddled field preparation. Grass species comprised 7% to 35% of the weed flora following puddling, but 56% to 64% of the weed flora following dry field preparation; conversely, broadleaf and sedge species were relatively more abundant following puddling than following dry field preparation.

Time of flooding and microtopography may also strongly affect the abundance and growth of weeds in rice fields. In an experiment conducted with irrigated rice during the dry season at Los Baños, Philippines, density and biomass of the annual grass *Leptochloa chinensis* were nil when rice was flooded five days after seeding, but increased progressively to 72 plants m^{-2} and 46 g m^{-2}, respectively, as flooding was delayed until 20 days after planting (Moody & Drost, 1983). Under rainfed conditions, establishment of *L. chinensis* seedlings began in higher areas of rice fields that dried during drought periods; as water levels receded across the field, establishment of the weed followed (Moody & Drost, 1983). Thus early, sustained, and uniform flooding can be important for control of *L. chinensis*.

Although water-seeding and maintenance of flooded conditions may have substantial value as components of weed management strategies for rice, heavy reliance on these methods has led to a clear demonstration of how weed floras can shift and adapt to the stresses placed upon them. Rice culture in California using water-seeding and continuous flooding began in the 1920s as a method to control severe infestations of *Echinochloa crus-galli*, which is well adapted to dry-seeding (Seaman, 1983). However, the change to water-seeded rice culture for control of *E. crus-galli* has increased the abundance of weed species that were previously unimportant in dry-seeded fields. *Echinochloa oryzoides* and *E. phyllopogon* have become important weeds and have largely replaced *E. crus-galli*. The former two species have large seeds which allow germination and emergence through flood water up to 30 cm deep; *E. crus-galli* has smaller seeds and is unable to emerge through deep water (Barrett, 1983; Seaman, 1983). *Echinochloa oryzoides* and *E. phyllopogon* are only partially controlled by continuous flooding, and California farmers have become greatly reliant on herbicides to control these and other species adapted to water-seeding and flooded conditions (Seaman, 1983; Hill *et al.*, 1990). Reliance on herbicides has resulted, in turn, in problems with water contamination (Cornacchia *et al.*, 1984) and herbicide-resistant weed genotypes (Pappas-Fader *et al.*, 1993). To improve weed control and prevent the development of adapted sets of weed species, rice growers in California are now advised to consider alternating between water-seeding and dry-seeding practices (Williams et al., 1992).

Spatial variation in moisture availability

Localized placement of water using drip irrigation technology offers considerable opportunities for weed management in arid environments. Access to water can be largely restricted to the crop root zone, thus minimizing water application to weeds growing between crop rows. Figure 5.2 illustrates how selective placement of irrigation water affected weed management in tomato production systems in California (Grattan, Schwankl & Lanini, 1988). The site used for this experiment is characterized by little or no precipitation during the summer months when many vegetable crops, including tomato, are produced. Three irrigation systems were compared: sprinkler irrigation, which spread water uniformly over the entire plot; furrow irrigation, which concentrated water between crop rows; and buried drip irrigation, which concentrated water directly beneath the crop (Figure 5.2a, b). Seeds of two annual weed species, *Amaranthus retroflexus* and *Echinochloa crus-galli*, were sown on all plots before irrigation treatments began. Weed growth and crop yield were measured in plots not treated with herbicides (Figure 5.2c, e) and in those treated with napropamide and pebulate (Figure 5.2d, f).

In the absence of herbicides, weed biomass production between crop rows in the sprinkler and furrow irrigation systems was >3.5 Mg ha^{-1} (Figure 5.2c), and tomato yield was reduced by weed competition (Figure 5.2e, f). In contrast, when buried drip irrigation was used, weed biomass was <0.2 Mg ha^{-1}, even in the absence of herbicides (Figure 5.2c), and crop yield was unaffected by weed competition (Figures 5.2e, f). Grattan, Schwankl & Lanini (1988) noted that initial costs of materials and installation for drip irrigation are high, but are offset in subsequent years by reduced traffic demands in the field, labor savings, higher water-use efficiency, and excellent control of many annual weed species.

Seed placement can affect a seed's access to soil moisture, the timing of germination and seedling emergence, and the outcome of crop–weed interactions. In a review of weed management tactics for rain-fed agronomic crops in Nebraska, Bender (1994, p. 37) identified planting conditions as optimum when enough moisture is present below the soil surface to germinate the crop, but when the surface is sufficiently dry to prevent germination of weeds until the next rain. In this case, seed placement in deeper, moister soil provides a competitive advantage to the crop by permitting it to emerge before the weeds. A similar approach can be used in irrigated systems, where water can be

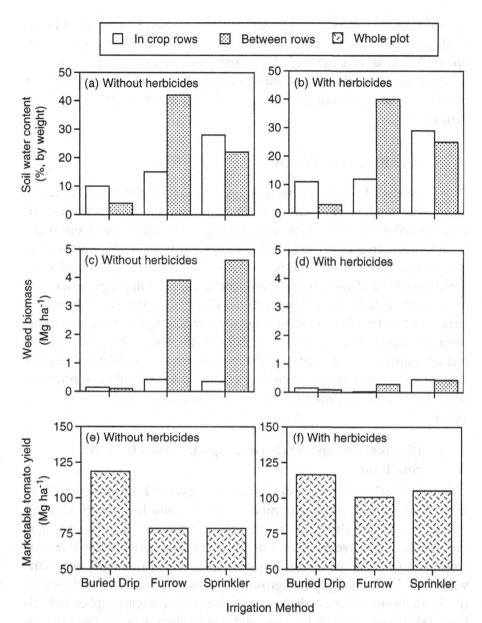

Figure 5.2 Effects of irrigation methods and herbicides (napropamide and pebulate) on soil moisture content (a and b), weed biomass (c and d), and marketable tomato yield (e and f) in a field experiment conducted in California, USA. Tomato was direct-seeded in the center of 1.5-m wide beds. All treatments were sprinkler irrigated until tomato was 23 cm tall, and cultivated and hand-weeded for 7 weeks after planting. Moisture content was measured in the top 2.5 cm of soil, one day after irrigation. (After Grattan, Schwankl & Lanini, 1988.)

applied to moisten the soil profile before planting a crop. Surface soil can then be allowed to dry and weeds can be killed by shallow tillage (Kempen, 1987, pp. 37, 53). Seeds of large-seeded crop species capable of emergence from a depth of several centimeters are subsequently planted into moist soil below the dry surface, promoting crop emergence before the next weed cohort germinates.

Fertility management

At least 14 mineral elements are essential for the growth and development of higher plants (Marschner, 1995). Applications of mineral elements to soil, especially nitrogen (N), phosphorus (P), potassium (K), and sulfur (S), often improve crop yield. Fertilization consequently plays a key role in crop production. Nitrogen is the nutrient whose supply most often limits the growth and yield of agricultural plants, and it is applied in the greatest quantities as synthetic fertilizer. Consequently, a large body of scientific literature has focused on the effects of N on crops and weeds and its fate in the environment. In recent years, increasing concerns over the energy costs and environmental impacts of synthetic fertilizers have led to greater interest in alternative nutrient sources, including crop residues, animal manures, composts, food-processing wastes, and sewage sludge (Parr, Miller & Colacicco, 1984).

Differential responses between species to soil fertility conditions

For both weeds and crops, increased uptake of P, K, and especially N can promote greater stem extension, branching, and leaf area production (Marschner, 1995). While these responses are important for increasing light interception, photosynthesis, and dry-matter accumulation when weeds or crops grow in single-species stands, such responses are particularly important when they differ among species growing in mixtures. Many weed species are considerably more effective than crops in capturing nutrients applied in fertilizers (Alkämper, 1976; DiTomaso, 1995), and increases in soil fertility may alter canopy relations in weed–crop mixtures in favor of the weed component. Such a situation is illustrated in Figure 5.3, which shows the effect of applying ammonium sulfate to mixtures of barley and *Brassica hirta*. Application of N fertilizer increased barley's green surface area (leaves and stems) by 10%, but increased the weed's surface area by 706%. The large increase in the weed's surface area was particularly marked in upper canopy levels, where the greatest proportion of photosynthetically active radiation was absorbed. In this

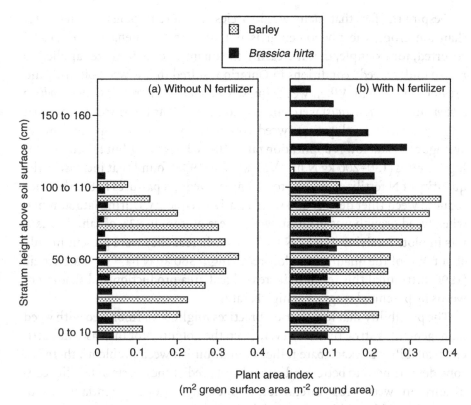

Figure 5.3 Green surface area of barley and *Brassica hirta* grown in mixture without (a) and with (b) N fertilizer. Data were collected in 10-cm strata, 160 days after planting, in a field experiment conducted in California, USA. Ninety kg N ha^{-1} (as ammonium sulfate) was applied to N fertilized treatments at planting and again 94 days later. (M. Liebman, unpublished data.)

experiment, N application greatly increased the weed's ability to shade the crop.

Not surprisingly, when weeds exhibit stronger height and leaf area responses to fertilizer than do crops, fertilizer application may have a neutral or even negative effect on crop yields. In Oregon, Appleby, Olson & Colbert (1976) observed that wheat yields were no higher or were slightly reduced by N application when the crop grew in association with high densities of *Lolium multiflorum*. In northern California, Carlson & Hill (1985) found that application of N fertilizer to wheat infested with *Avena fatua* increased crop yield only when *A. fatua* density was < 1.6% of the total weed plus crop density. At high weed densities, N application increased *A. fatua* panicle production by as much as 140%, and decreased wheat yield by as much as 49%, compared with unfertilized treatments.

Despite the fact that many weed species are more responsive to fertilizer than are crops, the phenomenon is not universal. Tollenaar et al. (1994) reported, for example, that increasing the quantity of N fertilizer applied to maize under weedy conditions in Ontario resulted in less weed biomass and greater maize yield. When weeds (mostly Amaranthus retroflexus, Chenopodium album, and Setaria viridis) were allowed to establish at the three- to four-leaf stage of maize development, weed competition reduced crop yield by an average of 31% at low N application rates (10–80 kg N ha^{-1}), but by only 13% at high N rates (130–200 kg N ha^{-1}). McKenzie (1996) found that increasing the quantity of N fertilizer applied to perennial ryegrass pastures in South Africa reduced weed tiller density and weed relative frequency. During late summer, when weed growth was greatest, weeds were present in 82% of sample quadrats in plots receiving 120 kg N ha^{-1} yr^{-1}, whereas they were present in only about 45% of the quadrats in plots receiving >360 kg N ha^{-1} yr^{-1}. McKenzie (1996) attributed this result to increased leaf area production and shading of weeds by perennial ryegrass at high N rates.

The possibility that fertilization practices might be harmonized with weed management is attractive. However, given the potential for increases in nutrient availability to exacerbate rather than diminish weed problems, there is a considerable need to better understand and predict the effects of fertility conditions on weed–crop interactions. Modeling is one potentially useful approach to address this issue. Models can be used to predict the performance of plant species in mixture based on knowledge of how the individual species respond to variations in environmental conditions when grown in pure stands. Models can also be used to generate testable hypotheses for experimental work.

Kropff and his co-workers examined weed-crop competition using the INTERCOM model, which incorporates information concerning nutrient supply and uptake, light interception, photosynthesis, and root and shoot growth (Kropff & van Laar, 1993). Under high nutrient conditions, the model indicated that height growth and leaf area production are critical factors determining the outcome of competition between species such as sugar beet and Chenopodium album (Kropff et al., 1993). Under low nutrient conditions, the model predicted that competitive dominance of one species over another is favored by morphological features that confer greater rates of nutrient capture (e.g., longer and denser root systems) and physiological features that confer greater rates of biomass production per unit of captured nutrient (e.g., photosynthetic C assimilation via the C_4 rather than C_3 pathway) (Kropff, 1993). Similar effects have been predicted by the ALLOCATE model developed by Tilman (1988).

It is clear that INTERCOM, ALLOCATE, and other models can be useful tools for examining the outcome of interactions among a range of weed–crop combinations grown in different soil environments, and for predicting morphological or physiological traits that can contribute to a crop's ability to tolerate or suppress weeds. However, as the following discussion indicates, consideration of several additional types of information would add to the value of models in developing weed management strategies. Of particular importance are the potential effects of spatial and temporal variation in nutrient availability, and qualitative differences between different nutrient sources.

Spatial variation in nutrient availability

Placement of fertilizer in bands close to crop rows is a method of concentrating nutrients for use by crops; it may also reduce nutrient availability to weeds not growing near the bands. Banding fertilizers within the row of such crops as bean, soybean, peanut, wheat, alfalfa, and rice has been shown not only to increase crop yield compared to broadcast applications, but also to reduce weed density and biomass (DiTomaso, 1995). Advantages of fertilizer banding are generally more pronounced for nutrients applied in a deep band (e.g., 5–7 cm below seed level) than for nutrients applied in a band on the soil surface (DiTomaso, 1995), perhaps because crop seedlings tend to emerge from deeper in the soil profile than do weed seedlings.

The success of fertilizer banding as a weed management strategy may depend on background levels of soil fertility. For example, in field experiments conducted for three years on two soil types in Denmark, Rasmussen, Rasmussen & Petersen (1996) observed that banding N fertilizer 5 cm below spring barley rows decreased weed biomass by an average of 55% and increased barley grain yield by an average of 28%, compared with broadcast fertilizer application. Comparison of results from the two soil types indicated that reductions in weed growth and increases in crop yield due to fertilizer banding were greater on a low-fertility coarse sand than on a more fertile sandy loam. In field experiments conducted on a silt loam in eastern Washington (USA), banding N fertilizer 5 cm below winter wheat rows generally increased wheat grain yield, but had little or no effect on biomass production by the dominant grass weeds (*Bromus tectorum* and *Aegilops cylindrica*) infesting the crop (Cochran, Morrow & Schirman, 1990). Cochran, Morrow & Schirman (1990) noted that background fertility levels at the site may have been sufficient to prevent major N deficits for weeds and that weeds growing close to fertilizer bands may have had access to the applied N. The researchers suggested that greater weed suppression would be expected when a lack of fertility at the soil surface resulted in greater nutrient deficits for weeds. Special attention to weed suppression near fertilizer bands may also be an important component of success with this approach.

Temporal variation in nutrient availability

Because weed and crop species can differ substantially in their abilities to absorb nutrients at different growth stages, the timing of fertilizer application can have strong effects on weed and crop performance. In cases where the peak period of nutrient absorption by a crop occurs after the period of maximum nutrient absorption by an associated weed, delayed application of fertilizer may starve the weed of nutrients during critical initial growth stages and better match the timing of crop nutrient demand. This hypothesis was tested by Alkämper, Pessios & Long (1979) in pot experiments in which maize was grown with *Sinapis arvensis* (*Brassica kaber*) or *Chenopodium album*, and different rates of NPK fertilizer were applied either in a single dose at planting, or with one-half the total application at planting plus one-half at maize ear emergence. Delayed fertilizer application increased crop biomass as much as 70% and reduced weed biomass as much as 50%, compared with early application of the same total quantity of fertilizer.

Delayed fertilizer application also improved rice yields in field experiments in which the crop was heavily infested with *Echinochloa crus-galli* (Smith & Shaw, 1966). Application of N in two split doses at 8 and 12 weeks after crop emergence increased yield of rice 30% to 80% compared with an earlier set of split applications at 3 and 8 weeks after emergence. Although weed biomass values in the different treatments of the experiment were not reported, the observed yield increases with delayed fertilization were attributed to a reduction in weed competition.

Both crop and weed responses to different times of N application were measured in experiments conducted by Angonin, Caussanel & Meynard (1996) with winter wheat and the annual dicot *Veronica hederifolia*. Effects of applying 60 kg N ha^{-1} at the tillering (earlier) or stem elongation (later) stages of wheat development were compared. Over a wide range of weed densities, weed biomass production was more than twice as high with earlier N application than with later N application. The weed species had no effect on N uptake and yield of wheat when fertilizer was applied at the stem elongation stage, but it significantly reduced N uptake and yield of wheat when fertilizer was applied at the tillering stage. The investigators noted that because of its short stature and relatively early growth and development, the major competitive effect of *V. hederifolia* occurs early in the growing season. Later N application can reduce the negative effect of *V. hederifolia* on wheat yield by benefiting the crop more than the weed.

In cases where weed species are more synchronous with crops in their patterns of nutrient capture, delayed fertilization may provide no advantage with regard to weed suppression and crop yield. In field experiments with winter wheat infested by the winter annual grass *Bromus tectorum*, Ball, Wysocki & Chastain (1996) observed that delaying application of N fertilizer until spring had either no effect or a stimulatory effect on weed biomass production, and either no effect or a negative effect on wheat yield, compared with N application at planting. Nitrogen fertilizer increased wheat yield without increasing *B. tectorum* biomass only when it was applied during the fallow period preceding crop production. The latter result was also observed by Anderson (1991), who attributed it to two phenomena: movement of fertilizer N into lower layers of the soil profile before growth of the crop and weed began, and differences in their rooting habits; the shallow-rooted weed was unable to extract fertilizer N at lower depths, while the deeper-rooted crop gained access to that source of nutrients. Ball, Wysocki & Chastain (1996) concluded that pre-planting N application may limit growth of *B. tectorum* and benefit the crop, but may also increase N leaching and water contamination. Other approaches for managing soil fertility and *B. tectorum* are therefore needed.

Synthetic nitrogen source effects

The form in which nutrients are provided can have differential effects on weed and crop performance. An illustration of this phenomenon can be seen in the results of Teyker, Hoelzer & Liebl (1991), who fertilized maize and *Amaranthus retroflexus* with nitrate or ammonium N sources. Shoot weight of the crop was unaffected by N source (Figure 5.4a), but use of ammonium (with the addition of a nitrification inhibitor) reduced the weed's shoot weight by 75%, compared with the nitrate N source (Figure 5.4b). Although the investigators did not examine the effects of different synthetic N sources on competitive interactions between the crop and weed, and did not compare growth of the two species over the same time interval, the observed effects suggest that use of ammonium N with a nitrification inhibitor could greatly enhance maize's ability to suppress growth of *A. retroflexus*.

Differences in synthetic N sources may also affect the species composition of weed communities. Pysek & Leps (1991) compared the weed floras of fields in the Czech Republic that had been cropped with barley for seven years and fertilized with ammonium sulfate (AS), calcium-ammonium nitrate (CAN), or

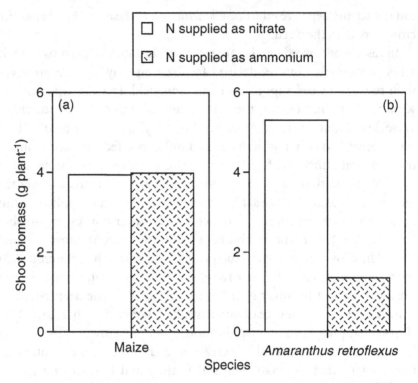

Figure 5.4 Shoot dry weight of maize and *Amaranthus retroflexus* as influenced by form of N fertilizer. Nitrogen was supplied as either calcium nitrate or ammonium sulfate with a nitrification inhibitor. Individual maize plants were grown in pots for 23 days before harvest; *A. retroflexus* was grown at a density of 2 plants per pot for 37 days before harvest. (After Teyker, Hoelzer & Liebl, 1991.)

a mixture of urea and ammonium nitrate (UAN). Total amounts of N applied in the different fertilizers were equal, but distinct differences were observed between treatments in the abundance of various weed species. *Galeopsis tetrahit*, *Veronica persica*, *Thlaspi arvense*, and *Stellaria media* were all relatively abundant in fields fertilized with AS or CAN, but were not detected in fields treated with UAN. Conversely, *Apera spica-venti* was observed in <8% of the quadrats sampled in fields fertilized with AS or CAN, but in 40% of the quadrats from the area fertilized with UAN. Although most of the weed species were similarly abundant in the AS and CAN treatments, *Fallopia convovulus* was present in 79% of the CAN quadrats, but only 41% of the AS quadrats. Total weed infestation, expressed as the estimated amount of weed cover in the sample quadrats, was least in the UAN treatment, greatest in the CAN treatment, and almost as high in the AS treatment. Statistical analyses indicated that the weed community was influenced both directly by fertilizer treatments and indirectly by crop canopy production and consequent competition with the crop. Although soil conditions were not monitored in this experiment, the different fertilizer sources may have resulted in contrasting soil pH levels and differences in the availability of non-N nutrients. Additional experiments focused on soil chemical conditions would be useful for understanding mechanisms through which fertility sources may affect weed community composition.

Competition between legumes and non-legumes

Many legume crops are able to satisfy a sizable proportion of their N requirements using atmospheric N_2 fixed by symbiotic bacteria on their roots. Most non-leguminous species lack access to this N source and are limited instead to the use of soil N. Because of this physiological difference, legumes may compete strongly with non-legumes and yield well under conditions of low soil N availability. In experiments in which soil N was manipulated by adding ammonium nitrate, the grass species *Lolium rigidum* overtopped, shaded, and suppressed the growth of subterranean clover under high soil N conditions; in contrast, clover grew vigorously and dominated the mixed species canopy under low soil N conditions (Stern & Donald, 1962). Similarly, pea grown with *Brassica hirta* produced 185% more biomass when ammonium sulfate fertilizer was not applied than when it was (Figure 5.5a), whereas *B. hirta* in mixture with pea produced 69% less biomass without ammonium sulfate than with it (Figure 5.5b). These patterns reflect direct effects of soil N conditions on crop and weed growth, as well as indirect effects that soil N conditions triggered through their influence on interspecific competition. A comparison of biomass values from pure stands and mixtures shows, for example, that competition from pea reduced the growth of *B. hirta* 54% under low soil N conditions, but had no significant effect on the weed's growth under high soil N conditions (Figure 5.5b).

Such data suggest that minimizing soil inorganic N levels may be one approach for enhancing the performance of certain legume crops growing in association with nitrophilous weeds and for placing additional stress upon the weeds. The approach would not be effective for varieties or species of legumes that fix little atmospheric N and depend heavily on soil inorganic N to produce adequate yields, e.g., short-season cultivars of common bean, *Phaseolus vulgaris* (Laing, Jones & Davis, 1984). To determine which legumes and weeds might be most amenable to management through manipulation of soil N levels would require inoculation of the legumes with appropriate strains of bacterial symbionts, and measurement of crop and weed growth under contrasting soil N conditions. Much in the same way that Tilman (1988,

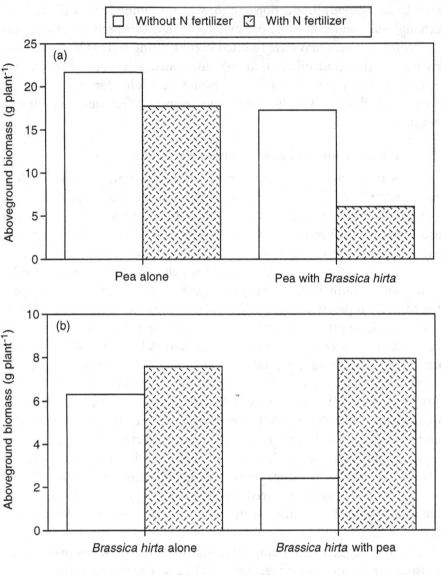

Figure 5.5 Above-ground biomass of pea (a) and *Brassica hirta* (b) grown in pure stands and mixtures, with and without N fertilizer, at 160 days after planting in a field experiment conducted in California, USA. Ninety kg N ha^{-1} (as ammonium sulfate) was applied to N fertilized treatments at planting and again 94 days later. (M. Liebman, unpublished data.)

pp. 19–24) predicts that competitive success between plant species will be a function of which grows best at the lowest levels of nutrient availability, those legumes that are best able to maintain growth at the lowest soil inorganic N levels should be most effective in suppressing the growth of nitrophilous weed species. Samson (1991) suggested that this approach was effective for weed management in soybean production systems and noted that winter rye and other cover crops preceding soybean could be used to sequester and temporarily immobilize soil inorganic N. We examine the impacts of cover crops on weed dynamics in more detail in the following pages.

Crop residue management

Crop residue constitutes about 65% of the total organic materials applied to land in the USA (Parr, Miller & Colacicco, 1984), and its composition and management are important factors affecting soil chemical, physical, and biological characteristics. Through its effects on soil, crop residue also can affect weed and crop germination, survival, growth, and competitive ability. In general, it appears that detrimental effects of crop residue are greater for small-seeded species than larger-seeded species. Because seeds of most major crops are one to three orders of magnitude larger than the weeds with which they regularly compete, residue management offers important opportunities for weed suppression (Mohler, 1996).

Taking advantage of these opportunities requires solving some considerable technical challenges, and manipulating and balancing a complex set of interacting ecological processes. Under some circumstances crop residue can promote rather than inhibit weeds, and suppress rather than enhance crop establishment and growth. Moreover, to generate enough residue to influence weeds appreciably, additional crops may need to be grown within a rotation sequence, increasing the complexity of the cropping system. Unlike certain ecologically based management practices that can be implemented with only modest changes to a farmer's existing cropping system, using residue for weed management can require substantial system redesign.

Green manures and mulches

Crops whose intended purpose is to alter soil characteristics rapidly and significantly can have particularly marked effects on weeds. We use the term *cover crop* for species grown expressly to add organic matter, maintain or increase nutrient availability, improve soil physical properties, prevent erosion, and, in some cases, reduce problems with soil-borne pathogens

(Hargrove, 1991; Sarrantonio, 1994). Cover crops are often N_2-fixing legumes, but other species, such as grasses and crucifers, are also used.

Cover crop residue can be managed in two distinct ways. Incorporation of cover crops into soil through various forms of tillage is called *green manuring*, a practice that has been used by farmers throughout the world for millennia (Pieters, 1927, pp. 10–16, 238–311). Alternatively, cover crop residue can be retained on the soil surface through no-tillage and zone tillage techniques and used as *mulch*. Other materials, such as sawdust (Obiefuna, 1986), food and distillation wastes (Singh, Singh & Singh, 1991), and sewage sludge (Roe, Stoffella & Bryan, 1993), also can be used as mulch, but the costs of purchasing and transporting them onto farm fields tend to restrict their use to high-value crops or situations where application to land is less expensive or more environmentally benign than alternative disposal methods. Cover cropping provides a means of inexpensively producing mulch *in situ*.

Green manures and mulch crops can be grown (i) when land would otherwise lie fallow, (ii) in mixtures with "main crops" grown for cash, fodder, and food; or (iii) as substitutes for main crops during normal seasons of crop production (Chapter 7) (Sarrantonio, 1992). The approach used depends upon crop characteristics and the environment. For example, in colder regions, such as the northeastern and north-central USA, sufficient time is usually available to establish winter wheat or rye cover crops after harvesting maize grown for silage, but not after maize grown for grain. Other cover crops, such as hairy vetch and clover species, must be planted before September if they are to survive until the following spring. Consequently, in many cropping systems used in short-season areas, a winter cover crop can be used successfully only if it is planted into a preceding main crop.

Broadcasting cover crop seeds on the soil surface to establish them during the growth of main crops generally results in poor stands, though occasional success has been reported with winter rye (Mohler, 1991; Johnson, DeFelice & Helsel, 1993). Reasons for poor establishment of surface-sown seeds include consumption by insects, mollusks, and small mammals (Figure 5.6) (Mohler, unpublished data), and failure to imbibe sufficient water. To improve establishment, cover crop seeds can be incorporated into soil during the last cultivation of main crops (Scott & Burt, 1987). Burial of cover crop seeds hides them from seed predators and, by speeding germination and establishment, shortens the period of vulnerability to herbivores (Figure 5.6). Cover crops planted at last cultivation rarely compete significantly with the associated main crops, although cover crop seedlings may die if main crops cast dense shade.

In temperate regions, cover crops used as green manures typically are killed

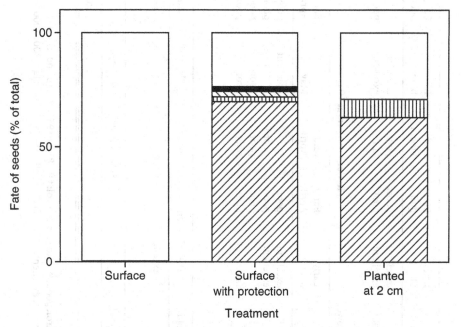

Figure 5.6 Fate of winter wheat seeds sown into conventionally tilled standing maize on 18 and 20 September 1990, and evaluated on 9 October 1990. Seeds were either placed on the soil surface, placed on the surface but protected with cloth covered screen to exclude mammals and surface dwelling macroinvertebrates, or planted at 2 cm depth. Partial exclosure treatments (data not shown) indicated that slugs and earthworms were the principal seed predators. (C. L. Mohler, unpublished data.)

by tillage several weeks before planting cash, fodder, and food crops. When used as mulches, cover crops often are killed with non-residual herbicides such as glyphosate or paraquat shortly before planting the next crop. Alternatively, some cover crops can be killed by mowing after they have reached a certain stage of maturity (Table 5.1). Creamer *et al.* (1995) developed an undercutter that killed winter rye, hairy vetch, and crimson clover cover crops by severing roots with minimal soil disturbance. Ideally, a mulch crop would grow during the fallow season and then die back naturally as the main

Table 5.1. *Effect of growth stage or time of treatment on mechanical control of cover crops*

Hoffman, Regnier & Cardina (1993)[a]

	Mower			Roller			Roller with blades		
	Early bud	Mid bloom	Late bloom	Early bud	Mid bloom	Late bloom	Early bud	Mid bloom	Late bloom
Hairy vetch, 1990	poor	fair	excellent	poor	poor	excellent	–	–	–
Hairy vetch, 1991	poor	excellent	excellent	poor	fair	excellent	–	good	excellent

Dabney, Buehring & Reginelli (1991)[a]

	Flail mower			Roller + coulters at 10 cm			Roller + coulters at 20 cm		
	Early April	Mid April	Early May	Early April	Mid April	Early May	Early April	Mid April	Early May
Hairy vetch	fair	fair	excellent	poor	fair	excellent	poor	poor	excellent
Crimson clover	fair	fair	excellent	poor	poor	fair	poor	poor	poor
Subterranean clover	poor	poor	fair	poor	poor	poor	poor	poor	poor
Berseem clover	poor	poor	fair	poor	poor	poor	poor	poor	poor

Wilkins & Bellinder (1996)[b]

1992

	First node	Second node	In boot	75% head	100% head	Milky kernel
Wheat	poor	fair		good		excellent
Rye	poor	good	excellent			

1993

	First node	Second node	In boot	Flag leaf	100% head	Watery kernel
Wheat	poor		fair	fair	good	excellent
Rye		fair	fair		good	excellent

Notes:
[a] Ratings based on percentage killed: excellent, ≥98% control; good, 95% to 98% control; fair, 85% to 95% control; poor, <85% control.
[b] Ratings based on amount of regrowth 4 weeks after treatment: excellent, ≤50 kg ha^{-1}; good, 50–200 kg ha^{-1}; fair, 200–1000 kg ha^{-1}; poor, >1000 kg ha^{-1}.

crop was established. One of the few examples of such a system involves use of certain cultivars of subterranean clover in a rather limited area of the eastern USA (Chapter 7) (Enache & Ilnicki, 1990; Ilnicki & Enache, 1992). Breeding efforts might increase the number of early maturing cover crops that could be used for mulch systems in other regions.

In warm regions with enough moisture to support the production of two or more crops per year, more options exist for the use of cover crops before, during, and after the production of main crops (Pieters, 1927, pp. 283–311; Miller, Graves & Williams, 1989; Reijntjes, Haverkort & Waters-Bayer, 1992, pp. 168–73). The cover species may be killed or suppressed by tillage, herbicides, or slashing with hand tools.

Slash/mulch (*tapado*) systems are used in many tropical countries to produce bean, maize, taro, and other staple foods in residues of forest, scrub, or weedy fallow vegetation (Thurston, 1997). The mulches used in such systems apparently suppress weeds considerably (Moreno & Sánchez, 1994; Thurston, 1997, pp. 26, 45–8), but these effects have not yet been studied systematically. Efforts by both farmers and researchers are under way to improve slash/mulch systems by introducing legume cover crops into fallow vegetation (Buckles *et al.*, 1994; Madrigal, 1994; Moreno & Sánchez, 1994; see also Chapters 3 and 7).

Allelopathy

Many plant species produce and release chemicals that are toxic to other plants, a phenomenon referred to as *allelopathy*. Allelochemicals may also be produced by microbes that transform plant products during residue decomposition. Living crops can have direct allelopathic effects on weeds (see Chapter 6), and live and decomposing weeds can reduce crop performance (Bhowmik & Doll, 1982, 1984; Putnam & Weston, 1986), but the most important application of allelopathy involves the use of crop residue to suppress weed germination, establishment, and growth.

Studies of allelopathic effects of crop residue on weed and crop species typically comprise a description of the symptoms and injuries present in target plants exposed to residue or extracts from it, isolation of the putative causal agent(s), and application of the isolated agent(s) to healthy plants to determine whether similar damaging effects can be reproduced. Using chemical isolation and bioassay techniques in laboratory and glasshouse experiments, a number of classes of chemicals have been identified as allelopathic agents. Those found frequently include alkaloids, coumarins, cyanogenic glucosides, flavonoids, phenolic acids, polyacetylenes, quinones, and terpenoids (Einhellig & Leather, 1988; Worsham, 1989; Rice, 1995).

The phytotoxicity of rye residue and its extracts has been particularly well studied. Allelopathic effects of this species have been attributed to β-phenyllactic acid and β-hydroxybutyric acid (Shilling, Liebl & Worsham, 1985; Shilling et al., 1986), and to 4-dihydroxy-1,4(2H)-benzoxazin-3-one (DIBOA), 2(3H)-benzoxazolinone (BOA), and related benzoxazolinone compounds (Barnes & Putnam, 1986, 1987; Barnes et al., 1987; Pérez & Ormeño-Núñez, 1993). Soil microbes can transform BOA into compounds that are considerably more phytotoxic than the parent chemical (Chase, Nair & Putnam, 1991; Gagliardo & Chilton, 1992).

Sorghum residue can suppress a range of plant species, an effect attributed to sorgoleone (a long-chain hydroquinone), dhurrin (a cyanogenic glucoside), and several other compounds (Weston, 1996). Allelopathic potential has been demonstrated for residues or extracts of a range of other crop species, including alfalfa (Waller, 1989; Chung & Miller, 1995), barley (Overland, 1966; Liu & Lovett, 1993a, 1993b), berseem clover (Bradow & Connick, 1990), rapeseed (Wanniarachchi & Voroney, 1997), crimson clover (White, Worsham & Blum, 1989; Creamer et al., 1996b), hairy vetch (White, Worsham & Blum, 1989), oat (Fay & Duke, 1977; Putnam & DeFrank, 1983; Putnam, DeFrank & Barnes, 1983), pea (Kimber, 1973; Cochran, Elliot & Papendick, 1977), red clover (Chang et al., 1969), sweetclover (McCalla & Duley, 1948; Guenzi & McCalla, 1962), sunflower (Leather, 1983), and wheat (Liebl & Worsham, 1983; Shilling, Liebl & Worsham, 1985).

Although many species have been shown to produce allelopathic compounds, demonstrating that allelopathy is responsible for weed suppression under field conditions is technically more difficult. Residue additions may change nutrient, temperature, moisture, and light conditions, and the possible effects of these factors must be distinguished from those of residue-derived chemicals. Cochran, Elliot & Papendick's (1977) approach using soil taken from beneath residue and Creamer et al.'s (1996b) approach using leached and unleached residues offer potential methods for studying allelopathy under field conditions.

Small-seeded weed and crop species appear especially susceptible to allelochemicals. Although large-seeded crops often show susceptibility in laboratory bioassays (Guenzi & McCalla, 1962; Kimber, 1973; Cochran, Elliot & Papendick, 1977; Chase, Nair & Putnam, 1991), they appear to be relatively insensitive in the field. Putnam & DeFrank (1983) found, for example, that residues of several grain crops substantially reduced emergence of lettuce, radish, tomato, and a mixture of small-seeded weed species, whereas they increased emergence of cucumber, pea, and snap bean. The differential suppression of smaller-seeded species by allelopathic substances released from residue may be

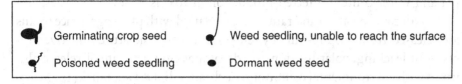
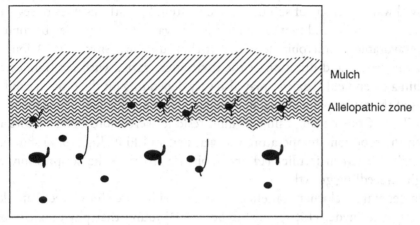

Figure 5.7 Phytotoxins leaching from a surface layer of organic mulch diffuse only a short distance into the soil. Large crop seeds commonly are planted below the toxic layer, and thus germinate in a chemically safe environment. Weed seeds that germinate within the allelopathic zone frequently are poisoned, whereas most weeds that germinate below the toxic layer exhaust seed reserves before reaching the surface.

the result of two processes. First, at least from germination until emergence, the surface-to-volume ratio of a small-seeded species is usually greater, and therefore its exposure per unit mass to allelopathic substances in the soil is also greater. Second, when residue is used as a mulch, the allelopathic toxins are released onto the soil surface and may not diffuse very deeply into the soil profile. Barnes & Putnam (1986) showed that percent germination and root elongation of several species decreased as the layer of soil separating seeds from rye residue decreased from 15 to 0 mm. To have any potential for emergence, a small-seeded crop or weed must germinate near the soil surface, but under an allelopathic mulch, this is where the toxins are most concentrated. In contrast, large-seeded crops are planted more deeply, and thus germination and initial root growth may occur in a less toxic environment (Figure 5.7). Both hypotheses require testing by careful experimentation.

Production and decomposition of allelopathic compounds is highly variable. Mwaja, Masiunas & Weston (1995) found that rye produced lower concentration of BOA and DIBOA in high-fertility conditions relative to medium- and low-fertility conditions. Similarly, Patrick & Koch (1958) found that soil pH and water content affected production of toxins in several species, and that toxin concentration and rate of release varied with plant age. Since toxins are released from crop residue by temperature- and moisture-sensitive processes of leaching, volatilization and decomposition, the rate of release also varies with environmental conditions. Cochran, Elliot & Papendick (1977) bioassayed water extracts of soil taken weekly from beneath residues of several crop species and found that toxin production was usually preceded by conditions favorable for microbial activity. Patrick, Toussoun & Snyder (1963) found that the degree and duration of the toxic effect varied with location, even within a given field. Toxicity may also vary depending on the growth conditions of the target plant. Thus, Bradow (1993) found that ketones of the sort volatilized from decomposing legume residue were more toxic to cotton when the root zone temperature was warmer, and Einhellig (1986) showed synergism between ferulic acid and moisture stress in the suppression of sorghum seedling growth.

In general, allelochemical effects of crop residue are short lived. Patrick, Toussoun & Snyder (1963) and Kimber (1973) found that phytotoxicity of several crop species declined markedly after two to three weeks of decomposition. The transitory nature of allelopathy and its dependence on soil and weather conditions represent major challenges to the use of crop residues for weed management. Fortunately, the toxic properties of crop residues are only one aspect of their inhibitory action on weeds. Even for allelopathically active materials such as rye residue, physical modifications of the environment can be responsible for a large portion of the weed suppression obtained (Teasdale & Mohler, 1992).

Nutrient availability from crop residue

Patterns of nutrient release from crop residue are important for weed management because, as explained below, they may affect weed density, the timing of weed emergence, and interactions between crops and weeds.

Decomposition of crop residue and subsequent changes in soil nutrient status are determined by multiple factors, including residue age and quality (e.g., C:N ratio, lignin and polyphenol contents), loading rate, temperature and moisture conditions, soil aeration and pH, and soil microbial, meso- and macrofaunal populations (Fox, Myers & Vallis, 1990; Honeycutt, Potaro & Halteman, 1991; Palm & Sanchez, 1991; Honeycutt et al., 1993; Honeycutt, Clapham & Leach, 1994; Killham, 1994). In general, crop residue retained on the soil surface decomposes and releases nutrients more slowly than residue incorporated into soil (Blevins, Smith & Thomas, 1984; Wilson & Hargrove, 1986; Sarrantonio & Scott, 1988; Dou, Fox & Toth, 1995). Application of crop residue with a C:N ratio >30:1 (e.g., cereal straw) typically results in temporary immobilization of nutrients by soil microbes before nutrients are released in plant-available, inorganic forms, whereas application of residue with a C:N ratio <20:1 (e.g., immature legume material) increases soil concentrations of plant-available nutrients as soon as environmental conditions allow enough microbial activity (Stevenson, 1986, pp. 164–6). Because of its critical importance in plant nutrition, much of the research relevant to residue decomposition and nutrient release has focused on N. However, P and S release from crop residue can follow temporal patterns similar to N (Stevenson, 1986, pp. 268, 301).

Because residue decay and nutrient transformations require time, it is not surprising that increases in soil inorganic nutrient concentrations often occur more slowly following crop residue application than synthetic fertilizer application. For example, in a maize field in Kentucky treated with ^{15}N-labeled ammonium nitrate and hairy vetch residue, Varco et al. (1993) found the percentage of soil inorganic ^{15}N derived from the fertilizer exceeded that from the green manure for 30 days after treatments were applied in one year, and for 45 days in a second year. Similarly, in a rice field in California, Westcott & Mikkelsen (1987) observed that soil inorganic N levels were lower for the first 48 days following incorporation of purple vetch, compared to application of ammonium sulfate containing the same amount of N.

Data concerning plant N uptake also suggest that crop residue can function as a slow-release nutrient source, compared to synthetic fertilizer applied in a single dose at the start of the growing season. In a greenhouse pot experiment using ammonium sulfate and *Sesbania aculeata* residue labeled with ^{15}N, Azam, Malik & Sajjad (1985) observed that after five weeks of growth maize had recovered 20% of the labeled N from the fertilizer, but only 5% from the green manure. In a field experiment conducted in a semiarid area of South Australia, Ladd & Amato (1986) found that 17% of the ^{15}N label in a *Medicago littoralis* green manure was taken up by a wheat crop, whereas 62% of the label remained in the soil organic fraction. In contrast, an average of 47% of the labeled N in urea, ammonium sulfate, and potassium nitrate fertilizers was taken up by wheat, and only 29% remained in the soil organic fraction. Similar results were obtained for comparisons of N dynamics following application of green manure and synthetic N fertilizer in wheat fields in Alberta and Saskatchewan (Janzen et al., 1990), and maize fields in Pennsylvania (Harris et al., 1994).

Release of nutrients from crop residue is not always a slow process, however. Luna-Orea, Wagger & Gumpertz (1996) measured nutrient release from two legume cover crops (*Desmodium adscendens* and *Pueraria phaseoloides*) in the Bolivian Amazon and reported that >50% of the N, P, K, and Mg contained in 12-month-old plants was released within four weeks after they were slashed and placed on the soil surface. In an experiment conducted in the north-central USA (Wisconsin), Stute & Posner (1995) observed that red clover and hairy vetch green manures released half of their N within four weeks after incorporation and increased soil inorganic N concentrations to levels similar to those obtained from ammonium nitrate fertilizer applied at 179 kg N ha^{-1}.

Because the rate and total amount of germination of certain weed species are positively correlated with soil nitrate concentration (Henson, 1970; Roberts & Benjamin, 1979; Taylorson, 1987; Karssen & Hilhorst, 1992), the use of crop residues that function as slow-release alternatives to early, pulsed application of synthetic N fertilizer may delay weed emergence and reduce weed density. Conversely, crop residues that release N quickly should not have this weed-suppressive effect. Field testing of these hypotheses is needed.

Increased reliance on decomposing crop residues rather than fertilizer applied at planting may also affect post-emergence weed management. Greater seed reserves convey greater tolerance of nutrient deficits and other stresses during early growth (Westoby, Leishman & Lord, 1996) and hence low availability of nutrients from crop residue early in the growing season may retard seedling growth of small-seeded weeds, but have a neutral effect on the early growth of large-seeded crops. Alternatively, if residue decomposes quickly and provides large amounts of nutrients at appropriate weed growth stages, weed problems might be worse with residue than with synthetic fertilizer, particularly if the fertilizer were applied in a split and delayed manner in narrow bands close to the crop row. Collaborative research involving soil, crop, and weed scientists is needed to resolve these issues and better manipulate residue effects on soil fertility to the advantage of crops and detriment of weeds.

Residue effects on temperature and moisture

Crop residue used as mulch substantially decreases maximum daily soil temperatures (Mitchell & Teel, 1977; Bristow, 1988; Fortin & Pierce, 1991; Teasdale & Mohler, 1993). Differences between maximum surface soil temperature in mulched and unmulched plots is greatest on hot, sunny days when the soil is dry, and under such conditions differences as great as 14 °C have been reported (Bristow, 1988). More typically, the difference between maxima in mulched and unmulched conditions is 2–5 °C. When weather is relatively cool, lower maximum temperature in a mulched field will tend to retard the emergence of both weeds and crops. Under hot conditions, lower maximum temperature may prevent some weed species from entering secondary dormancy (Forcella et al., 1997).

However, since mulch has either little effect on daily minimum soil tem-

Table 5.2. *Mean daily maximum and minimum soil temperature and amplitude of temperature fluctuation at 5 cm depth under hairy vetch mulch*

Year	Mulch biomass[a] (g m^{-2})	Soil temperature[b] (°C)		
		Maximum	Minimum	Amplitude
1989	0	30.2	20.6	9.6
	462	27.6	21.4	6.2
	924	26.9	21.1	5.8
1990	0	26.3	15.5	10.9
	319	23.7	17.0	6.7
	638	23.6	16.3	7.3
1991	0	34.0	18.8	15.3
	375	27.9	19.4	8.5

Notes:
[a] Biomass levels correspond to 0%, 100%, and 200% of the biomass produced by the winter cover crop in 1989 and 1990, and 0% and 100% in 1991.
[b] Mean of 7 days in 1989, 22 days in 1990, and 28 days in 1991.
Source: Teasdale & Mohler (1993), Teasdale & Daugherty (1993).

peratures or tends to cause modest night-time warming (Bristow, 1988; Fortin & Pierce, 1991; Teasdale & Mohler, 1993; Creamer *et al.*, 1996a), its principal effect on temperature conditions is to decrease the amplitude of diurnal fluctuations. For example, Teasdale & Mohler (1993) and Teasdale & Daughtry (1993) observed mean maximum–minimum differences at 5 cm depth of 6–8 °C with mulch and fluctuations of 10–15 °C without mulch (Table 5.2). Creamer *et al.* (1996a) reported similar differences at 10 cm depth, with a mean amplitude of 5 °C beneath mulch and 10 °C for bare soil (computed from their Figure 3).

For comparison, Totterdell & Roberts (1980) found that although a 5 °C temperature fluctuation promoted germination of *Rumex crispus* and *R. obtusifolius*, 10–15 °C was required for 100% germination. Similarly, Benech-Arnold *et al.* (1988) reported only 9% germination of after-ripened *Sorghum halapense* seeds exposed to 4 °C fluctuations for 20 cycles but 35% germination for those exposed to 10 °C fluctuations. Thus, although the magnitude of temperature fluctuations under mulch is sufficient to break dormancy of some seeds of species sensitive to temperature fluctuations, the decrease in amplitude by mulch can lower the percentage germination substantially.

Mulch decreases convection, which decreases the gradient in partial pressure of water vapor between the soil and the general atmosphere. Together with lower soil temperatures, this reduces evaporation from the soil surface and keeps soil moist for a longer period (Griffith, Mannering & Box, 1986; Teasdale & Mohler, 1993). These factors may facilitate the germination of weed seeds and probably reduce the number of germinants that die due to desiccation before establishment. Conversely, because light is absorbed by residue and convective dissipation of heat is reduced by the boundary layer of still air around mulch particles, high temperatures that could be detrimental to seedling survival can occur within the mulch layer itself. Mulch element temperatures over 50 °C have been recorded (Bristow, 1988; Wagner-Riddle, Gillespie & Swanton, 1996). Although data on weeds are lacking, Smith (1951) found that heat injury by girdling of stems was the principal cause of mortality for young white pine seedlings establishing in pine litter.

Most crop species are subtropical or tropical in origin and require warm soil for germination and establishment. As a result, mulch may delay crop establishment in regions where the soil is cool at planting (Kaspar, Erbach & Cruse, 1990; Burgos & Talbert, 1996). Moreover, growth may continue to be slower throughout the season, resulting in delayed maturity and reduced crop yield and quality (Fortin & Pierce, 1991; Mohler, 1995; Burgos & Talbert, 1996). Vidal & Bauman (1996) observed that spring frost damaged soybeans in a mulched treatment whereas plants in bare ground treatments escaped, presumably because the mulch insulated the crop from warmth emanating from the soil. Wicks, Crutchfield & Burnside (1994) found that maize yield increased with rate of wheat straw up to 4.4 Mg ha^{-1} due to increased soil moisture retention, but the trend reversed at higher rates due to slower maturation caused by cooler soil. Delays in maturity can be especially serious for horticultural crops, for which earlier production may mean greater market value. To address this problem, residue might be incorporated or moved from a strip directly over the crop row to create bare, warm soil conditions that promote crop emergence and growth; mulch could be retained over soil in inter-row areas to retard weed emergence.

In contrast with the problems experienced in cool conditions, germination of crop seeds may be hastened by mulch in warm, dry conditions due to increased moisture retention. In hot climates mulch may also increase crop production through increased water availability and the reduced root respiration that occurs in cooler soil (Midmore, Roca & Berrios, 1986; Obiefuna, 1986; Daisley et al., 1988). However, a mulch grown *in situ* may reduce soil moisture for the subsequent main crop (Liebl et al., 1992).

Residue effects on light

Light extinction by crop residue on the soil surface follows Beer's law. That is, the percentage of the photon flux density transmitted through residue usually follows an equation of the form

$$Y = 100\, e^{-kX}$$

where X is the residue biomass in g m^{-2}, and k is a fitted constant (Figure 5.8a) (Teasdale & Mohler, 1993; C. L. Mohler, unpublished data). Although this equation adequately describes the reduction in light by any given type of residue, the coefficient, k, varies between residue types, and through time as residue decays. A somewhat more general description of the relation between light level and amount of residue is given by the same equation, but with residue quantified as surface area per unit of ground area (Figure 5.8b) (i.e., m^2 m^{-2} – Teasdale & Mohler, 2000). Crop canopies differentially filter out the germination promoting red wavelengths (see Chapter 2), but dead mulch materials have only a minor effect on light quality (Teasdale & Daughtry, 1993; Teasdale & Mohler, 1993).

Although the mean light transmittance through crop residue is highly predictable, light level at the soil surface under residue varies greatly. Relatively high light levels are found at some locations even under a thick layer of mulch (Table 5.3) (Teasdale & Mohler, 1993). Thus, cover crops used as mulch can reduce the light-cued germination of many surface and shallowly buried weed seeds, but sufficient light to stimulate germination of some seeds will penetrate through all but the heaviest mulch layers. Seeds of most crop species do not require light for germination and are capable of high percentage germination beneath mulch.

In those microsites where weed seedlings are shaded by residue, seedlings may become etiolated as they attempt to extend photosynthetic surfaces above the mulch layer. Consequently, weed seedling growth in mulch may be slowed by shading effects on photosynthesis, as well as increased metabolic costs of extra stem material. In contrast, the larger seed size of most crops relative to weeds conveys a greater ability to grow up through mulch without exhausting seed reserves or becoming excessively etiolated.

Residue effects on herbivores and pathogens

A wide diversity of organisms potentially damaging to weeds can be promoted by crop residue. Mollusk populations increase under crop residue at the soil surface due to decreased desiccation and increased food supply, and may create problems for crop production in no-till, high-residue systems (Edwards, 1975). However, damage to weed seedlings by mollusks in high-residue conditions can also be considerable. C. L. Mohler (unpublished data) measured seedling survival of three weed species in tilled and untilled maize. In one of the two years of the study, survival of *Digitaria sanguinalis* from emergence until flowering was 60% in conventionally tilled maize, but only 16% in no-till maize. Systematic recording of the presence of slime trails and the types of damage sustained by each plant prior to death indicated that mollusk

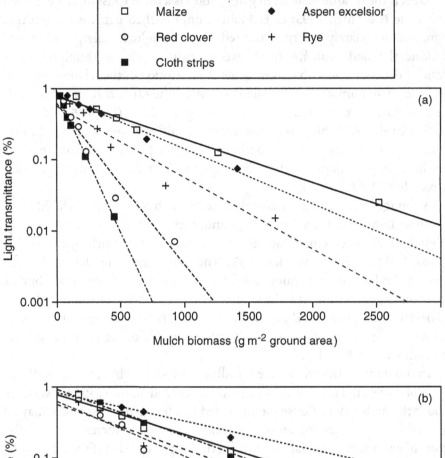
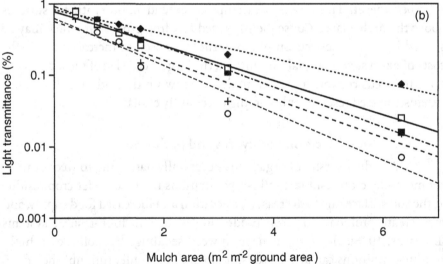

Figure 5.8 Light transmittance through five mulch materials as a function of mulch biomass (a) and mulch area (b). (C. L. Mohler, unpublished data.)

Table 5.3. *Frequency of locations under mulch that received photosynthetically active radiation at various levels relative to full sunlight*

	Hairy vetch		Rye	
Mulch biomass[a] (g m^{-2})	319	638	246	738
Bare ground (%)	19	4	37	3
Mean PPFD[b] (%)	24	7	34	5
Radiation relative to full sunlight				
<0.1%	0	25	0	6
0.1%–1%	0	19	0	42
1%–10%	40	31	25	35
10%–25%	27	19	19	10
25%–50%	17	4	31	6
>50%	17	2	25	0

Notes:
[a] Biomass levels correspond to 100% and 200% of the biomass produced by the winter cover crop.
[b] Photosynthetic photon flux density.
Source: Teasdale & Mohler (1993).

grazing was the primary cause of mortality. However, the effect was insignificant in the drier year of the study, and *Amaranthus retroflexus* and *Abutilon theophrasti* were not significantly affected by mollusks in either year. In another study, survival of *D. sanguinalis* was not affected by residue (Mohler & Callaway, 1992). Thus, the effect of mollusks on weed seedlings can be sporadic.

Earthworms remain closer to the soil surface under crop residue due to cooler, wetter soil conditions. Earthworms have been shown to consume grass seeds, and only part of these are egested in a viable condition (McRill & Sagar, 1973; McRill, 1974; Grant, 1983). Earthworms can also move weed seeds downward in the soil profile (van der Reest & Rogaar, 1988), making seedling emergence less likely. They also kill seedlings of both grass and broadleaf species by pulling young shoots into their burrows (Shumway & Koide, 1994). Other weed seed consumers that may be promoted by crop residue include carabid beetles, ants, crickets, and small mammals (Chapter 8) (Lund & Turpin, 1977; Risch & Carroll, 1986; Brust & House, 1988).

Since all of the organisms discussed above are generalist feeders that can threaten crops under some circumstances, careful management is required to exploit their weed control potential. Nevertheless, a species that finds a crop and a weed equally palatable may have a more damaging impact on the weed if it has a smaller seed than the crop. The larger seed size of the crop allows fast growth through the mulch layer, where herbivory is most intense. In contrast, smaller-seeded weeds may have their shoot apexes exposed to intense herbivory for several days while they attempt to accumulate sufficient photosynthate to complete growth through the mulch.

Grain straw and killed legume cover crops have been shown to promote populations of several disease-causing organisms, including species of *Pseudomonas, Rhizoctonia*, and *Pythium* (Stroo, Elliot & Papendick, 1988; Rickerl *et al.*, 1992; Rothrock *et al.*, 1995), and these can pose some threat to crops. Attack by damping off fungi sometimes also destroys substantial numbers of weed seedlings in high-residue systems (C. L. Mohler, personal observation). Whether disease-causing organisms promoted by residue can be manipulated to selectively control weeds without damaging crops remains to be determined. Factors that could be manipulated include type of mulch, crop planting date, direct seeded versus transplanted crops, and choice of crop cultivar. Separating the effects of reduced photosynthesis, etiolation, and increased humidity on disease susceptibility of particular weeds and crops could provide mechanistic insights into how to use these organisms successfully for weed management.

Residue effects on weed and crop performance

Data from studies investigating the combined chemical, physical, and biological impacts of crop residue on weeds and crops lead to two general conclusions. First, residue has the potential to suppress weeds while having a neutral or positive effect on crops, though this outcome is by no means universal. Second, weed responses to residue depend on the quantity of residue applied, whether or not it is incorporated into the soil, and the biology of the particular species involved.

An example of the successful use of soil-incorporated residue for weed management was reported by Boydston & Hang (1995), who measured weed growth in potato following bare fallow, sudangrass, and rapeseed green manure treatments. The sudangrass treatment was tilled in the fall, whereas fallow and rapeseed treatments were tilled the following spring, several weeks or several days before planting potato. Weed density and biomass (mostly the broadleaf annual species *Chenopodium album* and *Amaranthus retroflexus*) were strongly reduced in the rapeseed treatment compared with the bare fallow and sudangrass treatments (Figure 5.9). Total yield of potato tubers was 10% to 18% higher when the crop grew after rapeseed rather than after fallow, whether or not weeds were present. Sudangrass green manure raised potato yield 13% in one year, but tended to reduce yield in a second year, an effect the investigators attributed to leaching losses of N that occurred before potato could use nutrients released from the incorporated residue.

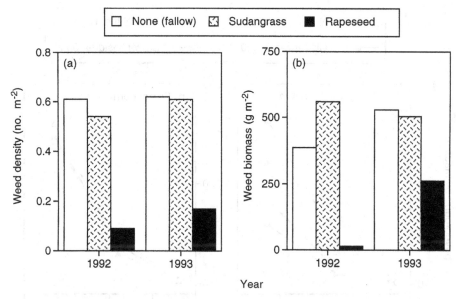

Figure 5.9 Mid-season density of weeds above the potato canopy (a) and final biomass of weeds in potato crops (b) following fallow, sudangrass, and rapeseed green manure treatments in a field experiment conducted in Washington, USA. (After Boydston & Hang, 1995.)

Rapeseed and many other cruciferous species contain glucosinolate compounds that hydrolyze to isothiocyanates. These have potent inhibitory effects on plant growth and seed germination (Brown & Morra, 1995). Boydston & Hang (1995) suggested that isothiocyanates were responsible for weed suppression by rapeseed residue, and that the residue's positive effect on potato growth may have been due to improved crop nutrition and suppression of soil-borne pathogens. They also suggested that potato's tolerance of rapeseed allelochemicals may have resulted from its large propagule size; small-seeded crops might be susceptible to rapeseed residue.

The ability of a legume green manure to suppress weed growth while supplying N to a crop was examined in field experiments investigating effects of crimson clover residue and ammonium nitrate fertilizer on *Chenopodium album* and sweet corn (Dyck & Liebman, 1994; Dyck, Liebman & Erich, 1995). Live clover plants were incorporated into soil shortly before planting *C. album* and sweet corn, and the synthetic N source was applied immediately after planting. Based on measurements of N uptake, the estimated fertilizer equivalency value of the clover green manure for the succeeding sweet corn crop was about 55 kg N ha^{-1} (Dyck, Liebman & Erich, 1995).

When *C. album* and sweet corn were grown in single-species stands at fixed densitites, biomass production by the weed species was significantly reduced

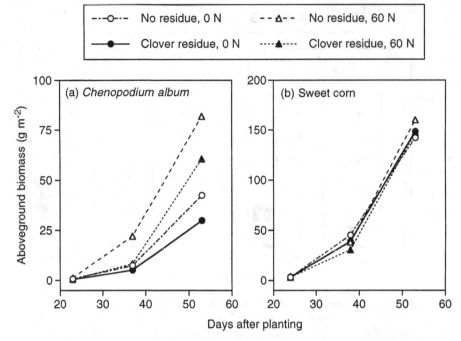

Figure 5.10 Effects of crimson clover residue and ammonium nitrate fertilizer on aboveground biomass production of *Chenopodium album* (a) and sweet corn (b) grown in pure stands in a field experiment conducted in Maine, USA. 0 and 60 indicate fertilizer rates in kg N ha^{-1}. (After Dyck & Liebman, 1994.)

by clover residue, but increased by N fertilizer (Figure 5.10a). At the end of the experiment, 53 days after planting, *C. album* biomass was 64% lower in the treatment receiving clover residue but no fertilizer compared with the treatment receiving 60 kg N ha^{-1} of synthetic fertilizer but no clover residue. In contrast, sweet corn biomass production was unaffected by clover residue or N fertilizer (Figure 5.10b), perhaps because of high background soil fertility or too short a growth period. Incorporation of clover residue had no significant effects on soil moisture and temperature, and no disease symptoms were observed (Dyck & Liebman, 1994).

Chenopodium album and sweet corn also were grown in competition at fixed densities in a two-year field experiment (Dyck, Liebman & Erich, 1995). Biomass of the weed species was 39% lower, on average, in a treatment receiving crimson clover green manure, but no synthetic fertilizer, than in a treatment receiving 45 kg N ha^{-1} as ammonium nitrate, but no clover residue. In contrast, when grown in competition with *C. album*, sweet corn biomass was 20% higher in the clover treatment than the synthetic fertilizer treatment. Comparison of sweet corn biomass from plots with and without *C. album* indicated that weed competition reduced crop growth by an average of 8% in the clover treatment, but by 28% in the synthetic fertilizer treatment. Thus, *C. album* was smaller and inflicted less damage on sweet corn when green manure, rather than synthetic fertilizer, was used as the major N source. Dyck, Liebman & Erich (1995) attributed these results to two factors: (i) lower levels of soil inorganic N present in the green manure treatment during the early portion of the growing season, which retarded growth of the weed but had little effect on the crop, and (ii) selective phytotoxic effects of clover residue on the weed.

In contrast to the reasonably high level of *C. album* suppression obtained from fresh crimson clover in the experiments described above, less effective and less consistent suppression was obtained from winter-killed, partially decomposed clover residue in an experiment with *C. album* and maize (Dyck & Liebman, 1995). Dyck & Liebman (1994) observed that fresh crimson clover residue incorporated into soil reduced *C. album* emergence, whereas Blum *et al.* (1997) found that it stimulated emergence of three other weed species. The factors responsible for these different outcomes might include differences in residue age, chemical composition, time of incorporation, environmental conditions, and weed identity, but more research is needed to understand their actual importance, especially if useful management practices are to be identified and damaging practices avoided.

The effects of crop residue used as mulch have been clarified by examining dose–response relationships (Figure 5.11) (Mohler & Teasdale, 1993; Teasdale & Mohler, 2000). For some weed species, such as *Amaranthus retroflexus*, *Chenopodium album*, and *Panicum capillare*, emergence declined monotonically as mulch rate increased (Figure 5.11). For others, such as *Abutilon theophrasti*, *Rumex crispus*, *Stellaria media*, and *Taraxacum officinale*, a low rate of mulch increased weed emergence. Because most effects of mulch on weed germination are negative, the enhancement of emergence at low mulch rates was probably due to improved water uptake by seeds in the moister environment under the mulch. Probably some of the difference between species in the shape of the mulch dose–response curve was due to propagule size, since surface-to-volume ratio favors the water uptake of small propagules relative to larger ones.

When soil moisture conditions are favorable for seed germination, the proportion of seedlings emerging through mulch relative to the number emerging in the unmulched condition, E, can usually be well described by a negative exponential curve (Teasdale & Mohler, 2000):

$$E = e^{-b \cdot \mathrm{MAI}}$$

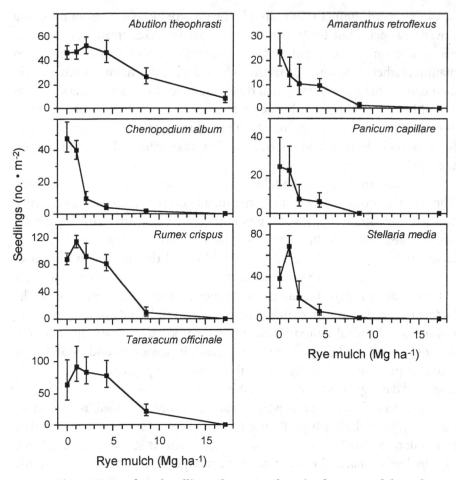

Figure 5.11 Number of seedlings of seven weed species that emerged through various rates of rye mulch in a field experiment conducted in New York, USA. (After Mohler & Teasdale, 1993.)

where MAI is mulch area index (i.e., mulch surface area per unit ground area) and b is a fitted constant. The magnitude of the exponential parameter, b, depends on the plant species and type of mulch. Most of the differences in weed emergence through the seven mulch materials investigated by Teasdale & Mohler (2000) could be accounted for by the proportion of solid volume in the mulch. That is, mulch materials with a high proportion of solid (e.g., bark chips, maize stalks) were more suppressive for a given MAI than were mulch materials with a low proportion of solid (e.g., oak leaves). This result may relate to the diffusion of reflected light into the mulch mass since the fraction of volume that was solid correlated with the light extinction coefficient of the mulches. Despite the well characterized allelopathic properties of rye (see section "Allelopathy" above), weed emergence through rye straw was no less than would have been expected for a non-phytotoxic mulch with rye's proportion of solid volume (Teasdale & Mohler, 2000). Rye roots have a disproportionate percentage of the plant's allelopathic activity (L. A. Weston, personal communication) and this may partially explain the apparent similarity between rye straw and other mulch materials in this and earlier studies (Teasdale & Mohler, 1992).

Most of the variation between plant species in ability to penetrate mulch is related to seed mass (Mohler, 2000). This allows good emergence of large-seeded crops like maize and soybean through mulch rates that severely suppress small-seeded weeds. Typically 5–10 Mg ha^{-1} of mulch is required to get substantial reduction in density of the small-seeded species (<2 mg) that dominate most weed communities; in some years, even higher mulch rates may be insufficient (Figure 5.11) (Mohler & Teasdale, 1993; Buhler, Mester & Kohler, 1996; Vidal & Bauman, 1996). Because fallow season cover crops typically only produce 2–4 Mg ha^{-1} of top growth, a mulch grown *in situ* should not be expected to provide adequate weed control by itself. Moreover, high mulch rates are likely to interfere with crop production. Wicks, Crutchfield & Burnside (1994) found, for example, that 6.8 Mg ha^{-1} of wheat straw gave better suppression of grass weeds than lower rates, but maximum maize yield was obtained at 4.4 Mg ha^{-1}. These considerations indicate that mulch should be considered as one component of an integrated weed management strategy rather than as a direct substitute for cultivation or herbicides.

The many studies using no-till planting into mulches grown *in situ* have largely supported these observations. In most of these studies mulch provided substantial suppression of annual weeds, but some additional weed management, usually in the form of herbicides, was necessary to obtain acceptable weed control and crop yield (Crutchfield, Wicks & Burnside, 1985; Shilling *et al.*, 1986; Wallace & Bellinder, 1992; Johnson, DeFelice & Helsel, 1993; Curran, Hoffman & Werner, 1994; Brecke & Shilling, 1996; Burgos & Talbert, 1996; Yenish, Worsham & York, 1996). In a few studies where mulch rate was low, the mulch had little effect on the weeds (Eadie *et al.*, 1992; Lanfranconi, Bellinder & Wallace, 1993). However, some studies have found that mulch alone provided adequate weed control in most or all location-years tested (Liebl *et al.*, 1992; Hoffman, Regnier & Cardina, 1993; Masiunas, Weston & Weller, 1995; Creamer *et al.*, 1996a; Smeda & Weller, 1996). Of these, all but Hoffman, Regnier & Cardina (1993) used rye or a mixture including rye, which supports the case for the effectiveness of rye's allelopathic properties under field conditions. In some cases, good weed control was probably also a consequence of low weed pressure (Hoffman, Regnier & Cardina 1993), or exceptionally heavy cover crop production (10–14 Mg ha^{-1} – Creamer *et al.*, 1996*b*).

Adequate weed control from mulch alone has often been observed when the mulch was applied from an outside source (Midmore, Roca & Berrios, 1986; Daisley *et al.*, 1988; Niggli, Weibel & Gut, 1990; Singh, Singh & Singh, 1991; Roe, Stoffella & Bryan, 1993; Davis, 1994), usually because of high application rates. Nevertheless, several studies found that supplemental weeding was needed, particularly when the mulch was applied at rates similar to those used in studies with *in situ* production (Obiefuna, 1986; Baryeh, 1987; Okugie & Ossom, 1988).

Effectiveness of residue for weed suppression often declines substantially after four to six weeks (Wallace & Bellinder, 1992; Creamer *et al.*, 1996*b*; Smeda & Weller, 1996; Vidal & Bauman, 1996), probably due to loss of mass through decomposition, and the breakdown of allelopathic compounds. In addition, condensed mulches that tend to retain water, such as sawdust, compost, and rotted baled hay, may provide a seed bed for the establishment of wind-borne seeds (Niggli, Weibel & Gut, 1990). This is rarely a problem with loose straw mulches that dry quickly following rain.

The foregoing discussion suggests three reasons why mulch-based weed management is a more viable strategy in tropical and warm temperate regions than in cool temperate regions. First, a warm fallow season provides better opportunities for production of more mulch biomass. Second, a greater range of cover crops are winter-hardy in warmer regions. Third, in cooler climates, mulch will generally lower soil temperatures to the detriment of the crop, whereas in hotter regions, lowered soil temperatures will pose less of a problem, and may be advantageous to crop production (Midmore, Roca & Berrios, 1986; Obiefuna, 1986).

Toward the integration of weed and soil management

Examples have been presented throughout this chapter illustrating how manipulations of soil temperature, moisture, nutrient, and residue conditions can alter weed density, growth, competitive ability, and community composition. In some cases the outcomes obtained are beneficial; in others they are undesirable. If soil management practices are to be used to regulate weeds in a consistently successful manner, a better understanding of the mechanisms is needed.

Serious relevant technical challenges confront researchers studying soil–weed relationships: fine roots are hard to recover, soil spatial heterogeneity introduces variability into field experiments, and soil chemical characteristics and microbial communities change rapidly and therefore require intensive sampling. Important questions remain unanswered concerning the necessary scope of investigations. For example, because a weed's response to resource conditions can depend on its genotype (Garbutt & Bazzaz, 1987; Tardif & Leroux, 1992), how large a pool of genotypes must be used to assess predictable species-level behavior? If the answer is that many genotypes must be tested in a range of environments, a significant research effort will be required.

Despite these challenges, a focus on soil management for weed regulation provides valuable opportunities to improve overall agroecosystem health. In particular, such a focus encourages scientists and farmers to integrate weed management with strategies for soil improvement and conservation. Soil solarization, flooding, banded nutrient applications, cover cropping with allelopathic species, and other practices clearly can play desirable roles in regulating weed populations, but the soil is more than just a medium in which to suppress and kill weeds – it is a resource that must be enhanced and protected for long-term crop production. The use of organic materials as soil amendments and the development of weed management machinery for mulch systems are two areas where weed and soil management could be integrated and simultaneously improved.

Soil amendment systems

Farmers, soil scientists, and agronomists have long recognized that regular addition of organic materials can markedly improve soil quality (Magdoff, 1992, pp. 23–38). As noted previously, cover crop residues can promote crop growth and yield by improving soil fertility, water availability, and aeration. Animal manures, composts, and other organic materials offer similar benefits and can be used in various combinations with cover crops to provide diversified soil amendment systems adapted to local conditions. Amending soil with organic materials may also improve crop performance by reducing pest pressures. Organic matter amendments have been shown to render crops less attractive to insect pests (Phelan, Mason & Stinner, 1995), and reduce crop disease problems by promoting better soil structure, more vigorous root systems, and greater populations of soil organisms that antagonize, outcompete, parasitize, or consume crop pathogens (van Bruggen, 1995).

Although the relationship between soil quality and weed dynamics has received little attention from researchers, data from a cropping systems study in northern Maine suggest that organic amendments can confer weed management benefits (Gallandt et al., 1998a), as well as improve soil physical and chemical characteristics (Gallandt *et al.*, 1998*b*). In this experiment, potato was planted in a two-year rotation sequence with either barley grown for grain or a mixture of oat, pea, and hairy vetch grown for green manure. In an "amended" soil management treatment, beef manure, cull potato compost, and low rates of synthetic fertilizer were applied to potato following the green manure crop. In an "unamended" soil management treatment, a high rate of synthetic fertilizer but no beef manure or compost was applied to potato following barley. Three contrasting weed management treatments were used during the potato phase of the green manure–potato-and-barley–potato rotations: full labeled rates of herbicides (metribuzin and paraquat) and zero or one cultivation other than standard hilling operations ("conventional"); half-rates of herbicides and one cultivation ("reduced input"); and one to three cultivations but no herbicides ("mechanical").

Applications of organic amendments significantly increased soil organic matter and water-stable aggregate content, cation exchange capacity, available P, K, Mg, and Ca, and potato leaf area and tuber yield (Gallandt *et al.*, 1998*a*, 1998*b*). By the fourth and fifth years of the study (1994 and 1995), weed growth in potato was strongly affected by an interaction between weed management and soil management systems (Figure 5.12) (Gallandt *et al.*, 1998*a*). When herbicides were applied (i.e., the conventional and reduced input treatments), weed growth in potato was minimal regardless of the soil management treatment. However, when herbicides were not applied (i.e., the mechanical treatment), significant differences in weed growth were evident between soil management treatments: amendments reduced weed biomass by 72% to 77%. Reductions in weed biomass occurred even when weed densities in the contrasting soil management treatments were equal, leading Gallandt *et al.* (1998*a*) to conclude that improvements in soil quality due to use of organic amendments promoted a more vigorous potato crop that was better able to compete with weeds.

Before generalizations can be drawn, the impact of organic matter amendments on crop–weed interactions needs to be examined in other cropping systems. Research is also needed to understand how soil amendments affect weed seed survival, seedling recruitment, and growth. Amendment-related factors that might affect weed dynamics include changes in communities of microbes and insects that attack weed seeds and seedlings, alterations of soil physical properties influencing safe sites for weed germination and establishment, increased concentrations of amendment-derived phytotoxins and growth stimulants, shifts in the timing of nutrient availability, and differential responses between crop and weed species to these factors (Gallandt, Liebman & Huggins, 1999; Liebman & Davis, 2000). The diversity of issues

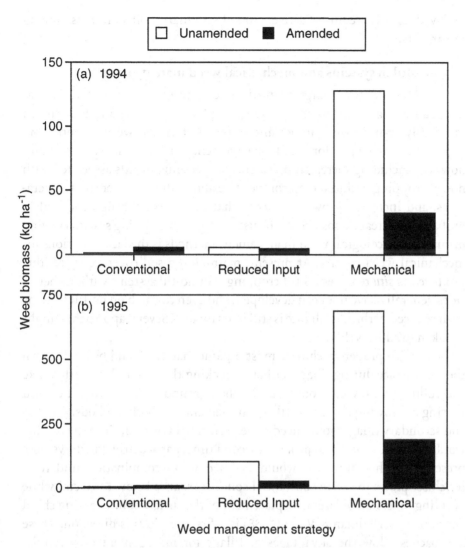

Figure 5.12 Mid-season weed biomass in 1994 (a) and 1995 (b) in potato crops managed with herbicides applied at full labeled rates and zero or one cultivation ("conventional"); with herbicides applied at half-rates and one cultivation ("reduced input"); or with one to three cultivations but no herbicides ("mechanical"). Potato in the "unamended" treatment followed a barley grain crop and received a high rate of synthetic fertilizer; potato in the "amended" treatment followed a pea–oat–vetch green manure crop and received compost, beef manure, and a low rate of synthetic fertilizer. The experiment was conducted in Maine, USA. (After Gallandt *et al.*, 1998*a*.)

involved clearly requires active cross-disciplinary collaborations among researchers.

Mulch systems and mechanical weed management

Maintenance of large quantities of crop residue on the soil surface has major advantages for soil conservation (Langdale *et al.*, 1994), and as discussed previously, can provide important opportunities for weed suppression. Significant challenges for weed management can occur in such systems, however, including increased prevalence of perennial weeds associated with no-till cropping, reduced crop competitive ability due to lower soil temperatures, and increased growth of weeds that do emerge through the mulch. Probably the greatest obstacle to the use of mulch in cropping systems relying primarily on ecological weed management methods is the lack of options for mechanical weed control. At present, maintaining soil coverage by residue produced *in situ* requires no-till cropping, but no-till systems without herbicides generally have not been developed, and even research on no-till systems with reduced herbicide reliance is still in its infancy. Several approaches might be taken to address this challenge.

First, with proper machinery most organic material could be retained on the soil surface during tillage, either by picking the material up with a rake and rolling or blowing it onto freshly tilled ground, or by picking it up and passing or blowing it over the tillage implement. For both methods, primary and secondary tillage would need to be performed together, since the residue would be more difficult to pick up a second time, particularly if furrows were present. Another approach would be to develop combination undercutter/chisel plows to create substantial soil disturbance below 5–10 cm, while avoiding inversion or lateral displacement of the surface soil. The wing chisel developed by Heilman & Valco (1988) is a first step in this direction. These approaches allow the advantages of tillage for management of perennial weeds, while still retaining large quantities of residue on the soil surface for soil conservation and suppression of annual weeds.

Second, although continuous no-till cropping on a large scale is probably impossible without herbicides, growing specific crops without tillage and herbicides may be feasible within a rotating tillage regime. Peters, Bohlke & Janke (1990) found that maize planted without tillage into mowed hairy vetch had adequate weed control and was as productive as conventionally planted maize, provided the ground was tilled before planting the vetch. Thus, even if tillage is considered necessary for suppression of perennial weeds, the time of tillage perhaps can be shifted from spring-planted row crops, where the potential for soil erosion is substantial, to fall-seeded cover crops, where erosion is less of a problem. Such systems also provide an additional option for shifting tillage from spring to fall in regions where spring-sown row crops predominate (see Chapter 7).

Third, crops can be planted without tillage into killed cover crops, and then cultivated with high-residue cultivators after the crop is established. The principal soil conservation advantage of no-tillage crop production occurs between planting and canopy closure, since cover crops can be used to protect soil during the preceding fallow season regardless of the tillage regime. After a main crop is established in a cover crop mulch, its leaves and roots increasingly protect the soil, and inter-row cultivation presents only a minor risk of erosion. With regard to weed control, the principal point of mulch is to suppress weeds in the crop row. In many studies in which weed control by mulch alone was considered inadequate, overall weed control would have been satisfactory if weeds in the inter-row area had been eliminated by cultivation.

As discussed in Chapter 4, several types of implements are available that can cultivate without jamming on dense residue, though none is ideal with regard to soil movement and stability in hard, untilled ground (C. L. Mohler, J. Mt. Pleasant & J. C. Frisch, unpublished data). Few studies have examined cultivation in mulch systems, but Liebman et al. (1995) found that two inter-row cultivations provided adequate weed suppression in dry bean after planting without tillage into herbicide-killed rye provided field margins were mowed to prevent seed production by *Taraxacum officinale*. Cultivation of untilled fields covered by mulch remains a challenge and continued work on this problem by agricultural engineers and farmers is highly desirable.

Mowing is an alternative approach to cultivation for controlling weeds between rows in mulch systems. Although mowing generally provides only short-term weed suppression, this may be adequate in a vigorously growing crop, especially if weed pressure has been reduced by the mulch. Baryeh (1987) found inter-row mowing was as effective as hoeing or hand pulling at reducing weed biomass and preserving maize yield. However, mowing significantly speeded decomposition of the dead velvet bean mulch relative to the other weeding methods.

Machinery can be used to address the problem of too much mulch for optimal crop emergence and growth, as well as the converse problem of insufficient mulch for adequate weed suppression. Several types of "trash wheels" are currently marketed to remove residue from the crop row. These decrease problems that arise when attempting to no-till plant through heavy residue, and allow quicker warming of the soil in spring-planted crops in cool climates (Swan, Kaspar & Erbach, 1996). Their performance in the heavy residue from killed cover crops needs to be explored.

In warm climates, and for crops that can emerge through dense residue, trailing wheels that brush residue into the crop row behind the planter unit would allow concentration of mulch in the zone that is most difficult to weed mechanically. A three-fold concentration of mulch over the row should be possible, and in many cases this would increase the mulch from an ineffective to an effective rate.

Finally, better methods could be developed for moving crop residues on the farm. Moving mulch materials is rarely cost-effective for producing field crops, but may be practical for some horticultural crops. Some fruit and vegetable growers have fields that are unsuited to intensive management of high-value crops. These could be planted to forage species and mowed periodically to provide material for soil improvement and weed suppression on the intensively managed fields. Care would need to be practiced to prevent introduction of weed seeds with the mulch. However, few weeds of perennial forages go to seed in late spring when the mulch would be most useful for production of summer crops. The biggest hurdle to moving mulch is that machines currently available for spreading bulk materials are not well adapted to spreading mulch between established plants. In some tropical regions, forest leaf litter is moved to nearby fields (Wilken, 1977), and simple human- or animal-powered machines could greatly reduce the labor involved. Mechanical management of mulch will likely be a fertile field for collaboration between farmers, agricultural engineers, weed scientists, and ecologists in the coming years.

REFERENCES

Alkämper, J. (1976). Influence of weed infestation on effect of fertilizer dressings. *Pflanzenschutz-Nachrichten Bayer*, **29**, 191–235.

Alkämper, J., Pessios, E., & Long, D. V. (1979). Einfluss der dungung auf die Entwicklung und Nahrstoffaufnahme verschiedener Unkrauter in Mais. *Proceedings of the European Weed Research Society (1979)*, 181–92.

Anderson, R. L. (1991). Timing of nitrogen application affects downy brome (*Bromus tectorum*) growth in winter wheat. *Weed Technology*, **5**, 582–5.

Angonin, C., Caussanel, J. P., & Meynard, J. M. (1996). Competition between winter wheat and *Veronica hederifolia*: influence of weed density and the amount and timing of nitrogen application. *Weed Research*, **36**, 175–87.

Appleby, A. P., Olson, P. D., & Colbert, D. R. (1976). Winter wheat yield reduction from interference by Italian ryegrass. *Agronomy Journal*, **68**, 463–6.

Azam, F., Malik, K. A., & Sajjad, M. I. (1985). Transformations in soil and availability to plants of ^{15}N applied as inorganic fertilizer and legume residues. *Plant and Soil*, **86**, 3–13.

Ball, D. A., Wysocki, D. J., & Chastain, T. G. (1996). Nitrogen application timing effects on downy brome (*Bromus tectorum*) and winter wheat (*Triticum aestivum*) growth and yield. *Weed Technology*, **10**, 305–10.

Barnes, J. P., & Putnam, A. R. (1986). Evidence for allelopathy by residues and aqueous extracts of rye (*Secale cereale*). *Weed Science*, 34, 384–90.

Barnes, J. P., & Putnam, A. R. (1987). Role of benzoxazinones in allelopathy by rye (*Secale cereale* L.). *Journal of Chemical Ecology*, 13, 889–906.

Barnes, J. P., Putnam, A. R., Burke, B. A., & Aasen, A. A. (1987). Isolation and characterization of allelochemicals in rye herbage. *Phytochemistry*, 26, 1385–90.

Barrett, S. C. H. (1983). Crop mimicry in weeds. *Economic Botany*, 37, 255–82.

Baryeh, E. A. (1987). Supplemental mechanical weed control for maize–cowpea rotation in mucuna mulch. *Agriculture International*, 39, 167–71.

Bell, C. E., Elmore, C. L., & Durazo, A III. (1988). Soil solarization for weed management in vegetables. In *Global Perspectives on Agroecology and Sustainable Agricultural Systems, Proceedings of the 6th International Scientific Conference of the International Federation of Organic Agriculture Movements*, ed. P. Allen and D. Van Dusen, pp. 475–9. Santa Cruz, CA: University of California Agroecology Program.

Bender, J. (1994). *Future Harvest: Pesticide-Free Farming*. Lincoln, NE: University of Nebraska Press.

Benech-Arnold, R. L., Ghersa, C. M., Sanchez, R. A., & Garcia Fernandez, A. E. (1988). The role of fluctuating temperatures in the germination and establishment of *Sorghum halepense* (L.) Pers.: regulation of germination under leaf canopies. *Functional Ecology*, 2, 311–18.

Bhowmik, P. C., & Doll, J. D. (1982). Corn and soybean response to allelopathic effects of weed and crop residues. *Agronomy Journal*, 74, 601–6.

Bhowmik, P. C., & Doll, J. D. (1984). Allelopathic effects of annual weed residues on growth and nutrient uptake of corn and soybeans. *Agronomy Journal*, 76, 383–8.

Blevins, R. L., Smith, M. S., & Thomas, G. W. (1984). Changes in soil properties under no-tillage. In *No-Tillage Agriculture*, ed. R. E. Phillips and S. H. Phillips, pp. 190–230. New York: Van Nostrand Reinhold.

Blum, U., King, L. D., Gerig, T. M., Lehman, M. E., & Worsham, A. D. (1997). Effects of clover and small grain cover crops and tillage techniques on seedling emergence of some dicotyledonous weed species. *American Journal of Alternative Agriculture*, 12, 146–61.

Boydston, R. A., & Hang, A. (1995). Rapeseed (*Brassica napus*) green manure crop suppresses weeds in potato (*Solanum tuberosum*). *Weed Technology*, 9, 669–75.

Bradow, J. M. (1993). Inhibitions of cotton seedling growth by volatile ketones emitted by cover crop residues. *Journal of Chemical Ecology*, 19, 1085–108.

Bradow, J. M., & Connick, W. J. (1990). Volatile seed germination inhibitors from plant residues. *Journal of Chemical Ecology*, 16, 645–66.

Brecke, B. J., & Shilling, D. G. (1996). Effect of crop species, tillage, and rye (*Secale cereale*) mulch on sicklepod (*Senna obtusifolia*). *Weed Science*, 44, 133–6.

Bristow, K. L. (1988). The role of mulch and its architecture in modifying soil temperature. *Australian Journal of Soil Research*, 26, 269–80.

Brown, P. D., & Morra, M. J. (1995). Glucosinolate-containing plant tissues as bioherbicides. *Journal of Agricultural and Food Chemistry*, 43, 3070–4.

Brust, G. E., & House, G. J. (1988). Weed seed destruction by arthropods and rodents in low-input soybean agroecosystems. *American Journal of Alternative Agriculture*, 3, 19–25.

Buckles, D., Ponce, I., Sain, G., & Medina, G. (1994). Cowardly land becomes brave: the use and diffusion of fertilizer bean (*Mucuna deeringianum*) on the hillsides of Atlantic Honduras. In *Tapado Slash/Mulch: How Farmers Use It and What Researchers Know About It*, ed. H. D. Thurston, M. Smith, G. Abawi & S. Kearl, pp. 247–61. Ithaca, NY: Cornell International Institute for Food, Agriculture and Development.

Buhler, D. D., Mester, T. C., & Kohler, K. A. (1996). The effect of maize residues and tillage on emergence of *Setaria faberi*, *Abutilon theophrasti*, *Amaranthus retroflexus* and *Chenopodium album*. *Weed Research*, **36**, 153–65.

Burgos, N. R., & Talbert, R. E. (1996). Weed control and sweet corn (*Zea mays* var. *rugosa*) response in a no-till system with cover crops. *Weed Science*, **44**, 355–61.

Carlson, H. L., & Hill, J. E. (1985). Wild oat (*Avena fatua*) competition with spring wheat: effects of nitrogen fertilization. *Weed Science*, **34**, 29–33.

Chang, C.-F., Suzuki, A., Kumai, S., & Tamura, S. (1969). Chemical studies on "clover sickness." II. Biological functions of isoflavonoids and their related compounds. *Agricultural and Biological Chemistry*, **33**, 398–408.

Chase, W. R., Nair, M. G., & Putnam, A. R. (1991). 2,2'-oxo-1,1'-azobenzene: selective toxicity of rye (*Secale cereale* L.) allelochemicals to weed and crop species: II. *Journal of Chemical Ecology*, **17**, 9–19.

Chung, I. M., & Miller, D. A. (1995). Natural herbicide potential of alfalfa residue on selected weed species. *Agronomy Journal*, **87**, 920–5.

Cochran, V. L., Elliot, V. F., & Papendick, R. I. (1977). The production of phytotoxins from surface crop residues. *Soil Science Society of America Journal*, **41**, 903–8.

Cochran, V. L., Morrow, L. A., & Schirman, R. D. (1990). The effect of N placement on grass weeds and winter wheat responses in three tillage systems. *Soil and Tillage Research*, **18**, 347–55.

Cornacchia, J. W., Cohen, D. B., Bowes, G. W., Schnagl, R. J., & Montoya, B. L. (1984). *Rice Herbicides: Molinate and Thiobencarb*, Special Project Report 84-4. Sacramento, CA: California State Water Research Control Board.

Creamer, N. G., Plassman, B., Bennett, M. A., Wood, R. K., Stinner, B. R., & Cardina, J. (1995). A method for mechanically killing cover crops to optimize weed suppression. *American Journal of Alternative Agriculture*, **10**, 157–62.

Creamer, N. G., Bennett, M. A., Stinner, B. R., & Cardina, J. (1996a). A comparison of four processing tomato production systems differing in cover crop and chemical inputs. *Journal of the American Society for Horticultural Science*, **121**, 559–68.

Creamer, N. G., Bennett, M. A., Stinner, B. R., Cardina, J., & Regnier, E. E. (1996b). Mechanisms of weed suppression in cover crop-based production systems. *HortScience*, **31**, 410–13.

Crutchfield, D. A., Wicks, G. A., & Burnside, O. C. (1985). Effect of winter wheat (*Triticum aestivum*) straw mulch level on weed control. *Weed Science*, **34**, 110–14.

Curran, W. S., Hoffman, L. D., & Werner, E. L. (1994). The influence of a hairy vetch (*Vicia villosa*) cover crop on weed control and corn (*Zea mays*) growth and yield. *Weed Technology*, **8**, 777–84.

Dabney, S. M., Buehring, N. W., & Reginelli, D. B. (1991). Mechanical control of legume cover crops. In *Cover Crops for Clean Water*, ed. W. L. Hargrove, pp. 146–7. Ankeny, IA: Soil and Water Conservation Society.

Daisley, L. E. A., Chong, S. K., Olsen, F. J., Singh, L., & George, C. (1988). Effects of surface-applied grass mulch on soil water content and yields of cowpea and eggplant in Antigua. *Tropical Agriculture (Trinidad)*, **65**, 300–4.

Davis, J. M. (1994). Comparison of mulches for fresh-market basil production. *HortScience*, **29**, 267–8.

DiTomaso, J. M. (1995). Approaches for improving crop competitiveness through the manipulation of fertilization strategies. *Weed Science*, **43**, 491–7.

Dou, Z., Fox, R. H., & Toth, J. D. (1995). Seasonal soil nitrate dynamics in corn as affected by tillage and nitrogen source. *Soil Science Society of America Journal*, **59**, 858–64.

Dyck, E., & Liebman, M. (1994). Soil fertility management as a factor in weed control: the effect of crimson clover residue, synthetic nitrogen fertilizer, and their interaction on emergence and early growth of lambsquarters and sweet corn. *Plant and Soil*, **167**, 227–37.

Dyck, E., & Liebman, M. (1995). Crop–weed interference as influenced by a leguminous or synthetic fertilizer nitrogen source. II. Rotation experiments with crimson clover, field corn, and lambsquarters. *Agriculture, Ecosystems and Environment*, **56**, 109–20.

Dyck, E., Liebman, M., & Erich, M. S. (1995). Crop–weed interference as influenced by a leguminous or synthetic fertilizer nitrogen source. I. Doublecropping experiments with crimson clover, sweet corn, and lambsquarters. *Agriculture, Ecosystems and Environment*, **56**, 93–108.

Eadie, A. G., Swanton, C. J., Shaw, J. E., & Anderson, G. W. (1992). Integration of cereal cover crops in ridge-tillage corn (*Zea mays*) production. *Weed Technology*, **6**, 553–60.

Edwards, C. A. (1975). Effects of direct drilling on the soil fauna. *Outlook on Agriculture*, **8**, 243–4.

Egley, G. H. (1983). Weed seed and seedling reductions by soil solarization with transparent polyethylene sheets. *Weed Science*, **31**, 404–9.

Einhellig, F. A. (1986). Interactions among allelochemicals and other stress factors of the plant environment. In *Allelochemicals: Role in Agriculture and Forestry*, ed. G. R. Waller, pp. 343–57. Washington, DC: American Chemical Society.

Einhellig, F. A., & Leather, G. R. (1988). Potentials for exploiting allelopathy to enhance crop production. *Journal of Chemical Ecology*, **14**, 1829–44.

Elmore, C. L., Roncoroni, J. A., & Giraud, D. D. (1993). Perennial weeds respond to control by soil solarization. *California Agriculture*, **47**, 19–22.

Enache, A. J., & Ilnicki, R. D. (1990). Weed control by subterranean clover (*Trifolium subterraneum*) used as a living mulch. *Weed Technology*, **4**, 534–8.

Ewel, J., Berish, C., Brown, B., Price, N., & Raich, J. (1981). Slash and burn impacts on a Costa Rican wet forest site. *Ecology*, **62**, 816–29.

Fay, P. K., & Duke, W. B. (1977). An assessment of allelopathic potential in *Avena* germplasm. *Weed Science*, **25**, 224–8.

Forcella, F., Wilson, R. G., Dekker, J., Kremer, R. J., Cardina, J., Anderson, R. L., Alm, D., Renner, K. A., Harvey, R. G., Clay, S. & Buhler, D. D. (1997). Weed seed bank emergence across the corn belt. *Weed Science*, **45**, 67–76.

Fortin, M.-C., & Pierce, F. J. (1991). Timing and nature of mulch retardation of corn vegetative development. *Agronomy Journal*, **83**, 258–63.

Fox, R. H., Myers, R. J. K., & Vallis, I. (1990). The nitrogen mineralization rate of legume residues in soil as influenced by their polyphenol, lignin, and nitrogen contents. *Plant and Soil*, **129**, 251–9.

Gagliardo, R. W., & Chilton, W. S. (1992). Soil transformation of 2(3H)-benzoxazolone of rye into phytotoxic 2-amino-3H-phenoxazin-3-one. *Journal of Chemical Ecology*, **18**, 1683–91.

Gallandt, E. R., Liebman, M., & Huggins, D. R. (1999). Improving soil quality: implications for weed management. *Journal of Crop Production*, **2**, 95–121.

Gallandt, E. R., Liebman, M., Corson, S., Porter, G. A., & Ullrich, S. D. (1998a). Effects of pest and soil management systems on weed dynamics in potato. *Weed Science*, **46**, 238–48.

Gallandt, E. R., Mallory, E. B., Alford, A. R., Drummond, F. A., Groden, E., Liebman, M., Marra, M. C., McBurnie, J. C., & Porter, G. A. (1998b). Comparison of alternative pest and soil management strategies for Maine potato production systems. *American Journal of Alternative Agriculture*, **13**, 146–61.

Gamliel, A., & Stapleton, J. J. (1997) Improvement of soil solarization with volatile compounds generated from organic amendments. *Phytoparasitica*, **25** (Supplement), 31S–38S.

Garbutt, K., & Bazzaz, F. A. (1987). Population niche structure: differential response of *Abutilon theophrasti* progeny to resource gradients. *Oecologia*, **72**, 291–6.

Giovanninni, G., Benvenuti, S., Lucchesi, S.,, & Giachetti, M. (1993). Weed reduction by burning straw and stubble in the field: positive responses and potential hazards. In *Soil Biota, Nutrient Cycling, and Farming Systems*, ed. M. G. Paoletti, W. Foissner & D. C. Coleman, pp. 279–85. Boca Raton, FL: Lewis Publishers.

Grant, J. D. (1983). The activities of earthworms and the fates of seeds. In *Earthworm Ecology*, ed. J. E. Satchell, pp. 107–22. London: Chapman & Hall.

Grattan, S. R., Schwankl, L. J., & Lanini, W. T. (1988). Weed control by subsurface drip irrigation. *California Agriculture*, **42**, 22–4.

Griffith, D. R., Mannering, J. V., & Box, J. E. (1986). Soil and moisture management with reduced tillage. In *No-Tillage and Surface-Tillage Agriculture: The Tillage Revolution*, ed. M. A. Sprague & G. B. Triplett, pp. 19–57. New York: John Wiley.

Grubb, P. J. (1977). The maintenance of species-richness in plant communities: the importance of the regeneration niche. *Biological Reviews*, **52**, 107–45.

Guenzi, W. D., & McCalla, T. M. (1962). Inhibition of germination and seedling development by crop residues. *Soil Science Society of America Proceedings*, **26**, 456–8.

Hargrove, W. L. (ed.) (1991). *Cover Crops for Clean Water*. Ankeny, IA: Soil and Water Conservation Society.

Harper, J. L. (1977). *Population Biology of Plants*. London: Academic Press.

Harris, G. H., Hesterman, O. B., Paul, E. A., Peters, S. E., & Janke, R. R. (1994). Fate of legume and fertilizer nitrogen-15 in a long-term cropping systems experiment. *Agronomy Journal*, **86**, 910–15.

Heilman, M. D., & Valco, T. D. (1988). Wing-chisel plow for in-row conservation tillage. *Agronomy Journal*, **80**, 1009–11.

Henson, I. E. (1970). The effects of light, potassium nitrate and temperature on the germination of *Chenopodium album*. *Weed Research*, **10**, 27–39.

Hill, J. E., Roberts, S. R., Bayer, D. E., & Williams, J. F. (1990). Crop response and weed control from new herbicide combinations in water-seeded rice (*Oryza sativa*). *Weed Technology*, **4**, 838–42.

Hoffman, M. J., Regnier, E. E, & Cardina, J. (1993). Weed and corn (*Zea mays*) responses to a hairy vetch (*Vicia villosa*) cover crop. *Weed Technology*, **7**, 594–9.

Honeycutt, C. W., Clapham, W. M., & Leach, S. S. (1994). A functional approach to efficient nitrogen use in crop production. *Ecological Modelling*, **73**, 51–61.

Honeycutt, C. W., Potaro, L. J., & Halteman, W. A. (1991). Predicting nitrate formation from soil, fertilizer, crop residue, and sludge in thermal units. *Journal of Environmental Quality*, **20**, 850–6.

Honeycutt, C. W., Potaro, L. J., Avila, K. L., & Halteman, W. A. (1993). Residue quality, loading rate, and soil temperature relations with hairy vetch (*Vicia villosa*) residue carbon, nitrogen, and phosphorus mineralization. *Biological Agriculture and Horticulture*, **9**, 181–99.

Horowitz, M., Regev, Y., & Herzlinger, G. (1983). Solarization for weed control. *Weed Science*, **31**, 170–9.

Ilnicki, R. D., & Enache, A. J. (1992). Subterranean clover living mulch: an alternative method of weed control. *Agriculture, Ecosystems and Environment*, **40**, 249–64.

Janzen, H. H., Bole, J. B., Biebderbeck, V. O., & Slinkard, A. E. (1990). Fate of N applied as green manure or ammonium sulphate fertilizer to soil subsequently cropped with spring wheat in three sites in western Canada. *Canadian Journal of Soil Science*, **70**, 313–23.

Johnson, G. A., DeFelice, M. S., & Helsel, Z. R. (1993). Cover crop management and weed control in corn (*Zea mays*). *Weed Technology*, **7**, 425–30.

Karssen, C. M., & Hilhorst, H. W. M. (1992). Effect of chemical environment on seed germination. In *Seeds: The Ecology of Regeneration in Plant Communities*, ed. M. Fenner, pp. 327–348. Wallingford, UK: CAB International.

Kaspar, T. C., Erbach, D. C., & Cruse, R. M. (1990). Corn response to seed-row residue removal. *Soil Science Society of America Journal*, **54**, 1112–17.

Katan, J. (1987). Soil solarization. In *Innovative Approaches to Plant Disease Control*, ed. I. Chet, pp. 77–105. New York: John Wiley.

Kempen, H. M. (1987). *Growers' Weed Management Guide*. Fresno, CA: Thomson Publications.

Killham, K. (1994). *Soil Ecology*. Cambridge, UK: Cambridge University Press.

Kimber, R. W. L. (1973). Phytotoxicity from plant residues. II. The effect of time of rotting straw from some grasses and legumes on the growth of wheat seedlings. *Plant and Soil*, **38**, 347–61.

Kropff, M. J. (1993). Mechanisms of competition for nitrogen. In *Modelling Crop–Weed Interactions*, ed. M. J. Kropff & H. H. van Laar, pp. 77–82. Wallingford, UK: CAB International.

Kropff, M. J., & van Laar, H. H. (eds) (1993). *Modelling Crop–Weed Interactions*. Wallingford, UK: CAB International.

Kropff, M. J., Van Keulen, N. C., Van Laar, H. H., & Schnieders, B. J. (1993). The impact of environmental and genetic factors. In *Modelling Crop–Weed Interactions*, ed. M. J. Kropff & H. H. van Laar, pp. 137–48. Wallingford, UK: CAB International.

Kumar, B., Yaduraju, N. T., Ahuja, K. N., & Prasad, D. (1993). Effect of soil solarization on weeds and nematodes under tropical Indian conditions. *Weed Research*, **33**, 423–9.

Ladd, J. N., & Amato, M. (1986). The fate of nitrogen from legume and fertilizer sources in soils successively cropped with wheat under field conditions. *Soil Biology and Biochemistry*, **18**, 417–25.

Laing, D. R., Jones, P. G., & Davis, J. H. C. (1984). Common bean (*Phaseolus vulgaris* L.). In *The Physiology of Tropical Field Crops*, ed. P. R. Goldsworthy & N. M. Fisher, pp. 305–51. New York: John Wiley.

Lanfranconi, L. E., Bellinder, R. R., & Wallace, R. W. (1993). Grain rye residues and weed control strategies in reduced tillage potatoes. *Weed Technology*, **7**, 23–8.

Langdale, G. W., Alberts, E. E., Bruce, R. R., Edwards, W. M., & McGregor, K. C. (1994).

Concepts of residue management: infiltration, runoff, and erosion. In *Crops Residue Management*, ed. J. L. Hatfield & B. A. Stewart, pp. 109–24. Boca Raton, FL: Lewis Publishers.

Leather, G. R. (1983). Sunflowers (*Heliothis annuus*) are allelopathic to weeds. *Weed Science*, **31**, 37–42.

Liebl, R. A., & Worsham, A. D. (1983). Inhibition of pitted morning glory (*Ipomoea lacunosa* L.) and certain other weed species by phytotoxic components of wheat (*Triticum aestivum* L.) straw. *Journal of Chemical Ecology*, **9**, 1027–43.

Liebl, R. A., Simmons, F. W., Wax, L. M., & Stoller, E. W. (1992). Effect of rye (*Secale cereale*) mulch on weed control and soil moisture in soybean (*Glycine max*). *Weed Technology*, **6**, 838–46.

Liebman, M., & Davis, A. S. (2000). Integration of soil, crop, and weed management in low-external-input farming systems. *Weed Research*, **40**, 27–47.

Liebman, M., Corson, S., Rowe, R. J., & Halteman, W. A. (1995). Dry bean responses to nitrogen fertilizer in two tillage and residue management systems. *Agronomy Journal*, **87**, 538–46.

Liu, D. L., & Lovett, J. V. (1993*a*). Biologically active secondary metabolites of barley. I. Developing techniques and assessing allelopathy in barley. *Journal of Chemical Ecology*, **19**, 2217–30.

Liu, D. L., & Lovett, J. V. (1993*b*). Biologically active secondary metabolites of barley. II. Phytotoxicity of barley allelochemicals. *Journal of Chemical Ecology*, **19**, 2231–44.

Luna-Orea, P., Wagger, M. G., & Gumpertz, M. L. (1996). Decomposition and nutrient release dynamics of two tropical legume cover crops. *Agronomy Journal*, **88**, 758–64.

Lund, R. D., & F. T. Turpin. (1977). Carabid damage to weed seeds found in Indiana cornfields. *Environmental Entomology*, **6**, 695–8.

Madrigal, R. R. Q. (1994). Some potential for leguminous species to help recover stability in agroecosystems. In *Tapado Slash/Mulch: How Farmers Use It and What Researchers Know About It*, ed. H. D. Thurston, M. Smith, G. Abawi & S. Kearl, pp. 223–31. Ithaca, NY: Cornell International Institute for Food, Agriculture and Development.

Magdoff, F. (1992). *Building Soils for Better Crops: Organic Matter Management*. Lincoln, NE: University of Nebraska Press.

Marschner, H. (1995). *Mineral Nutrition of Higher Plants*, 2nd edn. London: Academic Press.

Masiunas, J. B., Weston, L. A., & Weller, S. C. (1995). The impact of rye cover crops on weed populations in a tomato cropping system. *Weed Science*, **43**, 318–23.

McCalla, T. M., & Duley, F. L. (1948). Stubble mulch studies: effect of sweetclover extract on corn germination. *Science*, **108**, 163.

McKenzie, F. R. (1996). Influence of applied nitrogen on weed invasion of *Lolium perenne* pastures in a subtropical environment. *Australian Journal of Experimental Agriculture*, **36**, 657–60.

McRill, M. (1974). The ingestion of weed seed by earthworms. In *Proceedings of the 12th British Weed Control Conference*, pp. 519–24. London: British Crop Protection Council.

McRill, M., & Sagar, G. R. (1973). Earthworms and seeds. *Nature*, **243**, 482.

McWhorter, C. G. (1972). Flooding for johnsongrass control. *Weed Science*, **20**, 238–41.

Midmore, D. J., Roca, J., & Berrios, D. (1986). Potato (*Solanum* spp.) in the hot tropics. III.

Influence of mulch on weed growth, crop development, and yield in contrasting environments. *Field Crops Research*, **15**, 109–24.

Miller, P. R., Graves, W. L., & Williams, W. A. (1989). *Cover Crops for California Agriculture*, Leaflet no. 21471. Oakland, CA: University of California, Division of Agriculture and Natural Resources.

Mitchell, W. H., & Teel, M. R. (1977). Winter-annual cover crops for no-tillage corn production. *Agronomy Journal*, **69**, 569–73.

Mohler, C. L. (1991). Effects of tillage and mulch on weed biomass and sweet corn yield. *Weed Technology*, **5**, 545–52.

Mohler, C. L. (1995). A living mulch (white clover) / dead mulch (compost) weed control system for winter squash. *Proceedings of the Northeastern Weed Science Society*, **49**, 5–10.

Mohler, C. L. (1996). Ecological bases for the cultural control of annual weeds. *Journal of Production Agriculture*, **9**, 468–74.

Mohler, C. L. (2000). Seed size controls the ability of seedlings to emerge through rye mulch. *Weed Science Society of America Abstracts*, **40**, 98.

Mohler, C. L., & Callaway, M. B. (1992). Effects of tillage and mulch on the emergence and survival of weeds in sweet corn. *Journal of Applied Ecology*, **29**, 21–34.

Mohler, C. L., & Teasdale, J. R. (1993). Response of weed emergence to rate of *Vicia villosa* and *Secale cereale* L. residue. *Weed Research*, **33**, 487–99.

Moody, K. (1991). Weed management in rice. In *CRC Handbook of Pest Management in Agriculture*, vol. 3, 2nd edn, ed. D. Pimentel, pp. 301–28. Boca Raton, FL: CRC Press.

Moody, K., & Drost, D. C. (1983). The role of cropping systems on weeds in rice. In *Proceedings of the Conference on Weed Control in Rice*, 31 August–4 September 1981, pp. 73–86. Los Baños, Philippines: International Rice Research Institute.

Moreno, R., & Sánchez, J. F. (1994). The effect of using mulches with intercropping. In *Tapado Slash/Mulch: How Farmers Use It and What Researchers Know About It*, ed. H. D. Thurston, M. Smith, G. Abawi & S. Kearl, pp. 191–206. Ithaca, NY: Cornell International Institute for Food, Agriculture and Development.

Mwaja, V. N., Masiunas, J. B., & Weston, L. A. (1995). Effects of fertility on biomass, phytotoxicity, and allelochemical content of cereal rye. *Journal of Chemical Ecology*, **21**, 81–96.

Niggli, U., Weibel, F. P., & Gut, W. (1990). Weed control with organic mulch materials in orchards. Results from 8-year field experiments. *Acta Horticulturae*, **285**, 97–102.

Obiefuna, J. C. (1986). The effect of sawdust mulch and increasing levels of nitrogen on the weed growth and yield of false horn plantains (*Musa* ABB). *Biological Agriculture and Horticulture*, **3**, 353–9.

Okugie, D. N., & Ossom, E. M. (1988). Effect of mulch on the yield, nutrient concentration and weed infestation of the fluted pumpkin, *Telfairia occidentalis* Hook. *Tropical Agriculture (Trinidad)*, **65**, 202–4.

Overland, L. (1966). The role of allelopathic substances in the "smother crop" barley. *American Journal of Botany*, **53**, 423–32.

Palm, C. A., & Sanchez, P. A. (1991). Nitrogen release from the leaves of some tropical legumes as affected by their lignin and polyphenolic contents. *Soil Biology and Biochemistry*, **23**, 83–8.

Pappas-Fader, T., Cook, J. F., Butler, T., Lana, P. J., & Hare, J. (1993). Resistance of California arrowhead and smallflower umbrella sedge to sulfonylurea herbicides. *Proceedings of the Western Society of Weed Science*, **46**, 47.

Parr, J. F., Miller, R. H., & Colacicco, D. (1984). Utilization of organic materials for crop production in developed and developing countries. In *Organic Farming: Current Technology and Its Role in a Sustainable Agriculture*, ed. D. F. Bezdicek, J. F. Power, D. R. Keeney & M. J. Wright, pp. 83–96. Madison, WI: American Society of Agronomy, Crop Science Society of America, and Soil Science Society of America.

Patrick, Z. A., & Koch, L. W. (1958). Inhibition of respiration, germination, and growth by substances arising during the decomposition of certain plant residues in the soil. *Canadian Journal of Botany*, **36**, 621–47.

Patrick, Z. A., Toussoun, T. A., & Snyder, W. C. (1963). Phytotoxic substances in arable soils associated with decomposition of plant residues. *Phytopathology*, **53**, 152–61.

Pérez, F. J., & Ormeño-Núñez, J. (1993). Weed growth interference from temperate cereals: the effect of a hydroxamic-acids-exuding rye (*Secale cereale* L.) cultivar. *Weed Research*, **33**, 115–19.

Peters, S., Bohlke, M., & Janke, R. (1990). *A Comparison of Conventional and Low Input Cropping Systems with Reduced Tillage*. Kutztown, PA: Rodale Research Center.

Phelan, P. L., Mason, J. F., & Stinner, B. R. (1995). Soil-fertility management and host preference by European corn borer, *Ostrinia nubilalis* (Hubner), on *Zea mays* L.: a comparison of organic and conventional chemical farming. *Agriculture, Ecosystems and Environment*, **56**, 1–8.

Pieters, A. J. (1927). *Green Manuring: Principles and Practices*. New York: John Wiley.

Purseglove, J. W. (1985). *Tropical Crops: Monocotyledons*. New York: Longman.

Putnam, A. R., & DeFrank, J. (1983). Use of phytotoxic plant residues for selective weed control. *Crop Protection*, **2**, 173–81.

Putnam, A. R., DeFrank, J., & Barnes, J. P. (1983). Exploitation of allelopathy for weed control in annual and perennial cropping systems. *Journal of Chemical Ecology*, **9**, 1001–10.

Putnam, A. R., & Weston, L. A. (1986). Adverse impacts of allelopathy in agricultural systems. In *The Science of Allelopathy*, ed. A. R. Putnam & C.-S. Tang, pp. 43–56. New York: John Wiley.

Pysek, P., & Leps, J. (1991). Response of a weed community to nitrogen fertilization: a multivariate analysis. *Journal of Vegetation Science*, **2**, 237–44.

Ransom, J. K., & Oelke, E. A. (1983). Cultural control of common waterplantain (*Alisma triviale*) in wild rice (*Zizania palustris*). *Weed Science*, **31**, 562–6.

Rasmussen, K., Rasmussen, J., & Petersen, J. (1996). Effects of fertilizer placement on weeds in weed harrowed spring barley. *Acta Agriculturae Scandinavica, Section B, Soil and Plant Science*, **46**, 192–6.

Reijntjes, C., Haverkort, B., & Waters-Bayer, A. (1992). *Farming for the Future: An Introduction to Low-External-Input and Sustainable Agriculture*. London: Macmillan.

Rice, E. L. (1995). *Biological Control of Weeds and Plant Diseases: Advances in Applied Allelopathy*. Norman, OK: University of Oklahoma Press.

Rickerl, D. H., Curl, E. A., Touchton, J. T., & Gordon, W. B. (1992). Crop mulch effects on *Rhizoctonia* soil infestation and disease severity in conservation-tilled cotton. *Soil Biology and Biochemistry*, **24**, 553–7.

Risch, S. J., & Carroll, C. R. (1986). Effects of seed predation by a tropical ant on competition among weeds. *Ecology*, **67**, 1319–27.

Roberts, E. H., & Benjamin, S. K. (1979). The interaction of light, nitrate and alternating temperature on the germination of *Chenopodium album, Capsella bursa-pastoris* and *Poa annua* before and after chilling. *Seed Science and Technology*, **7**, 379–92.

Roe, N. E., Stoffella, P. J., & Bryan, H. H. (1993). Municipal solid waste compost suppresses weeds in vegetable crop alleys. *HortScience*, **28**, 1171–2.

Rothrock, C. S., Kirkpatrick, T. L., Frans, R. E., & Scott, H. D. (1995). The influence of winter legume cover crops on soilborne plant pathogens and cotton seedling diseases. *Plant Disease*, **79**, 167–71.

Rubin, B., & Benjamin, A. (1984). Solar heating of the soil: involvement of environmental factors in the weed control process. *Weed Science*, **32**, 138–42.

Samson, R. A. (1991). The weed suppressing effects of cover crops. In *Proceedings of the 5th Annual REAP Conference*, pp. 11–22. Ste. Anne de Bellevue, Quebec: Resource Efficient Agricultural Production (REAP).

Sarkar, P. A., & Moody, K. (1983). Effects of stand establishment techniques on weed population in rice. In *Proceedings of the Conference on Weed Control in Rice*, 31 August–4 September 1981, pp. 57–71. Los Baños, Philippines: International Rice Research Institute.

Sarrantonio, M. (1992). Opportunities and challenges for the inclusion of soil-improving crops in vegetable production systems. *HortScience*, **27**, 754–8.

Sarrantonio, M. (1994). *Northeast Cover Crop Handbook*. Emmaus, PA: Rodale Institute.

Sarrantonio, M., & Scott, T. W. (1988). Tillage effects on availability of nitrogen to corn following a winter green manure crop. *Soil Science Society of America Journal*, **52**, 1661–8.

Scott, T. W., & Burt, R. F. (1987). Use of red clover in corn polyculture systems. In *The Role of Legumes in Conservation Tillage Systems*, ed. J. F. Power, pp. 101–3. Ankeny, IA: Soil Conservation Society of America.

Seaman, D. E. (1983). Farmers' weed control technology for water-seeded rice in North America. In *Proceedings of the Conference on Weed Control in Rice*, 31 August–4 September 1981, pp. 167–76. Los Baños, Philippines: International Rice Research Institute.

Shilling, D. G., Liebl, R. A., & Worsham, A. D. (1985). Rye (*Secale cereale* L.) and wheat (*Triticum aestivum* L.) mulch: the suppression of certain broadleaved weeds and the isolation and identification of phytotoxins. In *The Chemistry of Allelopathy: Biochemical Interactions Among Plants*, ed. A. C. Thompson, pp. 243–71. Washington, DC: American Chemical Society.

Shilling, D. G., Jones, L. A., Worsham, A. D., Parker, C. E., & Wilson, R. F. (1986). Isolation and identification of some phytotoxic compounds from aqueous extracts of rye (*Secale cereale* L.). *Journal of Agricultural and Food Chemistry*, **34**, 633–8.

Shumway, D. L., & Koide, R. T. (1994). Seed preferences of *Lumbricus terrestris* L. *Applied Soil Ecology*, **1**, 11–15.

Singh, A., Singh, K., & Singh, D. V. (1991). Suitability of organic mulch (distillation waste) and herbicides for weed management in perennial aromatic grasses. *Tropical Pest Management*, **37**, 162–5.

Smeda, R. J., & Weller, S. C. (1996). Potential of rye (*Secale cereale*) for weed management in transplanted tomatoes (*Lycopersicon esculentum*). *Weed Science*, **44**, 596–602.

Smith, D. M. (1951). *The Influence of Seedbed Conditions on the Regeneration of Eastern White Pine*, Bulletin no. 545. New Haven, CT: Connecticut Agricultural Experiment Station.

Smith, R. J. Jr., & Shaw, W. C. 1966. *Weeds and Their Control in Rice Production*, Handbook no. 292. Washington, DC: US Department of Agriculture.

Standifer, L. C., Wilson, P. W., & Porche-Sorbet, R. (1984). Effects of solarization on soil weed seed populations. *Weed Science*, **32**, 569–73.

Stapleton, J. J., & DeVay, J. E. (1986). Soil solarization: a non-chemical approach for management of plant pathogens and pests. *Crop Protection*, **5**, 190–8.

Stapleton, J. J., & DeVay, J. E. (1995). Soil solarization: a natural mechanism of integrated pest management. In *Novel Approaches to Integrated Pest Management*, ed. R. Reuvani, pp. 309–22. Boca Raton, FL: Lewis Publishers.

Stern, W. R., & Donald, C. M. (1962). Light relationships in grass–clover swards. *Australian Journal of Agricultural Research*, **13**, 599–614.

Stevenson, F. J. (1986). *Cycles of Soil: Carbon, Nitrogen, Phosphorus, Sulfur, Micronutrients*. New York: Wiley-Interscience.

Stroo, H. F., Elliott, L. F., & Papendick, R. I. (1988). Growth, survival and toxin production of root-inhibitory pseudomonads on crop residues. *Soil Biology and Biochemistry*, **20**, 201–7.

Stute, J. K., & Posner, J. L. (1995). Synchrony between legume nitrogen release and corn demand in the upper Midwest. *Agronomy Journal*, **87**, 1063–9.

Swan, J. B., Kaspar, T. C., & Erbach, D. C. (1996). Seed-row residue management for corn establishment in the northern US Corn Belt. *Soil and Tillage Research*, **40**, 55–72.

Tardif, F. J., & Leroux, G. D. (1992). Response of three quackgrass biotypes to nitrogen fertilization. *Agronomy Journal*, **84**, 366–70.

Taylorson, R. B. (1987). Environmental and chemical manipulation of weed seed dormancy. *Reviews in Weed Science*, **3**, 135–54.

Teasdale, J. R., & Daughtry, C. S. T. (1993). Weed suppression by live and desiccated hairy vetch (*Vicia villosa*). *Weed Science*, **41**, 207–12.

Teasdale, J. R., & Mohler, C. L. (1992). Weed suppression by residue from hairy vetch and rye cover crops. In *Proceedings of the 1st International Weed Control Congress*, vol. 2, pp. 516–18.

Teasdale, J. R., & Mohler, C. L. (1993). Light transmittance, soil temperature, and soil moisture under residue of hairy vetch and rye. *Agronomy Journal*, **85**, 673–80.

Teasdale, J. R., & Mohler, C. L. (2000). The quantitative relationship between weed emergence and the physical properties of mulches. *Weed Science*, **48**, 385–92.

Teyker, R. H., Hoelzer, H. D., & Liebl, R. A. (1991). Maize and pigweed response to nitrogen supply and form. *Plant and Soil*, **135**, 287–92.

Thurston, H. D. (1997). *Slash/Mulch Systems: Sustainable Methods for Tropical Agriculture*. Boulder, CO: Westview Press.

Tilman, D. (1988). *Plant Strategies and the Dynamics and Structure of Plant Communities*. Princeton, NJ: Princeton University Press.

Tollenaar, M., Nissanka, S. P., Aguilera, A., Weise, S. F., & Swanton, C. J. (1994). Effect of weed interference and soil nitrogen on four maize hybrids. *Agronomy Journal*, **86**, 596–601.

Totterdell, S., & Roberts, E. H. (1980). Characteristics of alternating temperatures which stimulate loss of dormancy in seeds of *Rumex obtusifolius* L. and *Rumex crispus* L. *Plant, Cell and Environment*, **3**, 3–12.

Uhl, C., Clark, K., Clark, H., & Murphy, P. (1981). Early plant succession after cutting and burning in the upper Rio Negro region of the Amazon Basin. *Journal of Ecology*, **69**, 631–49.

van Bruggen, A. H. C. (1995). Plant disease severity in high-input compared to reduced-input and organic farming systems. *Plant Disease*, **79**, 976–84.

van der Reest, P. J., & Rogaar, H. (1988). The effect of earthworm activity on the vertical distribution of plant seeds in newly reclaimed polder soils in the Netherlands. *Pedobiologia*, **31**, 211–18.

Varco, J. J., Frye, W. W., Smith, M. S., & MacKown, C. T. (1993). Tillage effects on legume decomposition and transformation of legume and fertilizer nitrogen-15. *Soil Science Society of America Journal*, **57**, 750–6.

Vidal, R. A., & Bauman, T. T. (1996). Surface wheat (*Triticum aestivum*) residues, giant foxtail (*Setaria faberi*) and soybean (*Glycine max*) yield. *Weed Science*, **44**, 939–43.

Wagner-Riddle, C., Gillespie, T. J., & Swanton, C. J. (1996). Rye mulch characterization for the purpose of microclimatic modelling. *Agricultural and Forest Meteorology*, **78**, 67–81.

Wallace, R. W., & Bellinder, R. R. (1992). Alternative tillage and herbicide options for successful weed control in vegetables. *HortScience*, **276**, 745–9.

Waller, G. R. (1989). Allelochemical action of some natural products. In *Phytochemical Ecology: Allelochemicals, Mycotoxins, and Insect Pheromones and Allomones*, ed. C. H. Chou & G. R. Waller, pp. 129–54. Taipei, Taiwan: Academia Sinica.

Wanniarachchi, S. D., & Voroney, R. P. (1997). Phytotoxicity of canola residues: release of water-soluble phytotoxins. *Canadian Journal of Soil Science*, **77**, 535–41.

Westcott, M. P., & Mikkelsen, D. S. (1987). Comparison of organic and inorganic nitrogen sources for rice. *Agronomy Journal*, **79**, 937–43.

Westoby, M., Leishman, M., & Lord, J. (1996). Comparative ecology of seed size and dispersal. *Philosophical Transactions of the Royal Society of London, Series B*, **351**, 1309–18.

Weston, L. A. (1996). Utilization of allelopathy for weed management in agroecosystems. *Agronomy Journal*, **88**, 860–6.

White, R. H., Worsham, A. D., & Blum, U. (1989). Allelopathic potential of legume debris and aqueous extracts. *Weed Science*, **37**, 674–9.

Wicks, G. A., Crutchfield, D. A., & Burnside, O. C. (1994). Influence of wheat (*Triticum aestivum*) straw mulch and metolachlor on corn (*Zea mays*) growth and yield. *Weed Science*, **42**, 141–7.

Wilken, G. C. (1977). Integrating forest and small-scale farm systems in middle America. *Agro-Ecosystems*, **3**, 291–302.

Wilkins, E. D., & Bellinder, R. R. (1996). Mow-kill regulation of winter cereals for spring no-till crop production. *Weed Technology*, **10**, 247–52.

Williams, J., Wick, C., Scardaci, S., Klonsky, K., Chaney, D., Livingston, P., & Tourte, L. (1992). *Sample Costs to Produce Organic Rice, Water Seeded, in the Sacramento Valley.* Davis, CA: University of California Cooperative Extension.

Wilson, D. O., & Hargrove, W. L. (1986). Release of nitrogen from crimson clover residue under two tillage systems. *Soil Science Society of America Journal*, **50**, 1251–4.

Worsham, A. D. (1989). Current and potential techniques using allelopathy as an aid in weed management. In *Phytochemical Ecology: Allelochemicals, Mycotoxins, and Insect Pheromones and Allomones*, ed. C. H. Chou & G. R. Waller, pp. 275–91. Taipei, Taiwan: Academia Sinica.

Yenish, J. P., Worsham, A. D., & York, A. C. (1996). Cover crops for herbicide replacement in no-tillage corn (*Zea mays*). *Weed Technology*, **10**, 815–21.

Young, F. L, Ogg, A. G. Jr., & Dotray, P. A. (1990). Effect of postharvest field burning on jointed goatgrass (*Aegilops cylindrica*) germination. *Weed Technology*, **4**, 123–7.

CHARLES L. MOHLER

6

Enhancing the competitive ability of crops

Introduction

Many cultural practices, including crop density, arrangement, planting date and choice of cultivar affect the crop's ability to compete with weeds. However, most recommendations for the planting of crops are based on the assumption that weeds are absent. This is a result of the scientific and economic context in which recommendations are developed. Variety trials, fertility rate trials, and many other agronomic experiments are usually run in weed-free conditions to avoid the confounding effect of weed competition. For the agronomist or horticultural scientist, keeping a particular experiment free of weeds is a practical possibility. Given the high spatial and temporal variability in density and composition of weed communities, a weed-free trial may also be the easiest way to generate results that are applicable over a wide area. In addition, weeds generally decrease yield regardless of other parameters. Consequently, weeds are usually excluded from experiments unless they are specifically the object of investigation. However, weed-free fields are rarely practical on the farm, and as explained in the following sections, the presence of weeds generally changes the optimal choices for cultural practices relative to those developed in weed-free conditions.

The central thesis of this chapter is that *the density, arrangement, cultivar, and planting date of the crop that maximize the rate at which the crop occupies space early in the growing season usually minimize competitive pressure of weeds on the crop.* These cultural factors also have other effects on the competitive balance between weeds and crops. For example, allelopathic cultivars sometimes reduce weed biomass. However, the physical occupation of three-dimensional space by the crop, and the preemption of resources that this allows, is central to the operation of cultural weed control strategies discussed here and in the following chapter.

Crop density

Many studies have demonstrated that weed biomass and other measures of weed abundance usually decrease as crop density increases (Table 6.1) (Mohler, 1996). The pattern is remarkably consistent: of the 91 cases found in the literature, only six failed to show decreasing weediness with increasing crop density. Since most crops are sown in rows, variation in density involved variation in crop arrangement in most of these studies. The effects of crop arrangement in those studies in which arrangement was varied at constant crop density are discussed in the following section.

The role of crop density in weed management is well illustrated by a study of the interactions between safflower and *Setaria viridis* (Blackshaw, 1993). Safflower was sown in early May at six rates that resulted in densities ranging from 12 to 192 plants m^{-2}, and plots were maintained in either weedy or weed-free conditions. *S. viridis* biomass declined with increasing crop density (Figure 6.1a). Probably because safflower has a larger seed than *S. viridis*, the crop was initially taller than the weed, and maintained this height difference throughout the season (Figure 6.1b). As a consequence, by early July the crop substantially shaded the weed, and the degree of shading increased with crop density (Figure 6.1c). Due to the increased competitive pressure exerted by the crop at higher densities, the plateau in yield was reached at substantially higher crop density in the weedy, relative to the weed-free, condition (Figure 6.1d). Although the optimal crop density was much higher in the presence of weeds, even at very high density yield was still less in the weedy condition.

The mathematical analysis of how plant density affects the competition between species of annual plants has received considerable attention (De Wit, 1960; Håkansson, 1983; Spitters, 1983; Firbank & Watkinson, 1985). A brief introduction to this work helps explain the conditions under which crop density is most useful for weed management. Frequently, biomass of a crop, Y_C, can be expressed as

$$Y_C = N_C / (a + bN_C) \tag{6.1}$$

where N_C is the density of the crop (Shinozaki & Kira, 1956; Harper, 1977, p. 156; Håkansson, 1983). As N_C becomes large, Y_C rises to an asymptote at $1/b$ (Figure 6.2a). In the absence of weeds, the change in Y_C with density is negligible beyond a certain range, and consequently the flat portion of the Y_C curve at high density is sometimes referred to as the "law of constant final yield" (Kira, Ogawa & Shinozaki, 1953; Harper, 1977, p. 154). Now suppose the crop competes with a weed that is very similar to it in all respects. Then,

$$Y_C = N_C / [a + b(N_C + N_W)]$$
$$Y_W = N_W / [a + b(N_C + N_W)]$$
$$Y_C + Y_W = (N_C + N_W) / [a + b(N_C + N_W)]$$

Table 6.1. *Response of weeds and crop yield to increase in crop density*

Crop	Weed response to crop density[a,b]	Yield response to crop density		Reference
		Weed-free	Weeds present[c]	
Barley	−	+	++	Mann & Barnes (1947)
	−	+	++	Mann & Barnes (1949)
	−		+	Pfeiffer & Holmes (1961)
	−		V(+)	Bate, Elliott & Wilson (1970)
	V(−)	V(+)	+	Cussans & Wilson (1975)
	V(−)	0	0	Kolbe (1980)
	−	+	++	Håkansson (1983)
	−	+	++	Håkansson (1991)
	−	V	V(+)	Evans et al. (1991)
	−		+	Barton, Thill & Shafii (1992)
	−		+	Kirkland (1993)
	−	0	+	Doll, Holm & Søgaard (1995)
	−	+	++	Doll (1997)
Bean	−	+	++	Malik, Swanton & Michaels (1993)
Bean, snap		V	+	Williams et al. (1973)
Bluegrass, Kentucky	−		+	Parr (1985)
Cabbage	−	++	+	Weaver (1984)
Cotton		0	+	Rogers, Buchanan & Johnson (1976)
	−	V	V	Street et al. (1981)
Cowpea	V(−)		0	Nangju (1978)
	−		+	Brar, Gill & Randhawa (1984)
Cucumber		+	+	Staub (1992)
Fescue, red	−		+	Parr (1985)
Flax seed	−		V(+)	Robinson (1949)
	−	V(+)	+	Gruenhagen & Nalewaja (1969)
	−	+	++	Stevenson & Wright (1996)
	0		0	Blackshaw et al. (1999)
Lentil	−	V	V	Boerboom & Young (1995)
	V(−)	V	V	Ball, Ogg & Chevalier (1997)
Lupin	0		V(I)	Walton (1986)
	0	+	+	Putnam et al. (1992)
Maize	−	V(+)	V(+)	Nieto & Staniforth (1961)
		+	++	Williams et al. (1973)
	−		V(+,I)	Choudhary (1981)
	−	V(+)	+	Weil (1982)
	−		+	Ghafar & Watson (1983)
	−		+	Brar, Gill & Randhawa (1984)
	−	0	0	Forcella, Westgate & Warnes (1992)
	−	+	++	Tollenaar et al. (1994a)
	V(−)	0	V(+)	Teasdale (1995)
	−	[0]		Murphy et al. (1996)
	−	V		Teasdale (1998)
Oat	−		+	Bula, Smith & Miller (1954)
Onion		+[d]	++[d]	Williams et al. (1973)
Pea	−			Marx & Hagedorn (1961)
	−	I	I	Lawson (1982); Lawson & Topham (1985)
	V(−)	V(+)	+	Wall, Friesen & Bhati (1991)

Table 6.1. (*cont.*)

Crop	Weed response to crop density[a,b]	Yield response to crop density		Reference
		Weed-free	Weeds present[c]	
Pea, field	−	+	++	Townley-Smith & Wright (1994)
	−	V(+)	V(+)	Boerboom & Young (1995)
Peanut	−	+	++	Buchanan & Hauser (1980)
Ryegrass, perennial	−		+	Parr (1984, 1985)
Pigeonpea	0	0	0	Díaz-Rivera *et al.* (1985)
Rapeseed	−	+	++	O'Donovan (1994)
	−		+	Anderson & Bengtsson (1992)
Rhizoma peanut		I	+	Canudas *et al.* (1989)
Rice	−	+	++	Smith (1968)
	−	I,0	I,+	Akobundu & Ahissou (1985)
	−	+	++	Pantone & Baker (1991)
Safflower	−	+	++	Blackshaw (1993)
Sorghum		V(0)	+	Wiese *et al.* (1964)
Soybean	−	V(+)	+	Weber & Staniforth (1957)
	−	V(0)	V(+)	Staniforth (1962)
	−		+	Wax & Pendleton (1968)
	−	+	++	Kust & Smith (1969)
	−		+	McWhorter & Barrentine (1975)
	−	0	+	Felton (1976)
	−	V(−)	V(+)	Nangju (1978)
	V(−)	V	V	Weaver (1986)
	−	0	+	Howe & Oliver (1987)
	−		+	McWhorter & Sciumbato (1988)
	V(−)	0	+	Mickelson & Renner (1997)
	0			Pitelli, Charudattan & Devalerio (1998)
Sweet potato	−		V(I)	Ambe (1995)
Timothy	−		+	Parr (1985)
Wheat	−	V(+)	+	Radford *et al.* (1980)
	−		+	Shaktawat (1983)
	V(−)		V(+)	Vander Vorst, Wicks & Burnside (1983)
		V(+)	+	Carlson & Hill (1985)
	−	V(0)	+	Medd *et al.* (1985)
	−	V	+	Moss (1985)
	−	V(+)	+	Skorda & Efthimiadis (1985)
	−		+	Kukula (1986)
	−	I	+	Martin, Cullis & McNamara (1987)
	−		+	Koscelny *et al.* (1990)
	−		V(I)	Samuel & Guest (1990)
	−	V(0)	V(+)	Koscelny *et al.* (1991)
	−	−	V	Appleby & Brewster (1992)
	−		+[e]	Johri, Singh & Sharma (1992)
	−	0	0	Justice *et al.* (1993)

Table 6.1. (cont.)

Crop	Weed response to crop density[a,b]	Yield response to crop density Weed-free	Yield response to crop density Weeds present[c]	Reference
	0		+	Teich et al. (1993)
	−		+	Christensen, Rasmussen & Olesen (1994)
	V(−)	0	+	Justice et al. (1994)
	−	0		Doll, Holm & Søgaard (1995)
	−		+	Lemerle et al. (1996)
	−	V(0)		Anderson (1997)
	−	0	+	Tanji, Zimdahl & Westra (1997)
	−	−	+	Hashem, Radosevich & Roush (1998)
	V(−)	+		Blackshaw et al. (1999)

Notes:
[a] Studies were included only if weed response to crop density was measured, or yield response to crop density was measured with and without weeds. For studies in which all treatment series included weed control measures, the density series with the least effective weed control was used for evaluation of the weed response to crop density and the yield response to density in the presence of weeds.
[b] −, the measure of weed abundance investigated in the study (usually biomass) decreased as crop density increased; V(−), weed response to crop density varied among treatments, years, cultivars, sites, or experiments, but the tendency was to decrease as crop density increased; 0, no systematic change in weed measure as crop density increased.
[c] +, yield increased as crop density increased; ++, yield increased as crop density increased, and percentage increase was greater than in the corresponding treatment series without weeds; −, yield decreased as crop density increased; 0, no systematic change in yield as crop density increased; I, yield was maximum at intermediate crop density; V, yield response varied among treatments, years, cultivars, sites, or experiments. V with +, −, 0 or I shown in parentheses indicates that the response varied, but that an overall tendency was reasonably unambiguous. A blank indicates that no information was given on crop response to density in that condition.
[d] Yield of onions increased with density but percentage of onions >7.6 cm diameter decreased with density.
[e] Nutrient uptake by the crop rather than yield.
Source: Expanded from Mohler (1996).

(Figure 6.2a). Setting the density of the weed at various fixed values generates a series of curves for the response of weed biomass to crop density (Figure 6.2b). In each case, the yield of the crop is complementary to the yield of the weed and the biomass of the weed declines with increasing crop density. However, note that at any given crop density, the slope of the weed biomass curve is greater when density of the weed is high. Thus, *the suppression of weeds and increase in crop yield from an incremental increase in crop density increases with the density of weeds.*

The applicability of this conclusion to real crop–weed systems has been confirmed by factorial experiments (Håkansson, 1983; Cousens, 1985;

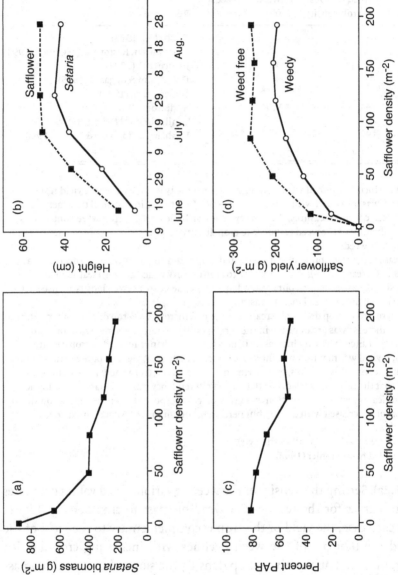

Figure 6.1 Effects of safflower density on competition between safflower and *Setaria viridis*, 1991. (a) Effect of safflower density on *S. viridis* biomass. (b) Height of safflower and *S. viridis*. (c) Percentage of photosynthetically active radiation (PAR) reaching the top of the *S. viridis* on 3 July. (d) Safflower seed yield in relation to density, with and without presence of *S. viridis*. (Drawn from data provided by R. E. Blackshaw – see Blackshaw, 1993).

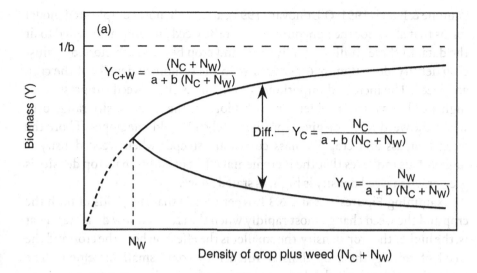
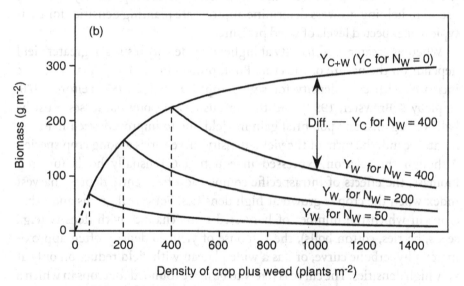

Figure 6.2 Biomass of crop and weed in response to their combined density ($N_C + N_W$) when the crop and weed are highly similar. (a) The model, with N_w fixed at an arbitrary value. (b) An example using barley as both crop and weed, with N_w fixed at 50, 200, and 400 plants m^{-2}, $a = 0.61$ and $b = 0.00293$. (Redrawn from Håkansson, 1983.)

Pantone & Baker, 1991; O'Donovan, 1994), although more complicated model forms that allow for the nonequivalence of the weed and crop are needed to fit the data. In these studies, usually weed and crop biomass are plotted against crop density alone (Figure 6.3) rather than the combined density of the crop and weed. The increased importance of crop density for weed suppression at high weed densities is evident in such plots: in the crop density range over which the weed-free crop gives approximately full yield, the slopes of both the weed biomass and crop biomass curves are steeper when weed density is higher. This indicates that the incremental effect of additional crop density is greater when weed density is high, as stated above.

Examining Figures. 6.2 and 6.3 further reveals that the yields of both the crop and the weed change most rapidly when the crop is at low densities. That is, the higher the crop density, the smaller is the effect on both the crop and the weed of raising the crop density further by some small increment. The optimal crop density is the density at which a further increment of seed costs more than the expected increase in yield is worth. Charts similar to Figure 6.3 showing the relation of yield and weed biomass to crop density could be useful in helping growers determine appropriate planting densities for their typical or expected levels of weed pressure.

Whether greater productivity at higher crop density results in greater yield depends on the nature of the crop. First, probability of lodging and disease increases with crop density for some crops (Hartwig, 1957; Felton, 1976; Appleby & Brewster, 1992), and the seriousness of potential losses must be balanced against any potential gain in yield due to improved weed management. Second, the form of the yield–density curve varies among crop species. Although the relation expressed in equation 6.1 usually holds for crop biomass, the effects of intraspecific competition frequently depress harvest index when the crop is grown at high densities. For cereal grains and other crops in which the number of harvested units changes with density (e.g., seeds, berries, cotton bolls), the response of yield to density often approximates a hyperbolic curve, or has a wide plateau with yield reduction only at very high densities. The curve is likely to be more rounded for crops in which a substantial reproductive structure has to be built in order to produce seeds (e.g., maize, sunflower). Finally, for root crops and the many vegetable crops for which small produce is unmarketable, the curve of yield in response to density is usually quite peaked. At high density, resources are divided among many small individuals, each of which makes one small root or fruit (Willey & Heath, 1969). For this latter type of crop, density is likely to have limited usefulness as a weed management strategy, and few density/weed management studies exist for these crops (Table 6.1). For many seed crops, however, the

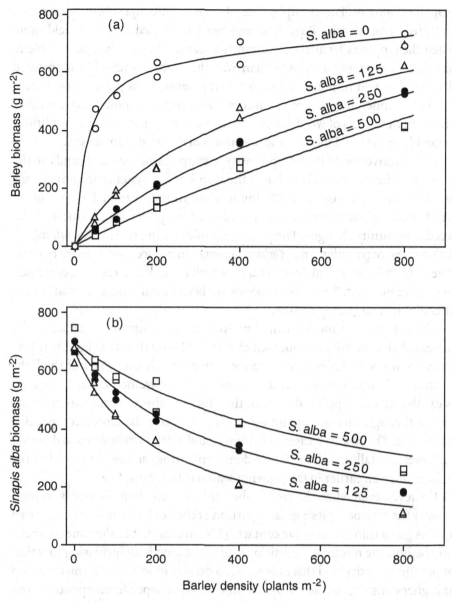

Figure 6.3 Biomass of (a) barley and (b) *Sinapis alba* grown in competition at several densities. (Drawn from data in Håkansson, 1983.)

response of yield to density is strictly increasing in the presence of weeds, although yield shows a peaked curve in weed-free conditions (Table 6.1). Thus, the response of yield to density is often qualitatively different in the presence of weeds.

Other cultural practices interact with crop density in affecting weed and crop development. For example, Gruenhagen & Nalewaja (1969) showed that at a given crop density *Polygonum convolvulus* decreased flax seed yield more when the crop was fertilized, apparently because the weed was better able to use nitrogen for plant growth than was the crop. However, at high weed density, the yield response to increased crop density was greater under fertilized than under unfertilized conditions. This result, combined with considerations on the effects of weed density discussed above, indicates that conditions favorable to the weed increase the usefulness of elevated crop density.

The effectiveness of high crop density in suppressing weeds depends on the biology of the weed as well as that of the crop, but few systematic comparisons are available. Håkansson (1983) found that although total biomass of all weeds declined as cereal density increased, the species composition of the weed community changed. The percentage of climbing species like *Bilderdykia (Polygonum) convolvulus* and *Galium aparine* increased with crop density, whereas *Sonchus asper* and *Brassica napus*, which begin life as rosettes, decreased in relative biomass. Thus, some species are better than others at resisting the effects of increased crop density.

The competitive mechanisms involved in the suppression of weeds by increased crop density are not well explored. Competition for light is inherently asymmetric: taller plants receive a disproportionate share of the light relative to their leaf areas (Weiner & Thomas, 1986; Weiner, 1990). If the crop is capable of overtopping the weed, then causing this to occur earlier in the season through increased planting density will give the crop a competitive advantage. This was the case for safflower and *Setaria viridis* discussed above: safflower was taller, and the higher-density plantings achieved a given level of light extinction earlier in the growing season (Blackshaw, 1993).

However, even when the crop is shorter in stature than the weeds, at high density the crop occupies a greater portion of the land area at the time the two species grow into competitive contact. This insures that at the time competition begins, the resource acquisition rate of the crop is an increasing function of planting density, and this may lead to a greater biomass of the mature crop at higher densities. Greater crop height due to intraspecific competition, and greater total root density and leaf area index may also improve the crop's relative performance at higher density. Experiments in which crop density is manipulated by transplanting before and after the onset of competition could falsify the hypothesis that early preemption of space governs the observed effects of crop density. Experiments in which above- and below-ground competition between the crop and weed is regulated by barriers over a range of densities could indicate the relative importance of above- and below-ground resource capture.

The above discussion begs the question of whether the yield advantage of increased crop density in the presence of weeds is greater for competitive crops or for relatively noncompetitive crops. The general relationships involved can be visualized by means of two hypothetical examples (Mohler, 1996). Suppose a field has a weed density of 100 plants m^{-2}. Further suppose that it is planted with a crop having a yield potential of 1000 kg ha^{-1} in weed-free conditions, and that the crop is usually planted at 100 plants m^{-2}. Thus, doubling planting density from 100 to 200 m^{-2} will increase the proportion of crop plants from 50% to 66.7%. The effect of such a change in crop density can be explored using a replacement series diagram in which yields of the two species are plotted against their relative densities (De Wit, 1960). If the crop is a better competitor than the weed, it will have a convex curve, whereas the weed will have a concave curve (Figure 6.4a). Figure 6.4 is drawn such that the relative percentages of the crop and weed are in the usual linear scale. However, since planting density does not affect initial density of the weed, weed density is fixed at 100 m^{-2}, and crop density therefore varies hyperbolically. For simplicity, competition is assumed to follow a simple replacement process in which the two species have equal growth potential and combined yield of the crop and weed is the same for all mixtures.

If the crop is competitive (Figure 6.4a), doubling density increases yield from 750 to 850 kg ha^{-1}. Even this is less than the potential yield of 1000 kg ha^{-1} possible in weed-free conditions, but if control measures that allow full yield are unavailable or prohibitively expensive, as in the case of safflower discussed above (Blackshaw, 1993), the increased planting density may be a viable tactic. Nevertheless, the percentage increase in yield due to elevated planting density is moderate – only 13% more than could be obtained by planting at the usual density.

The same replacement series curves can be inverted to examine the interaction between a highly competitive weed and a less competitive crop (Figure 6.4b). In this case, doubling crop density increases yield from 250 to 410 kg ha^{-1}, an increase of 64%. Although the percentage increase in yield is large, even the yield at twice normal density is so far below the crop's potential that either other weed management methods would be needed or the crop would probably not be grown. Low yield of noncompetitive crops under weedy conditions may be the reason why most studies of crop density involve large-seeded species (Table 6.1) that have an initial size advantage over weeds during establishment (see Chapter 2).

Note, however, that combining increased crop density with other weed control methods can be very effective when the crop is a poor competitor. For example, if the weed is thinned from 100 to 50 plants m^{-2} at the outset, then

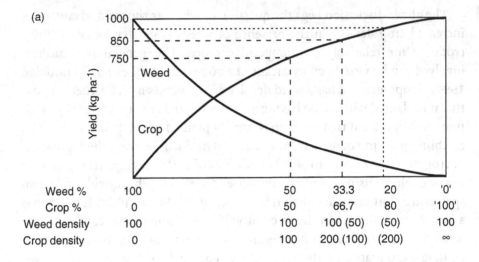

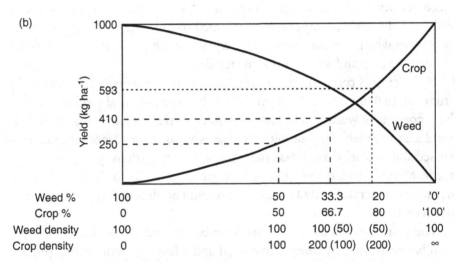

Figure 6.4 Replacement series diagrams showing the effect of changing crop density on crop yield when (a) the crop is more competitive than the weed and (b) the crop is less competitive than the weed. The curves show the yield response of the crop and weed. Numbers in parentheses correspond to the situation where half the weeds are removed randomly at or before emergence. (Modified from Mohler, 1996.)

100 crop plants m^{-2} corresponds to a mixture with 33.3% weed and 66.7% crop (Figure 6.4b). In this case, doubling crop density to 100 plants m^{-2} has an even greater effect on yield (183 kg ha^{-1} increase rather than the 160 kg ha^{-1} increase observed at the higher weed density). This occurs because the interaction shifts rightward onto the more steeply rising portion of the curve.

Lowering weed density also has a beneficial effect when the crop is a good competitor. When the crop is a competitive species, the combined effects of thinning weeds and increasing crop density are smaller than when the crop is a poor competitor, but yield may approach that for the weed-free condition (Figure 6.4a). Thus, enhanced crop density is likely to be an effective component of many integrated weed management programs.

Although increased planting density is clearly advantageous when weeds are present, exactly how much density should be increased depends on both the nature of the crop and the density of weeds present. Consequently, specific recommendations are difficult to obtain. In general, (i) a moderate increase in density above the accepted optimum for weed-free conditions is usually beneficial for those crops whose weed-free yield is relatively insensitive to density, (ii) an increase in density of 20% to 100% is likely to improve weed control measurably, and (iii) density increases greater than 50% to 100% are likely to result in lodging, disease, and other problems. Even for growers who often achieve good weed control, some increase in density may provide relatively cheap insurance against failures of other weed management measures.

Crop spatial arrangement

Row spacing and random versus regular planting patterns

Both theoretical and empirical studies indicate that crop planting pattern has a substantial effect on the competitive balance between crops and weeds. Fischer & Miles (1973) explored the effects of planting pattern using a stochastic model in which plants grew radially until every point in the field was occupied by either a crop plant or a weed without overlap. They assumed that the weeds were randomly positioned by a Poisson process, and examined a variety of plant arrangements for the crop. Several of their results warrant comment here. First, they showed that the percentage of the field surface occupied by weeds increased as the "rectangularity" of the crop increased (Figure 6.5), where rectangularity is the row spacing, b, divided by the within-row plant spacing, a. Second, random sowing of the crop was always inferior to a square lattice planting, and in fact, was equivalent to a rectangular planting arrangement in which $b/a = 3.537$ (Figure 6.5). Although a triangular lattice was superior to a square lattice, the difference was too small to have practical importance in field conditions. Finally, they found that grouping crop plants into clusters greatly decreased their ability to occupy space relative to the weeds. Although the percentage of space occupied by weeds in the model of Fischer & Miles (1973) varied with crop density, and with the relative emergence times and growth rates of the weeds and crops, the relative ranking

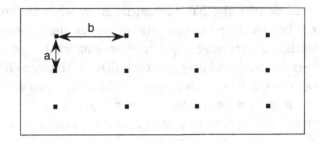

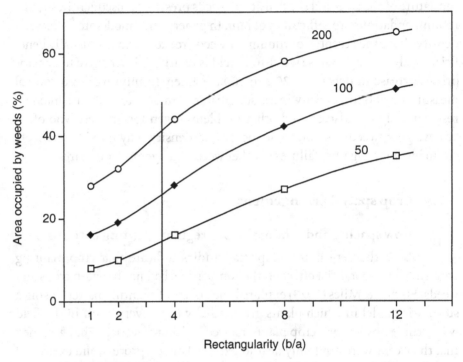

Figure 6.5 Effect of crop pattern on percentage of land area occupied by weeds in the model of Fischer & Miles (1973). Crop density was 300 m^{-2}, and curves are shown for weed densities of 50, 100, and 200 m^{-2}. Growth rate and emergence date were the same for the crop and the weeds. The vertical line shows the degree of rectangularity equivalent to a random (Poisson) placement of the crop plants ($b/a = 3.537$). (Drawn from data in Fischer & Miles, 1973.)

in the performance of different crop planting patterns remained the same. In an experiment with domestic oat and *Avena fatua*, Regnier & Bakelana (1995) confirmed Fischer & Miles's (1973) assumption of radial expansion of plant canopies, and found that the weeds grew larger as rectangularity increased.

Despite the strong theoretical basis for an inverse relation between row spacing and weed growth, the effect of row spacing on weeds in the field has been inconsistent. Although a majority of studies have shown that narrow row spacing suppresses weeds, a large minority of studies have shown no effect or an inconsistent effect across several experiments (Table 6.2). Variability in the effect of row spacing on weeds in the literature largely consists of variation among studies on a given species rather than variation among species (Table 6.2).

Plants tend to distribute their leaves into areas of high light and away from zones of high leaf area index, so that the horizontal distribution of leaves tends to be more uniform than the planting arrangement. This effect could explain some cases in which weed growth failed to respond to row spacing. However, several studies have shown that percent canopy cover develops faster, and light penetration through the canopy is less when rows are more closely spaced (Teasdale & Frank, 1983; Murdock, Banks & Toler, 1986; Yelverton & Coble, 1991; Murphy et al., 1996). Decreased light penetration with narrow rows results both from higher leaf area index (Murphy et al., 1996), and from a more efficient arrangement of leaves that increases light interception per unit leaf area (Flénet et al., 1996).

The studies in Table 6.2 found that narrowing row spacing in the presence of weeds increased crop yield in a slim majority of cases (27 out of 49). However, several studies were unclear as to whether the increased yield was due to weed control or would have occurred even in weed-free conditions. Although many studies showed no yield response to row spacing, only three observed a decrease in yield with narrower rows (Table 6.2).

Some cases in which weed biomass did not decrease with closer row spacing may be related to nutrient management. Banding fertilizer next to the crop row creates asymmetric competition for nutrients: on average, a unit of root growth by the crop early in the season allows greater access to the fertilizer than does a unit of root growth by the weeds because the fertilizer is closer to the crop. However, as the rows are placed closer together, this relative advantage declines.

Many cases in which weed biomass or crop yield in weedy conditions did not respond to row spacing can probably be explained by the height of the weed species relative to the crop. Note in this regard that the model of Fischer & Miles (1973) concerned only the occupation of ground area and did not include the relative height of the crop and weeds. In contrast, Schnieders et al. (1999) used a model to explore effects of row spacing on competition between two species that differed only in height. When the crop was potentially twice as tall as the weed, closer row spacing increased crop yield. When stature of the two species was equal, row spacing had a much smaller effect. Finally, when the weed was taller than the crop, crop yield declined at closer row spacing

Table 6.2. *Response of weeds and crop to decrease in crop row spacing at constant crop density*

Crop	Weed response to narrower rows[a]	Yield response to narrower rows		Reference
		Weed-free	Weeds present	
Barley	V		V(+)	Bate, Elliott & Wilson (1970)
	V(+)	0	0	Cussans & Wilson (1975)
	V(+)	0	0	Barton, Thill & Shafii (1992)
	−		+	Kirkland (1993)
Bean	−	V	+	Malik, Swanton & Michaels (1993)
Bean, snap	V(−)		V(+)	Teasdale & Frank (1983)
Cereals[b]	−		+	Håkansson (1984)
Cotton	V(+,0)	++	+	Miller, Carter & Carter (1983)
	−		+	Brown, Whitwell & Street (1985)
Flax seed		0	+	Alessi & Power (1970)
	−	+	+	Stevenson & Wright (1996)
	0		0	Blackshaw et al. (1999)
Lupin	−	+	+	Putnam et al. (1992)
Maize	V(−,I)		V	Choudhary (1981)
	0		V	Weil (1982)
	−		+	Harvey & McNevin (1990)
	−	[+]		Murphy et al. (1996)
	0		V(−)	Johnson, Hoverstad & Greenwald (1998)
	0	0		Teasdale (1998)
	−			Rambakudzibga (1999)
Oat	−		+	Pendleton & Dungan (1953)
Peanut	−	V	+	Colvin et al. (1986)
Pearl millet		+	−	Limon-Ortega, Mason & Martin (1998)
Pigeonpea	0	0	0	Díaz-Rivera et al. (1985)
Rapeseed	0	0	0	O'Donovan (1994)
Safflower	V(0)	0	0	Blackshaw (1993)
Sorghum	0[c]	−[c]	V[c]	Burnside, Wicks & Fenster (1964)
	−	V	V(+)	Wiese et al. (1964)
	−	V	V(+)	Holland & McNamara (1982)
		+	+	Limon-Ortega, Mason & Martin (1998)
Soybean	V(−)[c]	V(+)[c]	V(+)[c]	Burnside & Colville (1964)
	0	0	0	Peters, Gebhardt & Stritzke (1965)
	V(−)	0	0	Burnside (1979)
	−	0	+	Walker et al. (1984)
	−			Murdock, Banks & Toler (1986)
	−	0	+	Patterson et al. (1988)
	−	+[d]	++[d]	Légère & Schreiber (1989)
	V		0	Shaw, Smith & Snipes (1989)
	−			Yelverton & Coble (1991)
	V(−)			Mulugeta & Boerboom (1999)
Sunflower		+	V(0)	Woon (1987)
Sweet corn	−		+	Harvey & McNevin (1990)

Table 6.2. (cont.)

Crop	Weed response to narrower rows[a]	Yield response to narrower rows		Reference
		Weed-free	Weeds present	
Wheat	−		+	Shaktawat (1983)
	V(−)		0	Vander Vorst, Wicks & Burnside (1983)
	−		+	Kukula (1986)
	V(I)	+	0	Koscelny et al. (1990)
	V(−)	(+)	V(+)	Koscelny et al. (1991)
		+	+	Solie et al. (1991)
	−		+[e]	Johri, Singh & Sharma (1992)
	V(0)	0	0	Justice et al. (1993)
	−		0	Teich et al. (1993)
	V(+)	V(−)	0	Justice et al. (1994)
	+	−	−	Hashem, Radosevich & Roush (1998)
	V		0	Blackshaw et al. (1999)

Notes:
[a] Symbols are as in Table 6.1. Studies were included only if weed response to row spacing was measured, or yield response to row spacing was measured with and without weeds. For studies in which all treatment series included weed control measures, the row spacing series with the least effective weed control was used for evaluation of the weed response and the yield response in the presence of weeds.
[b] Summarizes 27 experiments on barley, wheat, oat, and rye.
[c] Widest row spacing was ignored because it alone was cultivated.
[d] Crop leaf area index rather than yield.
[e] Nutrient uptake by the crop rather than yield.

because weeds between rows covered the crop earlier. Teasdale & Frank (1983) had a related result in a field experiment with snap bean: the decrease in weed fresh weight with narrower row spacing was larger and more consistent when weed emergence was delayed for several weeks since this gave the crop a size advantage in competition for light. Similarly, Harvey & McNevin (1990) found that improvement of *Panicum miliaceum* control and sweet corn yield by narrow rows was greater when the crop was planted in May rather than April, possibly because of faster crop growth later in the season. These findings indicate that crop arrangement should be considered in the context of integrated weed management programs rather than as a stand-alone control tactic.

The relation between row spacing and weed management is considerably more complex than an analysis of weed abundance and yield may indicate. Prior to the widespread use of herbicides, row spacing in most row crops was determined largely by the need to cultivate between the rows. The change from horse-drawn to tractor-drawn cultivators allowed reduction in row spacing from 1 m or more to around 0.75 m. More recently, herbicides have allowed much narrower row spacings, particularly in soybean where narrow inter-rows often increase yields substantially even in weed-free conditions (Cooper, 1977; Parker, Marchant & Mullinix, 1981; Beatty, Eldridge & Simpson, 1982; Boquet, Koonce & Walker, 1982). Most of the work on row spacing postdates the widespread use of herbicides, and much of this work was stimulated by liberation from the planting constraints imposed by traditional inter-row cultivation methods. However, the advent of improved technology for mechanical weed management, including a new generation of harrows, and cultivator guidance systems coupled with tools that work close to crop rows, has opened possibilities for cultivation of narrowly spaced rows. To avoid crushing part of the crop with tractor tires during cultivation, the planter needs to leave wider inter-rows in the drive tracks. However, for wide planters and cultivators, the drive track inter-rows constitute a small percentage of the field. Adapting cultivation systems to narrow rows is as yet largely unexplored.

Effects of row orientation on light penetration into crop canopies

Row orientation in orchards and row crops affects the percentage of incident light that penetrates through crop canopies, particularly when the canopy is not closed. Because a variety of factors including row spacing, height and shape of the canopy, leaf area density, and time of year affect light interception, most of the work to date has emphasized mathematical modeling rather than empirical measurements. Several of these models agree in showing that throughout most of the temperate growing season light interception in orchards and row crops is greater when rows run N–S rather than E–W (Cain, 1972; Jackson & Palmer, 1972; Mutsaers, 1980; Palmer, 1989). Depending on the shape of the row canopy, the effects can be substantial (e.g., 47% vs. 38% interception – Palmer, 1977). Late in the temperate growing season, E–W-oriented rows intercept more light. A model by Schnieders *et al.* (1999) showed that a hypothetical crop in the Netherlands was most productive when rows were oriented NW–SE because this put rows perpendicular to the sun during the time of day when combined direct and indirect light intensity was highest.

As latitude decreases from 55° to 25°, the benefit of N–S row orientation increases (Cain, 1972; Jackson & Palmer, 1972; Mutsaers, 1980; Palmer, 1989). From 25° to 5° the difference between N–S and E–W orientation near the summer solstice decreases with latitude, but the portion of the year during which N–S orientation is advantageous increases from six months to nine months (Mutsaers, 1980). At the equator, N–S orientation is advantageous except in January and December, and again during June and July.

Effects of row orientation on weed management have received little direct attention, but several studies provide relevant information. In a seven-year study in Illinois, Pendleton & Dungan (1958) found that oat consistently yielded more when planted in N–S rather than E–W rows (mean increase of 75–276 kg ha^{-1}, depending on row spacing). Light levels in N–S-oriented inter-rows were lower than in E–W inter-rows early and late in the day when light levels were most likely to be limiting, and clover interseeded into the oats had lower density in the N–S-oriented plots, especially in the crop rows. In a study of interseeding cover crops into maize, Larson & Willis (1957) demonstrated that with E–W-oriented rows, a zone of high average light intensity on the south side of the row extended to near the base of the crop. With N–S-oriented rows, shading was more symmetrical. This is critical since weeds near the crop row are more competitively damaging and more difficult to control by cultivation than weeds in the inter-row (see Chapter 4). The model of Schnieders et al. (1999) found little effect of row orientation on competition between two species planted in parallel rows. However, their two species were given similar characteristics so this result may not represent systems in which the crop and weed differ in initial size and growth rate (see Chapter 2).

Although the effects of crop orientation on weeds remains to be verified, and are likely to be small relative to the benefits of other cultural practices, use of a N–S row orientation may increase the efficiency of the crop and suppress weeds with no extra expense to the grower. However, the benefits are unlikely to compensate for potential problems if field shape and topography are not conducive to N–S planting.

Crop genotype

Crop genotype affects interaction of crops with weeds

The role of crop genotype in weed management has received growing attention over the past 30 years. Callaway (1992) reviewed literature on crop varietal tolerance to weeds, and Callaway & Forcella (1993) assessed the prospects for breeding improved weed tolerance.

Following the lead of plant ecologists working on problems of competitive interaction between plants (Goldberg & Landa, 1991), Callaway (1992) distinguished between *competitive effect*, or the ability of the crop to suppress weeds, and *competitive response*, or the ability of the crop to avoid being suppressed.

Competitive effect can be measured as reduction in weed performance (e.g., biomass, cover) in the presence of a given cultivar relative to a weedy check treatment or to a standard cultivar. Competitive response can be measured as percentage reduction in yield relative to a weed-free control treatment. Although both concepts are important attributes of a crop's competitive ability, neither measures the attribute of greatest interest to a grower, namely the ability of the variety to produce a high yield despite weed competition. Thus, for example, Nangju (1978) found that 'Jupiter' soybean at low density allowed only 58% the weed biomass of 'Bossier' soybean (competitive effect), and had a yield loss due to weeds of 42% compared with 'Bossier's 53% (competitive response). However, 'Bossier' had a several-fold higher yield, both with and without the presence of weeds. Although neither competitive effect nor competitive response provides an adequate guide for choosing cultivars, they are both useful in identifying crop characteristics that correlate with competitive ability. Potentially, these characteristics could be bred into high-yielding lines to create new cultivars with high yield in the presence of weeds (Callaway & Forcella, 1993). Although competitive effect, competitive response, and yield under competition are not always correlated, they often are. In the discussion below, the term 'competitive ability' is used to encompass all three concepts.

Crop varietal differences contributing to competitive ability (of one sort or another) have been identified in a wide range of crops (Table 6.3). Some studies have failed to find differences in competitive ability among cultivars (e.g., Staniforth, 1962; Bridges & Chandler, 1988; Glaz, Ulloa & Parrado, 1989; Yelverton & Coble, 1991), but these all dealt with crops for which other studies have shown differences. Thus, differences in competitive potential among cultivars appear to exist for most crop species, although few data are available on tree crops. Not surprisingly, most work has focused on major crops, especially wheat and soybean (Table 6.3).

Variation in competitive ability of cultivars between experiments conducted in different years or at different locations is often large (Fiebig, Shilling & Knauft, 1991; Lemerle, Verbeek & Coombes, 1995; Cousens & Mokhtari, 1998; Ogg & Seefeldt, 1999). This is a problem commonly encountered in screening accessions for any desirable characteristic. However, competitive ability may be an inherently more complex characteristic than, say, monoculture yield or resistance to a fungal disease. Not only can several different types of competitive ability be identified (see above), but also, competition between plants involves several different resources, and varies with the identity of the weed species and the phenologies of the weed and crop. Although these factors can be controlled in experiments, use of results from

Table 6.3. *Crops with genotypic variation in competitive ability*

Crop	References
Alfalfa	Hycka & Benitez-Sidón (1979); Bittman, Waddington & McCartney (1991)
Barley	Kolbe (1980); Moss (1985); Siddiqi et al. (1985); Richards (1989); Richards & Davies (1991); Satorre & Snaydon (1992); Dhaliwal, Froud-Williams & Caligari (1993); Richards & Whytock (1993); Christensen (1994); Doll (1997)
Bean	Barreto (1970); Wilson, Wicks & Fenster (1980); Valverde & Araya (1986); Malik, Swanton & Michaels (1993); Urwin, Wilson & Mortensen (1996); Ngouajio, Foko & Fouejio (1997)
Carrot	William & Warren (1975)
Cotton	Chandler & Meredith (1983)
Cowpea	Nangju (1978); Remison (1978)
Guineagrass	Monzote, Funes & Díaz (1979)
Lupin	Walton (1986)
Maize	Staniforth (1961); Cadag & Mercado (1982); Woolley & Smith (1986); Ford & Mt. Pleasant (1994); Tollenaar et al. (1994b); Lindquist & Mortensen (1998); Lindquist, Mortensen & Johnson (1998)
Mungbean	Moody (1978)
Oat	Richards (1989); Satorre & Snaydon (1992); Lemerle, Verbeek & Coombes (1995)
Pea	Liebman (1989); Wall, Friesen & Bhati (1991)
Peanut	Colvin et al. (1985); Fiebig, Shilling & Knauft (1991)
Pigeonpea	Díaz-Rivera et al. (1985)
Potato	Sweet & Sieczka (1973); Yip, Sweet & Sieczka (1974); Selleck & Dallyn (1978); Nelson & Giles (1989)
Rapeseed	Lemerle, Verbeek & Coombes (1995)
Rice	Smith (1974); Kawano, Gonzalez & Lucena (1974); Akobundu & Ahissou (1985); Kwon, Smith & Talbert (1991); Stauber, Smith & Talbert (1991); Garrity, Movillon & Moody (1992); Fischer et al. (1995); Fischer, Ramírez & Lozano (1997); Johnson et al. (1998); Olofsdotter et al. (1999)
Ryegrass	Gibeault (1986); Sugiyama (1998)
Safflower	Paolini et al. (1998)
Sorghum	Guneyli, Burnside & Nordquist (1969); Burnside & Wicks (1972)
Soybean	Wax & Pendleton (1968); Burnside (1972, 1979); McWhorter & Hartwig (1972); McWhorter & Barrentine (1975); Burnside & Moomaw (1984); Rose et al. (1984); Murdock, Banks & Toler (1986); James, Banks & Karnok (1988); Monks & Oliver (1988); Callaway & Forcella (1993); Shilling et al. (1995); Bussan et al. (1997); Shaw, Rankins & Ruscoe (1997)
Squash	Stilwell & Sweet (1974)
Sugarcane	Arévalo, Cerrizuela & Olea (1978); Millhollon (1988)
Tall fescue	Forcella (1987)
Wheat	Appleby, Olson & Colbert (1976); Blackshaw, Stobbe & Sturko (1981); Kreuz (1982); Challaiah et al. (1983, 1986); Flood & Halloran (1984); Moss (1985); Sechniak, Lyfenko & Pika (1985); Wicks et al. (1986); González Ponce (1988); Ramsel & Wicks (1988); Richards (1989); Koscelny et al. (1990); Balyan et al. (1991); Kirkland & Hunter (1991); Richards & Davies (1991); Thompson, Gooding & Davies (1992); Valenti & Wicks (1992); Gooding, Thompson & Davies (1993); Richards & Whytock (1993); Christensen, Rasmussen & Olesen (1994); Wicks et al. (1994); Huel & Hucl (1996); Lemerle et al. (1996); Anderson (1997); Cosser et al. (1997); Cousens & Mokhtari (1998); Hucl (1998); Ogg & Seefeldt (1999); Seavers & Wright (1999); Seefeldt, Ogg & Hou (1999)

those experiments for weed management may prove challenging. A possible next step following screening studies would be evaluation of promising varieties in the context of integrated weed management on farms.

Many characteristics correlate with ability to tolerate or competitively suppress weeds. Since cultivars usually differ in many characteristics simultaneously, in some cases the attributes apparently distinguishing cultivars of high and low competitive ability may not be the ones that actually confer competitive ability. The problem is especially great for the many studies in which a small number of very different cultivars are compared. Forcella (1987) correctly pointed out that the only conclusive way to test the competitive effectiveness of a characteristic is to compare near isogenic lines that differ only with regard to that characteristic. A few other studies have also used this approach (Flood & Halloran, 1984; Seefeldt, Ogg & Hou, 1999). More work of this sort is badly needed. Nevertheless, cultivar screening trials are still useful in identifying existing varieties that can contribute to weed management immediately. These trials are most useful for identifying potential competitive traits when the various states of each character are represented in a wide range of genetic backgrounds (e.g., Garrity, Movillon & Moody, 1992; Lemerle et al., 1996), or when characteristics vary within a well defined class of cultivars (e.g., high-yielding semidwarfs – Fischer, Ramírez & Lozano, 1997)

Characteristics conferring competitive ability appear to differ between cereal grains and row crops. Row crop plants often do not contact weeds in the inter-rows until several weeks after emergence. Consequently, rapid growth and early canopy closure should provide increased competitive ability in row crops. In fact, several characters related to early growth rate have been found to correlate with competitive ability in row crops. These include large seed size in bean (Valverde & Araya, 1986), rapid emergence in sorghum, potato, and soybean (Guneyli, Burnside & Nordquist, 1969; Yip, Sweet & Sieczka, 1974; Rose et al., 1984), high early growth rate in sorghum, bean, and safflower (Guneyli, Burnside & Nordquist, 1969; Malik, Swanton & Michaels, 1993; Paolini et al., 1998), and rapid canopy closure in potato and soybean (Wax & Pendleton, 1968; Sweet & Sieczka, 1973; Yip, Sweet & Sieczka, 1974; Rose et al., 1984). Plant height and high leaf area index also correlate with competitive ability in row crops (soybean, cowpea – Nangju, 1978; mungbean – Moody, 1978; bean – Urwin, Wilson & Mortensen, 1996; safflower – Paolini et al., 1998). These characters allow the crop to overtop and shade the weeds. Indeterminate varieties of bean (Barreto, 1970; Wilson, Wicks & Fenster, 1980; Malik, Swanton & Michaels, 1993), cowpea (Remison, 1978), squash (Stilwell & Sweet, 1974), and cucumber (Staub, 1992) tend to be better competitors than determinate varieties. Indeterminate varieties of these crops usually have a vining habit that allows them to crawl over weeds and completely fill inter-row areas relatively quickly. Indeterminate growth may also allow crops to continue active competition with weeds through more of the season.

In contrast with row crops, height is the most frequent characteristic correlated with competitive ability in cereals (e.g., Appleby, Olson & Colbert, 1976; Ahmed & Hoque, 1981; Challaiah et al., 1986; Balyan et al., 1991; Garrity, Movillon & Moody, 1992; Lemerle et al., 1996; Hucl, 1998; Seefeldt, Ogg & Hou, 1999). This is reasonable since cereals are often no taller than the weeds with which they compete, and the near vertical orientation of much of the leaf surface in these crops allows light penetration deep into the crop canopy. Consequently, even when the taller cultivar does not completely overtop the competing weeds, at least the more elevated distribution of leaf area causes greater interception of light by the crop and more shade on the weeds. Greater light interception by taller cultivars has been demonstrated in rice and wheat (Jones, Zimmermann & Dall'Acqua, 1979; Gooding, Thompson & Davies, 1993). Unfortunately, short-statured cereal varieties often yield more. Indeed, increasing harvest index by decreasing stature has been used to breed higher-yielding cultivars in several cereal crops (Evans, 1980).

Nevertheless, some studies have found cereal cultivars that combine high yield with competitive ability (Fischer, Ramírez & Lozano, 1997; Hucl, 1998). High-yielding competitive varieties are possible because several characters other than height also contribute to competitive ability. Several studies suggest that high leaf area index or high biomass contribute to competitive ability in cereals (Kawano, Gonzalez & Lucena, 1974; Challaiah et al., 1983; Sechniak, Lyfenko & Pika, 1985; Balyan et al., 1991; Garrity, Movillon & Moody, 1992; Fischer et al., 1995; Huel & Hucl, 1996; Fischer, Ramírez & Lozano, 1997). These characteristics are frequently correlated with plant height, but they can also vary substantially within a height class (Sechniak, Lyfenko & Pika, 1985; Fischer, Ramírez & Lozano, 1997). Tillering is one route to high leaf area index and biomass in cereals. Several studies have found a correlation between tillering and competitiveness (Kreuz, 1982; Challaiah et al., 1983, 1986; Valenti & Wicks, 1992), although others have looked for this and found none (Kawano, Gonzalez & Lucena, 1974; Moss, 1985; Huel & Hucl, 1996). Challaiah et al. (1983) found that several of their highest-yielding cultivars also had high percentage light interception and correspondingly low weed densities. Lanning et al. (1997) demonstrated variation in light interception and suppression of *Avena fatua* among barley varieties of similar height. Recent work has shown that, as with row crops, early growth rate and characters relating to leaf area expansion rate appear to contribute to competitive ability in rice and wheat (Johnson et al., 1998; Ogg & Seefeldt, 1999; Seefeldt,

Ogg & Hou, 1999). Correlation between early ground cover and weed suppression has been demonstrated for barley and wheat varieties with a limited range of heights (Richards, 1989; Richards & Whytock, 1993; Huel & Hucl, 1996).

Little work has focused on characteristics conferring competitiveness in forage crops. Monzote, Funes & Díaz (1979) found that the tallest of 17 cultivars of guineagrass had the lowest infestation of weeds. Similarly, Black (1960) found that tall-petioled varieties of subterranean clover outcompeted short-petioled varieties in diallel competition experiments. Forcella (1987) compared isogenic strains of tall fescue differing in leaf-area expansion rate and found that faster expanding strains were more competitive. However, Sugiyama (1998) concluded that the greater competitive ability of tetraploid cultivars of perennial ryegrass relative to diploid cultivars could not be attributed to differences in early growth rate.

Few studies have addressed below-ground characteristics that relate to varietal differences in competitive ability. Satorre & Snaydon (1992) found that although root competition between *Avena fatua* and spring barley was more important than shoot competition, varieties differed more in above-ground competitive ability. Kawano, Gonzalez & Lucena (1974) found that rice cultivars with a high responsiveness of yield to nitrogen fertilization were poor competitors. In their experiment, traits of the 25 cultivars tested were correlated such that cultivars were either adapted to low intensity (low N, low density, no weeding) or to high intensity (high N, high density, weed-free) agronomic conditions, but not to mixed conditions (e.g., high N, no weeding). Whether this represents inherent trade-offs between characters, or merely the history of plant breeding in rice is unclear. Siddiqi *et al.* (1985) found substantial differences in the potassium uptake efficiency of barley cultivars that related to biomass accumulation under competition with *Avena fatua*. 'Fergus' produced a high biomass when grown at high potassium levels in monoculture, even though its shoot potassium concentration was relatively low. Related to this, potassium utilization efficiency of 'Fergus' was high when stressed for potassium in monoculture and when competing with *Avena fatua*. This apparently enabled it to attain substantially higher biomass in mixtures under low potassium conditions than could the other cultivars. Experiments designed to separate above- and below-ground competitive effects in herbaceous plants have usually found that below-ground competition was highly important (Donald, 1958; Aspinall, 1960; Martin & Snaydon, 1982; Satorre & Snaydon, 1992). Consequently, additional work comparing crop varieties with respect to competition for below ground resources would be useful.

Wu *et al.* (1999) reviewed studies showing variation in allelopathic potential among cultivars. For example, Fay & Duke (1977) screened 3000 oat accessions for presence of scopoletin, a compound known to have allelopathic effects. All accessions appeared likely to contain some scopoletin, but four studied in detail contained three- to four-fold more scopoletin than a standard commercial cultivar. Seedlings of *Brassica kaber* (*Sinapis arvensis*) grown with one of these varieties were stunted, twisted, and chlorotic. Putnam & Duke (1974) screened 540 cucumber accessions for allelopathic affects on *Brassica hirta* and *Panicum miliaceum*. One of these reduced *P. miliaceum* growth by 87% in a controlled environment chamber. It also had a substantial but short-lived effect on emergence and growth of several weed species in field trials (Lockerman & Putnam, 1979). Other studies have also found variation in allelopathic potential among accessions of crop species (sunflower – Leather, 1983; soybean – Rose et al., 1984; coffee – Waller et al., 1986; sweet potato – Harrison & Peterson, 1986; rice – Dilday et al., 1991; Olofsdotter et al., 1999). Nicol et al. (1992) found eight-fold variation in the concentration of the allelopathic compound DIMBOA among 47 cultivars of wheat. Wu et al. (1999) concluded that although the potential for development of allelopathic crops is substantial, more research on the genetic control of allelopathy is needed before breeding programs can be initiated. Crop allelopathy is most likely to provide effective weed management if combined with other competitive characteristics (Olofsdotter et al., 1999).

Highly competitive varieties often have multiple characteristics that contribute to competitive ability. This is illustrated by a study on potato (Yip, Sweet & Sieczka, 1974). The investigators compared four cultivars, including 'Katahdin', which was at the time the leading cultivar in the northeastern USA. Weed control differed substantially between cultivars. In particular 'Green Mountain' was highly competitive against a range of weed types (Figure 6.6). As a result, the yield of 'Green Mountain' with one cultivation plus hilling was as good as with an effective herbicide plus hilling (Figure 6.7). The greater competitiveness of 'Green Mountain' could be attributed to (i) its more rapid emergence and canopy closure, (ii) its extensive branching and dense foliage, and (iii) its maintenance of a closed canopy for a greater portion of the season (Table 6.4). Root competition was unimportant (Sweet, Yip & Sieczka, 1974). The characteristics just listed allowed 'Green Mountain' to suppress weed growth early in the season, compete effectively with the weeds that did establish, and maintain that competitive pressure through the critical part of the growing season. Together they allowed 'Green Mountain' to maintain yield under a weed pressure that substantially reduced yields in the less competitive cultivars (Figure 6.7). Although, 'Katahdin' produced a higher yield in weed-free conditions (Figure 6.7), the characteristics of 'Green

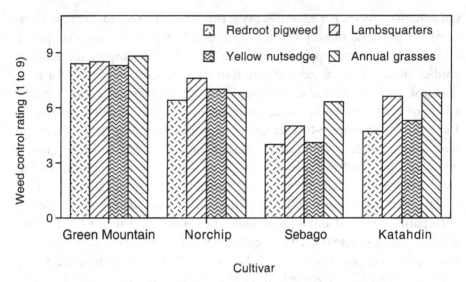

Figure 6.6 Weed control ratings (1 = no control, 9 = full control) in four potato cultivars in mid August. Weed management consisted of one inter-row cultivation plus hilling. (Drawn from data in Yip, Sweet & Sieczka, 1974.)

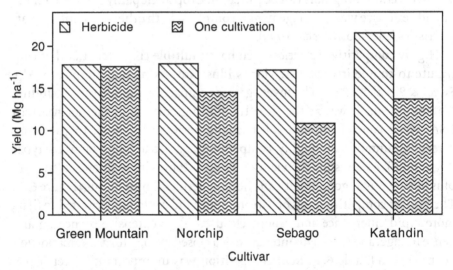

Figure 6.7 Tuber yield for four potato cultivars receiving either herbicide (linuron) plus hilling, or one cultivation plus hilling. (Redrawn from Yip, Sweet & Sieczka, 1974.)

Table 6.4. *Characteristics of four potato cultivars*

Cultivar	Emergence	Canopy closure	Falling over	Branches per seed piece	Plant size	Foliage density	Maturity
Green Mountain	6/14	7/5	7/28	15.2	Very large and spreading	Dense	Very late
Norchip	6/15–16	7/7	7/28	4.7	Medium-large and upright	Dense	Early
Sebago	6/16–17	7/9	7/11	6.0	Large, erect to spreading	Sparse	Very late
Katahdin	6/16–17	7/8	7/11	3.1	Medium to large and spreading	Sparse	Late

Source: Yip, Sweet & Sieczka (1974).

Mountain' that appear to give it greater competitive ability are not inherently inimical to yield. Thus, the potential exists for breeding cultivars that are high yielding under weed-free conditions, but also compete vigorously against weeds when they are present.

Breeding for competitive ability

Despite the many studies showing cultivar differences in competitive ability against weeds, work on improving the ability of crops to tolerate or suppress weeds is still in its infancy. Callaway & Forcella (1993) described an experiment designed to test the potential for improving the competitive ability of soybean. For three years they grew four to eight cultivars of soybean in weedy and weed-free plots, and at several dates during the season recorded a wide range of characteristics having potential effects on competitive ability. The character most consistently related to weed biomass was leaf area per plant (LAP). End-of-season weed biomass had a strong negative correlation with early to midseason LAP but not late season LAP. Thus, competitive ability appeared to result from high leaf area expansion rate.

Based on the preliminary trials, Callaway & Forcella chose a poorly competitive, low LAP cultivar which was adapted to the region ('Evans'), and a competitive, high LAP but nonadapted cultivar ('Gnome') for breeding material. These were crossed, and the F_1 progenies selfed. The resulting F_2 progenies were grown out and again selfed, and the leaf area of each plant was calculated. High and low LAP individuals were chosen as seed sources for the F_3 generation. F_3 lines with maturities at least as early as 'Evans', with yields as high as 'Evans', and whose LAP was at least one Least Significant Difference unit ($p = 0.05$) greater than 'Evans' were chosen. The competitive ability of the F_4 generation of these selected lines was tested by growing them with and without competition from *Amaranthus retroflexus* and *Setaria faberi*. One of the selected lines allowed only 62% as much growth of weeds as 'Evans', and had a yield 30% higher than 'Evans' in the weedy treatment. Since the intent of the study was to explore the feasibility of breeding for competitiveness against weeds rather than the introduction of a new cultivar, a full agronomic evaluation of the line was not attempted. However, the study does indicate that increasing the competitive ability of soybean without sacrificing yield in pure culture may be possible.

Callaway & Forcella (1993) also discussed several technical problems relating to the breeding of crops for competitive ability against weeds. They suggested that selection with respect to characters correlated with competitive ability (e.g., LAP) is more efficient than selection on competitive ability itself early in a breeding program when many lines are being examined. After these have been narrowed down to a few promising candidates, actual testing against weeds becomes more practical.

Choice of the weed or weeds against which the crop lines are to be tested is a critical issue. If the crop responds similarly to most of the weeds with which it occurs, with differences in response primarily due to weed biomass rather than taxon, then the choice of weed species is not especially critical and a standard "tester" species can be chosen. However, the limited evidence currently available indicates that crop response to weeds is taxon dependent (McWhorter & Hartwig, 1972; Monks & Oliver, 1988; Wilson & Wright, 1990; Bussan et al., 1997). Fortunately, statistical procedures are available that allow efficient choice of selection environments, in this case weed taxa (Zobel, Wright & Gauch, 1988; Crossa, Gauch & Zobel, 1990).

Phenology

Weed-free periods and weed infestation periods

The timing of crop emergence, growth, and maturation (*phenology*) relative to competing weeds has a large impact on crop production. Both empirical studies (Mann & Barnes, 1947; Bowden & Friesen, 1967; Håkansson, 1986) and simulation models (Kropff et al., 1993) have demonstrated that an advantage of even a few days can greatly shift the competitive balance between crops and weeds. Early in the life of an annual plant, biomass and leaf area tend to increase exponentially (Shinozaki & Kira, 1956). Consequently, a head start of a few days can greatly affect the relative sizes of the two species at the point when they grow into competitive contact. If a slightly larger size allows the crop to shade the weed at the time their canopies meet, then this size difference early in the season is likely to compound further.

Many studies in a wide range of annual crops have used this principle to demonstrate that an initial weed-free period is sufficient to obtain full crop yield (Zimdahl, 1980). Characteristically, as the initial weed-free period grows longer, yield rises to an asymptote (Figure 6.8). Thus, after some threshold date, weeds no longer have a measurable effect on the current crop. As discussed in Chapter 2, seed production by the weeds may cause problems in future years, but conceptually that is a separate issue.

Minimum initial weed-free periods result from the preemptive nature of competition between plants. Once the crop begins to shade the weeds, weed growth rate will be reduced. If the crop's head start is sufficiently great, the weeds will remain suppressed in a subcanopy position and will be unable to

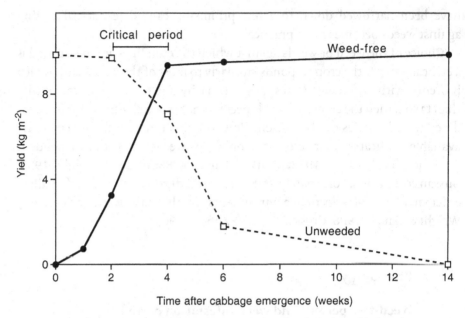

Figure 6.8 Effect of initial weed-free period and period of weed infestation on yield of cabbage. (Drawn from data in Miller & Hopen, 1991.)

compete effectively for light during the remainder of the crop's growing season. The length of initial weed-free period required for negligible yield loss depends on the initial growth rates of the crop and weed and on how the weed's growth rate and shoot architecture change in shade (see Chapter 2). Although the rate of uptake of water and nutrients per unit root surface may not be equal in the crop and weed, the total uptake for each species is necessarily proportional to their respective root surface areas. Providing the crop with an initial period of development prior to germination or sprouting of the weed thus allows proportionally more below-ground resources to flow to the crop (Mann & Barnes, 1947). However, in most cases competition for light is probably more important in making initial weed-free periods effective. Rates of water and nutrient uptake per unit of root surface usually do not depend on the relative size of the competing root systems. In contrast, a unit of leaf area for the taller competitor is far more effective at capturing sunlight than a unit of leaf area on the shorter species (Donald, 1963; Liebman, 1989).

Pre-emergence herbicides and most mechanical weeding methods are aimed at giving the crop a comparatively weed-free period for initial growth. After the herbicide dissipates, or following lay-by cultivation, weeds may sprout, but they have a limited effect on a well-established crop. Similarly, mulches of organic residue frequently lose effectiveness after a few weeks due to decomposition of biomass and breakdown of allelopathic toxins, but mulches may still be effective weed management tools if they suppress weeds through the minimum initial weed-free period (see Chapter 5). Deeply planting large-seeded crops into moist soil while the surface soil is too dry for germination of small-seeded weeds can also give the crop a head start over weeds (see Chapter 5).

The effect of an initial weed-free period on yield is far from constant for a given crop. In particular, a given weed-free period often allows greater yield loss when weed density is higher (Miller & Hopen, 1991; Baziramakenga & Leroux, 1994). Mechanistic simulations have demonstrated the same result (Weaver, Kropff & Groeneveld, 1992; Kropff, Lotz & Weaver, 1993), and these have, in addition, shown that the minimum initial weed-free period increases with the maximum height of the weed species present. In a field experiment, Dawson (1965) showed that yield response of sugar beet to initial weed-free period varied with weed species. Walker et al. (1984) found that in some trials minimum weed-free period for full yield was four weeks when soybean was planted with 40 cm or 80 cm between rows but only two weeks with 20 cm between rows. Barreto (1970) and Ngouajio, Foko & Fouejio (1997) found varietal differences in common bean yield response to weed-free period. In general, factors that increase the intensity of weed competitive pressure tend to lengthen the initial weed-free period required to avoid yield loss, although in some cases the effect is negligible.

Many studies have demonstrated a complementary phenomenon in which the crop does not suffer much if weeds are allowed to grow early in the season but are removed later (Zimdahl, 1980). In this case, the relation of crop yield to period of weed removal typically follows a falling sigmoidal curve (Figure 6.8). Essentially, the crop can tolerate the presence of weeds until the canopies and root systems of the two species are sufficiently extensive to reduce resource availability.

Maximum tolerated infestation periods are primarily useful in timing the application of post-emergence herbicides or hand-weeding with a hoe or machete. The optimal time for a single post-emergence treatment will be a compromise between the timing that allows the most weeds to emerge and thereby be exposed to the treatment, and the timing that minimizes crop loss by competition. However, when substantial weed emergence occurs throughout much of the cropping season, multiple treatments applied to successive cohorts of small weeds may be cheaper or allow fewer escapes.

The effect of period of infestation on yield depends on weed density (Bowden & Friesen, 1967; McNamara, 1976; Weaver, Kropff & Groeneveld, 1992; Baziramakenga & Leroux, 1994). In the simulation of Weaver, Kropff &

Groeneveld (1992), the maximum period of infestation tolerated by the crop was also greatly shortened by increased moisture stress. Cultivar comparisons in rice and bean noted that shorter-statured varieties suffered greater yield losses earlier in the season as the period of weed infestation increased (Barreto, 1970; Kwon, Smith & Talbert, 1991). Taken together these studies indicate that factors that increase the competitive ability of weeds relative to a crop tend to decrease the period of infestation that the crop can tolerate. However, Weaver (1984) found that the maximum tolerated period of infestation was shorter in narrow-row/high-density cabbage and cucumbers, probably because competition occurred earlier at higher crop densities.

As just discussed, functions relating yield to initial weed-free period and initial period of weed infestation depend on properties of the weeds (e.g., species, density), the crop (e.g., density, row spacing, cultivar), and the environment (e.g., fertility, soil moisture, temperature). Consequently, actually using information on the maximum tolerated infestation period and the minimum weed-free period to determine management strategies is a complex problem. Exploration of crop and weed phenology using mechanistic competition models offers hope for realistically addressing the complexity of interacting factors (Weaver, Kropff & Groeneveld, 1992; Kropff, Lotz & Weaver, 1993). In another approach, Dunan *et al.* (1995) modified the weed-free period and weed infestation period concepts to include the cost of alternative weed control options. They then used these relationships in a model that explicitly included weed density and competitive differences among weed species. In addition, they controlled a major source of environmental variation affecting the yield versus period of weed infestation relation by basing the timing of weed removal on cumulative heat units rather than days. Their simulations indicated that pre-emergence application of DCPA in onions was rarely cost effective. Such models provide the principal practical application for the weed-free period and weed infestation period concepts.

The time between the maximum tolerated period of infestation and the minimum initial weed-free period has been referred to as the "critical period for weed competition" (Figure 6.8) (Weaver, 1984; Dawson, 1986; Baziramakenga & Leroux, 1994; Ghosheh, Holshouser & Chandler, 1996; Singh *et al.*, 1996; Burnside *et al.*, 1998). It is the period during which weeds must be controlled to prevent yield reduction. A single weeding or nonresidual post-emergence herbicide application will not constitute sufficient weed management in systems that manifest a critical period. In situations where the tolerated period of infestation ends after the minimum initial weed-free period, the critical period is undefined, and a single weeding may be sufficient to avoid yield loss (Weaver, 1984; Van Acker, Swanton & Weise, 1993).

However, dependence of both the tolerated period of infestation and the minimum weed-free period on a wide variety of factors implies that application of the critical period concept to field situations requires both extensive data and careful judgement.

Initial size of crop at establishment

In some crops, possibilities exist for manipulating the size of the crop at planting time. Rice and many vegetable crops, including tomato, lettuce, leek, and the cole crops, are often planted in the field as transplants rather than seeds. Onion sets are essentially transplants, although they are dormant at planting time. Transplanting into ground that has been freshly tilled or from which the weeds have been eliminated by herbicides or flaming provides the crop with a substantial initial size advantage relative to weeds.

The advantages of transplanting for weed management are perhaps so obvious that they have not been explored much experimentally. Based on earlier experiments (Weaver & Tan, 1983, 1987), Weaver (1984) and Weaver, Kropff & Groeneveld (1992) concluded that the minimum weed-free period for transplanted tomatoes was about 35 days, versus about 70 days for direct-seeded tomatoes. Also, yield declined more slowly with increased period of infestation for transplanted tomatoes than for direct-seeded tomatoes. Hill, De Datta & Real (1990) analyzed several studies of *Echinochloa* spp. competition in transplanted and direct-seeded rice, and found that the transplanted rice was much more resistant to competition. For example, 3 *Echinochloa* m^{-2} resulted in 20% yield loss in direct-seeded rice, but 66 plants m^{-2} were required to produce the same yield loss in transplanted rice.

Most of the vegetable crops that are commonly transplanted are small-seeded species that establish slowly and remain noncompetitive for an extended period when growing from seeds. Early mechanical weeding in the crop row for these species is essentially impossible, and even hand-weeding may cause damage to delicate young plants. Consequently, organic growers usually transplant these crops even though effective direct-seeding equipment is available. If some herbicides are lost, either due to lack of reregistration or the development of herbicide resistance by major weeds, conventional growers may increase use of transplanting as well.

Some limited possibilities also exist for the manipulation of seed size. Kolbe (1980) screened winter barley to obtain two seed size fractions, and found that plots planted with the large seeds had higher yield and lower weed cover than plots planted with small seeds. Screening to obtain large seeds is a possibility in any crop with significant variation in seed size and in which a market exists for the smaller seed grades that will not be planted. Although screening to obtain larger seed would probably increase the cost of the seed, this might be offset somewhat by increased ease of removing weed seeds during production of certified seed. Using only large seeds is impractical for hybrid crops where the seed must be specially grown, and in forages and vegetables where no alternative market exists for the seed. Potato seed tubers are sometimes available in size classes, and the larger grades are likely to be better suited to growers relying on ecological weed management procedures. Lugo, Ortiz & González (1997) varied set size and planting density of taro and showed that the resulting differences in leaf area index had a substantial effect on weed biomass.

Probably the most effective way to obtain the competitive benefit of large seed size would be to breed larger-seeded crops. This is probably possible without yield loss in most crops. Larger seed size would also have benefits for management of weeds with cultivation (see Chapter 4) and mulch (see Chapter 5).

Planting date and weed–crop competition

As discussed in Chapter 4, planting date often has a substantial effect on weed pressure experienced by the crop, because weeds that germinate before planting are usually destroyed by seedbed preparation. However, planting date also affects the competitive balance between the crop and the weeds that emerge after planting. This occurs because the growth rate of both the crop and the weeds depends on environmental conditions, particularly temperature and soil moisture, and these typically change through the season. In weed-free conditions, the optimal planting date for agronomic crops is often the earliest that will allow successful emergence, since a longer season allows greater accumulation of resources and hence greater yield (Gunsolus, 1990). However, seedlings emerging from early plantings may grow slowly until weather conditions become more benign. Hence, they may become overgrown with weeds that are better adapted to weather conditions at that time of year.

Since a field is usually infested with several weed species, each with its own particular phenology, the problem of choosing a planting date that maximizes yield in the presence of weeds is not simple. To a first approximation, the crop will be most competitive at the planting date for which its growth rate per unit ground area (g m^{-2} d^{-1}) shortly after emergence is greatest relative to the weeds that emerge at that same time. This occurs because the early growth rates of the crop and weeds largely determine which has the greater stature, leaf area, and root occupation of the soil when they come into competitive contact. Water and nutrient use efficiencies and architecture of both the crop and weeds may also depend on environmental conditions, but their variation within the season is likely to be much less than variation in growth rate.

The competitive balance between the crop and weed depends not just on the growth rates of individual crop and weed plants, but rather on the ability of the crop population as a whole to occupy space in the field faster than the weed community as a whole. Consequently, the optimal planting date will often depend on both potential weed density and the effectiveness of weed control measures. If weeds are partially controlled, then yield as a function of planting date will be maximum for some date that depends on the degree of control of the several weed species present in the field. Good control will shift the optimal planting date toward that for the weed-free condition, and poor control will shift the optimum date toward that for the unmanaged condition. Thus, the optimal planting date depends on the grower's projection of the degree of weed control.

Many of the spring-planted row crops grown in the temperate zone, including maize, sorghum, soybean, and common bean, are subtropical species with relatively high optimal growth temperatures. Consequently, early planting is likely to put them at a disadvantage relative to weeds. For example, Oliver (1979) found that soybean planted in May in Arkansas had greater yield than July-planted soybean when the fields were kept free of weeds. However, leaf area index at four weeks was about three-fold greater in July-planted soybean, and due to day length dependence, growth of *Abutilon theophrasti* was slower in July-planted crops. Consequently *A. theophrasti* caused little yield loss for the July-planted soybean, but substantial yield loss for those planted in May. Similarly, Rushing & Oliver (1998) found that fresh weight of *Xanthium stumarium* planted with soybean declined with progressively later planting dates; weed suppression by late soybean planting was particularly great if weed emergence was delayed by two weeks or more. In a Wisconsin study, Harvey & McNevin (1990) found that control of weedy *Panicum miliaceum* and yield of maize and sweet corn improved when crops were planted in mid May rather than late April. However, late planting of row crops does not always improve weed control or decrease yield loss from a given density of weeds (Klingaman & Oliver, 1994).

The situation with spring-planted cereals in the temperate zone is rather different than that described above for spring row crops. Spring cereals germinate at temperatures as low as 5 °C (Cornell University, 1987, p. 144), and are thus well adapted to cool, wet conditions early in the season. They generally get maximum yield when planted in late winter in milder climates (Wibberley, 1989, p. 87), or as early as the ground can be worked in regions where the soil is frozen much of the winter (Cornell University, 1987, p. 144). Yield losses become substantial as the season progresses, largely due to decreased tillering. Moreover, because few annual weeds are well adapted for germination and rapid growth very early in the temperate growing season, weed pressure tends to increase as planting is delayed. Similarly, weedy safflower in Italy yielded more when sown in February than when sown in March, possibly because weed emergence was slower with the earlier planting date (Paolini *et al.*, 1998).

Consequently, for spring cereals the yield loss in the presence of weeds due to delayed planting is often greater than the yield loss experienced in the weed-free condition. For example, Vezina (1992) obtained full yield without weed control when wheat was planted in late April in Quebec, but competition from *Echinochloa crus-galli* and *Setaria pumila* (*S. glauca*) reduced yields when planting was delayed until early to mid May. Weeds did not affect barley yield in the same study, though yield was lower when planting was delayed. Deschenes & St-Pierre (1980) found lower weed biomass and higher oat biomass in plots seeded 5 May relative to plots seeded 3 June in Quebec, though the difference in oat biomass was not always reflected in grain yield. Wheat subjected to competition from controlled densities of *Setaria viridis* in Manitoba tended to have greatest yield when planted in late May to very early June (Blackshaw, Stobbe & Sturko, 1981), and yields declined for planting dates later in June. However, in contrast to the studies just cited, for some cultivars in some years, the optimal planting date without *S. viridis* was earlier than when the weed was present.

Planting date studies on fall-sown crops are few and contradictory. Lutman & Dixon (1986) found that yield of fall-sown oilseed rape (*Brassica napus*) declined with later planting dates, and that the yield decline was more severe as volunteer barley density increased. Given that early-seeded fall cereals have a faster initial growth rate (Wibberly, 1989), early sowing may improve competitiveness of cereals as well. However, Melander (1995) found that *Alopecurus myosuroides* and *Apera spica-venti* were slower to emerge and less competitive in late-sown winter wheat and rye. These contrasting studies show the importance of understanding the phenology of particular weeds relative to the crops with which they compete. In any case, effects of planting date on the relative competitive ability of a crop needs to be considered in light of the greater density of fall germinating weeds when the crop is planted early (see Chapter 4).

Planting date also affects weed competition in tropical crops (Díaz-Rivera *et al.*, 1985), but extraction of general rules for these situations will have to await further research.

Conclusions

The several cultural practices discussed in this chapter are based on common underlying general principles.

First, *the presence of weeds changes the cultural conditions that result in maximum crop yield*. The optimal choice for crop density, cultivar and planting date very often depend on whether or not the grower can expect significant weed competition with the crop. If the grower has a management program that reduces weed pressure to negligible levels in most years, then recommendations based on weed-free field studies may be optimal. However, weeds often significantly affect crop productivity due to a difficult-to-control species, lack of adequate herbicides for the particular crop, reluctance to use herbicides, or unpredictable effects of weather. In these situations, a moderate increase in planting density is likely to be beneficial, and cultivar, planting date, and plant arrangement may need to be adjusted as well.

Second, *crop yield in the presence of weeds increases with the difference in the height, biomass, and leaf area of the crop relative to that of the weeds at the onset of competition*. Factors such as increased planting density, larger seeds or transplants, larger or more rapidly growing cultivars, and planting dates that maximize early season growth of the crop all give the crop a head start relative to the weeds. Consequently, when the crop and the weeds grow into competitive contact, these factors lead to greater proportional occupation of the soil surface by crop shoots and the soil volume by crop roots. Taller stature gives the crop an even greater advantage over the weeds due to the asymmetric nature of competition for light. As discussed in Chapters 2 and 5, agricultural weeds are sensitive to both shade and nutrient stress. To the extent the crop is in a better position to shade the weeds and deplete water and nutrients in the rhizosphere of the weeds at the onset of competition, the less it will suffer yield loss from competition with weeds.

REFERENCES

Ahmed, N. U., & Hoque, M. Z. (1981). Plant height as a varietal characteristic in reducing weed competition in rice. *International Rice Research Newsletter*, 6(3), 20.

Akobundu, O., & Ahissou, A. (1985). Effect of inter-row spacing and weeding frequency on the performance of selected rice cultivars on hydromorphic soils of West Africa. *Crop Protection*, 4, 71–6.

Alessi, J., & Power, J. F. (1970). Influence of row spacing, irrigation, and weeds on dryland flax yield, quality, and water use. *Agronomy Journal*, 62, 635–7.

Ambe, J. T. (1995). Effect of plant population density of sweet potato (*Ipomoea batates* (L.) LAM) on weed incidence and severity in Cameroon. *International Journal of Pest Management*, 41, 27–30.

Anderson, R. L. (1997). Cultural systems can reduce reproductive potential of winter annual grasses. *Weed Technology*, **11**, 608–13.

Andersson, B., & Bengtsson, A. (1992). Influence of row spacing, tractor hoeing and herbicide treatment on weeds and yield in winter oilseed rape (*Brassica napus* L.). *Swedish Journal of Agricultural Research*, **22**, 19–27.

Appleby, A. P., & Brewster, B. D. (1992). Seeding arrangement on winter wheat (*Triticum aestivum*) grain yield and interaction with Italian ryegrass (*Lolium multiflorum*). *Weed Technology*, **6**, 820–3.

Appleby, A. P., Olson, P. D., & Colbert, D. R. (1976). Winter wheat yield reduction from interference by Italian ryegrass. *Agronomy Journal*, **68**, 463–6.

Arévalo, R. A., Cerrizuela, E. A., & Olea, I. L. (1978). Recent advances in weed competition studies in sugarcane in Argentina. In *Proceedings of the 16th Congress, International Society of Sugar Cane Technologists*, vol. 2, ed. F. S. Reis & J. Dick, pp. 1227–38. São Paulo, Brazil: IMPRESS.

Aspinall, D. (1960). An analysis of competition between barley and white persicaria. II. Factors determining the course of competition. *Annals of Applied Biology*, **48**, 637–54.

Ball, D. A., Ogg, A. G. Jr., & Chevalier, P. M. (1997). The influence of seeding rate on weed control in small red lentil (*Lens culinaris*). *Weed Science*, **45**, 296–300.

Balyan, R. S., Malik, R. K., Panwar, R. S., & Singh, S. (1991). Competitive ability of winter wheat cultivars with wild oat (*Avena ludoviciana*). *Weed Science*, **39**, 154–8.

Barreto, A. (1970). Competencia entre frijol y malas hierbas. *Agricultura Técnica en México*, **2**, 519–26.

Barton, D. L., Thill, D. C., & Shafii, B. (1992). Integrated wild oat (*Avena fatua*) management affects spring barley (*Hordeum vulgare*) yield and economics. *Weed Technology*, **6**, 129–35.

Bate, P. G., Elliott, J. G., & Wilson, B. J. (1970). The effect of barley population and row width on the growth of *Avena fatua*, wild oat. In *Proceedings of the 10th British Weed Control Conference*, pp. 826–30. London: British Crop Protection Council.

Baziramakenga, R., & Leroux, G. D. (1994). Critical period of quackgrass (*Elytrigia repens*) removal in potatoes (*Solanum tuberosum*). *Weed Science*, **42**, 528–33.

Beatty, K. D., Eldridge, I. L., & Simpson, A. M. Jr. (1982). Soybean response to different planting patterns and dates. *Agronomy Journal*, **74**, 859–62.

Bittman, S., Waddington, J., & McCartney, D. H. (1991). Performance of alfalfa strains grown in mixture with smooth bromegrass as affected by management. *Canadian Journal of Plant Science*, **71**, 1029–37.

Black, J. N. (1960). The significance of petiole length, leaf area, and light interception in competition between strains of subterranean clover (*Trifolium subterraneum* L.) grown in swards. *Australian Journal of Agricultural Research*, **11**, 277–91.

Blackshaw, R. E. (1993). Safflower (*Carthamus tinctorius*) density and row spacing effects on competition with green foxtail (*Setaria viridis*). *Weed Science*, **41**, 403–8.

Blackshaw, R. E., Stobbe, E. H., & Sturko, A. R. W. (1981). Effect of seeding dates and densities of green foxtail (*Setaria viridis*) on the growth and productivity of spring wheat (*Triticum aestivum*). *Weed Science*, **29**, 212–17.

Blackshaw, R. E., Semach, G., Li, X., O'Donovan, J. T., & Harker, K. N. (1999). An integrated weed management approach to managing foxtail barley (*Hordeum jubatum*) in conservation tillage systems. *Weed Technology*, 13, 347–53.

Boerboom, C. M., & Young, F. L. (1995). Effect of postplant tillage and crop density on broadleaf weed control in dry pea (*Pisum sativum*) and lentil (*Lens culinaris*). *Weed Technology*, 9, 99–106.

Boquet, D. J., Koonce, K. L., & Walker, D. M. (1982). Selected determinate soybean cultivar yield responses to row spacings and planting dates. *Agronomy Journal*, 74, 136–8.

Bowden, B. A., & Friesen, G. (1967). Competition of wild oats (*Avena fatua* L.) in wheat and flax. *Weed Research*, 7, 349–59.

Brar, L. S., Gill, H. S., & Randhawa, S. S. (1984). Competing ability of maize and cowpeas with variable seed rates against *Trianthema monogyna*. *Indian Journal of Ecology*, 11, 113–16.

Bridges, D. C., & Chandler, J. M. (1988). Influence of cultivar height on competitiveness of cotton (*Gossypium hirsutum*) with johnsongrass (*Sorghum halepense*). *Weed Science*, 36, 616–20.

Brown, S. M, Whitwell, T., & Street, J. E. (1985). Common bermudagrass (*Cynodon dactylon*) competition in cotton (*Gossypium hirsutum*). *Weed Science*, 33, 503–6.

Buchanan, G. A., & Hauser, E. W. (1980). Influence of row spacing on competitiveness and yield of peanuts (*Arachis hypogaea*). *Weed Science*, 28, 401–9.

Bula, R. J., Smith, D., & Miller, E. E. (1954). Measurements of light beneath a small grain companion crop as related to legume establishment. *Botanical Gazette*, 115, 271–8.

Burnside, O. C. (1972). Tolerance of soybean cultivars to weed competition and herbicides. *Weed Science*, 20, 294–7.

Burnside, O. C. (1979). Soybean (*Glycine max*) growth as affected by weed removal, cultivar, and row spacing. *Weed Science*, 27, 562–5.

Burnside, O. C., & Colville, W. L. (1964). Soybean and weed yields as affected by irrigation, row spacing, tillage and amiben. *Weeds*, 12, 109–12.

Burnside, O. C., & Moomaw, R. S. (1984). Influence of weed control treatments on soybean cultivars in an oat–soybean rotation. *Agronomy Journal*, 76, 887–90.

Burnside, O. C., & Wicks, G. A. (1972). Competitiveness and herbicide tolerance of sorghum hybrids. *Weed Science*, 20, 314–16.

Burnside, O. C., Wicks, G. A., & Fenster, C. R. (1964). Influence of tillage, row spacing, and atrazine on sorghum and weed yields from nonirrigated sorghum across Nebraska. *Weeds*, 12, 211–15.

Burnside, O. C., Wiens, M. J., Holder, B. J., Weisberg, S., Ristau, E. A., Johnson, M. M., & Cameron, J. H. (1998). Critical periods for weed control in dry beans (*Phaseolus vulgaris*). *Weed Science*, 46, 301–6.

Bussan, A. J., Burnside, O. C., Orf, J. H., Ristau, E. A., & Puettmann, K. J. (1997). Field evaluation of soybean (*Glycine max*) genotypes for weed competitiveness. *Weed Science*, 45, 31–7.

Cadag, M. R. T., & Mercado, B. L. (1982). Note: competitive ability of six Philippine corn varieties against *Ipomoea triloba* L. *Philippine Agriculturalist*, 65, 297–300.

Cain, J. C. (1972). Hedgerow orchard design for most efficient interception of solar radiation: effects of tree size, shape, spacing, and row direction. *Search Agriculture*, 2, 1–14.

Callaway, M. B. (1992). A compendium of crop varietal tolerance to weeds. *American Journal of Alternative Agriculture*, **7**, 169–80.

Callaway, M. B., & Forcella, F. (1993). Crop tolerance to weeds. In *Crop Improvement for Sustainable Agriculture*, ed. M. B. Callaway & C. A. Francis, pp. 100–31. Lincoln, NE: University of Nebraska Press.

Canudas, E. G., Quesenberry, K. H., Solenberger, L. E., & Prine, G. M. (1989). Establishment of two cultivars of rhizoma peanut as affected by weed control and planting rate. *Tropical Grasslands*, **23**, 162–70.

Carlson, H. L., & Hill, J. E. (1985). Wild oat (*Avena fatua*) competition with spring wheat: plant density effects. *Weed Science*, **33**, 176–81.

Challaiah, Ramsel, R. E., Wicks, G. A., Burnside, O. C., & Johnson, V. A. (1983). Evaluation of the weed competitive ability of winter wheat cultivars. *North Central Weed Control Conference Proceedings*, **38**, 85–91.

Challaiah, Burnside, O. C., Wicks, G. A., & Johnson, V. A. (1986). Competition between winter wheat (*Triticum aestivum*) cultivars and downy brome (*Bromus tectorum*). *Weed Science*, **34**, 689–93.

Chandler, J. M., & Meredith, W. R. Jr. (1983). Yields of three cotton (*Gossypium hirsutum*) cultivars as influenced by spurred anoda (*Anoda cristata*) competition. *Weed Science*, **31**, 303–7.

Choudhary, A. H. (1981). Effects of population and inter-row spacing on yields of maize and control of weeds with herbicides in the irrigated savanna. *Experimental Agriculture*, **17**, 389–97.

Christensen, S. (1994). Crop weed competition and herbicide performance in cereal species and varieties. *Weed Research*, **34**, 20–36.

Christensen, S., Rasmussen, G., & Olesen, J. E. (1994). Differential weed suppression and weed control in winter wheat. *Aspects of Applied Biology*, **40**, 335–42.

Colvin, D. L., Wehtje, G. R., Patterson, M., & Walker, R. H. (1986). Weed management in minimum-tillage peanuts (*Arachis hypogaea*) as influenced by cultivar, row spacing, and herbicides. *Weed Science*, **33**, 233–7.

Cooper, R. L. (1977). Response of soybean cultivars to narrow rows and planting rates under weed-free conditions. *Agronomy Journal*, **69**, 89–92.

Cornell University (1987). *Cornell Field Crops and Soils Handbook*, 2nd edn. Ithaca, NY: Cornell Cooperative Extension, New York State College of Agriculture and Life Sciences.

Cosser, N. D., Gooding, M. J., Thompson, A. J., & Froud-Williams, R. J. (1997). Competitive ability and tolerance of organically grown wheat cultivars to natural weed infestations. *Annals of Applied Biology*, **130**, 523–35.

Cousens, R. (1985). An empirical model relating crop yield to weed and crop density and a statistical comparison with other models. *Journal of Agricultural Science (Cambridge)*, **105**, 513–21.

Cousens, R. D., & Mokhtari, S. (1998). Seasonal and site variability in the tolerance of wheat cultivars to interference from *Lolium rigidum*. *Weed Research*, **38**, 301–7.

Crossa, J., Gauch, H. G. Jr., & Zobel, R. W. (1990). Additive main effects and multiplicative interaction analysis of two international maize cultivar trials. *Crop Science*, **30**, 493–500.

Cussans, G. W., & Wilson, B. J. (1975). Some effects of crop row width and seedrate on competition between spring barley and wild oat *Avena fatua* L. or common couch

Agropyron repens (L) Beauv. In *Proceedings of the Symposium on Status, Biology and Control of Grassweeds in Europe*, pp. 77–86. Wageningen, Netherlands: European Weed Research Society.

Dawson, J. H. (1965). Competition between irrigated sugar beets and annual weeds. *Weeds*, **13**, 245–9.

Dawson, J. H. (1986). The concept of period thresholds. In *Proceedings of the European Weed Research Society Symposium 1986, Economic Weed Control*, pp. 327–31. Paris, France: European Weed Research Society.

De Wit, C. T. (1960). On competition. *Verslagen van Landbouwkundige Onderzoekingen*, **66.8**, 1–82.

Deschenes, J. M., & St-Pierre, C. A. (1980). Effets des températures du sol, des dates de semis et des mauvaises herbes sur les composantes du rendement de l'avoine. *Canadian Journal of Plant Science*, **60**, 61–8.

Dhaliwal, B. K., Froud-Williams, R. J., & Caligari, P. D. S. (1993). Variation in competitive ability of spring barley cultivars. *Aspects of Applied Biology*, **34**, 373–6.

Díaz-Rivera, M., Hepperly, P. R., Riveros, G., & Almodóvar-Vega, L. (1985). Weed–crop competition in pigeon peas in Puerto Rico. *Journal of Agriculture of the University of Puerto Rico*, **69**, 201–13.

Dilday, R. H., Nastasi, P., Lin, J., & Smith, R. J. Jr. (1991). Allelopathic activity in rice (*Oryza sativa* L.) against Ducksalad (*Heteranthera limosa* (Sw.) Willd.). In *Sustainable Agriculture for the Great Plains, Symposium Proceedings*, 19–20 January 1989, Fort Collins, Colorado, eds. J. N. Hanson, M. J. Shaffer, D. A. Ball & C. V. Cole, pp. 193–201. Beltsville, MD: US Department of Agriculture, Agricultural Research Service.

Doll, H. (1997). The ability of barley to compete with weeds. *Biological Agriculture and Horticulture*, **14**, 43–51.

Doll, H., Holm, U., & Søgaard, B. (1995). Effect of crop density on competition by wheat and barley with *Agrostemma githago* and other weeds. *Weed Research*, **35**, 391–6.

Donald, C. M. (1958). The interaction of competition for light and for nutrients. *Australian Journal of Agricultural Research*, **9**, 421–35.

Donald, C. M. (1963). Competition among crop and pasture plants. *Advances in Agronomy*, **15**, 1–118.

Dunan, C. M., Westra, P., Schweizer, E. E., Lybecker, D. W., & Moore, F. D. III (1995). The concept and application of early economic period threshold: the case of DCPA in onions (*Allium cepa*). *Weed Science*, **43**, 634–9.

Evans, L. T. (1980). The natural history of crop yield. *American Scientist*, **68**, 388–97.

Evans, R. M., Thill, D. C., Tapia, L., Shafii, B., & Lish, J. M. (1991). Wild oat (*Avena fatua*) and spring barley (*Hordeum vulgare*) density affect spring barley grain yield. *Weed Technology*, **5**, 33–9.

Fay, P. K., & Duke, W. B. (1977). An assessment of allelopathic potential in *Avena* germ plasm. *Weed Science*, **25**, 224–8.

Felton, W. L. (1976). The influence of row spacing and plant population on the effect of weed competition in soybean (*Glycine max*). *Australian Journal of Experimental Agriculture and Animal Husbandry*, **16**, 926–31.

Fiebig, W. W., Shilling, D. G., & Knauft, D. A. (1991). Peanut genotype response to interference from common cocklebur. *Crop Science*, **31**, 1289–92.

Firbank, L. G., & Watkinson, A. R. (1985). On the analysis of competition within two-species mixtures of plants. *Journal of Applied Ecology*, **22**, 503–17.

Fischer, A., Ramírez, H. V., & Lozano, J. (1997). Suppression of junglerice [*Echinochloa colona* (L.) Link] by irrigated rice cultivars in Latin America. *Agronomy Journal*, **89**, 516–21.

Fischer, A., Chatel, M., Ramirez, H., Lozano, J., & Guimaraes, E. (1995). Components of early competition between upland rice (*Oryza sativa* L.) and *Brachiaria brizantha* (Hochst. ex A. Rich) Stapf. *International Journal of Pest Management*, **41**, 100–3.

Fischer, R. A., & Miles, R. E. (1973). The role of spatial pattern in the competition between crop plants and weeds: a theoretical analysis. *Mathematical Biosciences*, **18**, 335–50.

Flénet, F., Kiniry, J. R., Board, J. E., Westgate, M. E., & Reicosky, D. C. (1996). Row spacing effects on light extinction coefficients of corn, sorghum, soybean, and sunflower. *Agronomy Journal*, **88**, 185–90.

Flood, R. G., & Halloran, G. M. (1984). Growth habit variation in hexaploid wheat (*Triticum aestivum*) and competition with annual ryegrass (*Lolium rigidum*). *Protection Ecology*, **6**, 299–305.

Forcella, F. (1987). Tolerance of weed competition associated with high leaf-area expansion rate in tall fescue. *Crop Science*, **27**, 146–7.

Forcella, F., Westgate, M. E., & Warnes, D. D. (1992). Effect of row width on herbicide and cultivation requirements in row crops. *American Journal of Alternative Agriculture*, **7**, 161–7.

Ford, G. T., & Mt. Pleasant, J. (1994). Competitive abilities of six corn (*Zea mays* L.) hybrids with four weed control practices. *Weed Technology*, **8**, 124–8.

Garrity, D. P., Movillon, M., & Moody, K. (1992). Differential weed suppression ability in upland rice cultivars. *Agronomy Journal*, **84**, 586–91.

Ghafar, Z., & Watson, A. K. (1983). Effect of corn (*Zea mays*) population on the growth of yellow nutsedge (*Cyperus esculentus*). *Weed Science*, **31**, 588–92.

Ghosheh, H. Z., Holshouser, D. L., & Chandler, J. M. (1996). The critical period of Johnsongrass (*Sorghum halepense*) control in field corn (*Zea mays*). *Weed Science*, **44**, 944–7.

Gibeault, V. A. (1986). The potential of turf cultivar selection to minimize problems in turf. *Proceedings, Annual California Weed Conference*, **38**, 157–61.

Glaz, B., Ulloa, M. F., & Parrado, R. (1989). Cultivation, cultivar, and crop age effects on sugarcane. *Agronomy Journal*, **81**, 163–7.

Goldberg, D. E., & Landa, K. (1991). Competitive effect and response hierarchies and correlated traits in the early stages of competition. *Journal of Ecology*, **79**, 1013–30.

González Ponce, R. (1988). Competition between *Avena sterilis* spp. *macrocarpa* Mo. and cultivars of wheat. *Weed Research*, **28**, 303–7.

Gooding, M. J., Thompson, A. J., & Davies, W. P. (1993). Interception of photosynthetically active radiation, competitive ability and yield of organically grown wheat varieties. *Aspects of Applied Biology*, **34**, 355–62.

Gruenhagen, R. D., & Nalewaja, J. D. (1969). Competition between flax and wild buckwheat. *Weed Science*, **17**, 380–4.

Guneyli, E., Burnside, O. C., & Nordquist, P. T. (1969). Influence of seedling characteristics on weed competitive ability of sorghum hybrids and inbred lines. *Crop Science*, **9**, 713–16.

Gunsolus, J. L. (1990). Mechanical and cultural weed control in corn and soybeans. *American Journal of Alternative Agriculture*, **5**, 114–19.

Håkansson, S. (1983). *Competition and Production in Short-Lived Crop–Weed Stands: Density Effects*, Department of Plant Husbandry, Report no. 127. Uppsala, Sweden: Swedish University of Agricultural Sciences.

Håkansson, S. (1984). Row spacing, seed distribution in the row, amount of weeds: influence on production in stands of cereals. In *Weeds and Weed Control, Proceedings of the 25th Swedish Weed Conference*, pp. 17–34. Uppsala, Sweden: Swedish University of Agricultural Sciences.

Håkansson, S. (1986). Competition between crops and weeds: influencing factors, experimental methods and research needs. In *Proceedings of the European Weed Research Society Symposium 1986, Economic Weed Control*, pp. 49–60. Paris, France: European Weed Research Society.

Håkansson, S. (1991). *Growth and Competition in Plant Stands*, Crop Production Science, no. 12. Uppsala, Sweden: Department of Crop Production Science, Swedish University of Agricultural Sciences.

Harper, J. L. (1977). *Population Biology of Plants*. London: Academic Press.

Harrison, H. F. Jr., & Peterson, J. K. (1986). Allelopathic effects of sweet potatoes (*Ipomoea batatas*) on yellow nutsedge (*Cyperus esculentus*) and alfalfa (*Medicago sativa*). *Weed Science*, **34**, 623–7.

Hartwig, E. E. (1957). Row width and rates of planting in the southern states. *Soybean Digest*, **17**, 13–14, 16.

Harvey, R. G., & McNevin, G. R. (1990). Combining cultural practices and herbicides to control wild-proso millet (*Panicum miliaceum*). *Weed Technology*, **4**, 433–9.

Hashem, A., Radosevich, S. R., & Roush, M. L. (1998). Effect of proximity factors on competition between winter wheat (*Triticum aestivum*) and Italian ryegrass (*Lolium multiflorum*). *Weed Science*, **46**, 181–90.

Hill, J. E., De Datta, S. K., & Real, J. G. (1990). *Echinochloa* competition in rice: a comparison of studies from direct-seeded and transplanted flooded rice. In *Proceedings of the Symposium on Weed Management*, Special Publication no. 38, ed. B. A. Auld, R. C. Umaly & S. S. Tjitrosomo, pp. 115–29. Bogor, Indonesia: Biotrop.

Holland, J. F., & McNamara, D. W. (1982). Weed control and row spacing in dry-land sorghum in northern New South Wales. *Australian Journal of Experimental Agriculture and Animal Husbandry*, **22**, 310–16.

Howe, O. W. III, & Oliver, L. R. (1987). Influence of soybean (*Glycine max*) row spacing on pitted morningglory (*Ipomoea lacunosa*) interference. *Weed Science*, **35**, 185–93.

Hucl, P. (1998). Response to weed control by four spring wheat genotypes differing in competitive ability. *Canadian Journal of Plant Science*, **78**, 171–3.

Huel, D. G., & Hucl, P. (1996). Genotypic variation for competitive ability in spring wheat. *Plant Breeding*, **115**, 325–9.

Hycka, M., & Benitez-Sidón, J. M. (1979). Algunas características de nuevos cultivares españoles de alfalfa. *Anales de la Estación Experimental de Aula Dei*, **14**, 558–74.

Jackson, J. E., & Palmer, J. W. (1972). Interception of light by model hedgerow orchards in relation to latitude, time of year and hedgerow configuration and orientation. *Journal of Applied Ecology*, **9**, 341–57.

James, K. L., Banks, P. A., & Karnok, K. J. (1988). Interference of soybean, *Glycine max*, cultivars with sicklepod, *Cassia obtusifolia*. Weed Technology, **2**, 404–9.

Johnson, D. E., Dingkuhn, M., Jones, M. P., & Mahamane, M. C. (1998). The influence of rice plant type on the effect of weed competiton on *Oryza sativa* and *Oryza glaberrima*. Weed Research, **38**, 207–16.

Johnson, G. A., Hoverstad, T. R., & Greenwald, R. E. (1998). Integrated weed management using narrow corn row spacing, herbicides, and cultivation. Agronomy Journal, **90**, 40–6.

Johri, A. K., Singh, G., & Sharma, D. (1992). Nutrient uptake by wheat and associated weeds as influenced by management practices. Tropical Agriculture (Trinidad), **69**, 391–3.

Jones, C. A., Zimmermann, F. J. P., & Dall'Acqua, F M. (1979). Light penetration in wide-row upland rice. Tropical Agriculture (Trinidad), **56**, 367–9.

Justice, G. G., Peeper, T. F., Solie, J. B., & Epplin, F. M. (1993). Net returns from cheat (*Bromus secalinus*) control in winter wheat (*Triticum aestivum*). Weed Technology, **7**, 459–64.

Justice, G. G., Peeper, T. F., Solie, J. B., & Epplin, F. M. (1994). Net returns from Italian ryegrass (*Lolium multiflorum*) control in winter wheat (*Triticum aestivum*). Weed Technology, **8**, 317–23.

Kawano, K., Gonzalez, H., & Lucena, M. (1974). Intraspecific competition, competition with weeds, and spacing response in rice. Crop Science, **14**, 841–5.

Kira, T., Ogawa, H., & Shinozaki, K. (1953). Intraspecific competition among higher plants. I. Competition–density–yield inter-relationships in regularly dispersed populations. Journal of the Institute of Polytechnics, Osaka City University, Series D, **4**, 1–16.

Kirkland, K. J. (1993). Weed management in spring barley (*Hordeum vulgare*) in the absence of herbicides. Journal of Sustainable Agriculture, **3**, 95–104.

Kirkland, K. J., & Hunter, J. H. (1991). Competitiveness of Canada prairie spring wheats with wild oat (*Avena fatua* L.). Canadian Journal of Plant Science, **71**, 1089–92.

Klingaman, T. E., & Oliver, L. R. (1994). Influence of cotton (*Gossypium hirsutum*) and soybean (*Glycine max*) planting date on weed interference. Weed Science, **42**, 61–5.

Kolbe, W. (1980). Effect of weed control on grain yield of different winter barley cultivars with reference to sowing time, seed rate and seed size, in long-term trials at Höfchen and Laacherhof experimental stations (1968–1980). Pflanzenschutz-Nachrichten Bayer, **33**, 203–19.

Koscelny, J. A., Peeper, T. F., Solie, J. B., & Solomon, S. G. Jr. (1990). Effect of wheat (*Triticum aestivum*) row spacing, seeding rate, and cultivar on yield loss from cheat (*Bromus secalinus*). Weed Technology, **4**, 487–92.

Koscelny, J. A., Peeper, T. F., Solie, J. B., & Solomon, S. G. Jr. (1991). Weeding date, seeding rate, and row spacing affect wheat (*Triticum aestivum*) and cheat (*Bromus secalinus*). Weed Technology, **5**, 707–12.

Kreuz, E. (1982). Der Einfluß von Getreidefruchtfolgen auf die Spätverunkrautung von Winterweizenbeständen auf Schwarzerde. Institut für Getreideforschung Bernburg-Hadmersleben der Akademie der Landwirtschaftswissenschaften der DDR, **203**, 113–24.

Kropff, M. J., Lotz, L. A. P., & Weaver, S. E. (1993). Practical applications. In Modeling Crop–Weed Interactions, ed. M. L. Kropff & H. H. van Laar, pp. 149–67. Wallingford, UK: CAB International.

Kropff, M. J., van Keulen, N. C., van Laar, H. H., & Schneiders, B. J. (1993). The impact of environmental and genetic factors. In *Modeling Crop–Weed Interactions*, ed. M. L. Kropff & H. H. van Laar, pp. 137–47. Wallingford, UK: CAB International.

Kukula, S. T. (1986). Weed management in dryland cereal production with special reference to the Near East. *FAO Plant Protection Bulletin*, **34**, 133–8.

Kust, C. A., & Smith, R. R. (1969). Interaction of linuron and row spacing for control of yellow foxtail and barnyardgrass in soybeans. *Weed Science*, **17**, 489–91.

Kwon, S. L., Smith, R. J. Jr., & Talbert, R. E. (1991). Interference durations of red rice (*Oryza sativa*) in rice (*O. sativa*). *Weed Science*, **39**, 363–8.

Lanning, S. P., Talbert, L. E., Martin, J. M., Blake, T. K., & Bruckner, P. L. (1997). Genotype of wheat and barley affects light penetration and wild oat growth. *Agronomy Journal*, **89**, 100–3.

Larson, W. E., & Willis, W. O. (1957). Light, soil temperature, soil moisture and alfalfa–red clover distribution between corn rows of various spacings and row directions. *Agronomy Journal*, **49**, 422–6.

Lawson, H. M. (1982). Competition between annual weeds and vining peas grown at a range of population densities: effects on the crop. *Weed Research*, **22**, 27–38.

Lawson, H. M., & Topham, P. B. (1985). Competition between annual weeds and vining peas grown at a range of population densities: effects on the weeds. *Weed Research*, **25**, 221–9.

Leather, G. R. (1983). Sunflowers (*Helianthus annuus*) are allelopathic to weeds. *Weed Science*, **31**, 37–42.

Légère, A., & Schreiber, M. M. (1989). Competition and canopy architecture as affected by soybean (*Glycine max*) row width and density of redroot pigweed (*Amaranthus retroflexus*). *Weed Science*, **37**, 84–92.

Lemerle, D., Verbeek, B., & Coombes, N. E. (1995). Losses in grain yield of winter crops from *Lolium rigidum* competition depend on crop species, cultivar and season. *Weed Research*, **35**, 503–9.

Lemerle, D., Verbeek, B., Cousens, R. D., & Coombes, N. E. (1996). The potential for selecting wheat varieties strongly competitive against weeds. *Weed Research*, **36**, 505–13.

Liebman, M. (1989). Effects of nitrogen fertilizer, irrigation, and crop genotype on canopy relations and yields of an intercrop/weed mixture. *Field Crops Research*, **22**, 83–100.

Limon-Ortega, A., Mason, S. C., & Martin, A. R. (1998). Production practices improve grain sorghum and pearl millet competitiveness with weeds. *Agronomy Journal*, **90**, 227–32.

Lindquist, J. L., & Mortensen, D. A. (1998). Tolerance and velvetleaf (*Abutilon theophrasti*) suppressive ability of two old and two modern corn (*Zea mays*) hybrids. *Weed Science*, **46**, 569–74.

Linquist, J. L., Mortensen, D. A., & Johnson, B. E. (1998). Mechanisms of corn tolerance and velvetleaf suppressive ability. *Agronomy Journal*, **90**, 787–92.

Lockerman, R. H., & Putnam, A. R. (1979). Evaluation of allelopathic cucumbers (*Cucumis sativus*) as an aid to weed control. *Weed Science*, **27**, 54–7.

Lugo, M. de L., Ortiz, C. E., & González, A. (1997). Leaf area index related to weed suppression in upland taro. *Journal of the Agricultural University of Puerto Rico*, **81**, 220–1.

Lutman, P. J. W., & Dixon, F. L. (1986). The effect of drilling date on competition between volunteer barley and oilseed rape. In *Economic Weed Control, Proceedings of the European Weed Research Symposium 1986*, 12–14 March 1986, Stuttgart-Hohenheim, pp. 145–52. [No city]: European Weed Research Society.

Malik, S. V., Swanton, C. J., & Michaels, T. E. (1993). Interaction of white bean (*Phaseolus vulgaris* L.) cultivars, row spacing, and seeding density with annual weeds. *Weed Science*, **41**, 62–8.

Mann, H. H., & Barnes, T. W. (1947). The competition between barley and certain weeds under controlled conditions. II. Competition with *Holcus mollis*. *Annals of Applied Biology*, **34**, 252–68.

Mann, H. H., & Barnes, T. W. (1949). The competition between barley and certain weeds under controlled conditions. III. Competition with *Agrostis gigantea*. *Annals of Applied Biology*, **36**, 273–81.

Martin, M. P. L. D., & Snaydon, R. W. (1982). Root and shoot interactions between barley and field beans when intercropped. *Journal of Applied Ecology*, **19**, 263–72.

Martin, R. J., Cullis, B. R., & McNamara, D. W. (1987). Prediction of wheat yield loss due to competition by wild oats (*Avena* spp.) *Australian Journal of Agricultural Research*, **38**, 487–99.

Marx, G. A., & Hagedorn, D. I. (1961). Plant population and weed growth relations in canning peas. *Weeds*, **9**, 494–6.

McNamara, D. W. (1976). Wild oat density and the duration of wild oat competition as it influences wheat growth and yield. *Australian Journal of Experimental Agriculture and Animal Husbandry*, **16**, 402–6.

McWhorter, C. G., & Barrentine, W. L. (1975). Cocklebur control in soybeans as affected by cultivars, seeding rates, and methods of weed control. *Weed Science*, **23**, 386–90.

McWhorter, C. G., & Hartwig, E. E. (1972). Competition of Johnsongrass and cocklebur with six soybean varieties. *Weed Science*, **20**, 56–9.

McWhorter, C. G., & Sciumbato, G. L. (1988). Effects of row spacing, benomyl, and duration of sicklepod (*Cassia obtusifolia*) interference on soybean (*Glycine max*) yields. *Weed Science*, **36**, 254–9.

Medd, R. W., Auld, B. A., Kemp, D. R., & Murison, R. D. (1985). The influence of wheat density and spatial arrangement on annual ryegrass, *Lolium rigidum* Gaudin, competition. *Australian Journal of Agricultural Research*, **36**, 361–71.

Melander, B. (1995). Impact of drilling date on *Apera spica-venti* L. and *Alopecurus myosuroides* Huds. in winter cereals. *Weed Research*, **35**, 157–66.

Mickelson, J. A., & Renner, K. A. (1997). Weed control using reduced rates of postemergence herbicides in narrow and wide row soybean. *Journal of Production Agriculture*, **10**, 431–7.

Miller, A. B., & Hopen, H. J. (1991). Critical weed-control period in seeded cabbage (*Brassica oleracea* var. *capitata*). *Weed Technology*, **5**, 852–7.

Miller, J. H., Carter, L. M., & Carter, C. (1983). Weed management in cotton (*Gossypium hirsutum*) grown in two row spacings. *Weed Science*, **31**, 236–41.

Millhollon, R. W. (1988). Differential response of sugarcane cultivars to johnsongrass competition. *Southern Weed Society Proceedings*, **41**, 100.

Mohler, C. L. (1996). Ecological bases for the cultural control of annual weeds. *Journal of Production Agriculture*, **9**, 468–74.

Monks, D. W., & Oliver, L. R. (1988). Interactions between soybean (*Glycine max*) cultivars and selected weeds. *Weed Science*, **36**, 770–4.

Monzote, M., Funes, F., & Díaz, L. E. (1979). Comparación de cultivares de *Panicum*. 2. Bajo condiciones de pastoreo. *Centro Agrícola Enero-Abril*, **6**, 101–8.

Moody, K. (1978). Weed control in mungbean. In *The 1st International Mungbean Symposium*, ed. R. Cowell, pp. 132–6. Taipei, Taiwan: Asian Vegetable Research and Development Center.

Moss, S. R. (1985). The influence of crop variety and seed rate on *Alopecurus myosuroides* competition in winter cereals. In *Proceedings of the 1985 British Crop Protection Conference – Weeds*, pp. 701–8. Croydon, UK: British Crop Protection Council.

Mulugeta, D., & Boerboom, C. M. (1999). Seasonal abundance and spatial pattern of *Setaria faberi*, *Chenopodium album*, and *Abutilon theophrasti* in reduced-tillage soybeans. *Weed Science*, **47**, 95–106.

Murdock, E. C., Banks, P. A., & Toler, J. E. (1986). Shade development effects on pitted morningglory (*Ipomoea lacunosa*) interference with soybeans (*Glycine max*). *Weed Science*, **34**, 711–17.

Murphy, S. D., Yakubu, Y., Weise, S. F., & Swanton, C. J. (1996). Effect of planting patterns and inter-row cultivation on competition between corn (*Zea mays*) and late emerging weeds. *Weed Science*, **44**, 865–70.

Mutsaers, H. J. W. (1980). The effect of row orientation, date and latitude on light absorption by row crops. *Journal of Agricultural Science (Cambridge)*, **95**, 381–6.

Nangju, D. (1978). Effect of plant density, spatial arrangement, and plant type on weed control in cowpea and soybean. In *Weeds and Their Control in the Humid and Subhumid Tropics*, ed. I. O. Akobundu, pp. 288–99. Ibadan, Nigeria: IITA.

Nelson, D. C., & Giles, J. F. (1989). Weed management in two potato (*Solanum tuberosum*) cultivars using tillage and pendimethalin. *Weed Science*, **37**, 228–32.

Ngouajio, M., Foko, J., & Fouejio, D. (1997). The critical period of weed control in common bean (*Phaseolus vulgaris* L.) in Cameroon. *Crop Protection*, **16**, 127–33.

Nicol, D., Copaja, S. V., Wratten, S. D., & Niemeyer, H. M. (1992). A screen of worldwide wheat cultivars for hydroxamic acid levels and aphid antixenosis. *Annals of Applied Biology*, **121**, 11–18.

Nieto, H. J., & Staniforth, D. W. (1961). Corn–foxtail competition under various production conditions. *Agronomy Journal*, **53**, 1–5.

O'Donovan, J. (1994). Canola (*Brassica rapa*) plant density influences tartary buckwheat (*Fagopyrum tataricum*) interference, biomass, and seed yield. *Weed Science*, **42**, 385–9.

Ogg, A. G., & Seefeldt, S. S. (1999). Characterizing traits that enhance the competitiveness of winter wheat (*Triticum aestivum*) against jointed goatgrass (*Aegilops cylindrica*). *Weed Science*, **47**, 74–80.

Oliver, L. R. (1979). Influence of soybean (*Glycine max*) planting date on velvetleaf (*Abutilon theophrasti*) competition. *Weed Science*, **27**, 183–8.

Olofsdotter, M., Navarez, D., Rebulanan, M., & Streibig, J. C. (1999). Weed-suppressing rice cultivars: does allelopathy play a role? *Weed Research*, **39**, 441–54.

Palmer, J. W. (1977). Diurnal light interception and a computer model of light interception by hedgerow apple orchards. *Journal of Applied Ecology*, **14**, 601–14.

Palmer, J. W. (1989). The effects of row orientation, tree height, time of year and latitude on light interception and distribution in model apple hedgerow canopies. *Journal of Horticultural Science*, **64**, 137–45.

Pantone, D. J., & Baker, J. B. (1991). Weed–crop competition models and response-surface analysis of red rice competition in cultivated rice: a review. *Crop Science*, **31**, 1105–10.

Paolini, R., Del Puglia, S., Principi, M., Barcellona, O., & Riccardi, E. (1998). Competition between safflower and weeds as influenced by crop genotype and sowing time. *Weed Research*, **38**, 247–55.

Parker, M. B., Marchant, W. H., & Mullinix, B. J. Jr. (1981). Date of planting and row spacing effects on four soybean cultivars. *Agronomy Journal*, **73**, 759–62.

Parr, T. W. (1984). The effects of seed rate on weed populations during the establishment of amenity turf. *Aspects of Applied Biology*, **5**, 117–25.

Parr, T. W. (1985). The control of weed populations during grass establishment by the manipulation of seed rates. In *Weeds, Pests and Diseases of Grassland and Herbage Legumes,* Monograph no. 29, ed. J. S. Brockman, pp. 20–8. Croydon, UK: British Crop Protection Council.

Patterson, M. G., Walker, R. H., Colvin, D. L., Wehtje, G., & McGuire, J. A. (1988). Comparison of soybean (*Glycine max*)–weed interference from large and small plots. *Weed Science*, **36**, 836–9.

Pendleton, J. W., & Dungan, G. H. (1953). Effect of different oat spacings on growth and yield of oats and red clover. *Agronomy Journal*, **45**, 442–4.

Pendleton, J. W., & Dungan, G. H. (1958). Effect of row direction on spring oat yields. *Agronomy Journal*, **50**, 341–3.

Peters, E. J., Gebhardt, M. R., & Stritzke, J. F. (1965). Interrelations of row spacings, cultivations and herbicides for weed control in soybeans. *Weeds*, **13**, 285–9.

Pfeiffer, R. K., & Holmes, H. M. (1961). A study of the competition between barley and oats as influenced by barley seedrate, nitrogen level and barban treatment. *Weed Research*, **1**, 5–18.

Pitelli, R. A., Charudattan, R., & Devalerio, J. T. (1998). Effect of *Alternaria cassiae*, *Pseudocercospora nigricans*, and soybean (*Glycine max*) planting density on the biological control of sicklepod (*Senna obtusifolia*). *Weed Technology*, **12**, 37–40.

Putnam, A. R., & Duke, W. B. (1974). Biological suppression of weeds: evidence for allelopathy in accessions of cucumber. *Science*, **185**, 370–2.

Putnam, D. H., Wright, J., Field, L. A., & Ayisi, K. K. (1992). Seed yield and water-use efficiency of white lupin as influenced by irrigation, row spacing, and weeds. *Agronomy Journal*, **84**, 557–63.

Radford, B. J., Wilson, B. J., Cartledge, O., & Watkins, F. B. (1980). Effect of wheat seeding rate on wild oat competition. *Australian Journal of Experimental Agriculture and Animal Husbandry*, **20**, 77–81.

Rambakudzibga, A. M. (1999). Aspects of the growth and development of *Cyperus rotundus* under arable crop canopies: implications for integrated control. *Weed Research*, **39**, 507–14.

Ramsel, R. E., & Wicks, G. A. (1988). Use of winter wheat (*Triticum aestivum*) cultivars and herbicides in aiding weed control in an ecofallow corn (*Zea mays*) rotation. *Weed Science*, **36**, 394–8.

Regnier, E. E., & Bakelana, K. B. (1995). Crop planting pattern effects on early growth and canopy shape of cultivated and wild oats (*Avena fatua*). *Weed Science*, **43**, 88–94.

Remison, S. U. (1978). The performance of cowpea (*Vigna unguiculata* (L.) Walp) as influenced by weed competition. *Journal of Agricultural Science (Cambridge)*, **90**, 523–30.

Richards, M. C. (1989). Crop competitiveness as an aid to weed control. In *Proceedings of the*

Brighton Crop Protection Conference – Weeds, pp. 573–8. Farnham, UK: British Crop Protection Council.

Richards, M. C., & Davies, D. H. K. (1991). Potential for reducing herbicide inputs/rates with more competitive cereal cultivars. In *Proceedings of the Brighton Crop Protection Conference – Weeds*, pp. 1233–40. Farnham, UK: British Crop Protection Council.

Richards, M. C., & Whytock, G. P. (1993). Varietal competitiveness with weeds. *Aspects of Applied Biology*, 34, 345–54.

Robinson, R. G. (1949). The effect of flax stand on yields of flaxseed, flax straw, and weeds. *Agronomy Journal*, 41, 483–4.

Rogers, N. K., Buchanan, G. A., & Johnson, W. C. (1976). Influence of row spacing on weed competition with cotton. *Weed Science*, 24, 410–13.

Rose, S. J., Burnside, O. C., Specht, J. E., & Swisher, B. A. (1984). Competition and allelopathy between soybeans and weeds. *Agronomy Journal*, 76, 523–8.

Rushing, G. S., & Oliver, L. R. (1998). Influence of planting date on common cocklebur (*Xanthium strumarium*) interference in early-maturing soybean (*Glycine max*). *Weed Science*, 46, 99–104.

Samuel, A. M., & Guest, S. J. (1990). Effect of seed rates and within crop cultivations in organic winter wheat. In *Crop Protection in Organic and Low Input Agriculture: Options for Reducing Agrochemical Usage*, ed. R. J. Unwin, pp. 49–54. Farnham, UK: British Crop Protection Council.

Satorre, E. H., & Snaydon, R. W. (1992). A comparison of root and shoot competition between spring cereals and *Avena fatua* L. *Weed Research*, 32, 45–55.

Schnieders, B. J., van der Linden, M., Lotz, L. A. P., & Rabbinge, R. (1999). A model for interspecific competition in row crops. In *A Quantitative Analysis of Inter-Specific Competition in Crops with a Row Structure*, ed. B. J. Schnieders, pp. 31–56. Wageningen, Netherlands: Agricultural University Wageningen.

Seavers, G. P., & Wright, K. J. (1999). Crop canopy development and structure influence weed suppression. *Weed Research*, 39, 319–28.

Sechniak, L. R., Lyfenko, S. F., & Pika, I. N. (1985). Selection as a means of improving weed control in winter wheat cultivation. *Vestnik Sel'skokhozyaistvennoi Nauki, Moscow, USSR*, 10, 81–5.

Seefeldt, S. S., Ogg, A. G. Jr., & Hou, Y. (1999). Near-isogenic lines for *Triticum aestivum* height and crop competitiveness. *Weed Science*, 47, 316–20.

Selleck, G. W., & Dallyn, S. L. (1978). Herbicide treatments and potato cultivar interactions for weed control. *Proceedings of the Northeastern Weed Science Society*, 32, 152–6.

Shaktawat, M. S. (1983). Grass weed control in wheat through agronomic technique. *Indian Journal of Agronomy*, 28, 408–11.

Shaw, D. R., Rankins, A. Jr., & Ruscoe, J. T. (1997). Sicklepod (*Senna obtusifolia*) interference with soybean (*Glycine max*) cultivars following herbicide treatments. *Weed Technology*, 11, 510–14.

Shaw, D. R., Smith, C. A., & Snipes, C. E. (1989). Sicklepod (*Cassia obtusifolia*) control in soybean (*Glycine max*) grown in rotations of 97- and 18-cm row spacings. *Weed Science*, 37, 748–52.

Shilling, D. G., Brecke, B. J., Hiebsch, C., & MacDonald, G. (1995). Effect of soybean (*Glycine max*) cultivar, tillage, and rye (*Secale cereale*) mulch on sicklepod (*Senna obtusifolia*). *Weed Technology*, 9, 339–42.

Shinozaki, K., & Kira, T. (1956). Intraspecific competition among higher plants. VII. Logistic theory of the C–D effect. *Journal of the Institute of Polytechnics, Osaka City University, Series D*, 7, 35–72.

Siddiqi, M. Y., Glass, A. D. M., Hsiao, A. I., & Minjas, A. N. (1985). Wild oat/barley interactions: varietal differences in competitiveness in relation to K^+ supply. *Annals of Botany*, **56**, 1–7.

Singh, M., Saxena, M. C., Abu-Irmaileh, B. E., Al-Thahabi, S. A., & Haddad, N. I. (1996). Estimation of critical period of weed control. *Weed Science*, **44**, 273–83.

Skorda, E. A., & Efthimiadis, P. G. (1985). Effect of wheat seedrate on *Avena ludoviciana* competition. In *Proceedings of the 1985 British Crop Protection Conference – Weeds*, pp. 709–14. Croydon, UK: British Crop Protection Council.

Smith, R. J. Jr. (1968). Weed competition in rice. *Weed Science*, **16**, 252–5.

Smith, R. J. Jr. (1974). Competition of barnyardgrass with rice cultivars. *Weed Science*, **22**, 423–6.

Solie, J. B., Solomon, S. G. Jr., Self, K. P., Peeper, T. F., & Koscelny, J. A. (1991). Reduced row spacing for improved wheat yields in weed-free and weed-infested fields. *Transactions of the American Society of Agricultural Engineers*, **34**, 1654–60.

Spitters, C. J. T. (1983). An alternative approach to the analysis of mixed cropping experiments. 1. Estimation of competition effects. *Netherlands Journal of Agricultural Science*, **31**, 1–11.

Staniforth, D. W. (1961). Responses of corn hybrids to yellow foxtail competition. *Weeds*, **9**, 132–6.

Staniforth, D. W. (1962). Responses of soybean varieties to weed competition. *Agronomy Journal*, **54**, 11–13.

Staub, J. E. (1992). Plant density and herbicides affect cucumber productivity. *Journal of the American Society of Horticultural Science*, **117**, 48–53.

Stauber, L. G., Smith, R. J. Jr., & Talbert, R. E. (1991). Density and spatial interference of barnyardgrass (*Echinochloa crus-galli*) with rice (*Oryza sativa*). *Weed Science*, **39**, 163–8.

Stevenson, F. C., & Wright, A. T. (1996). Seeding rate and row spacing affect flax yields and weed interference. *Canadian Journal of Plant Science*, **76**, 537–44.

Stilwell, E. K., & Sweet, R. D. (1974). Competition of squash cultivars with weeds. *Proceedings, Northeastern Weed Science Society*, **28**, 229–33.

Street, J. E., Buchanan, G. A., Crowley, R. H., & McGuire, J. A. (1981). Influence of cotton (*Gossypium hirsutum*) densities on competitiveness of pigweed (*Amaranthus* spp.) and sicklepod (*Cassia obtusifolia*). *Weed Science*, **29**, 253–6.

Sugiyama, S. (1998). Differentiation in competitive ability and cold tolerance between diploid and tetraploid cultivars in *Lolium perenne*. *Euphytica*, **103**, 55–9.

Sweet, R. D., & Sieczka, J. B. (1973). Comments on ability of potato varieties to compete with weeds. *Proceedings, Northeastern Weed Science Society*, **27**, 302–4.

Sweet, R. D., Yip, C. P., & Sieczka, J. B. (1974). Crop varieties: can they suppress weeds? *New York's Food and Life Sciences Quarterly*, **7**, 3–5.

Tanji, A., Zimdahl, R. L., & Westra, P. (1997). The competitive ability of wheat (*Triticum aestivum*) compared to rigid ryegrass (*Lolium rigidum*) and cowcockle (*Vaccaria hispanica*). *Weed Science*, **45**, 481–7.

Teasdale, J. R. (1995). Influence of narrow row/high population corn (*Zea mays*) on weed control and light transmittance. *Weed Technology*, **9**, 113–18.

Teasdale, J. R. (1998). Influence of corn (*Zea mays*) population and row spacing on corn and velvetleaf (*Abutilon theophrasti*) yield. *Weed Science*, **46**, 447–53.

Teasdale, J. R., & Frank, J. R. (1983). Effect of row spacing on weed competition with snap beans (*Phaseolus vulgaris*). *Weed Science*, **31**, 81–5.

Teich, A. H., Smid, A., Welacky, T., & Hamill, A. (1993). Row-spacing and seed-rate effects on winter wheat in Ontario. *Canadian Journal of Plant Science*, **73**, 31–5.

Thompson, A. J., Gooding, M. J., & Davies, W. P. (1992). Shading ability, grain yield and grain quality of organically grown cultivars of winter wheat. *Annals of Applied Biology*, **120** (Supplement: *Tests of Agrochemicals and Cultivars*, 13), 86–7.

Tollenaar, M., Dibo, A. A., Aguilera, A., Weise, S. F., & Swanton, C. J. (1994a). Effect of crop density on weed interference in maize. *Agronomy Journal*, **86**, 591–5.

Tollenaar, M., Nissanka, S. P., Aguilera, A., Weise, S. F., & Swanton, C. J. (1994b). Effect of weed interference and soil nitrogen on four maize hybrids. *Agronomy Journal*, **86**, 596–601.

Townley-Smith, L., & Wright, A. T. (1994). Field pea cultivar and weed response to crop seed rate in western Canada. *Canadian Journal of Plant Science*, **74**, 387–93.

Urwin, C. P., Wilson, R. G., & Mortensen, D. A. (1996). Late season weed suppression from dry bean (*Phaseolus vulgaris*) cultivars. *Weed Technology*, **10**, 699–704.

Valenti, S. A., & Wicks, G. A. (1992). Influence of nitrogen rates and wheat (*Triticum aestivum*) cultivars on weed control. *Weed Science*, **40**, 115–21.

Valverde, L. R., & Araya, R. (1986). Tolerancia a la competencia de las malezas en seis cultivares de *Phaseolus vulgaris* L. *Turrialba*, **36**, 59–64.

Van Acker, R. C., Swanton, C. J., & Weise, S. F. (1993). The critical period of weed control in soybean [*Glycine max* (L.) Merr.]. *Weed Science*, **41**, 194–200.

Vander Vorst, P. B., Wicks, G. A., & Burnside, O. C. (1983). Weed control in a winter wheat–corn–ecofarming rotation. *Agronomy Journal*, **75**, 507–11.

Vezina, L. (1992). Influence de la date de semis sur la compétition de peuplements de *Setaria pumila* et d'*Echinochloa crus-galli* avec l'orge et le blé de printemps. *Weed Research*, **32**, 57–65.

Walker, R. H., Patterson, M. G., Hauser, E., Isenhour, D. J., Todd, J. W., & Buchanan, G. A. (1984). Effects of insecticide, weed-free period, and row spacing on soybean (*Glycine max*) and sicklepod (*Cassia obtusifolia*) growth. *Weed Science*, **32**, 702–6.

Wall, D. A., Friesen, G. H., & Bhati, T. K. (1991). Wild mustard interference in traditional and semi-leafless field peas. *Canadian Journal of Plant Science*, **71**, 473–80.

Waller, G. R., Kumari, D., Friedman, J., Friedman, N. & Chou, C.-H. (1986). Caffeine autotoxicity in *Coffea arabica* L. In *The Science of Allelopathy*, ed. A. R. Putnam & C.-S. Tang, pp. 243–69. New York: John Wiley.

Walton, G. H. (1986). Comparison of crop density and herbicide use on the seed yield of lupins subjected to weed competition. *Journal of the Australian Institute of Agricultural Science*, **52**, 167–9.

Wax, L. M., & Pendleton, J. W. (1968). Effect of row spacing on weed control in soybeans. *Weed Science*, **16**, 462–4.

Weaver, S. E. (1984). Critical period of weed competition in three vegetable crops in relation to management practices. *Weed Research*, **24**, 317–25.

Weaver, S. E. (1986). Factors affecting threshold levels and seed production of jimsonweed

(*Datura stramonium* L.) in soyabeans (*Glycine max* (L.) Merr.). *Weed Research*, **26**, 215–23.

Weaver, S. E., & Tan, C. S. (1983). Critical period of weed interference in transplanted tomatoes (*Lycopersicon esculentum*): growth analysis. *Weed Science*, **31**, 476–81.

Weaver, S. E., & Tan, C. S. (1987). Critical period of weed interference in field-seeded tomatoes and its relation to water stress and shading. *Canadian Journal of Plant Science*, **67**, 575–83.

Weaver, S. E., Kropff, M. J., & Groeneveld, R. M. W. (1992). Use of ecophysiological models for crop–weed interference: the critical period of weed interference. *Weed Science*, **40**, 302–7.

Weber, C. R., & Staniforth, D. W. (1957). Competitive relationships in variable weed and soybean stands. *Agronomy Journal*, **49**, 440–4.

Weil, R. R. (1982). Maize–weed competition and soil erosion in unweeded maize. *Tropical Agriculture (Trinidad)*, **59**, 207–13.

Weiner, J. (1990). Asymmetric competition in plant populations. *Trends in Ecology and Evolution*, **5**, 360–4.

Weiner, J., & Thomas, S. C. (1986). Size variability and competition in plant monocultures. *Oikos*, **47**, 211–22.

Wibberley, E. J. (1989). *Cereal Husbandry*. Ipswich, UK: Farming Press.

Wicks, G. A., Nordquist, P. T., Hanson, G. E., & Schmidt, J. W. (1994). Influence of winter wheat (*Triticum aestivum*) cultivars on weed control in sorghum (*Sorghum bicolor*). *Weed Science*, **42**, 27–34.

Wicks, G. A., Ramsel, R. E., Nordquist, P. T., Schmidt, J.W., & Challaiah. (1986). Impact of wheat cultivars on establishment and suppression of summer annual weeds. *Agronomy Journal*, **78**, 59–62.

Wiese, A. F., Collier, J. W., Clark, L. E., & Havelka, U. D. (1964). Effect of weeds and cultural practices on sorghum yields. *Weeds*, **12**, 209–11.

Willey, R. W., & Heath, S. B. (1969). The quantitative relationships between plant population and crop yield. *Advances in Agronomy*, **21**, 281–321.

William, R. D., & Warren, G. F. (1975). Competition between purple nutsedge and vegetables. *Weed Science*, **23**, 317–23.

Williams, C. F., Crabtree, G., Mack, H. J., & Laws, W. D. (1973). Effect of spacing on weed competition in sweet corn, snap beans, and onions. *Journal of the American Society for Horticultural Science*. **98**, 526–9.

Wilson, B. J., & Wright, K. J. (1990). Predicting the growth and competitive effects of annual weeds in wheat. *Weed Research*, **30**, 201–11.

Wilson, R. G. Jr., Wicks, G. A., & Fenster, C. R. (1980). Weed control in field beans (*Phaseolus vulgaris*) in western Nebraska. *Weed Science*, **28**, 295–9.

Woolley, J. N., & Smith, M. E. (1986). Maize plant types suitable for present and possible bean relay systems in Central America. *Field Crops Research*, **15**, 3–16.

Woon, C. K. (1987). Effect of two row spacings and hemp sesbania competition on sunflower. *Journal of Agronomy and Crop Science*, **159**, 15–20.

Wu, H., Pratley, J., Lemerle, D., & Haig, T. (1999). Crop cultivars with allelopathic capability. *Weed Research*, **39**, 171–80.

Yelverton, F. H., & Coble, H. D. (1991). Narrow row spacing and canopy formation reduces weed resurgence in soybeans (*Glycine max*). *Weed Technology*, **5**, 169–74.

Yip, C. P., Sweet, R. D., & Sieczka, J. B. (1974). Competitive ability of potato cultivars with major weed species. *Proceedings, Northeastern Weed Science Society*, **28**, 271–81.

Zimdahl, R. L. (1980). *Weed–Crop Competition: A Review*. Corvallis, OR: International Plant Protection Center, Oregon State University.

Zobel, R. W., Wright, M. J., & Gauch, H. G. Jr. (1988). Statistical analysis of a yield trial. *Agronomy Journal*, **80**, 388–93.

MATT LIEBMAN AND CHARLES P. STAVER

7

Crop diversification for weed management

Introduction

One of the defining characteristics of an ecosystem is the diversity of plant species it contains. In agricultural systems, diversity of the dominant plant species – crops – can vary in both spatial and temporal dimensions. Crops can be sown in pure stands (*sole crops*), but can also be sown in multispecies mixtures (*intercrops* or *polycultures*), a practice that probably began with the development of tropical agriculture (Plucknett & Smith, 1986). Temporally, a crop can be sown continuously in the same field (*continuous monoculture*) or sown only intermittently, in sequence with other crops (*rotation*), a practice known from ancient Greece, Rome, and China (Karlen *et al.*, 1994). Rotation sequences often contain only food, feed, and fiber crops, but may also include cover crops to improve and conserve soil during seasons when "main" crops are absent. In temperate areas, rotation cycles typically extend over several years, with only annual changes of crops, but in areas with long or continuous growing seasons, farmers may plant a sequence of several crops within a single year (*multiple cropping*), or overlap the late growth period of one crop with the interplanting and early development of another (*relay cropping*).

Spatial and temporal diversity in agricultural systems may also result from growing trees and shrubs with herbaceous species (*agroforestry*). Orchard trees with a cover crop understory form one type of agroforestry system. Another type of agroforestry mimics the ecological process of succession: woody perennials are planted with short-lived herbaceous crops and after the herbaceous species senesce, the woody species persist, grow larger, and dominate the system until removed by disturbance or harvest (Hart, 1980). Mixtures of trees can be tended in a manner that resembles late stages of succession and merges agricultural crop production with forestry. Alternatively, trees in agroforestry systems can be pruned cyclically to arrest succession and create periodic opportunities for herbaceous crop production. Agroforestry practices were probably components of the earliest agricultural systems in the humid tropics and remain widespread there today (Plucknett & Smith, 1986).

Crop diversity in conventional, traditional, and organic farming systems

Diversity comprises both the number of species present in a given area (species richness) and the degree to which each species is equally abundant (evenness) (Pielou, 1977). Large differences in crop species diversity exist among conventional high-input, traditional, and organic farming systems. As explained later in this chapter, these differences strongly influence weeds and weed management.

Over the past half-century, crop diversity has declined precipitously in conventional high-input farming systems used in the USA and other industrialized countries. Large areas are now planted with only one or two annual crop species. This trend is exemplified by changes in the landscape of Illinois, where cropland occupies 66% of the state's total land area (United States Department of Agriculture, 1999). Maize and soybean were planted on 45% of Illinois' cropland in 1958 (United States Department of Agriculture, 1973), but on 86% of its cropland in 1997 (United States Department of Agriculture, 1999). Lost from the Illinois landscape during this period were large areas of pasture, hay, and small-grain cereals (Power & Follett, 1987; Bullock, 1992).

Not surprisingly, crop uniformity across broad landscapes is related to low diversity in individual fields over time. Clear illustrations of this point are provided by a 1991 survey of American cropping systems (Economic Research Service, 1992). In Iowa, where maize was planted on 5.1 million hectares in 1991, maize and soybean were the only two crops planted on 90% of that area in 1989 and 1990. Similarly, in Kansas, 74% of the 4.8 million hectares sown with wheat in 1991 was used only for wheat production or fallow in the two preceding years. Production of the same crop in the same field for at least three consecutive years was found on 86% of the land used for wheat in Oklahoma, 82% of the land used for cotton in Louisiana, 57% of the land used for soybean in Mississippi, and 55% of the land used for maize in Nebraska.

Reductions in crop diversity in the USA and other industrialized countries have been driven by a set of interrelated technical and socioeconomic factors that have encouraged farmers to seek higher profits through greater economies of scale, enterprise specialization, and increased substitution of purchased inputs for labor and management time. Cropping systems have become less diverse as farmers have become more reliant on synthetic pesticides and fertilizers, crop varieties responsive to chemical inputs, and larger, more costly machinery. Diversity has fallen as crop and livestock production have become separated onto different farms or into different regions, and as government programs have subsidized the production of some crops, but not others (Power & Follett, 1987). A similar decline in crop diversity has taken place in some developing countries through the expanded use of agrochemical inputs, "high response" crop varieties, and agricultural machinery (Chambers, 1990; Pretty, 1995). As in the industrialized countries, government policies and subsidies, extension education, and lending practices have contributed to increased crop specialization and, concomitantly, greater reliance on purchased production inputs.

In contrast to the low levels of crop diversity that have become characteristic of most industrialized farming systems, many farmers in developing countries choose to maintain traditional production methods characterized by high levels of crop diversity. Much of this diversity is the result of intercropping and multiple cropping practices, which are used to raise and stabilize crop yields with minimal reliance on purchased fertilizers, pesticides, and machinery (see section "Why farmers plant crops in mixtures" below). Farmers in the Latin American tropics plant more than 40% of their cassava, 60% of their maize, and 80% of their bean in intercrop mixtures (Francis, Flor & Temple, 1976; Leihner, 1983). West African farmers plant intercrops on more than 80% of the cultivated land (Steiner, 1984) and may sow as many as 12 crop species in a 7-m^2 area (Okigbo & Greenland, 1976). A survey of six villages in India indicated that up to 91% of the cropped area was sown in mixtures that contained as many as eight crop species (Jodha, 1981). Intercropping and multiple cropping are widely practiced in the Himalayan region (Ashby & Pachico, 1987) and have been described as "almost universal" in Chinese vegetable production systems (Wittwer, 1987).

Agroforestry practices can also increase crop diversity dramatically. *Kebuntalun* agroforestry systems in Java may contain 100 or more plant species used as ornamentals, medicines, cash crops, building materials, firewood, spices, vegetables, fruits, and other foods (Christanty *et al.*, 1986). Huastec farmers in northeastern Mexico manage mixtures of more than 300 species of herbs, shrubs, and trees used for construction materials, fencing, fuel, medicines, basketry, tools, soaps, dyes, musical instruments, feeds, fruits, vegetable greens, and cash crops, especially coffee (Alcorn, 1984). Such systems feature large numbers of woody perennial species, minimal soil disturbance, leaf litter accumulation, and long-term successional changes in plant species composition.

In industrialized countries, a high degree of crop diversity may still be found in organic farming systems. Diversity in these systems is considered crucial for encouraging ecological processes that suppress pests, maintain soil productivity, and reduce dependence on purchased inputs (Lampkin, 1990, pp. 125–60; Macey, 1992, pp. 80–105). In an area of Ohio where conventionally managed farms are dominated by continuous maize or a maize–soybean rotation, an organic farm uses a four-year rotation composed of maize, soybean, winter wheat, and red clover (National Research Council, 1989, pp. 253–65). In an area of Saskatchewan where conventional farmers typically plant spring wheat in alternate years with fallow, an organic farmer uses rotation sequences five or more years in length that include wheat, barley, flax, mustard, safflower, sweet clover, alfalfa, chickpea, lentil, and field pea (Matheson *et al.*, 1991, pp. 46–51). A German organic farm producing both crop and livestock products uses a 13-year rotation that includes forage grasses, clovers, alfalfa, maize, potato, carrot, barley, wheat, oat, rapeseed, mustard, field pea, and vetch (Kaffka, 1985). Intercropping forage grasses or legumes with cereal crops is common to each of these three organic farms.

Principles guiding crop diversification for weed management

Cropping systems create stresses, mortality risks, and growth and reproduction opportunities to which weeds respond. Different crops and their associated management practices have different effects. Light, water, and nutrient conditions that affect weed performance vary due to crop-specific differences in resource use and specialized fertilization and irrigation practices. Crop-specific tillage and cultivation practices produce variations in weed mortality, and in the nature and timing of weed germination cues. Damage to weeds by mowing and grazing occurs with certain types of crops, but not others. Finally, chemical, physical, and biological characteristics of soil change through the addition of different crop residues and herbicides. Compared with simple cropping systems, in which the same crop or type of crop is grown repeatedly with the same set of management practices, cropping systems that are diversified through rotation, intercropping, and agroforestry have a greater variety of factors acting on weed populations. Consequently, diversified cropping increases opportunities for weed management.

Existing information suggests two general principles for managing weeds through crop diversification:

1. *Weeds should be challenged with a broad range of stress and mortality factors through the use of crop sequences containing dissimilar species and disparate*

management practices. Diverse crop sequences repeatedly change the environment to which a weed community is adjusting. Over time, diverse sequences reduce weed density by creating inhospitable or fatal conditions throughout the life history of each weed species present. Sequences that include crops suited to different seasons of the year, and that alternate annual and perennial crops, can be especially effective for weed suppression.

2. *Crop mixtures and sequences should be designed to maximize capture of light, water, and nutrients by crops and preempt resources used by weeds.* Production of a single crop generally does not exhaust the resources necessary to support plant growth. Weeds are well adapted for rapid establishment at microsites where resources are available and, once established, compete with crops for resources. Because annual crop mixtures often exploit a greater range and quantity of resources than sole crops, they can be more effective for suppressing weeds through resource preemption. Weed establishment and growth also can be reduced by sowing cover crops that compete for resources during seasons when main crops are absent or dormant. In agroforestry systems, trees and shrubs can suppress infestations of both perennial and annual herbaceous weeds through greater capture of resources, especially light.

In the following sections of this chapter, we examine applications of these principles in a range of farming systems. The reader should note that rotation, intercropping, and agroforestry are discussed separately for ease of presentation, but that farmers often use these practices in various combinations to address weed management and other concerns.

Crop rotation

Why farmers use crop rotation

Because crop rotation can improve soil characteristics and reduce pest pressures, yields from rotation systems are often higher and more stable than those from continuous monocultures. Although market forces and government policies can make one- or two-crop systems profitable and attractive to farmers, at least in the short term (National Research Council, 1989, pp. 233–40), a growing number of analysts consider the environmental and economic costs of such systems to be unsustainable (Power & Follet, 1987; Reganold, Papendick & Parr, 1990; Brummer, 1998). Agronomic and biological advantages of crop rotation are likely to become increasingly important if protecting soil and water quality, minimizing agrichemical use, and reducing crop subsidy payments become policy priorities (Faeth *et al.*, 1991; Faeth, 1993).

Potential benefits of crop rotation depend on the choice of crops grown and their order within the rotation sequence. Pasture, hay, and cover crops can be especially useful for improving soil physical characteristics, including bulk density, aggregate stability, aeration, and water infiltration capacity (MacRae & Mehuys, 1985; Karlen et al., 1994). Pasture, hay, and cover crops can also reduce soil erosion (Gantzer et al., 1991; Karlen et al., 1994) and loss of nutrients through leaching (Jackson, Wyland & Stivers, 1993; McLenaghen et al., 1996). When legumes are included in crop sequences, biologically fixed nitrogen can become available to subsequent crops, increasing yields of non-legume species (Fox & Pielielek, 1988; Frye et al., 1988). Even when nutrients are supplied at high levels or N-supplying legumes are not included in the crop sequence, crops grown in rotations typically yield 5% to 15% more than crops grown continuously (Karlen et al., 1994). This effect appears to be related to changes in soil microbiology and biochemistry (Bullock, 1992). Rotation of host and non-host crops can reduce damage by insect pests and pathogens that have a narrow host range, limited dispersal ability, and low persistence in the soil or adjacent land (Sumner, 1982; Lashomb & Ng, 1984; Sturz & Bernier, 1987; McEwen et al., 1989).

Continuous production of a single crop and short sequences of crops with similar management practices promote the increase of weed species adapted to conditions similar to those used for producing the crops. The resulting weed flora can be highly competitive and difficult to control, even with chemical technology (Froud-Williams, 1988). In contrast, over the course of a diverse rotation employing crops with different planting and harvest dates, different growth habits and residue characteristics, and different tillage and weed management practices, weeds can be challenged with a wide range of stresses and mortality risks, and given few consistent opportunities for unchecked growth and reproduction (Figure 7.1) (Liebman & Dyck, 1993; Derksen, 1997; Liebman & Ohno, 1998). Consequently, crop rotation can be a powerful tool for weed management.

Variation in the timing of crop management practices

As noted in Chapter 2, many weed species exhibit characteristic pulses of germination and growth at particular times of year. Crop management practices that affect weeds, such as tillage and seedbed preparation, planting, fertilization, herbicide application, and cultivation, also have a marked periodicity. Interactions between the periodicity of weed emergence and crop-specific farming operations can generate distinct weed communities in different crops. In a study of weed dynamics during 20 years of arable cropping in England, Chancellor (1985) noted that spring-germinating weed

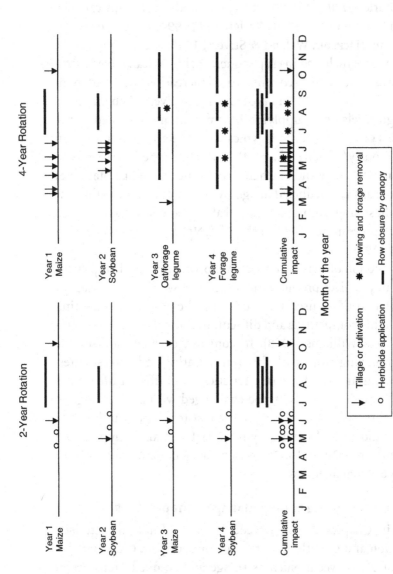

Figure 7.1 Stress and mortality factors affecting weeds in a simple two-year rotation and in a more diverse four-year rotation. Both rotations are suitable for the midwestern USA. The two-year rotation uses conventional weed management practices, whereas the four-year rotation uses practices commonly found on low-external-input farms. The forage legume (e.g., alfalfa or red clover) used in the four-year rotation is planted with oat, which is harvested for grain in July. As indicated by symbols shown on lines labeled "cumulative impact," the timing of stress and mortality factors is more varied and the duration of canopy cover is greater in the more diverse rotation. (Adapted from Kelner, Juras & Derksen, 1996, and Derksen, 1997.)

species, such as *Aethusa cynapium*, were most abundant in spring-sown crops (barley and potato), but fall-germinating weeds, such as *Poa annua*, were most abundant in fall-sown crops (winter barley, wheat, and oat).

Whereas continuous monocultures and short rotations of similar species select for weeds adapted to particular temporal niches, rotation of crops with different planting dates, contrasting growth periods, and differently timed management practices can disrupt selection for adapted weed species. In experiments conducted in Sweden, Håkansson (1982) observed that *Allium vineale* was an important weed in winter cereal crops sown repeatedly on the same land, but that it was unimportant in spring-sown cereals. The weight of new bulbs produced by *A. vineale* in spring-sown cereals was only 1% to 5% of that in winter cereals. This difference was attributed to different times of tillage for seedbed preparation coupled with seasonal changes in the weed's susceptibility to mechanical damage. Spring tillage preceding spring planting was most effective in killing *A. vineale* because at that time the plants had few reserves with which to recover from damage; tillage later in the year, after *A. vineale* had accumulated reserves, provided less effective control. The impact of spring tillage was persistent and created a rotation effect. When spring-sown crops occurred in more than 40% of a rotation, *A. vineale* populations were greatly reduced in subsequent winter cereals.

In the short term, the use of particular crops and management practices within a rotation can contribute to general reductions in weed density. Dotzenko, Ozkan & Storer (1969) found, for example, that weed seed density at the end of a four-year cropping cycle was lower when bean followed sugar beet than when maize or barley followed sugar beet. Much of this effect was attributed to differences among the crops in tillage and planting dates. Because bean had the latest planting date of the three crops preceding sugar beet, it allowed the highest proportion of weeds to emerge and be killed during seedbed preparation, and was most effective in limiting weed seed production. Whether this effect would have continued in future years is unknown, however. If weed species preadapted for late germination migrated into the field, or if resident weed species evolved late-germinating genotypes, differences among rotations in weed community composition might have become evident, and the general reduction in weed density obtained by rotating sugar beet with bean might have disappeared. Long-term experiments tracking the population dynamics of different weed species within the context of different crop rotations are needed to address this issue. In addition to fixed rotation treatments, such experiments might include flexible rotation treatments that would change in response to shifts in weed community composition.

Crop rotation and variations in soil conditions

As crops vary within a rotation sequence, changes occur in soil moisture and fertility conditions, residue cover, microtopography, and other soil properties that influence weed dynamics (see Chapter 5). Generally, a particular set of soil conditions will be favorable for some weed species, but less favorable or inhibitory for others. Crop sequences that involve large changes in soil conditions offer opportunities to disrupt selection for adapted weed species by destabilizing their environment.

This concept is exemplified by sequences that include both irrigated and rainfed crops. Rotation of wetland rice with dryland crops, such as maize, soybean, peanut, mungbean, sweet potato, and pasture species, reduces infestation by water-tolerant weeds. In an experiment in which no herbicides, cultivation, or hand-weeding were used during four succeeding cropping seasons, density of the perennial, water-tolerant weed *Scirpus maritimus* was 36% to 49% lower when wetland rice was grown in rotation with a dryland maize/mungbean intercrop than when wetland rice was grown continuously (Moody & Drost, 1983). Reductions in *S. maritimus* density were greatest during the dryland phase of the rotation, when soil moisture conditions were least favorable for the weed. Nonetheless, rotation of wetland and dryland crops in this experiment failed to fully suppress *S. maritimus* and did not adequately reduce annual weed densities, indicating the need for other weed management tactics to complement soil moisture management.

As discussed in Chapter 5, residues of various crop species have allelopathic properties that can influence weed germination, growth, and competitive ability. Integration of these effects into crop rotation strategies has not been addressed systematically, but opportunities appear to exist. For example, Einhellig & Rasmussen (1989) found that weed cover and biomass were lower following sorghum than following soybean, and attributed this effect to allelopathic properties of sorghum residue. Roder *et al.* (1989) obtained an average of 9% more soybean in rotation with sorghum than in continuous monoculture, and suggested that the yield advantage was due to greater soil water content early in the growing season. Thus, sequencing sorghum before soybean might provide both agronomic and weed management benefits. Other rotation sequences showing analogous crop yield and weed suppression benefits should be identified and investigated, with attention directed toward understanding the mechanisms and management of allelopathic interactions.

Crop rotation with and without herbicides

Results of an experiment conducted in Alberta by Blackshaw (1994) provide an exceptionally clear illustration of how weed management can

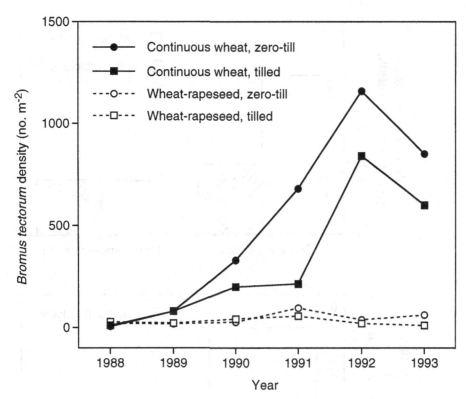

Figure 7.2 Density of *Bromus tectorum* plants in winter wheat grown in rotation with rapeseed or in continuous monoculture, with or without tillage, in an experiment conducted in Alberta, Canada. (Adapted from Blackshaw, 1994.)

benefit from rotating crops managed with different herbicides, as well as contrasting tillage and planting dates. Over a six-year period, densities of the winter annual grass *Bromus tectorum* were substantially lower in winter wheat grown in rotation with spring-sown rapeseed than in continuous winter wheat (Figure 7.2). Wheat yield was inversely proportional to *B. tectorum* density. Available herbicides were ineffective against *B. tectorum* in continuous wheat because of the weed's phenological and physiological similarity to the crop. Control of *B. tectorum* in the wheat–rapeseed rotation was attributed to spring cultivation and pre-plant herbicides, which killed the fall-germinating weed before rapeseed planting, and to selective post-emergence herbicides, which were tolerated by rapeseed, but effective against *B. tectorum*. Because seeds of *B. tectorum* have little dormancy and generally germinate in one to two years, suppression of the weed during the rapeseed phase of the rotation averted a problem in the subsequent wheat phase.

Although rotation of crops in conventionally managed cropping systems typically implies a rotation of herbicides as well, changes in weed density and

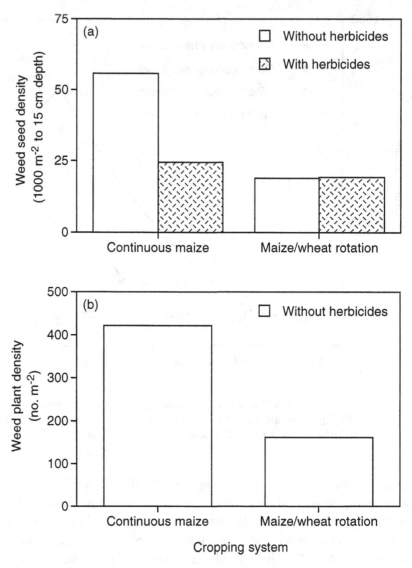

Figure 7.3 Weed seed density (a) and seedling density (b) in maize grown in rotation with winter wheat or in continuous monoculture, with or without herbicides, in an experiment conducted in the Tiber Valley, Italy. (Adapted from Covarelli & Tei, 1988.)

species composition can occur in response to variations in crop diversity *per se*, regardless of herbicide diversity. Covarelli & Tei (1988) compared weed complexes after six years of continuous maize or three cycles of a two-year winter wheat–maize rotation; the experiment included treatments in which both cropping systems were not sprayed with herbicides. Total weed seed density was 66% lower (Figure 7.3a) and seedling density 62% lower (Figure 7.3b) for maize in rotation than for continuous maize. Weed seed densities were as low in the wheat–maize rotation without herbicides as with herbicides. Although the continuous maize and wheat–maize systems contained equal numbers of weed species, they differed in the relative importance of the species. In continuous maize, the most abundant weed, *Echinochloa crus-galli* (a C_4 grass like maize), accounted for 71% of the total number of weed seedlings. In contrast, in the wheat–maize rotation, the most abundant species, *Amaranthus retroflexus*, accounted for only 39% of the total weed density, and *E. crus-galli* accounted for only 26% of the total. Impacts of crop rotation practices on weed species diversity are discussed in more detail in Chapter 10.

Effects of lengthening rotations and increasing diversity from two to three crops can be seen in a long-term study conducted by Schreiber (1992). The experiment included a zero-tillage, minimum weed management treatment that received herbicides at concentrations lower than commercially recommended. Subplots were left untreated with herbicides to allow fuller expression of possible crop rotation effects on weed dynamics. Density of the annual grass weed *Setaria faberi* was measured in continuous maize, maize rotated with soybean, and maize following winter wheat in a soybean–wheat–maize rotation. Over a seven-year period, mean density of *S. faberi* plants was highest in continuous maize, lowest in the three-year rotation, and intermediate in the two-year rotation. Thus, *S. faberi* density decreased as crop diversity increased. Perhaps more important from a farmer's viewpoint is that variance in the weed's density between years decreased as crop diversity increased. Consequently farmers using the three-year rotation could feel more confident that *S. faberi* density would not become unexpectedly high.

Schreiber (1992) attributed the marked reductions in *S. faberi* density in the three-year rotation to allelopathic effects of wheat straw. However, differences among the fall-sown crop (wheat) and the two spring-sown crops (maize and soybean) with regard to planting and harvest dates and other management practices may also have affected weed dynamics. Results of Schreiber's (1992) study do not resolve questions of how many and what types of crops should be included in a rotation for maximum weed suppression, but they indicate that further research addressing those questions would be valuable.

Perennial forage crops

Perennial forage crops constitute important components of many crop rotation systems and offer opportunities to suppress weeds through competition (Risser, 1969), mowing (Hodgson, 1958; Norris & Ayres, 1991), and grazing (Chapter 9) (Heard, 1963; Dowling & Wong, 1993). Sunlight filtered through a canopy of well-established forage plants inhibits the germination of many weed species due to changes in spectral composition (see Chapter 2). Reduced soil disturbance during the time when a perennial forage crop occupies a field also suppresses weed germination relative to tilled conditions (Roberts & Feast, 1973). For all of these reasons, few new weed seeds may be added to the soil seed bank during the forage phase of a crop rotation.

Weed seed survival in soil can be higher in untilled forages than tilled crops (Warnes & Andersen, 1984; Lueschen *et al.*, 1993), but the combined effects of reducing seed inputs and maintaining a moderate level of seed mortality can result in substantial reductions in weed density. Eighty-three percent of the Manitoba and Saskatchewan farmers surveyed by Entz, Bullied & Katepa-Mupondwa (1995) reported fewer weeds in cereals following perennial forages (especially alfalfa) than following cereal crops; 67% reported higher grain yields following forages than following cereal crops. The farmers noted that annual grass and annual broadleaf weed species were particularly well controlled through the use of perennial forages.

In field experiments, weed seed populations were found to decline by 99% for *Avena fatua* following three years of perennial ryegrass/white clover sod (Wilson & Phipps, 1985), 47% for *Abutilon theophrasti* following two years of alfalfa (Lueschen *et al.*, 1993), and 47% for *Brassica kaber* following one and a half years of smooth bromegrass (Warnes & Andersen, 1984). In each case, maintenance of the forage stands for longer periods of time had little additional effect on weed seed mortality. Inclusion of perennial forage crops in rotations should therefore be expected to reduce, but not eliminate, seed populations of many weeds typically found in arable crops.

If forage management practices are improperly timed relative to weed growth and reproduction, weed seed density may increase more during the production of forages than annual crops. In one of two years, Clay & Aguilar (1998) observed higher weed seed densities in soil following alfalfa than continuous maize, and attributed this difference to seed shed that occurred between the penultimate and final alfalfa harvests. However, weed seedling densities in maize were consistently lower following alfalfa than following maize (Clay & Aguilar, 1998), suggesting that seedling establishment may have been more important than seed survival for regulating weed density. Differences between the two cropping systems in soil chemical, physical, or biological characteristics may have been responsible for differences in weed emergence (see Chapter 5).

While perennial forage phases of a rotation may reduce infestations of certain weed species, they may also allow for increases in populations of others. Ominski, Entz & Kenkel (1999) measured weed densities in Manitoba wheat, oat, and barley fields that were occupied the previous year by either alfalfa hay or cereals. Densities of *Avena fatua*, *Circium arvense*, *Brassica kaber*, and

Galium aparine were lower following alfalfa than following cereal crops. This effect was attributed to alfalfa's competitive ability and to the cutting regime used for it, which largely prevented seed production by the weeds. In contrast, densities of *Taraxacum officinale* and *Thlaspi arvense* were greater in cereals following alfalfa than in continuous cereal production. Success of *Taraxacum officinale* in the alfalfa/cereal system was attributed to its prostrate growth habit, which allowed it to avoid defoliation during hay harvest, and to its extended period of germination in untilled soil, which permitted seedling establishment after the soil surface was exposed by hay removal in the summer or by crop dormancy in the fall. Success of *Thlaspi arvense* in the alfalfa/cereal system was attributed to its winter annual habit: it germinated in the fall when alfalfa was dormant and noncompetitive, and it resumed growth early in the spring prior to the resumption of alfalfa growth. The absence of tillage or herbicides during alfalfa production allowed both *Taraxacum officinale* and *Thlaspi arvense* to survive and produce seeds, and led to increased infestations of the two weed species in succeeding cereal crops.

Certain perennial grass weeds, such as *Elytrigia (Agropyron) repens*, are particularly well adapted to survival in sod environments (Sheaffer *et al.*, 1990). For this reason, their densities should be reduced as much as possible during annual crop phases, before forages are sown. Tillage and herbicides used to terminate the growth of forage stands are critical factors regulating the persistence of *E. repens* and other perennials in subsequent annual phases of a rotation. Impacts of different tillage practices on perennial weeds are examined in Chapter 4.

Cover crops within rotation systems

As discussed in Chapter 5, cover crops can be grown during periods when main crops are absent to suppress weed germination and growth through resource competition, allelopathy, and other processes. In effect, cover crops fill gaps in a cropping system that would otherwise be occupied by weeds. This type of niche preemption is illustrated in data from McLenaghen *et al.* (1996), who sowed five winter cover crops or let ground lie fallow after fall-plowing sod. The quantity of ground cover produced by weeds was inversely proportional to that produced by the cover crops (Figure 7.4). In the fallow treatment without a cover crop, weeds covered 52% of the ground area. In contrast, a white mustard cover crop produced 92% ground cover and reduced weed cover to just 4%. Rye produced 85% ground cover and permitted only 9% weed cover. At least part of the weed suppression observed in this study may have been due to competition for nitrogen, since most of the cover crops captured soil nitrate effectively (McLenaghen *et al.*, 1996).

Although this and other short-term experiments indicate that cover crops

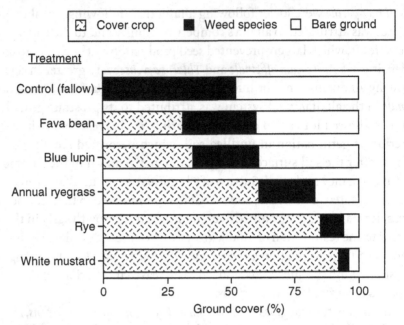

Figure 7.4 Ground cover produced by cover crops and weeds two months after seeding in an experiment conducted on the Canterbury Plains, New Zealand. (Adapted from McLenaghen et al., 1996.)

may play an important role in weed management, long-term impacts of cover cropping on weeds remain poorly understood. Multiyear field experiments comparing weed dynamics in rotations with and without cover crops are needed to address this issue. Models of weed population dynamics will also be useful for identifying which weed species may decline and which species may become more abundant with different cover crop management strategies.

Intercropping

Why farmers plant crops in mixtures

Intercropping is practiced widely where farmers seek (i) the highest combined yield of two or more crops per unit of land area, or (ii) the smallest risk of not meeting food or income requirements. Intercropping is particularly important where crop production and food security are challenged by a limited amount of arable land, and stressful soil, pest, and climatic conditions.

Intercrop yield performance is most often assessed by calculating a land equivalent ratio (LER), which indicates the total amount of land that must be sown in different sole crops to produce the same yields obtained from one land unit sown with an intercrop (Vandermeer, 1989). If LER>1, intercropping can produce a higher total yield from a given area than can sole cropping; when LER<1, sole cropping provides more yield per unit area.

The land use efficiency of intercropping sorghum with pigeonpea, a practice used extensively by Indian farmers, is illustrated by data from Natarajan & Willey (1980a). Sorghum yields under intercrop and sole crop conditions averaged 4190 and 4470 kg ha^{-1}, respectively; intercrop and sole-crop yields of pigeonpea averaged 690 and 1020 kg ha^{-1}, respectively. Based on these values, 0.94 ha of sole-cropped sorghum (4190 ÷ 4470) and 0.68 ha of sole-cropped pigeonpea (690 ÷ 1020) were needed to produce the same yields harvested from one hectare of the intercrop, and the LER value of the intercrop was 0.94+0.68=1.62. In this case, yield of each component of the mixture was reduced by competition from the associated crop, but total yield from intercropping, on a unit area basis, was 62% greater than from sole cropping. Although intercropping does not always provide yield advantages over sole cropping, LER values exceeding one have been reported for many different intercropping systems (Willey, 1979a; Ofori & Stern, 1987; Balasubramanian & Sekayange, 1990).

In some situations, farmers are interested primarily in obtaining full yield of one "main" crop, but sow other species into the main crop for additional benefits: more food and fodder, improved soil conservation, better weed control, or other purposes (Willey, 1979a). LER values typically are not calculated in these cases, but exceed 1 if the farmer manages the intersown minor crop(s) so that competition against the main crop is prevented. For reasons discussed below, main crop yields may even be enhanced by intersown species. For example, in an experiment conducted by Abraham & Singh (1984), average grain yield of sole-cropped sorghum was 1.97 Mg ha^{-1}, whereas yield of intercropped sorghum was 2.70 Mg ha^{-1}; intercropped cowpea provided an additional 0.40 Mg ha^{-1} of seed.

Greater yield stability may be a particularly important advantage of intercropping systems. Using data from 51 cropping systems trials conducted in India from 1972 to 1978, Rao & Willey (1980) calculated that for a disaster income level of 1000 rupees ha^{-1}, growing only sole-cropped pigeonpea would fail to produce enough revenue one year in five, growing only sole-cropped sorghum would fail one year in eight, growing both sorghum and pigeonpea as sole crops would fail one year in 13, but growing a sorghum/pigeonpea intercrop would fail just one year in 36. Using similar data, Trenbath (1983) calculated that to minimize a farm family's risk of failing to produce an adequate number of calories for subsistence, less land was needed with a sorghum/pigeonpea intercrop than with sorghum or pigeonpea sole crops. This was due to both higher yield and lower variance of yield.

Resource use and weed suppression

Weed suppression by intercrops is often, but not always, better than that obtained from sole crops (Liebman & Dyck, 1993). Greater land use efficiency, yield stability, and weed suppression of intercropping relative to sole cropping all appear to derive from complementary patterns of resource use and facilitative interactions between crop species.

Complementary use of resources occurs when the component species of an intercrop use qualitatively different resources, or use the same resources at different places or different times. Complementarity can exist between crops that differ in nitrogen sources (e.g., in mixtures of legumes and non-legumes), photosynthetic responses to varying light intensities (e.g., in mixtures of C_3 and C_4 species), rooting depth, peak periods of leaf area display and root activity, and other physiological, spatial, and temporal aspects of resource use (Snaydon & Harris, 1981; Ofori & Stern, 1987; Willey, 1990). In ecological terms, resource complementarity minimizes niche overlap and competition between crop species, and permits crop mixtures to capture a greater range and quantity of resources than can sole crops (Vandermeer, 1989).

Facilitative interactions occur when one crop species directly or indirectly aids the growth of another (Vandermeer, 1989). Taller crops can improve the growth and yield of shorter crops by reducing wind speed and improving water use efficiency (Radke & Hagstrom, 1976). Erect crops can improve the performance of climbing crops by providing physical support (Budelman 1990a, 1990b). Growth and yield of non-legume crops may be increased by N transfer from roots of associated legume species (Agboola & Fayemi, 1972; Eaglesham et al., 1981). Various crop species may also reduce the amount of insect and disease damage suffered by associated crops by (i) disrupting the ability of pests to locate, move to, feed upon, or infect host plants, or by (ii) providing favorable conditions for the proliferation and effective action of natural enemies of pests (Vandermeer, 1989; Trenbath, 1993; Altieri, 1994; Liebman, 1995).

Because complementary patterns of resource use and facilitative interactions between intercrop components can lead to greater capture of light, water, and nutrients, intercrops can be more effective than sole crops in preempting resources used by weeds and suppressing weed growth. Linkages between resource use and weed suppression can be seen in a study conducted by Abraham & Singh (1984), who planted sorghum alone and in mixture with different legume species (fodder cowpea, grain cowpea, soybean, mungbean, and peanut). Under both weed-free and weed-infested conditions, intercropping legumes with sorghum increased crop leaf area, light interception, macronutrient (N, P, and K) uptake, dry matter production, and seed yield above levels obtained from sole-cropped sorghum. Intercropping also reduced the amount of light available to weeds, decreased weed macronutrient uptake, and reduced weed dry matter production up to 76%, compared with the sorghum sole crop. Figure 7.5 illustrates these effects for the sorghum/fodder cowpea intercrop. Similar patterns of resource use, crop yield, and weed suppression have been observed for intercrops and sole crops of sorghum and pigeonpea (Natarajan & Willey, 1980a, 1980b; Shetty & Rao, 1981).

A resource-based approach for studying weeds in intercropping systems can provide mechanistic insights into why certain crop combinations are more weed-suppressive than others. For example, based on data presented in Tables 7 and 8 of Abraham & Singh's (1984) study, >90% of the variation among cropping systems in final weed biomass was predictable as a linear function of light penetration to ground level at 25 days after planting. That is, the mixtures of crop species that were most effective in suppressing weed growth were those that intercepted the most light early in the growing season. Efforts to improve weed suppression by sorghum-based intercropping systems might therefore focus on identifying management practices and cultivars that increase early canopy development and light interception. Similar resource-based approaches could be pursued for improving weed management in other systems.

Crop diversity and density

One of the important issues emerging from studies of intercrop/weed interactions is whether weeds are suppressed by increasing crop diversity *per se*, or by the combined effects of increasing crop diversity and density. Mixtures of crop species that complement or facilitate each other's use of resources are often sown at greater total densities than those used for sole crops. The most common way to do this is to add all or part of the normal sole-crop population density for one crop into the normal sole crop density of another crop. Thus, crop diversity and density effects are often confounded in intercropping experiments.

To separate the effects of crop diversity from those of crop density requires the use of experimental designs in which overall density is maintained constant between sole crop and intercrop treatments. Suppose, based on the study by Bulson, Snaydon & Stopes (1997), that the "normal" sowing densities for sole crops of wheat and fava bean are 250 and 50 plants m^{-2}, respectively. For a

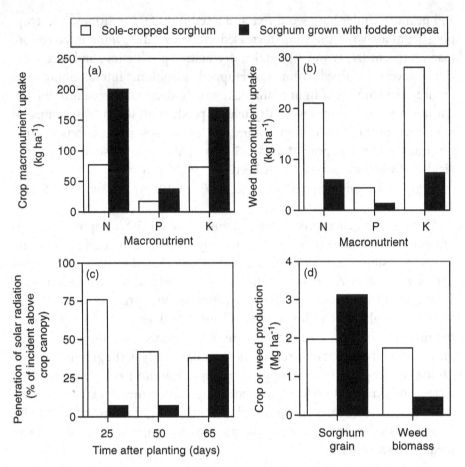

Figure 7.5 Resource use and dry matter production of sole-cropped sorghum, a sorghum/fodder cowpea intercrop, and associated weeds in an experiment conducted in New Delhi, India. Data are averaged over five weed management treatments (unweeded control, two hand-weedings, linuron, nitrofen, and fluchloralin) and two cropping seasons. (a) Uptake of N, P, and K by sole-cropped sorghum and the sorghum/cowpea intercrop. (b) Uptake of N, P, and K by weeds in the two cropping systems. (c) Penetration of solar radiation to ground level in sole-cropped and intercropped sorghum as a percentage of incident radiation above the canopy. The increase in radiation penetration at 65 days after planting resulted from harvest of cowpea at 60 days after planting. (d) Sorghum grain yield and weed above-ground biomass in sole-cropped and intercropped sorghum. In addition to sorghum grain, the intercrop produced cowpea fodder material. (Adapted from Abraham & Singh, 1984.)

specified set of growing conditions and genotypes, these normal density sole crops are expected to produce the highest yields from the least amount of seed. An "additive mixture" containing 250 wheat plus 50 fava bean plants m^{-2} could also be sown and would represent 200% of the normal population densities used for the sole crops. In contrast, a "replacement series mixture" containing equal proportions of the two crops at their normal densities would contain 125 wheat plus 25 fava bean plants m^{-2}. Such a mixture would represent the same total plant density as the two normal density sole crops, and comparisons of yield and weed suppression characteristics could be made between the intercrop and sole crops without confounding changes in crop diversity with changes in density.

Sharaiha & Gliessman (1992) found that yield advantages and superior weed control resulted directly from increases in crop diversity. The investigators grew sole crops and two-species mixtures of fava bean, pea, and lettuce in constant-density replacement series. The two legumes were direct-seeded; lettuce was transplanted. Intercrops were formed by substituting rows of other species for rows in the sole crops. All plots were weeded until 34 days after planting; thereafter, weeds were allowed to grow. Intercropping yield advantages were evident in LER values for fava bean/pea, fava bean/lettuce, and pea/lettuce intercrops that averaged 1.55, 1.48, and 1.26, respectively. Mean weed biomass in the fava bean/pea intercrop was 48% less than in the fava bean sole crop, and 40% less than in the pea sole crop. Mean weed biomass in the fava bean/lettuce intercrop was 30% and 32% less than in the respective sole crops; in the pea/lettuce intercrop, mean weed biomass was 6% and 23% less than in the respective sole crops.

Weed suppression advantages of intercrops over sole crops in constant-density replacement series were also observed by Fleck, Machado & De Souza (1984) in experiments with bean, sunflower, and maize. Intercrops were formed by substituting rows of one crop for another. Bean/sunflower and maize/sunflower intercrops were most successful in suppressing weed growth (Figure 7.6), an effect attributed to the greater canopy cover of those two intercrops, and to the ability of maize and sunflower to grow taller than the associated weeds.

In other studies, intercropping has failed to suppress weed growth more than sole cropping unless total crop density has been elevated above levels used for sole crops. Experiments with intercrops and sole crops of wheat and fava bean (Table 7.1)(Bulson, Snaydon & Stopes, 1997), barley and pea (Mohler & Liebman, 1987), sorghum and pigeonpea (Shetty & Rao, 1981), and pearl millet and peanut (Shetty & Rao, 1981) indicated that at a given overall crop density, weed biomass in the intercrop was always more than in

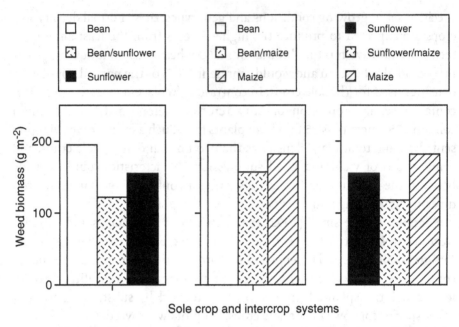

Figure 7.6 Above-ground biomass of weeds present in different cropping system treatments 87 days after crop emergence, in an experiment conducted in Rio Grande do Sul, Brazil. (Adapted from Fleck, Machado & De Souza, 1984.)

the sole-cropped cereal, but less than in the sole-cropped legume. It is unclear why intercropping failed to provide a weed control advantage over sole cropping in these studies, but *did* provide an advantage in the studies by Sharaiha & Gliessman (1992) and Fleck, Machado & De Souza (1984). To answer this question, the influence of environmental conditions and species-specific resource use and growth characteristics needs to be investigated in more detail.

In contrast to the variable results of studies testing the effects of crop diversity on weeds, studies focused on crop density consistently show that increasing intercrop density reduces weed biomass (Bantilan, Palada & Harwood, 1974; Shetty & Rao, 1981; Tripathi & Singh, 1983; Mohler & Liebman, 1987; Lanini *et al.*, 1991; Weil & McFadden, 1991; Bulson, Snaydon & Stopes, 1997). Increases in intercrop density can also have important beneficial effects on crop yield (Willey, 1979*b*). Bulson, Snaydon & Stopes (1997) observed that yields of wheat and fava bean sole crops failed to increase when densities exceeded 100% of recommended levels; in contrast, yield of a wheat/fava bean intercrop, expressed as a land equivalent ratio (LER), continued to rise when total crop density exceeded the level used for a normal density sole crop (Table 7.2). A wheat/fava bean intercrop with each component sown at 75% of its

Table 7.1. *Above-ground biomass of weeds (g m^{-2}) present in fava bean and wheat sole crops and intercrops sown at different densities in an experiment conducted in Berkshire, UK. Recommended densities for fava bean and wheat sole crops were 25 and 250 plants m^{-2}, respectively*

Fava bean density[a]	Wheat density[a]				
	0	25	50	75	100
0	–	302	146	97	124
25	398	168	148	96	93
50	346	162	133	80	100
75	284	138	151	75	36
100	169	117	72	83	62

Notes:
[a] Percentage of recommended sole crop density.
Source: Adapted from Bulson, Snaydon & Stopes (1997).

Table 7.2. *Land equivalent ratios (LER) of fava bean/wheat intercrops sown at different densities, based on seed yields of 3.7 Mg ha^{-1} for both fava bean and wheat sole crops grown at their respective recommended densities (25 and 250 plants m^{-2})*

Fava bean density[a]	Wheat density[a]			
	25	50	75	100
25	0.76	1.06	1.02	1.27
50	0.86	1.08	1.05	1.25
75	0.94	1.16	1.29	1.27
100	0.97	1.15	1.24	1.16

Notes:
[a] Percentage of recommended sole-crop density.
Source: Adapted from Bulson, Snaydon & Stopes (1997).

normal sole crop density (i.e., a 150% mixture) provided the highest LER value (1.29) obtained in the experiment. Similarly, intercrop yield advantages over sole crops rose with increasing density in mixtures of sorghum with pigeonpea (Shetty & Rao, 1981) and barley with pea (Mohler & Liebman, 1987).

The preceding discussion of how intercropping affects weeds and crop yields indicates that increases in crop diversity through intercropping may or may not provide weed control advantages over sole cropping. However, increases in crop density consistently reduce weed biomass in intercrops as they do in sole crops (see Chapter 6), and intercropping often allows use of a total crop density that would be pointless or problematic in a sole crop. Farmers seeking to maximize their yield per unit land area by sowing a high-density intercrop can gain weed suppression benefits, whether these result from increasing crop diversity or crop density.

Other management factors affecting weeds in intercrops

In addition to changes in crop density, other management factors, especially crop spatial arrangement, crop genotype, and soil fertility, can affect weed performance in intercrops. In sole cropping systems, soil and crop management factors may directly affect crop–weed competition. In intercropping systems, soil and crop management factors can also have indirect effects on weeds through their influence on crop–crop interactions (Vandermeer, 1989, pp. 127–40). As a consequence of these additional interactions, it is difficult to predict which management options will maximize the strength of intercrop competition against weeds, while reducing competition between crop components and maximizing their yield. Both empirical and theoretical research are needed to address these issues.

Crop spatial arrangement

In sole crop systems, competition among crop plants can be reduced and competition from crops against weeds can be increased by placing a given number of crop seeds or transplants in more equidistant arrangements (see Chapter 6). Generally this is accomplished by reducing distances between crop rows and increasing distances between plants within rows.

In intercrop systems, weed performance can also be affected by crop spatial arrangement, but optimum crop arrangements are not readily predictable, possibly because changes in distance-dependent interactions between crop species may alter their effects on weeds. Less weed growth was measured in rice/blackgram intercrops when rice was sown in paired rows rather than uniformly spaced rows (Sengupta, Bhattacharyya & Chatterjee, 1985), and in pigeonpea/sorghum intercrops when pigeonpea was sown in paired rather than uniform rows (Prasad, Gautam & Mohta, 1985). In contrast, in one year of an experiment in which pigeonpea was intercropped with urdbean, mungbean, soybean, cowpea, or sorghum, a uniform row arrangement of pigeonpea resulted in less weed growth than did a paired row arrangement; no difference was found in weed growth between uniform and paired row arrangements in a second year (Ali, 1988). Mechanistic modeling of interspecific interactions, using approaches such as those presented by Kropff & van Laar (1993), could prove useful in understanding how contrasting spatial arrangements affect crop–crop and crop–weed competition.

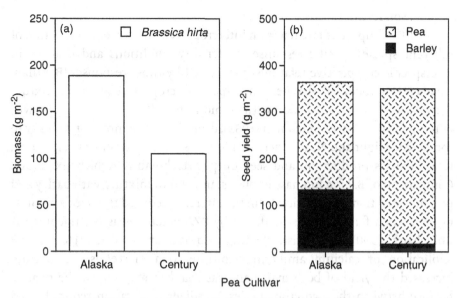

Figure 7.7 Effect of pea cultivar identity on (a) above-ground biomass of *Brassica hirta* and (b) seed yields of barley and pea in mixtures of the three species grown under well-watered, low nitrogen conditions, in an experiment conducted in California, USA. (Adapted from Liebman, 1989.)

Crop genotype

Similar to the situation for sole crops (see Chapter 6), large differences in weed growth have been observed between different cultivars of crops sown in mixtures. Weed biomass was 49% to 56% lower when sorghum was intercropped with a cowpea cultivar used for fodder than when it was intercropped with one used for seed production (Abraham & Singh, 1984). Samson, Foulds & Patriquin (1990) found significant differences in weed suppression ability between cultivars of both perennial and annual ryegrasses intercropped with maize. *Brassica hirta* biomass was 44% lower when barley was intercropped with a long-vined, large-leafed pea cultivar ('Century') rather than a shorter-vined, smaller-leafed pea cultivar ('Alaska') (Figure 7.7a) (Liebman, 1989).

Generally, intercrops are more weed-suppressive when they include cultivars that have rapid initial growth and dense, planiform canopies. Choosing a cultivar for its weed suppression characteristics may conflict with other objectives, however. Intercropping barley with 'Century' pea rather than 'Alaska' pea gave better suppression of *B. hirta* (Figure 7.7a) and increased pea seed yield by 45% (Figure 7.7.b), but also reduced barley seed yield by 88% (Figure 7.7b). In this case, better weed suppression and greater production of the legume would have to be weighed against the reduced yield of the cereal.

Soil fertility

The impact of fertilizers on intercrop–weed mixtures is a function of inherent species-specific responses to fertility conditions and changes in interspecific competitive relationships created by those responses (Bantilan, Palada & Harwood, 1974). When barley/pea intercrops were grown in mixture with *Brassica hirta*, fertilization with ammonium sulfate reduced the growth of pea relative to barley (Liebman, 1989), diminished the intercrop's ability to compete for light and N, and increased *B. hirta*'s photosynthetic rate, leaf area, total biomass production, and seed output (Liebman & Robichaux, 1990). Comparison of *B. hirta* biomass in pure stand and in mixture with barley and pea indicated that competition from the intercrop reduced the weed's biomass by 49% when fertilizer was used, but by 87% when it was not (Liebman & Robichaux, 1990). In contrast, Olasantan, Lucas & Ezumah (1994) found that application of calcium ammonium nitrate to a maize/cassava mixture increased the yield of both maize and cassava, but maintained the relative balance between the two crop species. Fertilizer application reduced weed biomass (a mixture of many grass, sedge, and broadleaf species) by increasing the intercrop's leaf area index and light interception ability.

Because many intercrops are composed of legumes, which use atmospheric N_2, and non-legumes, which use inorganic forms of N from the soil pool and fertilizer, a key research question is how to manage the N nutrition of different intercrop components so that total yield is optimized and weed growth is minimized. Can the location, timing, and form of fertilizer N application be adjusted to match the N requirements of non-legume crops, while withholding inorganic N from legumes and weeds? Research is also needed to understand how different types of N management affect crop–crop interactions in mixtures containing only non-leguminous species and how these relationships affect weed performance. Additional research is needed to determine whether management of non-N nutrients should differ between intercrops and sole crops.

Living mulches and smother crops

The preceding discussion has focused on how intercropping may affect weeds when each component crop is intended to be a source of food, feed, or fiber. Here we consider the impacts of intercropping with species that are intended mostly for other purposes, such as soil conservation, nitrogen fixation, and weed control, though they may produce food and other harvestable products incidentally. Used in this way, such species are "minor" components in mixture with "main crops," and are commonly called *living mulches* and *smother crops* (Shetty & Rao, 1981; Paine & Harrison, 1993; De Haan,

Sheaffer & Barnes, 1997; De Haan *et al.*, 1997). Living mulches are used to provide cover before, during, and after main crop production, and are often well established at the time main crops are sown; in contrast, smother crops are generally planted simultaneously with or after main crops, and are small or absent during early stages of main crop growth. The key management issue that must be addressed for both living mulches and smother crops is how they can suppress weed recruitment and growth without functioning themselves as weeds that compete against main crops.

When a main crop is planted into an established living mulch, competition against the main crop frequently occurs. Kurtz, Melsted & Bray (1952) planted maize into sods of several grass and legume species (smooth bromegrass, timothy, lespedeza, alfalfa, birdsfoot trefoil, ladino clover, or red clover) and found that all of the species reduced or eliminated maize grain yield, at least in some years. Application of N fertilizer and irrigation water only partially mitigated the problem. Other investigators working with maize and sweet corn production systems have used low doses of herbicides, repeated mowing, and partial incorporation into soil to manage, but not kill, grass and legume living mulches (Echtenkamp & Moomaw, 1989; Grubinger & Minotti, 1990; Mohler, 1991; Eberlein, Sheaffer & Oliveira, 1992; Fischer & Burrill, 1993). The results of such approaches have been inconsistent, however, and it appears that a considerable amount of fine-tuning will be required before most living mulch systems can be regulated predictably.

One living mulch system that may serve as a model for the successful development of others involves the use of subterranean clover in mixture with maize and other warm-season annual crops. Much of the success of this system derives from a fortuitous match between local climatic conditions and subterranean clover's ecophysiology, and from differences in the timing of growth between the living mulch and the main crops that prevent competitive reduction of main crop growth and yield. In New Jersey, Ilnicki & Enache (1992) observed that subterranean clover germinated during late summer, grew vegetatively until early winter, lay dormant during winter, resumed growth the following spring, and died in early summer after setting seed. For most of the summer, while maize and other summer annual crops made most of their growth, a dense mat of dead subterranean clover lay on the soil surface; late in the summer the regeneration and growth cycle began again as high densities of clover seedlings emerged from seeds that had been produced *in situ*.

Enache & Ilnicki (1990) found that maize planted without tillage and without herbicides into established subterranean clover produced as much or more biomass and grain as sole-cropped maize grown with herbicides, either with or without tillage. Late-season weed biomass in the maize/subterranean clover intercrop managed without herbicides or tillage was 54% to 90% lower than in sole-cropped maize managed with herbicides and conventional tillage, and 70% to 96% lower than in sole-cropped maize with herbicides but without tillage. Ilnicki & Enache (1992) attributed weed suppression by subterranean clover to the physical barrier created by the dead mulch layer and suggested that the mulch could have allelopathic effects. Competition by clover plants that germinated during the summer also may have been important for late-season weed suppression.

Subterranean clover is well suited for use as a living mulch with soybean, squash, cabbage, snap bean, and tomato in New Jersey (Ilnicki & Enache, 1992), but it is poorly adapted for many other environments because of insufficient winter hardiness, more synchronous and competitive growth with main crops, and other factors. The concept of temporal complementarity between living mulches and main crops is nonetheless intriguing and should be explored further.

Although smother crops planted at the same time as main crops can be used to suppress weeds, they can also reduce main crop yields unacceptably (De Haan et al., 1994; De Haan, Scheaffer & Barnes, 1997). However, certain species may provide weed control without a loss of main crop yield. This may be possible if (i) resource depletion by minor crops occurs at times when it has no effect on main crops; (ii) the minor crops avoid competition for N because they are legumes, or avoid competition for light and water because they are shorter than main crops (shading by the main crop decreases transpiration by the minor crop); and (iii) the rapid early growth or physiological shade tolerance of minor crops allows them to produce an understory canopy that suppresses weed germination and establishment.

Examples of smother crops that can be planted simultaneously with main crops to suppress weed growth while maintaining crop yields can be drawn from both tropical and temperate regions. In Andhra Pradesh, India, Shetty & Rao (1981) found that adding smother crops of cowpea or mungbean to main crops of sorghum or pigeonpea reduced early-season weed growth and allowed a reduction of hand-weeding from twice to once without a reduction in main crop yield. In Maine, USA, intercropping red clover with spring barley had no effect on barley yield but reduced weed biomass at the time of barley harvest and up to nine months thereafter (Figure 7.8) (M. Liebman, unpublished data). Similarly, in England, intercropping red clover with barley reduced growth of the perennial grass weed *Agropyron (Elytrigia) repens* from both seeds and rhizome fragments (Williams, 1972; Dyke & Barnard, 1976). Red clover has a low light compensation point of about 140 μmol s^{-1} m^{-2}

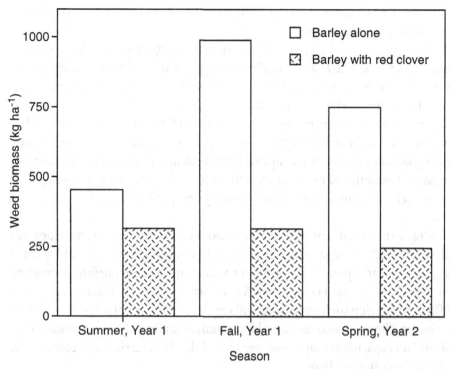

Figure 7.8 Weed biomass present in a barley sole crop and a barley/red clover intercrop at the time of barley harvest (summer, year 1) and in stubble at subsequent sampling dates (fall, year 1; spring, year 2), in an experiment conducted from 1988 to 1993 in Maine, USA. (M. Liebman, unpublished data.)

(Kendall & Stringer, 1985) and is thus well adapted to growing under barley. After the barley is harvested, the clover rapidly develops a dense canopy that shades weeds.

Delayed seeding of minor crops into established main crops offers a more predictably successful approach for reducing or eliminating competition against main crops while gaining some improvement in weed control. Because this type of intercropping is generally conducted after a final weeding operation, any weed control advantage it may provide derives from the combination of direct control measures with competition from the smother crop. An example of how such a system may work is shown in a study by Sengupta, Bhattacharyya & Chatterjee (1985). They compared weed growth (mostly grasses and sedges) and crop yields in an unweeded rice sole crop, an unweeded rice/blackgram intercrop with both crops sown simultaneously, and a rice/blackgram relay intercrop in which rice was sown first and blackgram was added 21 days later after one hand-weeding. Weed density and biomass were lowest and rice yield and monetary return were highest in the relay intercropping system.

Weed control can be improved in maize production systems by planting a smother crop at the last cultivation, about one month after sowing maize. Samson, Foulds & Patriquin (1990) found that late-season weed biomass could be reduced up to 63% by interseeding annual or perennial ryegrass into maize. Although relay intercropping with ryegrass had no effect on maize yield, it probably did reduce the number of weed propagules that would infest future crops. Samson, Foulds & Patriquin (1990) recommended against sowing ryegrass into established maize crops with grain yield potentials of >9 Mg ha^{-1}, since such vigorous crops would be overly competitive for light, water, and nutrients.

More extensive use of smother crops would be fostered by the development of short-lived, short-statured genotypes that suppress weed establishment during the early part of the growing season, but senesce before competing with main crops (De Haan, Scheaffer & Barnes, 1997; De Haan et al., 1997). Efforts to develop such genotypes will require the active participation of plant breeders. They will also require a better understanding of the ecophysiologies of minor crops, main crops, and weeds, and the characteristics of competitive interactions among them.

Changes in weed community composition due to intercropping

In contrast to crop rotation, which alters weed communities through the effects of different crops sequenced over multiple seasons, intercropping combines the effects of different crops within a single season. Will farmers encounter different species of weeds or different proportions of weed species in intercropping systems compared with sole cropping systems? If so, how might these differences be predicted? Two impacts of intercropping on weed community organization appear particularly important.

First, the dominant weed species found in intercrops may closely reflect the identities and proportions of the different component crops, each of which may have distinct associated weed communities. Shetty & Rao (1981) observed, for example, that the weed community of sole-cropped peanut was dominated by *Digitaria*, *Cyperus*, and *Celosia* spp., which produced about 3%, 40%, and 50%, respectively, of the total weed biomass. In contrast, the weed community of sole-cropped millet was a mixture of many species, with *Digitaria*, *Cyperus*, and *Celosia* spp. constituting about 25%, 15%, and 5%, respectively, of total weed biomass. *Celosia* and *Digitaria* grew taller than peanut, whereas *Cyperus* grew beneath the peanut canopy and was shaded by it. As more rows of peanut replaced millet to form various crop mixtures, both *Celosia* density and biomass increased markedly, but only within the peanut rows. These results suggest that the composition of weed communities in intercrops can be predicted from the proportions of the component crops.

Second, greater evenness in the relative importance of different weed species can be promoted by cropping practices that increase crop biomass and reduce total weed biomass. Whether or not the weed species found in an intercrop grow in more balanced proportions than those in sole crops will then depend on which cropping system provides the greatest degree of overall weed suppression. This point is illustrated by data from Mohler & Liebman (1987), who grew barley and pea sole crops and intercrops in replacement series at different densities. High-density crops produced more biomass than normal-density crops, and for a given density, crop biomass was greatest in sole-cropped barley, least in sole-cropped pea, and intermediate in the intercrop. For both intercrops and sole crops, increased planting density led to higher crop biomass production. These more productive crops had fewer weeds, lower total weed biomass, and reduced relative importance of the dominant weed species, *Amaranthus retroflexus* or *Brassica kaber*, depending on the site. Other weed species, such as *Ambrosia artemisiifolia*, also produced less biomass, but comprised a larger fraction of the weed community as total weed biomass decreased. Thus, with greater crop competition, the composition of the intercrop and sole crop weed communities shifted from dominance by *Amaranthus retroflexus* or *B. kaber* to more mixed assemblages. This shift implies that the weed communities were structured more by crop resource use than by crop diversity.

Given the general dearth of data concerning weed species abundance in intercrops, it is difficult to assess the prevalence of these impacts of intercropping on weed community organization. However, as noted in Chapter 10, a better understanding of how cropping practices affect weed communities is needed to anticipate shifts in the abundance of resident weed species and prevent invasions by dispersing weed species.

Agroforestry

Why farmers use agroforestry systems

The use of trees by temperate and tropical farmers ranges from forest fallow systems and traditional dooryard gardens to modern fruit orchards, plantation tree crops, and alley cropping. Depending on the species and management strategies employed, agroforestry systems may be labor efficient or labor intensive, highly dependent or largely independent of purchased inputs, and oriented toward maximizing income or minimizing risk. The defining characteristic of all of these systems is that they include woody perennials and herbaceous species that interact over periods ranging from months to years (Young, 1989, p. 11). Agroforestry systems are also characterized by a low frequency of tillage.

Woody perennials perform a variety of functions in agroforestry systems. Trees and shrubs interplanted with annual crops can provide farmers with fuel and construction materials. They can also be used to add organic matter to soil, thereby improving water holding and cation exchange capacities, and increasing microbial activity (Nair, 1984, pp. 31–46; Young, 1989, pp. 93–150). Interplanting deep-rooted tree species, such as *Erythrina* and *Inga* spp., with shallower-rooted crops, such as coffee, fosters nutrient recovery from lower soil horizons, and recycles nutrients through decomposing leaf litter and prunings (Aranguren, Escalante & Herrera, 1982). Leguminous trees can also fix atmospheric N and provide it to associated crops through decomposing residues (Young, 1989, pp. 130–43). The multistory canopies and deep roots of some agroforestry systems minimize soil erosion and reduce soil and leaf temperatures, which is particularly valuable in hot regions (Willey, 1975; Young, 1989, pp. 53–77). Shade cast by overstory trees can benefit the physiological performance of certain shorter tree and bush crops, such as coffee, tea, and cacao (Willey, 1975).

Because they are larger and longer-lived than annual crops, trees and shrubs more strongly affect their environment. Consequently, cropping systems that include trees offer unique opportunities and challenges for weed management.

Temporal patterns of weed infestation in agroforestry systems

As in annual crop systems, weed growth and interference in agroforestry systems depend on the presence of weed propagules, conditions for germination, and resources available for growth, all of which vary over time. During establishment and early growth phases, trees capture only a small fraction of available resources. Weeds present in the first months or even years after tree establishment are a function of the previous use of the site and the seed rain from surrounding vegetation. Annual crops are often intersown among trees during early stages of system development to capture as crop yield some of the resources that would otherwise be captured by weeds. This period may last from one to three or more years. As trees increase in size and ability to capture resources, they become less susceptible to interference from annual and perennial herbaceous weeds.

Once orchards and plantations are established and productive, the potential for weed infestation depends on the proportion of available resources captured by the trees. When trees achieve nearly complete resource use, as in dense plantations, weed problems are often minimal, though vines, parasitic canopy weeds, and undesirable trees can invade. In other cases, practical management factors such as access for annual harvest or wide spacing for fruit quality and pest control preclude maximal resource capture. This fosters weed persistence.

Following a multiyear period of peak production, orchards and plantations enter a period of replanting or renovation. Weed pressure and weed management strategies during renovation depend on resident weed vegetation, the soil seed bank, and inputs to the seed bank. Since many orchard and plantation systems have a productive life of 10 to 30 years or more, weedy vegetation in the last years before renovation represents the accumulated effectiveness of weed management over an extended period.

Other agroforestry systems, such as forest fallow and alley cropping, are cyclic, with regular, periodic shifts in the relative proportions of woody and non-woody components. When annual crops are produced in cyclic systems, trees are temporarily suppressed, which favors weed growth. Later, however, when tree growth occurs during fallow periods, weed growth and reproduction are reduced by competition from trees. This managed fluctuation is essential to maintaining both soil productivity and annual crop production.

From an ecological perspective, weed management in agroforestry systems differs from that in annual cropping systems because it often exploits successional phenomenon, i.e., long-term, directional changes in species composition and environmental conditions driven by species interactions. Additionally, because distances between woody perennials in agroforestry systems generally are farther than those between crop plants in annual systems, agroforestry provides more opportunities for spatially manipulating growth resources to the advantage of crops and detriment of weeds. Trees can also produce large quantities of leaves and branches that can be used as mulch for weed suppression. Integrating these processes to achieve the desired balance between tree growth, annual crop production, weed suppression, and soil improvement is the major management challenge that must be addressed in agroforestry systems.

Forest fallow systems and weed dynamics

Historically, in many tropical and temperate agricultural systems, crop fields were surrounded by forests, and trees re-established in fields after a few years of crop production. After several decades of growth, trees were cut and burned, and annual crops were planted again. In recent years, numerous factors, including human population increase and land conversion to export crops and pastures, have led to shorter fallow periods and a reduction in tree cover relative to cropped area.

Saxena & Ramakrishnan (1984) investigated weed responses to shortened periods of forest fallow in northeastern India, where the interval between successive crops has declined from 20 to 30 years to four to five years. In their study, weed seed density, weed plant density, and weed biomass were more than twice as high in fields cropped after four and six years of fallow in native vegetation than in fields cropped after 10 and 20 years of fallow. As a result, fields in the area generally were hand-hoed twice after long fallow periods, but required three or four hoeings after shorter fallow periods. Saxena & Ramakrishnan (1984) noted that repeated crop cycles with only four to six years of fallow resulted in an "arrested succession" in which herbaceous weedy species with high reproductive potential persisted through fallow periods. In contrast, with longer fallow periods, weedy species were replaced by bamboo, trees, and shade-tolerant herbs, and weed seeds and perennating vegetative parts were largely destroyed by decay, physiological exhaustion, and predation.

On river bottom lands in Amazonian Peru, four- to six-year periods of tree fallow are rotated with a sequence of short and medium cycle crops that are progressively more tolerant of weeds (Staver, 1989a). Indigenous farmers cut and burn trees and shrubs, and then plant maize and cassava. Plantain is added during the first six months of crop production and cassava may be planted a second time after the first crop is harvested. During the second and third years of cropping, which are dominated by plantain, weed biomass increases, insect pests of plantain multiply, and crop nutrient deficiencies become more prevalent (Staver, 1989a). Weeding frequency declines during the same period. Shrubs and trees begin to re-emerge in the field, although the harvest of plantain continues. As trees form a complete overhead canopy, herbaceous weeds begin to disappear. Staver (1991) found that herbaceous weed biomass in this system was eliminated by two to five years of fallow, but that readily germinable weed seeds in the soil continued to decline through 10 years of tree cover (Figure 7.9). Typically, fields are cleared again after fallowing for three to five years, even though viable weed seeds are still present in the soil. As a result, early weed control in the maize–cassava–plantain relay sequence is necessary to minimize crop yield reductions.

In this system, trees and shrubs act both as competitors with crops and as agents of soil restoration and weed control. In the early stages of crop production, trees and shrubs sprout from stumps and roots or germinate from the

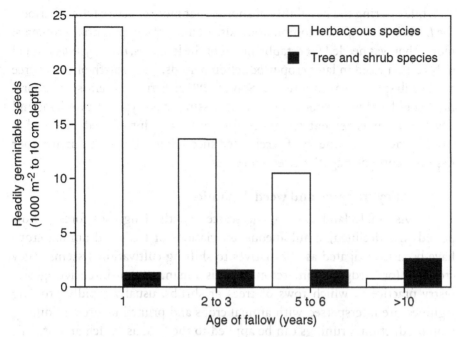

Figure 7.9 Density of readily germinable seeds in soils collected from fallows of different ages in the Palcazu Valley, Peru. (Adapted from Staver, 1991.)

seed bank, and are usually eliminated as weeds. If cropping is intensified either by longer cropping periods, more frequent weeding, or greater crop cover, tree regeneration is delayed (Staver, 1991). A balance exists between crop management practices, especially weeding, and the rate at which tree cover is re-established after cropping is discontinued.

To suppress the growth of herbaceous weeds and accelerate the rate at which woody perennial species come to dominate a field, Staver (1989b) examined the effects of planting a combination of *Inga edulis* and *Desmodium ovalifolium* into a cassava/plantain intercrop. *Inga edulis* is a fast-growing leguminous tree adapted to high light conditions; *Desmodium ovalifolium* is a slow-growing, woody, stoloniferous, shade-tolerant legume used for ground cover. Weed growth was suppressed both by presence of *I. edulis* and increasing *D. ovalifolium* cover (Staver, 1989b). The combination of *I. edulis* plus *D. ovalifolium* increased woody biomass production and plantain yields, but had no effect on cassava yields, compared with plots without the legumes. Three years after the first clearing, fallow vegetation was cut and burned again and maize was planted. Plots following *I. edulis* plus *D. ovalifolium* produced higher maize yields and less weed biomass than those following natural fallow (C. Staver, unpublished data).

While offering the possibility of shorter, but more effective fallow periods, the *I. edulis/D. ovalifolium* combination also demonstrated potential problems: more labor was needed for establishment of the legumes, *D. ovalifolium* tended to become a weed in later crop production periods, and growth of native tree and shrub species was suppressed (Staver, 1989*b*). Farmer–scientist collaborations could be highly productive for addressing these types of problems and developing management strategies for native and introduced plants that would provide the benefits of accelerated succession, while minimizing labor requirements and negative effects on crops.

Alley cropping and weed dynamics

As arable land has become scarce and the length of forest fallow periods has declined, simultaneous associations of trees and annual crops have been investigated as alternatives to shifting cultivation systems. Alley cropping (or "hedgerow intercropping") is a management-intensive agroforestry practice in which rows of trees or shrubs, usually rapidly growing legumes, are interspersed with annual crops and pruned before and during crop production. Prunings can be applied to the field as mulch and organic fertilizer, fed to livestock, or used as firewood. Farmers have been slow to adopt alley cropping, except for soil conservation purposes on sloped lands (Thurston, 1997), but experimental studies provide insights into how management practices and environmental conditions affect relationships among weeds, crops, and trees. Five points that emerge from alley cropping studies merit attention here.

First, during fallow periods preceding annual crop production, hedgerow trees and shrubs can suppress weeds through shading and other forms of competition. At the end of a four-month fallow preceding taro production in Western Samoa, Rosecrance, Rogers & Tofinga (1992) found photosynthetically active radiation (PAR) levels between *Calliandra calothyrsus* and *Gliricidia sepium* hedgerows were only 2% to 10% of values measured in control plots without the trees. In southwestern Nigeria, Anoka, Akobundu & Okonkwo (1991) observed that rhizome biomass of *Imperata cylindirica* declined by 96% three years after planting *G. sepium* hedgerows, and by 90% after planting *Leucaena leucocephala*, whereas it increased in a hedgerow-free control treatment. Death of *I. cylindrica* rhizomes was attributed to the nearly complete canopy cover produced by the trees during the rainy season. Yamoah, Agboola & Mulongoy (1986) compared different legumes for use as hedgerows in alley cropping systems and found that *Cassia siamea* intercepted more PAR and suppressed weed biomass production more effectively than *Flemingia congesta* or *G. sepium*. Greater PAR interception and weed suppression by *C. siamea* were attributed to its planophile leaves and extensive side branches.

Second, narrower spacing between hedgerows promotes greater weed suppression. Rosecrance, Rogers & Tofinga (1992) found weed coverage between rows of *Gliricidia sepium* and *Calliandra calothyrsus* during the last four months of taro production averaged 60% in 6-m alleys, but only 35% in 5-m alleys, and 15% in 4-m alleys. Jama, Getahun & Ngugi (1991) reported that shading by *Leucaena leucocephala* during fallow periods between cropping seasons reduced weed biomass by 74% in 8-m alleys, but by 84% in 4-m alleys, and 93% in 2-m alleys, compared to a hedgerow-free control treatment. Improvements in weed control and possible soil benefits from higher densities of trees must be weighed against the loss of crop production area, however.

Third, tree prunings applied to the soil surface can suppress weeds. Effects are species-specific and rate dependent. In Ivory Coast, Budelman (1988) found that at an application rate of 5 Mg ha^{-1}, leaf mulch of *Leucaena leucocephala* had no effect on weed biomass, whereas *Gliricidia sepium* mulch had some suppressive ability and *Flemingia macrophylla* mulch was strongly weed-suppressive. Leaf mulch decomposition was most rapid for *G. sepium*, slowest for *F. macrophylla*, and intermediate for *L. leucocephala*. The lack of weed suppression from *L. leucocephala*, in spite of its moderate decomposition rate, was attributed to its small leaflets, which separated from the rachis as soon as drying began after pruning. *Flemingia macrophylla* had a larger leaflet size and an intact, although curled, leaf after drying. Weed suppression by *F. macrophylla* increased as the amount of mulch material increased. Eleven weeks after mulching with *F. macrophylla*, weed biomass was reduced about 70% by a 3 Mg ha^{-1} application, about 85% by a 6 Mg ha^{-1} application, and virtually eliminated by a 9 Mg ha^{-1} application, compared to an unmulched control treatment. Budelman (1988) suggested that mulches were most effective against annual weed species, but also hindered the emergence of *Cyperus rotundus* shoots. They would not be effective against root and stump suckers, which can easily pass through a mulch layer.

Fourth, alley cropping can change weed community composition. Siaw, Kang & Okali (1991) observed that grass species were common in control plots, whereas grasses were uncommon in plots with *Leucaena leucocephala* and *Acioa barteri* hedgerows. They also found that *L. leucocephala* seedlings were increasingly common in hedgerow plots. Rippin *et al*. (1994) found that grass weeds were greatly reduced in *Erythrina poeppigiana* hedgerow and mulch plots and that the grass weed *Rottboellia exaltata* was absent from hedgerow plots.

Fifth, alley cropping must be evaluated in terms of its impacts not only on weed biomass and community composition, but also on crop production. Alley cropping is clearly better suited for some crops, soils, and climates than others, though more research is still needed to identify optimum combinations. Siaw, Kang & Okali (1991) reported that *Leucaena leucocephala* and *Acioa barteri* hedgerows increased cowpea yield by 20% to 57% and maize yield by 11% to 132% compared to plots without hedgerows. Jama, Getahun & Ngugi (1991) also obtained higher maize yields from mixtures with *L. leucocephala* than from maize sole crops. Budelman (1990b) reported yam yields were 135% higher from yam/*Gliricidia sepium* mixtures than from yam sole crops. Salazar, Szott & Palm (1993) found, however, that hedgerows of *Inga edulis*, *L. leucocephala*, and *Erythrina* spp. spaced 4 m apart reduced yield of rice growing within 1.5 m of the tree bases. Lal (1989) found that maize and cowpea yields were lower in associations with *L. leucocephala* and *G. sepium* at 2-m and 4-m spacing than in a no-tillage treatment without hedgerows, especially in seasons with below average rainfall. He attributed the negative effect of the hedgerows on maize and cowpea to competition for soil moisture.

Using herbaceous vegetation to manage weeds in tree crops

In systems where trees function as main crops, herbaceous species can be planted or allowed to volunteer from the soil seed bank to produce an understory layer that reduces soil erosion and compaction, adds organic matter, improves soil fertility, limits damage by insect and mite pests, and suppresses weeds (Altieri & Schmidt, 1985; Hogue & Neilsen, 1987; Bugg *et al.*, 1991; Bugg & Waddington, 1994; Prokopy, 1994).

In an early review of cover cropping practices for closed-canopy tropical plantation crops, such as coconut, rubber, and oil palm, Sampson (1928) noted that various legume species were useful for suppressing *Imperata cylindrica* and other weeds, while young trees were being established. Research focused on the use of two legume cover crops, *Pueraria phaseoloides* and *Desmodium ovalifolium*, in Honduran oil palm plantations indicated that they greatly reduced labor requirements for weed suppression (Centro Internacional de Información Sobre Cultivos de Cobertura, 1994). Obiefuna (1989) found that intercropping plantain with egusi melon, a fast-spreading vine, suppressed weed biomass production, reduced weeding requirements for young plantains, and increased plantain yield as much as 27% above the level obtained from a weed-free treatment without egusi melon. In addition to weed suppression, Obiefuna (1989) noted that intercropped egusi melon enhanced soil microbial activity and nutrient supply, moderated soil temperature, and increased soil moisture content, all of which could improve plantain performance.

For many tree crops, the canopy never closes, even when each individual plant is at mature size and in full production. A mosaic of light conditions exists on the ground, with shade patches beneath trees interspersed with light gaps between tree rows and sometimes within rows (Vandermeer, 1989, pp. 106–26). Capture of nutrients and water by trees is greatest beneath their shoots, where their roots are concentrated (Atkinson, 1980). Thus, resource use in open-canopy tree systems is spatially variable and incomplete: the tree crops fail to fully use light, water, and nutrients that are available in inter-row areas.

Patterns of resource use in open-canopy tree crops have led to the differentiation of close-to-tree and inter-row zones for weed and cover crop management (Glenn & Welker, 1989). Weeds or cover crops growing close to tree trunks can compete with trees for resources and reduce tree performance. In an experiment conducted in West Virginia, Welker & Glenn (1989) grew young peach trees within a tall fescue (*Festuca arundinacea*) sod and found that leaf nitrogen, canopy width, trunk diameter, and fruit yield of the trees increased as the size of vegetation-free areas surrounding the trees increased up to 9 m^2. To prevent this type of competitive stress, farmers can suppress vegetation close to trees by mowing, surface cultivating, grazing, mulching, or applying herbicides (Glenn & Welker, 1989). In contrast, weeds and cover crops growing farther away from trees pose little or no competitive threat to tree growth and production, and can be used to improve and conserve soil, and aid pest management. In California, Hendricks (1995) found that almond orchards with cover crops grown between tree rows had more soil organic matter, higher earthworm and predatory insect densities, reduced insect pest damage, and similar yields compared to conventionally managed almond orchards with bare soil. The major additional cost associated with cover cropping was for more irrigation water.

Perennial cover crops have been tested for weed control in established coffee plantations. In a region of Nicaragua with a five-month dry season, a three-year-old stand of sun-grown coffee was used to compare the use of two perennial legume cover crops (*Arachis pintoi* and *Desmodium ovalifolium*) against local grower weed management practices that combined mowing and herbicides (paraquat and simazine with spot applications of glyphosate) (Bradshaw, 1993; Bradshaw & Lanini, 1995). Both cover crops and weeds were removed manually from coffee rows in all treatments during the course of the experiment, and cover crops were hand-weeded for three months after they were sown. Weed biomass in cover cropped plots was lower than in plots managed with mowing and herbicides at three of four sampling dates; no difference was detected among treatments at a fourth sampling date, after growers had mowed twice and applied herbicides twice (Bradshaw & Lanini, 1995). No differences were detected in coffee growth and yield among the legume cover crop treatments, a weedy control, and the mowing plus herbicide treatment, though cover crops increased moisture stress in coffee leaves during the last month of the dry season (Bradshaw, 1993).

Selective weeding is a possible alternative to planting cover crops between rows of trees. In a five-year study initiated in newly planted coffee, Aguilar *et al.* (1997) compared conventional total weed suppression (using machete slashing and herbicide tank mixes) with selective weeding (using herbicides plus slashing or just slashing). The objective of the selective treatments was to suppress potentially competitive weeds in inter-row areas and promote the growth of weed species, such as *Oplismenus burmanii*, *Commelina difusa*, and *Drymaria cordata*, characterized by a low, creeping growth habit, and shallow root systems. These weeds were considered to be benign ground covers.

Results of the experiment indicated that selective weeding was effective in changing weed community composition, minimizing weed competition against the coffee crop, and promoting plant cover for soil conservation (Aguilar *et al.*, 1997). Before initiation of the experiment, ground cover weeds constituted 11% of the herbaceous vegetation cover, but by the fifth year they were 34% of the vegetation cover in the selective herbicide-plus-machete treatment, and 64% of the selective machete-only treatment. During the dry season, soil in selectively weeded plots was protected by a mulch of senesced ground-cover weeds, whereas soil in the conventional plots was largely bare. Selective weed management resulted in reduced coffee yield in the first year of production, but for each of the following two years and for the three-year total, yields did not differ among treatments. As compared with the conventional treatment, the selective herbicide-plus-machete and machete-only treatments increased labor costs 28% and 135%, but reduced herbicide use by 25% and 100%.

Using additional tree strata to manage weeds in tree crops

Agroforestry systems composed of mixtures of tree species are common in the humid tropics and often are characterized by multilayered canopies, very complete use of light, and few weeds (Christanty *et al.*, 1986). Such systems mimic the natural forests surrounding them (Ewel, 1986), and weeds that are able to grow in tree mixtures are most frequently vines and undesirable tree species rather than herbaceous annuals. Ewel (1986) and others have suggested that mixed-species tree farming is a sustainable agricultural system for high rainfall, low soil nutrient environments where farmers lack cash or credit to buy synthetic agrichemicals and other production inputs. Coffee-based systems illustrate how tree crops can be grown with other trees, and how weeds may be affected.

Overstory tree species are planted with coffee primarily to regulate the sunlight available for photosynthesis by coffee and to stabilize coffee yields over several-year periods (Carvajal, 1984; Kimemia & Njoroge, 1988). With little or no shade (80% to 100% of full sunlight), coffee must be fertilized more heavily to reach a greater yield potential. Muschler (1997) proposed that only in certain optimum growing conditions is the complete absence of shade advantageous for coffee production. Otherwise, some level of shade, depending on growing conditions, fertilizer inputs, and coffee varieties, should be included.

The use of shade trees in coffee orchards maintains or improves soil conditions by reducing soil temperatures and erosion, adding organic matter, recycling subsoil nutrients, and in the case of legumes, adding nitrogen. Shade trees also create less favorable conditions for weed growth, thereby reducing the need for hand-weeding or herbicides. Nestel & Altieri (1992) investigated weed biomass production and species composition in Mexican coffee orchards that differed in shade level and species mix. Measurements were made in plots of coffee grown in pure stand (unshaded), in mixture with trees of the genus *Inga* (monogeneric shade), and in mixture with trees in the genera *Inga*, *Citrus*, and *Musa* (polygeneric shade). Coffee trees were 8–12 years old and shade trees were 8–15 years old. Plots were cleared of weeds by hand at the start of both the wet and dry seasons, but no other weed control measures were used.

Weed biomass increased at a much lower rate in the two shade treatments than in the unshaded treatment. In the wet season, the unshaded treatment accumulated >2000 kg ha^{-1} of weed biomass, whereas weed biomass in the shaded treatments reached only 800–1000 kg ha^{-1} (Figure 7.10). In the dry season, weed biomass in the unshaded treatment rose to >1500 kg ha^{-1}, but reached only 200 kg ha^{-1} in monogeneric shade and 400 kg ha^{-1} in polygeneric shade (Figure 7.10). In shaded treatments, weeds in the Commelinaceae (*Tripogandra serrulata* and *Commelina erecta*) were most abundant, whereas without shade, weeds in the Asteraceae (*Galinsoga quadriradiata*, *Smallanthus maculatus*, *Bidens alba*, and *Melampodium microcephalum*) were most abundant. Nestel & Altieri (1992) noted that local growers considered weeds in the Asteraceae to be more difficult to control and more damaging to coffee than weeds in the Commelinaceae. The investigators proposed that differences among shade treatments in weed biomass and community composition could be related to solar radiation levels and allelopathic interactions. Water extracts from *Inga* and *Musa* leaves and roots have been reported to inhibit germination and root elongation of weeds in the Asteraceae and Gramineae, but to have less effect on those in the Commelinaceae (Anaya *et al.*, 1982).

The natural growth of most overstory tree species often results in greater than optimum shade levels in numerous patches within a multistory orchard. The regulation of shade levels through pruning as frequently as twice annually can improve the light environment for coffee production and generate additional leaf litter for localized weed control and ground cover. As was

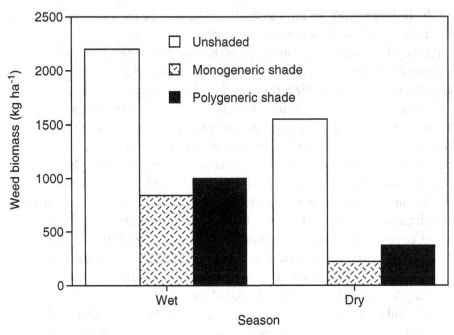

Figure 7.10 Weed biomass at the termination of wet season and dry season experiments in which coffee was grown without shade trees (unshaded), in the shade of *Inga* spp. (monogeneric shade), or in the shade of *Inga, Citrus,* and *Musa* spp. (polygeneric shade). The experiments were conducted in Veracruz, Mexico. (Adapted from Nestel & Altieri, 1992.)

noted for tree and shrub species used in alley cropping, mulches from shade tree species associated with coffee vary greatly in their resistance to decomposition and ability to suppress weeds. Rivas, Staver & Blanco (1993) found, for example, that leaf mulches of *Inga paterna, Simarouba glauca,* and *Clusia rosea* decomposed four times more slowly in shaded coffee than did a *Gliricidia sepium* mulch. Mulches of *I. paterna, S. glauca,* and *C. rosea* suppressed weed establishment up to two months after application.

Given the spatial variability that characterizes weeds, shade levels, and litter deposition, Staver, Bradshaw & Somarriba (1993) proposed that coffee growers should visualize a mosaic of different patches in their orchards. A well-managed orchard floor would have patches of naturally occurring ground-cover weed species, patches of leaf litter and pruned leaf mulch, and areas under coffee plants with minimal living ground cover. Partial shade, fluctuating throughout the year with pruning regimes, would moderate coffee physiological rates as well as limit weed growth. This approach could be applied to other tree crops adapted to production in similar light conditions, such as cacao.

Obstacles and opportunities in the use of crop diversification for weed management

This chapter has presented many examples of how crop rotation, intercropping, and agroforestry practices improve crop production and weed management. Despite the advantages offered by crop diversification, the trend in many industrialized nations and a growing number of developing nations is toward greater specialization and less crop diversity. Specialization in only one or two crops on a farm can lead to greater economies of scale, but attendant costs should not be ignored. Compared with more diversified systems, weed management in simplified cropping systems generally requires either more manual labor, more frequent use of mechanical control tactics, or greater quantities of herbicides.

How can crop diversification be better used to improve crop production and weed management? The first challenge is philosophical: agricultural practitioners and scientists must recognize the importance of ecology as well as technology in managing weeds. Successful management of weeds in multi-component cropping systems is a sophisticated activity that requires comprehensive knowledge of competitive and facilitative interactions, and factors that mediate those interactions. Applied community ecology must have as much importance and intellectual status as the discovery, synthesis, and use of herbicides.

A second challenge is scientific: crop rotation, intercropping, and agroforestry systems can promote crop performance and suppress weed germination, establishment, growth, competitive ability, and reproduction. However, information on the relevant mechanisms is inadequate. How can crop rotation be used to manipulate soil conditions so as to stimulate crop root growth while increasing weed seed mortality? What types of diversity within crop sequences best suppress particular species of weeds? What are the best ways to simultaneously increase the yield potential and weed suppression ability of intercrops? What chemical and physical characteristics of tree leaves and canopies are particularly desirable for suppressing weeds in agroforestry systems? What species and cultivars are best suited for use in rotation, intercropping, and agroforestry systems? Modeling will help address these questions (see, for example, Vandermeer, 1989; Jordan *et al.*, 1995), but models must be developed and validated with field data.

A third challenge is technical: realization of the full potential of intercropping and agroforestry systems will require development of a new generation of highly adaptable agricultural machinery. Existing farm machinery can be used to sow, maintain, and harvest many intercrops adapted to temperate regions, including soybean/wheat (Reinbott *et al.*, 1987), fava bean/wheat (Bulson, Snaydon & Stopes, 1997), alfalfa/oat (Hesterman *et al.*, 1992), and red clover/maize (Scott *et al.*, 1987). A lack of appropriate machinery for managing other intercrops may represent insufficient attention from agricultural engineers, not an inherent incompatibility of intercropping with mechanization (Cordero & McCollum, 1979; Vandermeer, 1989, pp. 199–201). Intercropping is most common on small, tropical farms with little access to improved farming technologies, but it is not restricted to such situations. For example, better animal-drawn weed control implements are being designed for use in intercropping systems (Anderson, 1981). Can appropriate technologies be developed for alley cropping systems to prevent competition from weed-suppressing trees against herbaceous crops, perhaps by root pruning? More attention from engineers will allow the potential of diversified cropping systems to be expressed more fully throughout the world.

A fourth challenge is social: systems for propagating farmer knowledge about diversified cropping systems are needed. In both temperate and tropical regions, farmers managing complex cropping systems often have site-specific knowledge about their impacts on weeds (Exner, Thompson & Thompson, 1996; Thurston, 1997). To better disperse farmer-generated knowledge, policy makers, educational institutions, and funding agencies need to support networks linking farmers, extension educators, and researchers.

A fifth challenge is political: obtaining the benefits of cropping system diversity on a large scale requires that national and regional policies promote crop diversity. If greater reliance on nonchemical weed management is deemed environmentally desirable, then government policies must be consistent with this goal. Government programs to stabilize farm income must promote longer crop rotations. Private and public lenders must be encouraged to recognize the value of crop diversification in stabilizing yields and in reducing challenges from weeds and other pests. In many cases, local and regional markets need to be developed for diversified products. The political work necessary to accomplish these and other related tasks is enormous, but so are the potential benefits to farmers and the environment.

REFERENCES

Abraham, C. T., & Singh, S. P. (1984). Weed management in sorghum–legume intercropping systems. *Journal of Agricultural Science (Cambridge)*, **103**, 103–15.

Agboola, A. A., & Fayemi, A. A. (1972). Fixation and excretion of nitrogen by tropical legumes. *Agronomy Journal*, **64**, 409–12.

Aguilar, A., Aguilar, V., Somarriba, S., & Staver, C. (1997). Selective weed management for soil conservation in young coffee: evaluation of herbicide-mechanical and mechanical systems with and without *Arachis pintoi*. In *Proceedings of 18th Latin American Coffee Symposium*, pp. 115–24. San Jose, Costa Rica: Interamerican Institute for Agricultural Sciences – Central American Program for Coffee Improvement (IICA – PROMECAFE). (in Spanish).

Alcorn, J. B. (1984). Development policy, forests, and peasant farms: reflections on Huastec-managed forests' contributions to commercial production and resource conservation. *Economic Botany*, **38**, 389–406.

Ali, M. (1988). Weed suppressing ability and productivity of short duration legumes intercropped with pigeonpea under rainfed conditions. *Tropical Pest Management*, **34**, 384–7.

Altieri, M. A. (1994). *Biodiversity and Pest Management in Agroecosystems*. Binghamton, NY: Haworth Press.

Altieri, M. A., & Schmidt, L. L. (1985). Cover crop manipulation in northern California orchards and vineyards: effects on arthropod communities. *Biological Agriculture and Horticulture*, **3**, 1–24.

Anaya, L. A., Ruy-Ocotla, G., Ortiz, L. M., & Ramos, L. (1982). Potencial alelopático de las principales plantas de un cafetal. In *Estudios Ecológicos en el Agroecosistema Cafetalero*, ed. E. Jimenez-Avila and A. Gomez-Pompa, pp. 85–94. Veracruz, Mexico: Instituto Nacional de Investigaciones sobre Recursos Bióticos.

Anderson, D. T. (1981). Seeding and interculture mechanization requirements related to intercropping in India. In *Proceedings of the International Workshop on Intercropping*, 10–13 January 1979, Hyderabad, India, ed. R. W. Willey, pp. 328–36. Patencheru, India: International Crops Research Institute for the Semi-Arid Tropics (ICRISAT).

Anoka, U. A., Akobundu, I. O., & Okonkwo, S. N. C. (1991). Effects of *Gliricidia sepium* (Jacq.) Steud and *Leucaena leucocephala* (Lam.) de Wit on growth and development of *Imperata cylindrica* (L.) Raeuschel. *Agroforestry Systems*, **16**, 1–12.

Aranguren, J., Escalante, G., & Herrera, R. (1982). Nitrogen cycle of tropical perennial crops under shade trees. 1. Coffee. *Plant and Soil*, **67**, 247–58.

Ashby, J. A., & Pachico, D. (1987). Agricultural ecology of the mid-hills of Nepal. In *Comparative Farming Systems*, ed. B. L. Turner II & S. B. Brush, pp. 195–222. New York, NY: Guilford Press.

Atkinson, D. (1980). The distribution and effectiveness of the roots of tree crops. *Horticultural Reviews*, **2**, 424–90.

Balasubramanian, V., & Sekayange, L. (1990). Area harvests equivalency ratio for measuring efficiency in multiseason cropping. *Agronomy Journal*, **82**, 519–22.

Bantilan, R. T., Palada, M., & Harwood, R. R. (1974). Integrated weed management. 1. Key factors affecting weed–crop balance. *Philippine Weed Science Bulletin*, **1**, 14–36.

Blackshaw, R. E. (1994). Rotation affects downy brome (*Bromus tectorum*) in winter wheat (*Triticum aestivum*). *Weed Technology*, **8**, 728–32.

Bradshaw, L. (1993). Perennial cover crops in Nicaraguan coffee orchards: mechanisms and manipulations of plant competition. PhD. thesis, University of California, Davis.

Bradshaw, L., & Lanini, W. T. (1995). Use of perennial cover crops to suppress weeds in Nicaraguan coffee orchards. *International Journal of Pest Management*, **41**, 185–94.

Brummer, E. C. (1998). Diversity, stability, and sustainable American agriculture. *Agronomy Journal*, **90**, 1–2.

Budelman, A. (1988). The performance of the leaf mulches of *Leucaena leucocephala*, *Flemingia macrophylla* and *Gliricidia sepium* in weed control. *Agroforestry Systems*, **6**, 137–45.

Budelman, A. (1990a). Woody legumes as live support systems in yam cultivation. 1. The tree–crop interface. *Agroforestry Systems*, **10**, 47–59.

Budelman, A. (1990b). Woody legumes as live support systems in yam cultivation. 2. The yam–*Gliricidia sepium* association. *Agroforestry Systems*, **10**, 61–9.

Bugg, R. L., & Waddington, C. (1994). Using cover crops to manage arthropod pests of orchards: a review. *Agriculture, Ecosystems and Environment*, **50**, 11–28.

Bugg, R. L., Sarrantonio, M., Dutcher, J. D., & Phatak, S. C. (1991). Understory cover crops in pecan orchards: possible management systems. *American Journal of Alternative Agriculture*, **6**, 50–62.

Bullock, D. G. (1992). Crop rotation. *Critical Reviews in Plant Science*, **11**, 309–26.

Bulson, H. A. J., Snaydon, R. W., & Stopes, C. E. (1997). Effects of plant density on intercropped wheat and field beans in an organic farming system. *Journal of Agricultural Science (Cambridge)*, **128**, 59–71.

Carvajal, J. (1984). *Cafeto: Cultivo y Fertilización*. Berne, Switzerland: International Potassium Institute.

Centro Internacional de Informacion Sobre Cultivos de Cobertura (CIDICCO) (1994). *La Utilización de Leguminosas de Cobertura en Plantaciones Perennes*, Noticias Sobre Cultivos de Cobertura no. 7. Tegucigalpa, Honduras: CIDICCO.

Chambers, R. J. H. (1990). Farmer-first: a practical paradigm for the third agriculture. In *Agroecology and Small Farm Development*, ed. M. A. Altieri & S. B. Hecht, pp. 237–44. Boca Raton, FL: CRC Press.

Chancellor, R. J. (1985). Changes in the weed flora of an arable field cultivated for 20 years. *Journal of Applied Ecology*, **22**, 491–501.

Christanty, L., Abdoellah, O. S., Marten, G. G., & Iskander, J. (1986). Traditional agroforestry in west Java: the *pekarangan* (homegarden) and *kebun-talun* (annual–perennial rotation) cropping systems. In *Traditional Agriculture in Southeast Asia*, ed. G. G. Marten, pp. 132–58. Boulder, CO: Westview Press.

Clay, S. A., & Aguilar, I. (1998). Weed seedbanks and corn growth following continuous corn or alfalfa. *Agronomy Journal*, **90**, 813–18.

Cordero, A., & McCollum, R. E. (1979). Yield potential of interplanted food crops in the southeastern US. *Agronomy Journal*, **71**, 834–42.

Covarelli, G., & Tei, F. (1988). Effet de la rotation culturale sur la flore adventice du maïs. In *8ième Colloque International sur la Biologie, l'Ecologie et la Systematique des Mauvaises Herbes*, vol. 2, pp. 477–84. Paris, France: Comité Français de Lutte contre les Mauvaises Herbes, and Leverkusen, Germany: European Weed Research Society.

Creamer, N. G., Plassman, B., Bennett, M. A., Wood, R. K., Stinner, B. R., & Cardina, J. (1995). A method of mechanically killing cover crops to optimize weed suppression. *American Journal of Alternative Agriculture*, **10**, 157–62.

De Haan, R. L., Sheaffer, C. C., & Barnes, D. K. (1997). Effect of annual medic smother plants on weed control and yield in corn. *Agronomy Journal*, **89**, 813–21.

De Haan, R. L., Wyse, D. L., Ehlke, N. J., Maxwell, B. D. & Putnam, D. H. (1994). Simulation of spring-seeded smother plants for weed control in corn (*Zea mays*). *Weed Science*, **42**, 35–43.

De Haan, R. L., Wyse, D. L., Ehlke, N. J., Maxwell, B. D., & Putnam, D. H. (1997). Spring-seeded smother plants for weed control in corn and other annual crops. In *Ecological Interactions and Biological Control*, ed. D. A. Andow, D. W. Ragsdale & R. F. Nyvall, pp. 178–94. Boulder, CO: Westview Press.

Derksen, D. A. (1997). Weeds. In *Zero Tillage: Advancing the Art*, ed. D. Domitruk & B. Crabtree, pp. 24–8. Brandon, MB: Manitoba–North Dakota Zero Tillage Farmers Association.

Dotzenko, A. D., Ozkan, M., & Storer, K. R. (1969). Influence of crop sequence, nitrogen fertilizer and herbicides on weed seed populations in sugar beet fields. *Agronomy Journal*, **61**, 34–7.

Dowling, P. M., & Wong, P. T. W. (1993). Influence of preseason weed management and in-crop treatments on two successive wheat crops. 1. Weed seedling numbers and wheat grain yield. *Australian Journal of Experimental Agriculture*, **33**, 167–72.

Dyke, G. V., & Barnard, A. J. (1976). Suppression of couchgrass by Italian ryegrass and broad red clover undersown in barley and field beans. *Journal of Agricultural Science (Cambridge)*, **87**, 123–6.

Eaglesham, A. R. J., Ayanaba, A., Ranga Rao, V., & Eskew, D. L. (1981). Improving the nitrogen nutrition of maize by intercropping with cowpea. *Soil Biology and Biochemistry*, **13**, 169–71.

Eberlein, C. V., Sheaffer, C. C., & Oliveira, V. F. (1992). Corn growth and yield in an alfalfa living mulch system. *Journal of Production Agriculture*, **5**, 332–9.

Echtenkamp, G. W., & Moomaw, R. S. (1989). No-till corn production in a living mulch system. *Weed Technology*, **3**, 261–6.

Economic Research Service (ERS) (1992). *Agricultural Resources: Inputs Situation and Outlook Report*, Publication AR-28. Washington, DC: ERS, US Department of Agriculture.

Einhellig, F. A., & Rasmussen, J. A. (1989). Prior cropping with sorghum inhibits weeds. *Journal of Chemical Ecology*, **15**, 951–60.

Enache, A. J., & Ilnicki, R. D. (1990). Weed control by subterranean clover (*Trifolium subterraneum*) used as a living mulch. *Weed Technology*, **4**, 534–8.

Entz, M. H., Bullied, W. J., & Katepa-Mupondwa, F. (1995). Rotational benefits of forage crops in Canadian prairie cropping systems. *Journal of Production Agriculture*, **8**, 521–9.

Ewel, J. J. (1986). Designing agricultural ecosystems for the humid tropics. *Annual Review of Ecology and Systematics*, **17**, 245–71.

Exner, D. N., Thompson, R. L., & Thompson, S. N. (1996). Practical experience and on-farm research with weed management in an Iowa ridge tillage-based system. *Journal of Production Agriculture*, **9**, 496–500.

Faeth, P. (ed.) (1993). *Agricultural Policy and Sustainability: Case Studies from India, Chile, the Philippines, and the United States*. Washington, DC: World Resources Institute.

Faeth, P., Repetto, R., Kroll, K., Dai, Q., & Helmers, G. (1991). *Paying the Farm Bill: US Agricultural Policy and the Transition to Sustainable Agriculture*. Washington, DC: World Resources Institute.

Fischer, A., & Burrill, L. (1993). Managing interference in a sweet corn–white clover living mulch system. *American Journal of Alternative Agriculture*, **8**, 51–6.

Fleck, N. G., Machado, C. M. N., & De Souza, R. S. (1984). Eficiência da consorciação de culturas no controle de plants daninhas. *Pesquisa Agropecuaria Brasileira (Brasília)*, **19**, 591–8.

Fox, R. H., & Pielielek, W. P. (1988). Fertilizer N equivalence of alfalfa, birdsfoot trefoil, and red clover for succeeding corn crops. *Journal of Production Agriculture*, **1**, 313–17.

Francis, C. A., Flor, C. A., & Temple, S. R. (1976). Adapting varieties for intercropped systems in the tropics. In *Multiple Cropping*, ed. R. I. Papendick, P. A. Sanchez & G. B. Triplett, pp. 235–53. Madison, WI: American Society of Agronomy – Crop Science Society of America – Soil Science Society of America.

Froud-Williams, R. J. (1988). Changes in weed flora with different tillage and agronomic management systems. In *Weed Management in Agroecosystems: Ecological Approaches*, ed. M. A. Altieri & M. Liebman, pp. 213–36. Boca Raton, FL: CRC Press.

Frye, W. W., Blevins, R. L., Smith, M. S., Corak, S. J., & Varco, J. J. (1988). Role of annual legume cover crops in efficient use of water and nitrogen. In *Cropping Strategies for Efficient Use of Water and Nitrogen*, ed. W. L. Hargrove, pp. 129–54. Madison, WI: American Society of Agronomy – Crop Science Society of America – Soil Science Society of America.

Gantzer, C. J., Anderson, S. H., Thompson, A. L., & Brown, J. R. (1991). Evaluation of soil loss after 100 years of soil and crop management. *Agronomy Journal*, **83**, 74–7.

Glenn, D. M., & Welker, W. V. (1989). Cultural practices for enhanced growth of young peach trees. *American Journal of Alternative Agriculture*, **4**, 8–11.

Grubinger, V. P., & Minotti, P. L. (1990). Managing white clover living mulch for sweet corn production with partial rototilling. *American Journal of Alternative Agriculture*, **5**, 4–12.

Håkansson, S. (1982). Multiplication, growth, and persistence of perennial weeds. In *Biology and Ecology of Weeds*, ed. W. Holzner & M. Numata, pp. 123–35. The Hague, Netherlands: Dr. W. Junk.

Hart, R. D. (1980). A natural ecosystem analog approach to the design of a successional crop system for tropical forest environments. *Biotropica*, **12** (Supplement), 73–82.

Heard, A. J. (1963). Weed populations on arable land after four-course rotations and after short leys. *Annals of Applied Biology*, **52**, 177–84.

Hendricks, L. (1995). Almond growers reduce pesticide use in Merced County field trials. *California Agriculture*, **49**, 5–10.

Hesterman, O. B., Griffin, T. S., Williams, P. T., Harris, G. H., & Christenson, D. R. (1992). Forage legume–small grain intercrops: nitrogen production and response of subsequent corn. *Journal of Production Agriculture*, **5**, 340–8.

Hodgson, J. M. (1958). Canada thistle (*Cirsium arvense* Scop.) control with cultivation, cropping, and chemical sprays. *Weeds*, **6**, 1–10.

Hogue, E., & Neilsen, G. (1987). Orchard floor vegetation management. *Horticultural Reviews*, **9**, 377–430.

Ilnicki, R. D., & Enache, A. J. (1992). Subterranean clover living mulch: an alternative method of weed control. *Agriculture, Ecosystems and Environment*, **40**, 249–64.

Jackson, L. E., Wyland, L. J., & Stivers, L. J. (1993). Winter cover crops to minimize nitrate losses in intensive lettuce production. *Journal of Agricultural Science (Cambridge)*, **121**, 55–62.

Jama, B., Getahun, A., & Ngugi, D. N. (1991). Shading effects of alley cropped *Leucaena leucocephala* on weed biomass and maize yield at Mtwapa, Coast Province, Kenya. *Agroforestry Systems*, **13**, 1–11.

Jodha, N. S. (1981). Intercropping in traditional farming systems. In *Proceedings of the*

International Workshop on Intercropping, 10–13 January 1979, Hyderabad, India, ed. R. W. Willey, pp. 282–91. Patencheru, India: International Crops Research Institute for the Semi-Arids Tropics (ICRISAT).

Jordan, N., Mortensen, D. A., Prenzlow, D. M., & Curtis-Cox, K. (1995). Simulation analysis of crop rotation effects on weed seedbanks. *American Journal of Botany*, **82**, 390–8.

Kaffka, S. R. (1985). Thirty years of energy use, nutrient cycling, and yield on a self-reliant dairy farm. In *Sustainable Agriculture and Integrated Farming Systems*, ed. T. C. Edens, C. Fridgen & S. L. Battenfield, pp. 143–58. East Lansing, MI: Michigan State University Press.

Karlen, D. L., Varvel, G. E., Bullock, D. G., & Cruse, R. M. (1994). Crop rotations for the 21st century. *Advances in Agronomy*, **53**, 1–45.

Kelner, D., Juras, L., & Derksen, D. (1996). *Integrated Weed Management: Making It Work on Your Farm*. Extension bulletin, Saskatchewan Agriculture and Food, Manitoba Agriculture, and Agriculture and Agri-Food Canada. Internet: http://gov.mb.ca/agriculture/crops/weeds/weedfacts/fba06s00.html.

Kendall, W. A., & Stringer, W. C. (1985). Physiological aspects of clover. In *Clover Science and Technology*, ed. N. L. Taylor, pp. 111–59. Madison, WI: American Society of Agronomy.

Kimemia, J., & Njoroge, J. (1988). Effect of shade on coffee: a review. *Kenya Coffee*, **53**, 387–91.

Kropff, M. J., & van Laar, H. H. (eds) (1993). *Modelling Crop–Weed Interactions*. Wallingford, UK: CAB International.

Kurtz, T., Melsted, S. W., & Bray, R. H. (1952). The importance of nitrogen and water in reducing competition between intercrops and corn. *Agronomy Journal*, **44**, 13–17.

Lal, R. (1989). Agroforestry systems and soil surface management of a tropical Alfisol. 1. Soil moisture and crop yields. *Agroforestry Systems*, **8**, 7–29.

Lampkin, N. (1990). *Organic Farming*. Ipswich, UK: Farming Press.

Lanini, W. T., Orloff, S. B., Vargas, R. N., Orr, J. P., Marble, V. L., & Grattan, S. R. (1991). Oat companion crop seeding rate effect on alfalfa establishment, yield and weed control. *Agronomy Journal*, **83**, 330–3.

Lashomb, J. H., & Ng, Y. S. (1984). Colonization by Colorado potato beetle, *Leptinotarsa decemlineata* (Say) (Coleoptera: Chrysomelidae), in rotated and nonrotated fields. *Environmental Entomology*, **13**, 1352–6.

Leihner, D. (1983). *Management and Evaluation of Intercropping Systems with Cassava*. Cali, Colombia: Centro Internacional de Agricultura Tropical (CIAT).

Liebman, M. (1989). Effects of nitrogen fertilizer, irrigation, and crop genotype on canopy relations and yields of an intercrop/weed mixture. *Field Crops Research*, **22**, 83–100.

Liebman, M. (1995). Polyculture cropping systems. In *Agroecology: The Science of Sustainable Agriculture*, 2nd edn, ed. M. A. Altieri, pp. 205–18. Boulder, CO: Westview Press.

Liebman, M., & Dyck, E. (1993). Crop rotation and intercropping strategies for weed management. *Ecological Applications*, **3**, 92–122.

Liebman, M., & Ohno, T. (1998). Crop rotation and legume residue effects on weed emergence and growth: applications for weed management. In *Integrated Soil and Weed Management*, ed. J. L. Hatfield, D. D. Buhler & B. A. Stewart, pp. 181–221. Chelsea, MI: Ann Arbor Press.

Liebman, M., & Robichaux, R. H. (1990). Competition by barley and pea against mustard:

effects on resource acquisition, photosynthesis and yield. *Agriculture, Ecosystems and Environment*, **31**, 155–72.

Lueschen, W. E., Andersen, R. N., Hoverstad, T. R., & Kanne, B. K. (1993). Seventeen years of cropping systems and tillage affect velvetleaf (*Abutilon theophrasti*) seed longevity. *Weed Science*, **41**, 82–6.

Macey, A. (ed.) (1992). *Organic Field Crop Handbook*. Ottawa, Ontario: Canadian Organic Growers.

MacRae, R. J., & Mehuys, G. R. (1985). The effect of green manuring on the physical properties of temperate-area soils. *Advances in Soil Science*, **3**, 71–93.

Matheson, N., Rusmore, B., Sims, J. R., Spengler, M., & Michalson, E. L. (1991). *Cereal–Legume Cropping Systems: Nine Farm Case Studies in the Dryland Northern Plains, Canadian Prairies and Intermountain Northwest*. Helena, MT: Alternative Energy Resources Organization.

McEwen, J., Darby, R. J., Hewitt, M. V., & Yeoman, D. P. (1989). Effects of field beans, fallow, lupins, oats, oilseed rape, peas, ryegrass, sunflowers and wheat on nitrogen residues in the soil and on the growth of a subsequent wheat crop. *Journal of Agricultural Science (Cambridge)*, **115**, 209–19.

McLenaghen, R. D., Cameron, K. C., Lampkin, N. H., Daly, M. L., & Deo, B. (1996). Nitrate leaching from plowed pasture and the effectiveness of winter catch crops in reducing leaching losses. *New Zealand Journal of Agricultural Research*, **39**, 413–20.

Mohler, C. L. (1991). Effects of tillage and mulch on weed biomass and sweet corn yield. *Weed Technology*, **5**, 545–52.

Mohler, C. L., & Liebman, M. (1987). Weed productivity and composition in sole crops and intercrops of barley and field pea. *Journal of Applied Ecology*, **24**, 685–99.

Moody, K., & Drost, D. C. (1983). The role of cropping systems on weeds in rice. In *Proceedings of the Conference on Weed Control in Rice*, 31 August–4 September 1981, pp. 73–86. Los Baños, Philippines: International Rice Research Institute.

Muschler, R. (1997). Shade or sun for sustainable coffee: a new angle on an old discussion. *Proceedings of the 18th Symposium of the Latin American Coffee Symposium*, pp. 471–6. San José, Costa Rica: Interamerican Institute for Agricultural Sciences – Central American Program for Coffee Improvement (IICA-PROMECAFE). (in Spanish).

Nair, P. K. R. (1984). *Soil Productivity Aspects of Agroforestry*. Nairobi, Kenya: International Council for Research on Agroforestry (ICRAF).

Natarajan, M., & Willey, R. W. (1980a). Sorghum–pigeonpea intercropping and the effects of plant population density. 1. Growth and yield. *Journal of Agricultural Science (Cambridge)*, **95**, 51–8.

Natarajan, M., & Willey, R. W. (1980b). Sorghum–pigeonpea intercropping and the effects of plant population density. 2. Resource use. *Journal of Agricultural Science (Cambridge)*, **95**, 59–65.

National Research Council (NRC) (1989). *Alternative Agriculture*. Washington, DC: National Academy Press.

Nestel, D., & Altieri, M. A. (1992). The weed community of Mexican coffee agroecosystems: effect of management upon plant biomass and species composition. *Acta Oecologica*, **13**, 715–26.

Norris, R. F., & Ayres, D. (1991). Cutting interval and irrigation timing in alfalfa: yellow foxtail invasion and economic analysis. *Agronomy Journal*, **83**, 552–8.

Obiefuna, J. C. (1989). Biological weed control in plantains (*Musa* AAB) with egusi melon (*Colocynthis citrullus* L.). *Biological Agriculture and Horticulture*, 6, 221–7.

Ofori, F., & Stern, W. R. (1987). Cereal–legume intercropping systems. *Advances in Agronomy*, 41, 41–90.

Ogg, A. G. Jr., & Dawson, J. H. (1984). Time of emergence of eight weed species. *Weed Science*, 32, 327–35.

Okigbo, B. N., & Greenland, D. J. (1976). Intercropping systems in tropical Africa. In *Multiple Cropping*, ed. R. I. Papendick, P. A. Sanchez & G. B. Triplett, pp. 63–101. Madison, WI: American Society of Agronomy – Crop Science Society of America – Soil Science Society of America.

Olasantan, F. O., Lucas, E. O., & Ezumah, H. C. (1994). Effects of intercropping and fertilizer application on weed control and performance of cassava and maize. *Field Crops Research*, 39, 63–9.

Ominski, P. D., Entz, M. H., & Kenkel, N. (1999). Weed suppression by *Medicago sativa* in subsequent cereal crops: a comparative survey. *Weed Science*, 47, 282–90.

Paine, L. K., & Harrison, H. (1993). The historical roots of living mulch and related practices. *HortTechnology*, 3, 137–43.

Pielou, E. C. (1977). *Mathematical Ecology*. New York: John Wiley.

Plucknett, D. L., & Smith, N. J. H. (1986). Historical perspectives on multiple cropping. In *Multiple Cropping Systems*, ed. C. A. Francis, pp. 20–39. New York: Macmillan.

Power, J. F., & Follett, R. F. (1987). Monoculture. *Scientific American*, 256(3), 78–86.

Prasad, K., Gautam, R. C., & Mohta, N. K. (1985). Studies on weed control in arhar and soybean as influenced by planting patterns, intercropping and weed control methods. *Indian Journal of Agronomy*, 30, 434–9.

Pretty, J. N. (1995). *Regenerating Agriculture: Policies and Practices for Sustainability and Self-Reliance*. Washington, DC: Joseph Henry Press.

Prokopy, R. (1994). Integration of orchard pest and habitat management: a review. *Agriculture, Ecosystems and Environment*, 50, 1–10.

Radke, J. K., & Hagstrom, R. T. (1976). Strip intercropping for wind protection. In *Multiple Cropping*, ed. R. I. Papendick, P. A. Sanchez & G. B. Triplett, pp. 201–22. Madison, WI: American Society of Agronomy – Crop Science Society of America – Soil Science Society of America.

Rao, M. R., & Willey, R. W. (1980). Evaluation of yield stability in intercropping: studies on sorghum/pigeonpea. *Experimental Agriculture*, 16, 105–16.

Reganold, J. P., Papendick, R. I., & Parr, J. F. (1990). Sustainable agriculture. *Scientific American*, 262(6), 112–20.

Reinbott, T. M., Helsel, Z. R., Helsel, D. G., Gebhardt, M. R., & Minor, H. C. (1987). Intercropping soybean into standing green wheat. *Agronomy Journal*, 79, 886–91.

Rippin, M., Haggar, J. P., Kass, D., & Kopke, U. (1994). Alley cropping and mulching with *Erythrina poeppigiana* (Walp.) O. F. Cook and *Gliricidia sepium* (Jacq.) Walp.: effects on maize/weed competition. *Agroforestry Systems*, 25, 119–34.

Risser, P. G. (1969). Competitive relationships among herbaceous grassland plants. *Botanical Review*, 35, 251–84.

Rivas, C., Staver, C., & Blanco, M. (1993). Leaf mulch from four shade tree species for weed control in shaded coffee. In *Proceedings of the 16th Latin American Coffee Symposium*, p. 83. Managua, Nicaragua: Interamerican Institute for Agricultural

Sciences – Central American Program for Coffee Improvement (IICA – PROMECAFE).

Roberts, H. A., & Feast, P. M. (1970). Seasonal distribution of emergence in some annual weeds. *Experimental Horticulture*, **21**, 36–41.

Roberts, H. A., & Feast, P. M. (1973). Emergence and longevity of seeds of annual weeds in cultivated and undisturbed soil. *Journal of Applied Ecology*, **10**, 133–43.

Roder, W., Mason, S. C., Clegg, M. D., & Kniep, K. R. (1989). Yield–soil water relationships in sorghum–soybean cropping systems with different fertilizer regimes. *Agronomy Journal*, **81**, 470–5.

Rosecrance, R. C., Rogers, S., & Tofinga, M. (1992). Effects of alley cropped *Calliandra calothyrsus* and *Gliricidia sepium* hedges on weed growth, soil properties, and taro yields in Western Samoa. *Agroforestry Systems*, **19**, 57–66.

Salazar, A., Szott, L. T., & Palm, C. A. (1993). Crop–tree interactions in alley cropping systems on alluvial soils of the Upper Amazon basin. *Agroforestry Systems*, **22**, 67–82.

Sampson, H. C. (1928). Cover crops in tropical plantations. *Tropical Agriculturalist (Ceylon)*, **71**, 153–70.

Samson, R. A., Foulds, C., & Patriquin, D. (1990). *Choice and Management of Cover Crop Species and Varieties for Use in Row Crop Dominant Rotations*. Ste. Anne de Bellevue, Quebec: Resource Efficient Agricultural Production (REAP) – Canada.

Saxena, K., & Ramakrishnan, P. (1984). Herbaceous vegetation development and weed potential in slash and burn agriculture (jhum) in N. E. India. *Weed Research*, **24**, 135–42.

Schreiber, M. M. (1992). Influence of tillage, crop rotation, and weed management on giant foxtail (*Setaria faberi*) population dynamics and corn yield. *Weed Science*, **40**, 645–53.

Scott, T. W., Mt. Pleasant, J., Burt, R. F., & Otis, D. J. (1987). Contributions of ground cover, dry matter, and nitrogen from intercrops and cover crops in a corn polyculture system. *Agronomy Journal*, **79**, 792–8.

Sengupta, K., Bhattacharyya, K. K., & Chatterjee, B. N. (1985). Intercropping upland rice with blackgram (*Vigna mungo* L.). *Journal of Agricultural Science (Cambridge)*, **104**, 217–21.

Sharaiha, R., & Gliessman, S. (1992). The effects of crop combination and row arrangement in the intercropping of lettuce, favabean and pea on weed biomass and diversity and on crop yields. *Biological Agriculture and Horticulture*, **9**, 1–13.

Sheaffer, C. C., Wyse, D. L., Marten, G. C., & Westra, P. H. (1990). The potential of quackgrass for forage production. *Journal of Production Agriculture*, **3**, 256–9.

Shetty, S. V. R., & Rao, A. N. (1981). Weed management studies in sorghum/pigeonpea and pearl millet/groundnut intercrop systems: some observations. In *Proceedings of the International Workshop on Intercropping*, 10–13 January 1979, Hyderabad, India, ed. R. W. Willey, pp. 238–48. Patencheru, India: International Crops Research Institute for the Semi-Arid Tropics (ICRISAT).

Siaw, D. E. K. A., Kang, B. T., & Okali, D. U. U. (1991). Alley cropping with *Leucaena leucocephala* (Lam.) De Wit and *Acioa barteri* (Hook. F.) Engl. *Agroforestry Systems*, **14**, 219–31.

Snaydon, R. W., & Harris, P. M. (1981). Interactions belowground: the use of water and nutrients. In *Proceedings of the International Workshop on Intercropping*, 10–13 January 1979, Hyderabad, India, ed. R. W. Willey, pp. 188–201. Patencheru, India: International Crops Research Institute for the Semi-Arids Tropics (ICRISAT).

Staver, C. (1989a). Why farmers rotate fields in maize–cassava–plantain bush fallow agriculture in the wet Peruvian Amazon. *Human Ecology*, **17**, 401–26.

Staver, C. (1989b). Shortened bush fallow rotations with relay-cropped *Inga edulis* and *Desmodium ovalifolium* in wet central Amazonian Peru. *Agroforestry Systems*, **8**, 173–96.

Staver, C. (1991). The role of weeds in the productivity of Amazonian bush fallow agriculture. *Experimental Agriculture*, **27**, 287–304.

Staver, C., Bradshaw, L., & Somarriba, S. (1993). Ground cover management in Central American shaded coffee: selective weeding, mulch, and perennial living covers. *Agronomy Abstracts* (American Society of Agronomy), **85**, 61.

Steiner, K. G. (1984). *Intercropping in Tropical Smallholder Agriculture with Special Reference to West Africa*. Eschborn, Germany: Deutsche Gesellschaft fur Technische Zusammenarbeit (GTZ).

Sturz, A. V., & Bernier, C. C. (1987). Survival of cereal root pathogens in the stubble and soil of cereal versus noncereal crops. *Canadian Journal of Plant Pathology*, **9**, 205–13.

Sumner, D. R. (1982). Crop rotation and plant productivity. In *CRC Handbook of Agricultural Productivity*, vol. 1, ed. M. Rechcigl, pp. 273–313. Boca Raton, FL: CRC Press.

Thurston, H. D. (1997). *Slash/Mulch Systems: Sustainable Methods for Tropical Agriculture*. Boulder, CO: Westview Press.

Trenbath, B. R. (1983). The dynamic properties of mixed crops. In *Frontiers of Research in Agriculture*, ed. S. K. Roy, pp. 265–86. Calcutta, India: Indian Statistical Institute.

Trenbath, B. R. (1993). Intercropping for the management of pests and diseases. *Field Crops Research*, **34**, 381–405.

Tripathi, B., & Singh, C. M. (1983). Weed and fertility management using maize/soyabean intercropping in the northwestern Himalayas. *Tropical Pest Management*, **29**, 267–70.

United States Department of Agriculture (USDA) (1973). *Monoculture in Agriculture: Extent, Causes, and Problems. Report of the Task Force on Spatial Heterogeneity in Agricultural Landscapes and Enterprises*. Washington, DC: US Department of Agriculture.

United States Department of Agriculture (USDA) (1999). *1997 Census of Agriculture*. Washington, DC: US Department of Agriculture, National Agricultural Statistics Service.

Vandermeer, J. H. (1989). *The Ecology of Intercropping*. Cambridge, UK: Cambridge University Press.

Warnes, D. D., & Andersen, R. N. (1984). Decline of wild mustard (*Brassica kaber*) seeds in soil under various cultural and chemical practices. *Weed Science*, **32**, 214–17.

Weil, R., & McFadden, M. E. (1991). Fertility and weed stress effects on performance of maize/soybean intercrop. *Agronomy Journal*, **83**, 717–21.

Welker, W. V., & Glenn, D. M. (1989). Sod proximity influences the growth and yield of young peach trees. *Journal of the American Horticultural Society*, **114**, 856–9.

Willey, R. W. (1975). The use of shade in coffee, cocoa, and tea. *Horticultural Abstracts*, **45**, 791–8.

Willey, R. W. (1979a). Intercropping: its importance and research needs. 1. Competition and yield advantages. *Field Crop Abstracts*, **32**, 1–10.

Willey, R. W. (1979b). Intercropping: its importance and research needs. 2. Agronomy and research approaches. *Field Crop Abstracts*, **32**, 73–85.

Willey, R. W. (1990). Resource use in intercropping systems. *Agricultural Water Management*, **17**, 215–31.

Williams, E. D. (1972). Growth of *Agropyron repens* seedlings in cereals and field beans. *Proceedings of the 11th British Weed Control Conference*, pp. 32–7. London: British Crop Protection Council.

Wilson, B. J., & Phipps, P. A. (1985). A long term experiment on tillage, rotation, and herbicide use for the control of *Avena fatua* in cereals. In *Proceedings of the 1985 British Crop Protection Conference – Weeds*, pp. 693–700. London: British Crop Protection Council.

Wittwer, S. (1987). Vegetable abundance: from yardlong cowpeas to bitter melons. In *Feeding a Billion: Frontiers of Chinese Agriculture*, ed. S. Wittwer, Y. Yu, H. Sun & L. Wang, pp. 253–70. East Lansing, MI: Michigan State University Press.

Yamoah, C. F., Agboola, A. A., & Mulongoy, K. (1986). Decomposition, nitrogen release and weed control by prunings of selected alley cropping shrubs. *Agroforestry Systems*, **4**, 239–46.

Young, A. (1989). *Agroforestry for Soil Conservation*. Wallingford, UK: CAB International.

MATT LIEBMAN

8

Managing weeds with insects and pathogens

Introduction

Weeds and other plant species are susceptible to attack by a diversity of invertebrate herbivores and pathogens. Virtually every plant organ provides a niche for some type of insect, mite, nematode, fungus, bacterium, or virus (Harper, 1977, p. 484). Protection of crop plants from these organisms is a major issue in crop production. Conversely, the promotion of herbivory and disease to suppress weed recruitment, growth, and reproduction is a major objective of biological control programs.

Biological control of weeds requires that sufficiently high densities of herbivores and pathogens are present when weeds are at susceptible developmental stages. For this to happen, herbivores and pathogens used as biological control agents must be well adapted to abiotic components of the environment, such as temperature and precipitation regimes (Crawley, 1986; Cullen, 1995). To control weeds effectively, they must also largely escape the effects of predation, parasitism, disease, competition, and chemical interference (Newman, Thompson & Richman, 1998).

Three approaches are used in efforts to regulate weed populations with herbivores and pathogens (Andow, Ragsdale & Nyvall, 1997). *Conservation* methods involve modifying the environment to retain or increase populations of resident control agents and intensify the damage they inflict on weeds. *Inoculation* methods involve introducing relatively small numbers of biological control agents that will suppress a target weed species as their populations establish, increase, and disperse. In many cases, organisms used as inoculative control agents are not native to the regions into which they are released and are collected from the original territory of their introduced host weed. *Inundation* methods involve introducing either native or exotic control agents in large numbers to suppress the target weed quickly. Organisms used as inundative control agents are not expected to persist for long periods or disperse long distances.

Usually, inoculative and inundative control methods are directed toward specific weed species and employ herbivores and pathogens with narrow host ranges (Boyetchko, 1997; McFayden, 1998). Both methods are generally best suited to weed species that dominate the agroecosystems they infest, and that can not be suppressed successfully or cost-effectively by other means. Weeds in rangeland, pastures, orchards, and arable cropland may be the targets of both inoculative and inundative biocontrol methods, although in practice inoculative release of highly dispersible agents is favored in extensive habitats (e.g., rangeland), whereas inundative release tends to be favored in intensively managed, short-duration crops (e.g., soybean). Of the three weed biocontrol approaches, conservation methods remain the least well understood and least frequently used. Much of the information relevant to conservation methods for weed biocontrol comes from studies conducted in arable cropland, but in theory other types of agroecosystems could be manipulated to enhance the effects of resident weed-attacking organisms.

The history of biological control indicates that three types of success are possible: complete, occasional, and partial (Andow, Ragsdale & Nyvall, 1997). In a few cases, a target weed species is constantly suppressed to very low population densities over a wide area and *complete control* is achieved. Complete control is demonstrated by a combination of several types of evidence: weed population densities at several locations decrease following introduction of the control agent; weed densities remain at a low level following the control agent's establishment; and weed survival or reproduction increases when weeds are artificially protected from the control agent (Smith & DeBach, 1942; McEvoy, Cox & Coombs, 1991; Andow, Ragsdale & Nyvall, 1997). More commonly, suppression of a target weed species is successful in some years but not others, a situation described as *occasional control*. Weed suppression by biocontrol agents also may not be particularly strong, a situation described as *partial control*. Occasional and partial control can be viewed as failures if the standard against which they are assessed is highly effective herbicides. Alternatively, they might be viewed as components of multitactic strategies that bring diverse and temporally variable sets of stresses to bear against weed communities (Liebman & Gallandt, 1997).

In this chapter we examine conservation, inoculation, and inundation approaches for managing weeds with insect herbivores and phytopathogens. We believe these approaches are best guided by two principles:

1. *The use of any one organism as a weed biocontrol agent should be integrated with the use of other weed-suppressive herbivores, pathogens, competitors, and management practices.*
2. *The possible impacts of introduced weed biocontrol agents on target and non-target organisms should be rigorously evaluated and openly discussed before the agents are released.*

The first principle is directed at intensifying the impact of biocontrol agents on target weed species, while the second is directed at using those agents prudently. Throughout this chapter, our intent is to better understand the ecological relationships underlying weed biocontrol, and to see where insect herbivores and phytopathogens might provide farmers with practical benefits in the near future.

Conservation of resident herbivores and pathogens

As noted in Chapters 2 and 5, weed density and growth can be reduced by insects and microorganisms already present in agricultural habitats and not intentionally introduced for weed suppression. In the absence of other weed management tactics, populations of these organisms generally are not capable of suppressing weeds to the point where crop yield loss is prevented. They may contribute, however, to the regulation of weed populations at several life stages. By better understanding the ecology of resident herbivore and pathogen species that attack weeds, it may be possible to identify management strategies that enhance their impact.

Weed seeds on the soil surface are vulnerable to predation by resident insects, rodents, and other organisms. Carabid beetles are highly active on the soil surface and may be particularly important as weed seed predators (Lund & Turpin, 1977). Cardina *et al.* (1996) put 50 seeds of *Abutilon theophrasti* in soil-filled petri dishes, placed them on the surface of Ohio corn fields, and observed rates of seed removal of 1% to 57% per day. They calculated that at the average removal rate of 11.2% per day, 80% of the seeds would be removed in four weeks. Pitfall trapping, field experiments with protective cages made with different mesh sizes, and laboratory feeding trials indicated that carabid beetles, mice, and slugs were responsible for weed seed removal and destruction. In North Carolina soybean fields, Brust & House (1988) observed carabid beetles, ants, crickets, and mice feeding on weed seeds. Over a five-week period, seeds of four broadleaf annual weeds, placed on cards at densities of 25 to 50 seeds per card, were removed at rates of 2.2% to 4.2% per day. In Maine potato and grain fields, carabid beetles were found to reduce or eliminate weed seedling emergence by consuming seed tissues directly, and by burying seeds in "caches" at depths from which seedlings were unable to reach the soil surface (Zhang, 1993; Hartke, Drummond & Liebman 1998). When seeds of *Echinochloa crus-galli* and *Brassica kaber* were placed for one-week periods in trays (73 seeds per tray) covered with wire mesh that excluded small mammals and birds but not insects, 40% to 95% of the seeds were attacked or removed (Zhang, 1993). Andersson (1998) reported that seeds of *Bilderdykia (Polygonum) convolvulus*, *Chenopodium album*, *Matricaria perforata*, *Polygonum lapathifolium*, and *Thlaspi arvense* were removed at rates of 20% to 90% by unidentified organisms in Swedish pastures and oat fields.

Natural populations of soil-borne fungi can reduce the germinability of buried weed seeds. In a Nebraska field experiment with weedy genotypes of *Sorghum bicolor*, Fellows & Roeth (1992) compared the germination of seeds that were treated or not treated with fungicides (carboxin and thiram) before burial for four to five months during late autumn and winter. In the first year of the experiment, 17% of the treated seeds versus 0.5% of the untreated seeds germinated after recovery from the soil. In a second year, which was drier, 40% of the treated seeds versus 17% of the untreated seeds germinated. The effects of soil-borne organisms on seed viability during warmer months of the year may be at least as great, but experimental studies are needed to quantify them (Kremer, 1993).

Resident soil-borne fungi may also inflict substantial damage on populations of emerged weeds. Lindquist *et al.* (1995*a*) measured survival of *Abutilon theophrasti* seedlings that emerged in a Minnesota soybean field at different times during two growing seasons and found that maximal survival rates after emergence were only 12% to 21%. They attributed much of the observed mortality to attack of the seedlings by an unidentified *Verticillium* species. Similarly, in an experiment in which *A. theophrasti* was grown at a fixed density and allowed to produce seed, Hartzler (1996) noted a 20-fold difference in seedling recruitment between two Iowa locations that differed in the degree of *Verticillium* infestation. Results of a multiyear simulation model that included temporally variable *Verticillium* infection indicated that attack of *A. theophrasti* by the fungus increased net financial return from a maize–soybean rotation up to 19% and decreased the number of years herbicide was necessary to control the weed up to 35% (Lindquist *et al.*, 1995*b*).

Resident insects can defoliate weeds and reduce their biomass production substantially, although such damage may be insufficient to limit crop yield loss. In California sugar beet fields, Norris (1997) observed that a sawfly, *Schizocerella pilicornis*, and a weevil, *Hypurus bertrandi*, often reduced leaf area and biomass of the weed *Portulaca oleracea* by 70% to 80%. In competition experiments, however, exposure of *P. oleracea* to *S. pilicornis* and *H. bertrandi* failed to limit the weed's negative impact on sugar beet yield in three of four years, compared to treatments in which *P. oleracea* was protected from insect attack by the systemic insecticide aldicarb. Insect attack of *P. oleracea* limited the weed's competitive effect on sugar beet in only one of four years. In that year, weed populations protected from herbivores by aldicarb reduced sugar beet yield by up to 40%, whereas yield loss due to competition from unprotected weeds did not exceed 15%. Norris (1997) concluded that *S. pilicornis* and *H. bertrandi* could greatly damage *P. oleracea* without providing sufficient biological control to reduce the need for other weed management techniques.

Why do high levels of damage to weeds by indigenous herbivores and pathogens fail to prevent yield loss? One explanation is that damage to the weed may occur too late in the crop's development to prevent yield reduction (Boyetchko, 1997). For this reason, it may be better to view indigenous herbivores and phytopathogens as agents for reducing weed growth and reproduction, rather than as therapeutic tools for crop protection.

How can background levels of herbivory and disease be managed to consistently subject weeds to stress? The first step is to identify systems in which resident natural enemies sometimes suppress weeds effectively. Next, experiments are needed to determine whether inoculation of fields with a regionally native natural enemy can lead to a new local population. The extent to which various taxa of weed natural enemies are dispersal-limited rather than habitat-limited is largely unknown at present. Ultimately, methods for increasing weed suppression by established natural enemy populations are needed. Existing data suggest five approaches may be particularly worthwhile in achieving the latter objective.

First, the use of pesticides that have detrimental effects on resident weed biocontrol agents should be minimized or eliminated. If not, broad-spectrum insecticides used to protect crops from insect attack may inadvertently decrease endemic biological weed control and increase requirements for weed management inputs (Norris, 1997). The impact of reducing pesticide use on weed seed predators is suggested by farming systems comparisons conducted in Switzerland. Pfiffner & Niggli (1996) reported that carabid beetle densities were twice as high in organic and biodynamic systems than in a conventional system, and attributed this effect to lower pesticide use, as well as greater ground cover and greater use of compost and organic soil amendments.

Second, to maximize the percentage of weed seeds that are destroyed by seed predators, post-harvest tillage should be delayed as long as possible (Cardina *et al.*, 1996). This would leave weed seeds on the soil surface where they are most vulnerable to attack by surface-searching insects, rodents, and birds.

Third, because higher levels of biological activity occur when crop residue is maintained near the soil surface, the use of minimum tillage techniques should be considered. Pitty, Staniforth & Tiffany (1987) studied fungal attack of *Setaria viridis* and *S. faberi* seeds under field conditions in Iowa and found that more seeds were killed by pathogens when maize and soybean plots were disturbed with shallow sweep cultivation rather than a sequence of moldboard plowing, disking, harrowing, and sweep cultivation. Derksen, Blackshaw & Boyetchko (1996) suggested that in semiarid farming regions of Canada, minimum tillage practices may enhance weed suppression by microorganisms favored by the cooler, moister conditions created by residues retained on the soil surface. Brust & House (1988) found carabid beetle density and weed seed removal were two to three times higher in a no-tillage soybean production system that maintained wheat straw on the soil surface compared with a conventionally tilled system without residue cover (Figure 8.1). The effects of minimum tillage practices on the soil environment, weeds, and their natural enemies are examined in more detail in Chapter 5.

Fourth, the impacts of native pathogens attacking weed seeds and other organs may be increased by using complementary weed management tactics, including inoculative or inundative releases of other biocontrol agents. Kremer & Spencer (1989) found that infection of *Abutilon theophrasti* seeds by fungi in the genera *Alternaria*, *Fusarium*, *Cladosporium*, and bacteria in the genera *Pseudomonas*, *Erwinia*, *Flavobacterium* increased greatly when the seed-piercing bug *Niesthrea lousianica* was released at experiment sites. Viability of *Abutilon theophrasti* seeds declined linearly as fungal infection increased. An average of 92% of *A. theophrasti* seeds were viable at sites where the insect was not released, whereas only 16% were viable where releases were made. More research needs to be directed at this type of synergistic interaction between indigenous and introduced biocontrol agents.

Finally, farm landscapes might be altered to provided better habitat for weed biocontrol agents. Carabid densities, for example, can be greatly increased by planting narrow strips of perennial grasses at 200-m intervals within arable fields (Wratten & van Emden, 1995). Research is needed to determine how grass strips and other forms of non-crop vegetation affect weed seed predation and weed dynamics.

Inoculative releases of control agents

Many weed species have been introduced into new regions by human activities (see Chapter 10). During the immigration process, introduced weed species may leave behind herbivores and pathogens that suppressed them to

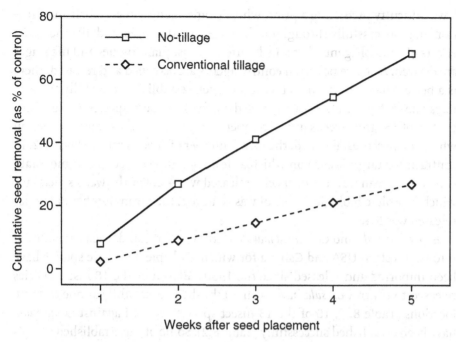

Figure 8.1 Cumulative consumption of weed seeds by seed feeders over a five-week period in no-tillage and conventional tillage soybean field plots in North Carolina. Results are expressed as percentages of control treatments that were protected from seed predators. (Adapted from Brust & House, 1988.)

low levels of abundance in their native range. Thus, one strategy for controlling a weed species that lacks natural enemies is to search its original homeland or climatically similar areas for organisms that feed upon and parasitize it, and release those organisms where their weed host was introduced without them. This strategy is often called "classical biological control."

A critical step in the process of screening exotic species for use as inoculative biocontrol agents is feeding and infection studies to ensure that non-target plant species will not be attacked. After determining that the candidate control agents attack only the targeted weed species, experimental colonies are then established to determine the agents' requirements and their potential impacts on the weed. If the agents can be established in new habitats and are effective in suppressing the target weed under field conditions, colonies will then be reared and distributed throughout the weed's range.

Use of inoculative control agents against exotic weed species

Typically, inoculative control efforts focused on a single weed species involve the release of multiple species of control agents, because (i) researchers have difficulty predicting a priori which single agent will establish and function most successfully throughout the target weed's range, and (ii) synergistic effects of control agents that attack different host plant tissues and life stages are desired. As more potential control agents are screened and released, there is a better chance of having at least one agent establish successfully. Andow, Ragsdale & Nyvall (1997) reviewed data for 74 insect species released for control of 35 target weeds in cool temperate regions. For those target weeds on which no insects established, the mean number of species released was 1.2. In contrast, for target weeds on which at least one insect species became established, the mean number of species released was 4.2. For the weed species for which complete biological control was achieved, the mean number of species released was 5.4.

Euphorbia esula and *Centaurea maculosa* are two rangeland weeds introduced into the western USA and Canada for which multiple herbivore species have been imported and released since the 1960s. Eleven of the 18 insect species released to control *E. esula* have been established successfully at one or more locations (Table 8.1); 10 of the 13 insect species released against *C. maculosa* have been established successfully (Table 8.2). Some of the established species inflict substantial damage to the targeted weeds under field conditions. Insects that are particularly effective for suppression of *E. esula* include flea beetles (*Aphthona cyparissiae*, *A. czwalinai*, *A. flava*, *A. lacertosa*, and *A. nigriscutis*) that mine the plant's roots as larvae and eat its leaves as adults; another beetle (*Oberea erythrocephala*) that bores into stems and root crowns; midges (*Spurgia capitigena* and *S. esulae*) that form galls in shoot tips and prevent flowering; and moths (*Hyles euphorbiae* and *Lobesia euphorbiana*) whose larvae defoliate the plant and kill stems (Julien & Griffiths, 1998; Swiadon, Drlik & Woo, 1998). Successfully established species that are particularly effective at suppressing *C. maculosa* include a root-mining moth (*Agapeta zoegana*); a beetle (*Cyphocleonus achates*) that kills small plants and shoots arising from stools of larger plants; and a moth (*Metzneria paucipunctella*) and two gallflies (*Urophora affinis* and *U. quadrifasciata*) that attack seed heads (Grossman, 1989; Julien & Griffiths, 1998).

Although insect herbivores have substantially reduced *Euphorbia esula* and *Centaurea maculosa* densities at some locations, they have failed to completely control the weeds throughout their full geographical ranges (Julien & Griffiths, 1998). However, control may become more widespread in the future as introduced control agents disperse, increase in number, and deplete weed seed banks in the soil (Harris, 1997). Control is also likely to improve as more is learned about the ecological characteristics and requirements of the herbivores.

Table 8.1. *Insects introduced into the United States and Canada for biological control of* Euphorbia esula *and related hybrids*

Insect species introduced	Country of origin	States and provinces with successful establishment
Aphthona abdominalis (Coleoptera: Chrysomelidae)	Italy	none reported
Aphthona cyparissiae (Coleoptera: Chrysomelidae)	Austria, Hungary, Switzerland	AB, BC, CO, IA, ID, MB, MN, MT, ND, NE, NM, NV, ON, OR, SD, SK, WA, WI, WY
Aphthona czwalinai (Coleoptera: Chrysomelidae)	Austria, Germany, Hungary, Russia	CO, IA, ID, MB, MN, MT, ND, NE, OR, SD, WA, WI, WY
Aphthona flava (Coleoptera: Chrysomelidae)	Hungary, Inner Mongolia, Italy	AB, BC, CO, IA, ID, MN, MT, ND, NS, ON, OR, SD, UT, WA, WI, WY
Aphthona lacertosa (Coleoptera: Chrysomelidae)	Austria, Hungary, Yugoslavia	AB, ID, MB, MT, ND, OR, SK, WA, WY
Aphthona nigriscutis (Coleoptera: Chrysomelidae)	Hungary	AB, BC, CO, IA, ID, MB, MN, MT, ND, NE, ON, OR, SD, SK, WA, WI, WY
Chamaesphecia astatiformis (Lepidoptera: Sesiidae)	Yugoslavia	none reported
Chamaesphecia crassicornis (Lepidoptera: Sesiidae)	Hungary	none reported
Chamaesphecia hungarica (Lepidoptera: Sesiidae)	Hungary, Yugoslavia	none reported
Chamaesphecia tenthrediniformis (Lepidoptera: Sesiidae)	Austria, Greece	none reported
Hyles euphorbiae (Lepidoptera: Sphingidae)	France, Germany, Hungary, Switzerland	ID, MT, ND, NE, ON, WY
Lobesia euphorbiana (Lepidoptera: Tortricidae)	Italy	BC, MB, ON, SK
Minoa murinata (Lepidoptera: Geometridae)	Germany	none reported
Oberea erythrocephala (Coleoptera: Cerambycidae)	Hungary, Italy, Switzerland	AB, MT, ND, OR
Pegomya curticornis (Diptera: Anthomyiidae)	Hungary	MB, SK
Pegomya euphorbiae (Diptera: Anthomyiidae)	Hungary	none reported
Spurgia capitigena (Diptera: Cecidomyiidae)	Italy	AB, MT, ND, SK
Spurgia esulae (Diptera: Cecidomyiidae)	Italy	ID, MB, MT, ND, NS, ON, SK

Source: Adapted from Julien & Griffiths (1998).

Table 8.2. *Insects introduced into the United States and Canada for biological control of* Centaurea maculosa

Insect species introduced	Country of origin	States and provinces with successful establishment
Agapeta zoegana (Lepidoptera: Tortricidae)	Austria, Romania, Hungary	BC, CO, ID, MN, MT, OR, SD, UT, WA, WI, WY
Bangasternus fausti (Coleoptera: Curculionidae)	Greece	none reported
Chaetorellia acrolophi (Diptera: Tephritidae)	Austria, Switzerland	MT, OR, WA, WY
Cyphocleonus achates (Coleoptera: Curculionidae)	Austria	BC, CO, MT, OR, UT, WA, WY
Larinus minutus (Coleoptera: Curculionidae)	Greece, Romania	BC, CA, CO, ID, MN, MT, NE, OR, SD, UT, WA, WY
Larinus obtusus (Coleoptera: Curculionidae)	Romania	BC, MT, WA
Metzneria paucipunctella (Lepidoptera: Gelechiidae)	Switzerland	BC, CO, ID, MT, OR, WA
Pelochrista medullana (Lepidoptera: Tortricidae)	Austria, Hungary, Romania	none reported
Pterolonche inspersa (Lepidoptera: Pterolonchidae)	Hungary	none reported
Sphenoptera jugoslavica (Coleoptera: Buprestidae)	Greece	BC, OR
Terellia virens (Diptera: Tephritidae)	Austria, Switzerland	MT, OR, WA
Urophora affinis (Diptera: Tephritidae)	Austria, France, Switzerland	AZ, BC, CA, CO, ID, MI, MN, MT, NV, NY, OR, QB, SD, UT, VA, WA, WY
Urophora quadrifasciata (Diptera: Tephritidae)	Ukraine	AB, BC, QB

Source: Adapted from Julien & Griffiths (1998).

When the match between a weed, its imported natural enemies, and the environment is correct, the results can be quite spectacular. In Australia, where various species of the New World cactus *Opuntia* escaped from cultivation and heavily infested rangeland, importation and release of insects for biological control were initiated in the 1920s (Grossman, 1989; Julien & Griffiths, 1998). Among the agents that proved effective were a moth (*Cactoblastis cactorum*) from Argentina, whose larvae mine and destroy pads of the cactus; scale insects (*Dactylopius austrinus, D. ceylonicus, D. opuntiae,* and *D. tomentosus*) from Argentina, Brazil, India, Sri Lanka, and the USA, which attack cactus pads and fruits; a bug (*Chelinidea tabulata*) from the USA, which attacks new shoots and fruits; and a beetle (*Archlagocheirus funeatus*) from Mexico, whose larvae feed in woody stem tissue and cause collapse of adult plants.

Within a decade of their release in Australia, a sufficient number of these agents were established to maintain effective biological control of *Opuntia* cacti, and millions of hectares once considered useless for grazing were rendered productive (Grossman, 1989). For certain control agents used against *Opuntia* spp., seasonal reinoculation is still necessary in some locations. In New South Wales, for example, low winter temperatures reduce populations of *Dactylopius austrinus* to a level at which no control of *Opuntia aurantiaca* is achieved. To overcome this, augmentative releases of *D. austrinus* are carried out in the spring by government employees (Tisdell, Auld & Menz, 1984).

Successful biological control of a weed through the introduction of multiple herbivore species attacking different life stages has also been accomplished in agricultural regions of South Africa infested by the South American tree *Sesbania punicea* (Moran, 1995). The bud-feeding weevil *Trichapion lativentre* was imported, released, and established in the 1970s, and found capable of reducing pod production of the target weed by >98%. The impact of *T. lativentre* was enhanced by introduction of a second weevil species, *Rhyssomatus marginatus*, whose larvae destroy the seeds of *S. punicea*. The combined action of the two agents lowered *S. punicea* fecundity by >99%, but little effect was evident on the density of mature plants. A third weevil, *Neodiplogrammus quadrivittatus*, which feeds in the branches and stems of mature host plants, was then introduced in 1986 from Brazil. Although it was ineffective against juvenile stages, it proved to be quite deadly against adult trees. Combination of the three control agents has been effective in reducing both seed production (and hence seedling recruitment) and standing biomass of the weed in the Southwest Cape Province (Moran, 1995).

The importance of *T. lativentre* in the biological control of *S. punicea* has been emphasized by results of an inadvertent, large-scale experiment. Hoffman & Moran (1995) found that organophosphate insecticide drift onto thickets of *S.*

punicea surrounding citrus orchards in the Oliphants River valley reduced populations of *T. lativentre* early in the growing season and almost eliminated the weevil's ability to suppress bud, flower, pod, and seed production by the tree. On farms in the same valley where insect pests of citrus were managed through biological control, or where infestations of *S. punicea* were remote from citrus orchards, *T. lativentre* populations were higher and *S. punicea* seed production was substantially reduced by the weevil.

McEvoy, Cox & Coombs (1991) used observations of natural population dynamics at a single site, data from a controlled field experiment involving population manipulations, and results of a regional survey to examine biological control of *Senecio jacobaea* by exotic insects in Oregon. The weed is a biennial or short-lived perennial that displaces desirable forage and is toxic to livestock. Its native range extends from Scandinavia south through Asia Minor, and from Great Britain east to Siberia, but it has spread to New Zealand, Australia, Argentina, and coastal areas of the USA and Canada. Starting in the 1960s, three insects native to France and Italy were introduced to control *S. jacobaea* in western Oregon: a moth, *Tyria jacobaeae*; a fly, *Botanophila seneciella* (formerly *Hylemia seneciella*); and a flea beetle, *Longitarsus jacobaeae* (McEvoy, Cox & Coombs, 1991). Larvae of *T. jacobaeae* will feed on vegetative stages of the weed but prefer flowering individuals; complete defoliation is common, but death is rare, and plants often develop secondary flowering shoots after the primary shoot is destroyed. Larvae of *B. seneciella* consume immature seeds and involucre bases. Adults of *L. jacobaeae* chew holes in the leaves and larvae tunnel into leaf petioles and roots.

Population dynamics of *Senecio jacobaea* were monitored in an abandoned pasture in western Oregon from 1981 through 1988 (McEvoy, Cox & Coombs, 1991). *Tyria jacobaeae*, *Longitarsus jacobaeae*, and *Botanophila seneciella* were all present as the result of deliberate introduction at the site or migration from other release points in the state, and each of the three insect species was observed to feed on *S. jacobaea* plants at the study site in at least one year. Density of the weed decreased dramatically between 1981 and 1988, from > 90% of total plant standing crop to < 1%. The reduction in weed biomass was matched by an increase in biomass of four perennial grass species. Seed densities of *S. jacobaea* in soil (to 10 cm depth) decreased in concert with the reduction in above-ground plant biomass, from 14 500 seeds m^{-2} in 1982 to 4800 seeds m^{-2} in 1988.

Further insight into the impacts of herbivory on *S. jacobaea* came from a field experiment in which high-density, mixed-age *S. jacobaea* populations were created by sowing seeds directly and transplanting additional small rosettes from a greenhouse (McEvoy, Cox & Coombs, 1991). Following an initial period in which plants in all plots were protected from herbivores with cages, two treatments were established. Plants in "unprotected" plots had the sides of their cages rolled up to expose them to herbivores; plants in "protected" plots remained caged and were treated with the insecticide carbofuran. High densities of *L. jacobaeae* were observed in exposed plots. Because densities of *T. jacobaeae* were lower than expected during the experiment, plants were artificially defoliated to match the effect of grazing larvae. In plots exposed to natural enemies (and their human mimics), *S. jacobaea* density, biomass, and reproductive output all declined within 18 months to <1% of populations protected from enemies.

A third line of evidence demonstrating the suppressive effects of insect biocontrol agents on *S. jacobaea* came from a regional survey conducted by the Oregon Department of Agriculture. *Senecio jacobaea* densities were found to decline >93% across 42 sites in western Oregon in the six years following introduction of *L. jacobaeae* (McEvoy, Cox & Coombs, 1991). Based on the combination of observational and experimental studies, McEvoy, Cox & Coombs (1991) concluded that introduced insect herbivores successfully controlled *S. jacobaea* in western Oregon.

The majority of inoculative biological control agents used against weeds have been insects. These have been dominated by Coleoptera (especially Chrysomelidae and Curculionidae), Lepidoptera (especially Pyralidae), Diptera (especially Tephritidae), and Hemiptera (especially Dacylopiidae and Tingidae) (Julien, Kerr & Chan, 1984; Julien & Griffiths, 1998). The preponderance of insect taxa as introduced weed biocontrol agents may be because they are particularly well adapted to persist, disperse, and attack their hosts in the rangeland and pasture environments typical of most weed biocontrol efforts. Nonetheless, certain microbial taxa have proven effective as inoculative agents for biological control of weeds in some environments. These taxa typically have the ability to withstand dry periods and disperse relatively long distances, often by wind (TeBeest, 1991; Watson, 1991). In Chile, for example, the rust fungus *Phragmidium violaceum* was introduced from Germany to control *Rubus constrictus* and *R. ulmifolius* on rangeland and pastures, and large reductions in weed density resulted in several areas (Hasan & Ayres, 1990; Julien & Griffiths, 1998). In Hawaii, importation and release of the fungus *Entyloma ageratinae* in combination with release of an imported fly, *Procecidochares alanai*, and a moth, *Oidaematophorus beneficus*, reduced densities of the weed *Ageratina riparia* to <5% of preintroduction levels at many locations (TeBeest, 1996; Julien & Griffiths, 1998).

The rust fungus *Puccinia chondrillina* was introduced from Mediterranean Europe into Australia in the early 1970s to control *Chondrilla juncea* in pastures and grain fields (Hasan & Ayres, 1990). Initially, introduction of *P. chondrillina* resulted in near elimination of the target weed at many locations. In subsequent years, however, a resurgence of the weed has taken place. Investigation of the situation revealed that three biotypes of *C. juncea* (narrow-leaf, intermediate-leaf, and broad-leaf) exist, each with different susceptibilities to the particular strain of fungus (IT 32) originally introduced from Italy to control the weed. The intermediate-leaf and broad-leaf biotypes are resistant to IT 32 and have replaced the formerly dominant, susceptible, narrow-leaf biotype (Chaboudez & Sheppard, 1995). Consequently efforts have been initiated to collect additional strains of the fungus in the Mediterranean basin capable of infecting the resurgent *C. juncea* biotypes in Australia (Hasan & Ayres, 1990). As this example points out, adequate attention to genetic variation within populations of weeds and the organisms that attack them is critical to the success of biological control efforts.

Use of inoculative control agents against native weed species

In addition to collecting and releasing exotic herbivores and pathogens to control introduced weed species, exotic natural enemies can be considered for release to control native weed species. In essence, this approach represents exposing a target weed species to herbivores and pathogens it has not yet encountered in its evolutionary history. In addition, it represents introducing herbivore and pathogen species where few indigenous natural enemies may be adapted to attack their populations (Hokkanen, 1986). The exploitation of evolutionarily new relationships for weed biocontrol remains largely untested, but it is useful to examine a case where the approach might be applied.

Lawton (1988) suggested that exotic herbivores might be introduced for control of *Pteridium aquilinum*, a fern native to Great Britain, but also to every continent except Antarctica. The fern is acutely poisonous and carcinogenic to livestock, serves as a reservoir for ticks that vector viral diseases of sheep, and aggressively invades pasture land. Costs of control, lost grazing, and stock poisoning in Great Britain are estimated at several million pounds per year (Lawton, 1988). While *P. aquilinum* can be controlled by repeated cutting or herbicide applications, these measures often are not cost-effective and relaxation of control may result in rapid resurgence of the problem.

In Britain, 27 native insects regularly attack the above-ground portions of *P. aquilinum* but fail to adequately control the plant because *they* are attacked by parasites, predators, and pathogens (Lawton, 1988). In an attempt to subject the weed to greater herbivore pressure, 12 insect and one mite species that feed on *P. aquilinum* were located in temperate regions of South Africa.

Two moth species appeared to be especially good candidates for control agents: *Conservula cinisigna*, which chews fronds of *P. aquilinum*, and an unidentified *Panotima* species, whose larvae mine fronds and then attack the rachis of the plant. Feeding experiments showed both species were virtually monophagous on the target weed and fed on it early in the season, when the plant is especially vulnerable to attack. Because the two potential control agents are ecologically and taxonomically distinct from resident herbivores of *P. aquilinum* in Great Britain, Lawton (1988) proposed that they might escape the effects of predation, parasitism, and disease from indigenous natural enemies. Nonetheless, despite the many characteristics of the two insects which suggested they might be successful in suppressing *P. aquilinum*, Lawton (1988) noted that their effectiveness could be demonstrated definitively only after field releases took place.

Choosing target weed species and protecting non-target species

At first thought, field release of an organism that may control a noxious weed seems a worthwhile endeavor that should be pursued as soon as the suitable agent is identified. Caution, consultation, and serious planning are advised, however. Organisms used for inoculative biocontrol releases are chosen because of their ability to reproduce and disperse, and once released they cannot be recalled. Thus, there is a distinct possibility that susceptible plant species will be suppressed outside of the areas originally designated for treatment with biocontrol agents. A very real possibility also exists that, through evolution, the range of host plants attacked by introduced control agents may shift or expand to include non-target species (Simberloff & Stiling, 1996).

In the case of *Pteridium aquilinum* in Great Britain, the target plant constitutes a major portion of the native vegetation over a large area, and the ecological impact of removing it needs to be considered very carefully before the release of exotic control agents takes place. The plant clearly has undesirable characteristics in pasture land, but what benefits would be lost due to its removal from pastures and other habitats? Are these more or less valuable than the benefits derived from the plant's suppression? Lawton (1988) noted that *P. aquilinum* may provide desirable habitat for native animal populations, and he proposed construction of an "environmental balance sheet" that would set out the pros and cons of *P. aquilinum* before exotic herbivores were released to control it. Such an assessment process has apparently worked in favor of the weed. MacFayden (1998) reported that after extensive agent testing, biocontrol efforts directed at *P. aquilinum* in Great Britain were ultimately abandoned because of regulatory requirements for costly field cage experiments and doubts over the wisdom of controlling a native species.

The difficulty and necessity of resolving social issues related to weed biocontrol are further illustrated by the controversy associated with release of several insects for control of *Echium plantagineum* in Australia (Tisdell, Auld & Menz, 1984). Depending on one's perspective, *E. plantagineum* may be called either "Paterson's curse" or "Salvation Jane." For Australian grain farmers and orchardists, the plant is a noxious weed. In contrast, for livestock producers, it is a useful (though cumulatively toxic) source of sheep fodder during drought periods. For honey producers, it is an important nectar source for bees. Failure to reconcile these opposing perspectives led farmers and others who derive benefits from *E. plantagineum* to obtain a court injunction and restraining order against the Commonwealth Scientific and Industrial Research Organization, the Australian government agency responsible for biocontrol releases. This type of conflict is expensive, time-consuming, and should be avoided if weed biocontrol programs are to gain strong public support. Open discussions of potential losses and gains are necessary to insure that the goals of weed biocontrol efforts are shared by all parties holding an interest in the targeted species.

The potential for weed biocontrol agents to have unintended impacts on non-target species is illustrated by the case of *Rhinocyllus conicus*, a weevil collected from France and Italy and released in the USA and Canada, starting in 1968, to control exotic thistles (Louda *et al.*, 1997). Originally targeted were weedy Eurasian species in the genus *Carduus*. Although feeding preference trials conducted before release of *R. conicus* indicated that its range of hosts included the native North American genera *Circium*, *Silybum*, and *Onopordum*, the weevil's stronger oviposition preference and more successful larval development on *Carduus* spp. were supposed to limit its use of native plants.

The real outcome of releasing *R. conicus* differed markedly from this expectation, however. The weevil fed on both native and exotic thistle species. Measurements made in national parks and conservation land in Nebraska, South Dakota, and Colorado indicated that seed production by five native thistle species dropped precipitously because of attack by the weevil (Louda *et al.*, 1997). For the native, sparsely distributed species *Circium canescens*, the percentage of flowerheads per plant infested by *R. conicus* rose at one location from none in 1992 to 58% in 1996. The average number of viable seeds produced by *C. canescens* flowerheads infested by weevils was only 14% of that produced by similar heads without insects or with exposure only to native insects. Because regeneration of native thistle populations is seed-limited, sizable reductions in seed production by *R. conicus* may reduce densities of these plants to the point their persistence is threatened.

McFayden (1998) reported six other cases in which release of weed biocontrol agents has resulted in damage to non-target plants. Included within her list is the release of the moth *Cactoblastis cactorum* in the West Indies. When *C. cactorum* was released in the 1950s to control the native cactus *Opuntia triacantha*, few problems were anticipated and few objections were raised. Since that time, however, the moth has spread naturally and through deliberate introductions throughout the Caribbean basin. In 1989, *C. cactorum* was found in the Florida Keys, where it now threatens the survival of the native cactus *Opuntia spinosissima*, which was already endangered by clearing and development. McFayden (1998) noted that "*C. cactorum* is likely to continue its spread westward into Mexico and the cactus country of the southwest USA, where its impact may be severe unless it is reduced by the effects of parasitism or by competition with similar native moths in the genus *Melitara*."

The lesson to be learned from these examples is that very thorough investigations of effects on potential alternative hosts are needed before exotic herbivores and pathogens are released to control weeds (Strong, 1997). Weeds clearly can wreak havoc with existing communities unless controlled effectively, but the biological agents released to control them may also severely affect the native biota (Simberloff & Stiling, 1996). Information concerning the full range of potential economic and ecological risks, costs, and benefits is needed before introductions of weed biocontrol agents can be assessed adequately (Harris *et al.*, 1985).

Inundative releases of control agents

Unlike inoculative biological weed control, in which herbivores and pathogen populations are expected to increase naturally and spread considerable distances from points of introduction, inundative biological control methods rely on human intervention to increase and disperse host-specific control agents onto target weed species. While inoculative methods are expected to act slowly, often over a period of years, inundative methods are expected to severely damage and kill nearly all susceptible weeds quickly, often over a period of a few days. The inundative approach is particularly well suited to annual cropping systems, in which weeds must be strongly suppressed early in the growing season to avoid crop yield loss due to competition.

So far, inundative methods for biological weed control have relied almost exclusively on native microorganisms. The bias toward microbes rather than insects reflects the greater ease and lower cost of raising the large populations needed for inundative releases, and the ability of certain microbial control agents to mimic the ease of application and efficacy of chemical herbicides. Greater reliance on native rather than exotic microbial taxa reflects greater ease of collection, fewer regulatory requirements, and better preadaptation of native species to local climatic conditions (Quimby & Walker, 1982). Although there are important exceptions, shoot-attacking pathogens are more commonly used than those attacking roots because of greater ease and uniformity of inoculum application, narrower host range, and more effective dispersal of reproductive propagules in air than in soil (Hasan & Ayres, 1990).

In theory the use of inundative control agents is expected to have fewer negative environmental impacts than the use of inoculative controls, since the former are less well adapted for persistence and dispersal without human intervention. That is not to say that mass releases of microorganisms to suppress weeds are without potential problems. TeBeest (1996) noted that "almost without exception, these pathogens can and do infect cultivated and horticulturally important plant species in controlled experiments." Toxic effects on livestock and people are another undesired but possible impact of using microbial control agents. *Phomopsis emicis*, a fungal pathogen considered for use in Australia as a biocontrol agent against the weed *Emex australis*, was found to produce large quantities of a compound poisonous to grazing animals (Auld & Morin, 1995). Thus, the host range and ecological impact of organisms used for inundative control programs must be very thoroughly studied, and prudence used in their deployment (Weidemann, 1991). Great care must be taken to separate target and non-target species in both space and time. No major accidents have yet been reported, but that may reflect limited use more than a set of safety factors inherent to the organisms used. In anticipation of greater use of microorganisms for controlling weeds and other pests, Canadian and American public regulatory agencies are now mandated to consider how microbial products affect the environment, human health, and food and feed quality (Makowski, 1997).

Mycoherbicides

More than 100 taxa of microorganisms have now been identified as potential candidates for the control of at least 100 weed species (Charudattan, 1991; Kennedy & Kremer, 1996). The most intensively studied of these are fungi (Watson, 1989). Typically, a fungus is collected and isolated from a diseased weed, tested for its pathogenicity via inoculation of healthy weeds, cultured and maintained on artificial media, and identified by taxonomic specialists. Host range of the pathogen is then determined, optimum conditions for infection and disease development are identified, and mechanisms of action of the pathogen are investigated. Next, particularly virulent strains of the pathogen are mass-produced, formulated to improve their infective ability, evaluated for efficacy on a field scale, and registered with regulatory agencies. Through this process, a number of fungi have been developed as sprayable, potentially marketable products, commonly called "mycoherbicides." These include *Alternaria cassiae* ('CASST'), for use against *Cassia obtusifolia*, *C. occidentalis*, and *Crotalaria spectabilis*; *Colletotrichum gloeosporioides* f.sp. *aeschynomene* ('Collego'), for use against *Aeschynomene virginica*; *Colletotrichum gloeosporioides* f.sp. *cuscutae* ('Lubao'), for use against *Cuscuta australis*, *C. chinensis*, and *C. maritima*; *Colletotrichum gloeosporioides* f.sp. *malvae* ('BioMal'), for use against *Malva pusilla*; *Phytophthora palmivora* ('DeVine'), for use against *Morrenia odorata*; and *Puccinia canaliculata* ('Dr. Biosedge'), for use against *Cyperus esculentus* (TeBeest, 1996; Boyetchko, 1997).

In nature, widespread infection of host plants by pathogens and subsequent disease development are relatively rare: inoculum levels may be too low, host plants may resist infection, and environmental conditions may not be conducive to infection, attack on host tissues, and propagule dispersal. The mycoherbicide approach to weed control overcomes most barriers to infection and disease development by applying large doses of inoculum (e.g., 10^{10}–10^{11} spores ha^{-1}) to susceptible life stages of target weeds using carrier formulations that promote attack (Watson, 1989; Auld & Morin, 1995). The timing of applications may be chosen to make use of environmental conditions favorable to plant disease, or environmental conditions may be modified directly. Because leaf wetness is an especially critical factor affecting infection and disease development by many foliar pathogens (TeBeest, 1991), delivery of a high concentration of inoculum in an optimally moist environment is desirable. Mabbayad & Watson (1995) found, for example, that increasing both the conidia concentration of an *Alternaria* species and the volume of water carrying the fungus improved suppression of the weed *Sphenoclea zeylanica* in rice fields (Table 8.3). The fungus was effective against both seedling and adult stages of the target weed.

By reducing weed density and suppressing weed biomass production, application of mycoherbicides can diminish the competitive effects of weeds on crops and increase crop yields. For example, in experiments conducted by Kempenaar, Horsten & Scheepens (1996), application of *Ascochyta caulina* spore suspensions improved the performance of maize and sugar beet grown with *Chenopodium album*. The fungus created necrotic lesions on leaves and stems of *C. album* and reduced *C. album* density up to 65%, but did not harm the two

Table 8.3. *Effect of* Alternaria *conidia concentration and spray volume on* Sphenoclea zeylanica *biomass and density two weeks after application in field experiments*

Conidia concentration (conidia mL^{-1})	Spray volume (mL 0.25 m^{-2})	Biomass (g 0.25 m^{-2})	Density (number 0.25 m^{-2})
Experiment 1			
0	50	11.2	201.8
7.0×10^5	50	0.2	16.3
1.8×10^6	50	0.04	5.3
Experiment 2			
0	50	17.1	81.5
8.0×10^3	50	13.7	93.0
3.0×10^4	50	8.0	79.5
2.6×10^5	50	0.2	4.3
Experiment 3			
0	0	17.6	145.0
1.6×10^6	12.5	3.0	40.7
1.6×10^6	25	0.8	9.3
1.6×10^6	50	0.2	1.3
Experiment 4			
0	0	28.2	169.3
4.8×10^5	12.5	10.5	89.7
4.8×10^5	50	2.1	33.3

Source: Adapted from Mabbayad & Watson (1995).

crop species. Competition from *C. album* reduced maize biomass 20% when *A. caulina* spores were not applied, but the weed had no effect on maize growth when sprayed with the fungus. Competition from *C. album* reduced sugarbeet biomass 80% in unsprayed plots, but reduced it only 20% to 60% when the mycoherbicide was applied.

Similarly, in experiments conducted by Paul & Ayres (1987), competition from *Senecio vulgaris* against lettuce was diminished or eliminated by application of *Puccinia lagenophorae* spores. In the absence of the pathogen, lettuce fresh weight was reduced 50% to 97% by competition from *S. vulgaris* sown at densities of 250 to 65 000 seeds m^{-2}. When *S. vulgaris* was infected by the fungus, however, the weed had no effect on lettuce yield until its sowing density was >4000 seeds m^{-2}. Even in treatments containing the highest *S. vulgaris* densities, lettuce yield in plots sprayed with *Puccinia lagenophorae* spores was higher than in unsprayed plots. Infection by *P. lagenophorae* had little effect on the density of *S. vulgaris* growing with lettuce, but reduced the weed's biomass by an average of 16%.

Table 8.4. *Effects of different spray applications on mortality of* Sesbania exaltata *and soybean yield in field trials*

Spray treatment[a]	Sesbania exaltata mortality (%)		Soybean yield (kg ha^{-1})
	1989	1990	1990
Conidia in water	10	40	2160
Conidia in invert emulsion[b]	95	97	2593
Invert emulsion only	8	15	1408
Acifluorfen[c]	96	98	2618
Untreated	3	2	1309

Notes:
[a] Conidia concentrations of *Colletotrichum truncatum* were 1×10^7 mL^{-1}; carrier volume for all spray treatments was 187 L ha^{-1}.
[b] The invert emulsion was prepared from paraffinic wax and oil, monoglyceride emulsifier, lanolin, and water.
[c] Acifluorfen was applied at 1.1 kg a.i. ha^{-1}.
Source: Adapted from Boyette *et al.* (1993).

Improvements in the efficacy of fungi as weed biocontrol agents can result from the use of two or more species in combination. Infection of *Senecio vulgaris* with the fungus *Botrytis cinerea* caused 10% mortality when *B. cinerea* was applied alone, but 100% mortality when it was applied after the weed species had been infected with *Puccinia lagenophorae* (Hallett, Paul & Ayres, 1990).

Improvements in formulation technologies also can markedly increase the efficacy of fungi used for inundative weed control. An illustrative example is provided by Boyette *et al.* (1993), who examined the effects of carrier formulation on infectiousness of *Colletotrichum truncatum* used to suppress *Sesbania exaltata*, a leguminous weed found in soybean, rice, and cotton in the southern USA. The fungus is highly virulent, host-specific, and can be mass-produced readily. Invert emulsions (water suspended in oil, rather than a standard emulsion of oil suspended in water) are thought to retard evaporation of water spray droplets, thereby reducing or eliminating the need for a 6- to 24-hour dew period for infection by fungi used as biocontrol agents. In field experiments, application of *C. truncatum* conidia in water (at 187 L ha^{-1}) killed an average of 25% of the *S. exaltata* population and increased soybean yield 65%, compared with an untreated control (Table 8.4). In contrast, application of *C. truncatum* conidia in an invert emulsion with the same carrier volume killed an average of 96% of the *S. exaltata* population and increased soybean yield 98%. The level of weed suppression and crop yield increase obtained from the invert emulsion formulation was similar to that obtained from the herbicide acifluorfen (Table 8.4).

A variety of other formulations have been examined as means to improve application and efficacy of fungi used as weed biocontrol agents (Boyette et al., 1996). For liquid formulations, which are used primarily to incite leaf and stem diseases, the focus has been on developing better surfactants, diluents, emulsions, and gels. *Colletotrichum gloeosporioides* f.sp. *malvae* has, for example, been formulated in a silica gel carrier to improve dispersion of the pathogen in water and enhance spray coverage of the target weed. For granular formulations, which appear best suited to deliver pathogens that attack at or below the soil surface, the focus has been on producing granules that buffer environmental extremes, serve as alternate food sources for pathogens (thus increasing their persistence), and minimize the likelihood of being washed away from the target area. Examples include cornmeal and sand mixed with *Fusarium solani* f.sp. *cucurbitae* for control of *Cucurbita texana*; vermiculite mixed with *Alternaria macrospora* for control of *Anoda cristata*; sodium alginate and kaolin mixed with *Fusarium laterium* for control of *Abutilon theophrasti*; and wheat gluten mixed with *Fusarium oxysporum* for control of *Sesbania exaltata* (Boyette et al., 1996).

In addition to the development of better formulation technologies for mycoherbicides, efforts have also been directed toward improving production and culture technologies. In general, three methods exist for producing fungal pathogens in large quantities. Certain fungi, such as *Puccinia* spp., require the use of living host plants. Others, such as *Alternaria* spp., are best adapted to solid substrate fermentations. Both of those methods are considered expensive and poorly suited to industrial-scale production. A large number of other fungi can be produced, however, by liquid culture fermentation, which is relatively inexpensive and already widely used for pharmaceutical and food products.

The possibility of improving liquid culture media for the production of mycoherbicides has been emphasized by Jackson et al. (1996), who investigated the effects of varying carbon concentrations in media used to produce microsclerotia of *Colletotrichum truncatum*. Microsclerotia of this fungus are considered to be more desirable than conidia because of their greater stability as dry preparations and their greater efficacy in controlling *Sesbania exaltata* when used as soil amendments. Jackson et al. (1996) found that media rich in carbon (>25 g L^{-1}) promoted production of microsclerotia, whereas media containing less carbon (<16 g L^{-1}) favored production of conidia. They concluded that a better understanding of how nutritional conditions in liquid culture media affect propagule formation, yield, efficacy, and stability will enhance the use of fungi and other microorganisms as biocontrol agents.

Deleterious rhizobacteria

Deleterious rhizobacteria (DRB) are a second group of microorganisms that have received attention as potential weed biocontrol agents. DRB colonize seeds, root mucigel, epidermal and cortical tissues, and intercellular spaces, and can reduce seed viability, seedling emergence, and plant growth through the release of toxic compounds (Kremer, 1993; Kennedy, 1997). Data from a number of researchers indicate that the effects of different DRB taxa on plants can be host- and rhizobacterial-isolate-specific (Boyetchko, 1996, 1997). Certain DRB can harm crop species, such as potato, sugar beet, wheat, citrus, and bean, but others have benign or neutral effects on crops and detrimental effects on weeds, such as *Abutilon theophrasti*, *Aegilops cylindrica*, *Amaranthus* spp., *Bromus japonicum*, *B. tectorum*, *Chenopodium* spp., *Datura stramonium*, *Ipomoea* spp., *Polygonum* spp., and *Xanthium canadense* (Kremer, 1993; Boyetchko, 1996; Kennedy & Kremer, 1996). Genera of DRB that may be useful for weed suppression include *Achromobacter*, *Alcaligenes*, *Arthrobacter*, *Citrobacter*, *Enterobacter*, *Erwinia*, *Flavobacterium*, *Klebsiella*, and *Pseudomonas* (Boyetchko, 1996; Kremer & Kennedy, 1996).

Most of the research work focused on DRB has been conducted in laboratories and glasshouses. In one of the few experiments examining the potential of DRB as selective agents for weed suppression under field conditions, Kennedy et al. (1991) tested whether an isolate of *Pseudomonas fluorescens* (strain D7) could inhibit the grass weed *Bromus tectorum* without harming winter wheat. *Bromus tectorum* is poorly controlled by conventional herbicides available for winter wheat production. It is often the dominant weed species in cereal fields of the western USA and Canada, where it causes losses estimated at $300 million per year (Skipper, Ogg & Kennedy, 1996; Kennedy, 1997). In laboratory experiments, strain D7 has been shown to produce toxins that inhibit *B. tectorum* but not wheat (Tranel, Gealy & Kennedy, 1993; Gurusiddaiah et al., 1994; Gealy et al., 1996). In the field, application of strain D7 to soil of plots sown with wheat and infested with *B. tectorum* reduced the weed's density up to 35%, reduced its late-season biomass up to 54%, and reduced its seed production up to 64%. At two of three field sites, application of strain D7 increased winter wheat yields 18% to 35%, an effect attributed to a reduction in *B. tectorum*'s ability to compete against the wheat crop.

In subsequent field experiments with strain D7, *B. tectorum* was suppressed in only one of 20 trials, and the bacterium consistently failed to increase winter wheat yield compared to uninoculated controls (Skipper, Ogg & Kennedy, 1996). At the present time, it appears that DRB have potential but lack consistency as agents for biological control of weeds under field conditions. One suggested avenue for addressing this problem is improvements in delivery systems, including the use of clay, peat, and alginate carriers and encapsulations (Boyetchko, 1996; Kremer & Kennedy, 1996). DRB can rapidly colonize crop residues, such as barley straw (Stroo, Elliot & Papendick, 1988), and such materials may serve as appropriate media for improved DRB application and management.

The integration of multiple stress factors

As discussed previously, the joint use of several insect or microbial biocontrol agents can result in better weed suppression than that obtained with a single organism. The integration of biocontrol agents with mechanical, chemical, and other biotic stress factors can also provide advantages over reliance on a single agent alone. For example, in three out of five experiments conducted by Klein & Auld (1996), wounding *Xanthium spinosum* by dragging steel mesh over it before applying *Colletotrichum orbiculare* spores caused greater mortality than did spores alone; mowing *X. spinosum* before applying spores of the fungus increased weed mortality in two out of three experiments. Application of the fungus *Puccinia canaliculata* in mixture with the herbicide paraquat provided 99% control of *Cyperus esculentus*, compared with 60% control with *P. canaliculata* alone and 10% control with paraquat alone (Phatak, Callaway & Vavrina, 1987). Similarly, application of the growth regulator thidiazuron in concert with the fungus *Colletotrichum coccodes* provided greater control of *Abutilon theophrasti* than did use of either stress factor separately (Hodgson *et al.*, 1988).

Plant competition can add greatly to or synergize the effects of pathogens and insects on weeds. Groves & Williams (1975) conducted a pot experiment to determine how infection by *Puccinia chondrillina* and competition from subterranean clover affected growth of the weed *Chondrilla juncea*. At 146 days after planting, weed biomass was reduced 51% by the pathogen alone, 69% by clover competition alone, and 94% by the combination of *P. chondrillina* infection and competition (Figure 8.2).

Cumulative effects of plant competition and insect attack on *Senecio jacobaea* were studied experimentally by McEvoy *et al.* (1993) in an Oregon pasture. At the start of the experiment, the pasture community was dominated by the perennial grasses *Holcus lanatus*, *Dactylis glomerata*, *Anthoxanthum odoratum*, and *Festuca arundinacea*; *S. jacobaea* represented only 0.1% of the above-ground dry mass of the community. Plots were then tilled to stimulate germination of *S. jacobaea*. To manipulate interspecific competition against the emerged *S. jacobaea* plants, other plants colonizing each plot were either manually removed,

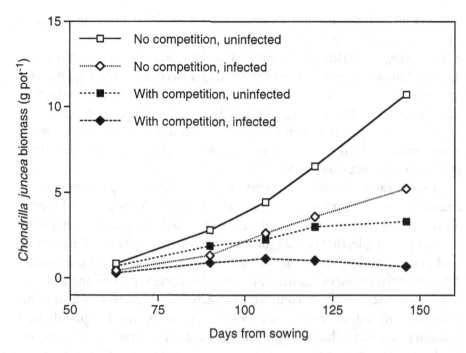

Figure 8.2 Effects of competition from subterranean clover and infection by *Puccinia chondrillina* on biomass production of *Chondrilla juncea* in a glasshouse pot experiment. (Adapted from Groves & Williams, 1975.)

clipped to a height of 5 cm, or left unaltered. Cages were used to regulate *S. jacobaea* exposure to two herbivores: the moth *Tyria jacobaeae*, and the flea beetle *Longitarsus jacobaeae*. In addition to caging, herbivores were removed from "protected" treatments using the insecticide rotenone. The moth and beetle fed on *S. jacobaea* at different times of year, so it was possible to vary attack by the two herbivore species independently, by opening and closing cages at different times.

Without competition from desired grasses, *L. jacobaeae* was incapable of eliminating *S. jacobaea* from the pasture. Plant competition alone eliminated actively growing *S. jacobaea* individuals in four years, but the combination of competition and herbivory by *L. jacobaeae* eliminated all *S. jacobaea* individuals except those in the buried seed bank within two years. *Tyria jacobaeae* reduced *S. jacobaea* seed production, but had no detectable effect on the weed's biomass production during four years of measurements.

Because both the total removal of competitors and their partial suppression by clipping helped maintain *S. jacobaea* populations, even when *L. jacobaeae* was present, this experiment strongly suggests the importance of integrating grazing practices with weed biocontrol agents. If grazing by livestock on desirable pasture species cripples their ability to compete against a target weed, biocontrol agents may be of little value. Conversely, if livestock grazing is regulated through timing and stocking densities to maintain a high degree of plant competition against the target weed, biocontrol agents may make valuable contributions toward weed suppression. Further increases in the strength of competition against weeds subjected to biocontrol efforts in rangelands and pastures might be gained by intentionally sowing desired plant species to increase their densities (Muller-Scharer & Schroeder, 1993; Jacobs, Sheley & Maxwell, 1996).

In some cases, the weed-suppressive effects of microorganisms used as inundative control agents may not be evident unless the target weed species grows in competition with other plants. DiTommaso, Watson & Hallett (1996) found, for example, that the fungus *Colletotrichum coccodes* had little impact on seed production by *Abutilon theophrasti* when the weed grew in pure stand. After an initial period of stunting due to disease, height growth and leaf production resumed rapidly. In contrast, when *A. theophrasti* grew in mixture with soybean, early-season growth suppression caused by the fungus allowed soybean to dominate the canopy, shade the weed, and prevent it from recovering later in the season. In two of three years, spraying weed–crop mixtures with *C. coccodes* reduced the weed's height 20–30 cm and diminished its seed production by an average of 60% compared with uninoculated plants. DiTommaso, Watson & Hallett (1996) noted that protocols for screening potential agents for weed biocontrol may overlook useful organisms if they do not include competition from other plant species within the set of experimental conditions.

Moving ahead with weed biocontrol

In this chapter we have seen that insect herbivores and phytopathogens can strongly affect the survival, growth, and competitive ability of weeds. Weed-attacking organisms can be conserved in fields where they occur already, added inoculatively where they are not yet present but can be established, or introduced in an inundative manner after mass-production in insectaries and microbiology laboratories. The conservation approach seems well suited to weed seed predators and might be extended, after more research, to better exploit the impacts of foliage and flower feeders and pathogens. Inoculative releases of biocontrol agents are currently best suited to introduced weeds of pasture and rangeland, whereas inundative releases are best suited to annual crops. Due to the need for fast action and cheap production, most inundative biocontrol of weeds is with microorganisms.

Despite their potential, herbivory and disease remain largely unexploited for weed management. Why is this so? How might the situation be improved? Attention to three sets of issues may improve the development and practical application of weed biocontrol.

First, an appropriate philosophical perspective is needed. If individual biocontrol agents are expected to act alone with the efficacy and broad target range of most herbicides, results will often be disappointing. If, on the other hand, biocontrol agents are viewed as stress factors that are most useful when integrated with a variety of other weed management tactics, many more opportunities for success may be encountered. For this to happen, closer integration of weed biocontrol work with other weed research and extension efforts is highly desirable.

A second set of issues is technical: weed biocontrol agents need to be produced in such a manner that they are readily accessible to farmers. Although many weed-suppressive organisms have been identified, studied, and formulated for application, very few have been brought into the marketplace. The costs of developing weed biocontrol agents are lower than those for synthetic herbicides, but commercialization is hindered by small market size due to the narrow spectrum of control, and the perception of relatively low profits (Auld & Morin, 1995). Consequently, most companies are reluctant to develop and register a product that controls only one weed species.

To overcome the inertia of the marketplace, research and oversight by publicly funded institutions are needed to aid individual farmers and cooperatives in culturing biocontrol agents for local use. Models for how this might be done already exist in certain parts of the world. Fermented food products are made in many homes and small-scale industries in Asia, and the modification of fermentation technologies to produce weed-suppressive microorganisms is possible (Mabbayad & Watson, 1995; Auld & Morin, 1995). In Cuba, fungi, bacteria, and insects are currently produced in small-scale facilities for deployment in local biocontrol efforts directed toward insect pests and crop pathogens (Perfecto, 1994). Similar approaches are needed for producing weed biocontrol agents in both industrialized and developing countries. Decentralized programs for producing weed biocontrol agents present challenges related to quality control, formulation, and storage, but many of those challenges can probably be overcome. Because of the potential risks to non-target organisms, national and regional oversight is needed when making choices about target weed species, herbivores, and phytopathogens in biocontrol efforts. But once risks, benefits, and costs are assessed rigorously and debated thoroughly, local efforts to implement weed biocontrol programs should be supported.

Finally, improved understanding is needed of the basic biology and ecology of herbivore–weed and pathogen–weed interactions. To avoid potentially irreversible damage to non-target organisms, more research needs to focus on factors determining host specificity, host recognition, and colonization, infection, and dispersal processes. How do weed genotypes differ in their susceptibility to different types of herbivory and disease? To identify particularly useful genotypes of control agents and to deploy them most effectively against target weeds that vary genetically, temporally, and spatially requires research focused on the population genetics and evolution of both the agents and their targets.

Suppression of a dominant weed species by biocontrol agents may increase the production of desirable plant species, but it might also result in increased growth of other weed species that had been only minor components of the community (Figure 8.3). Sequences of invasions by new plant species are also possible. Randall (1996) described the history of a county in Oregon in which *Hypericum perforatum* was suppressed by biocontrol agents, only to be replaced by *Senecio jacobaea*, which was in turn suppressed by biocontrol agents, but replaced by *Carduus pycnocephalus*. To prevent the substitution of one weed problem for another, more needs to be learned about how selective herbivores and pathogens and vegetation management practices affect the dynamics of multispecies plant communities.

Research should also focus on interactions between weed biocontrol agents and other organisms that can affect their performance through competition, chemical interference, predation, parasitism, and disease. To improve the efficacy of microbial weed biocontrol agents, we need to know how to predict and manipulate their relationships with other microorganisms inhabiting plant surfaces and soil. Tillage and crop residue management are particularly important for regulating microbial interactions (Derksen, Blackshaw & Boyetchko, 1996), and more interdisciplinary research is needed to determine how microbial communities can be manipulated to better reduce weed seed survival, seedling establishment, competitive ability, and reproduction. Similarly, there is an important need to identify importation strategies and habitat manipulations that minimize the effects of competitors, predators, parasites, and pathogens on insect herbivores used as weed biocontrol agents.

Research addressing these and other issues presents excellent opportunities for collaboration between plant, animal, and microbial ecologists, and between basic and applied biologists. The results of such collaborations should have real practical value in the development of the next generation of weed biocontrol strategies.

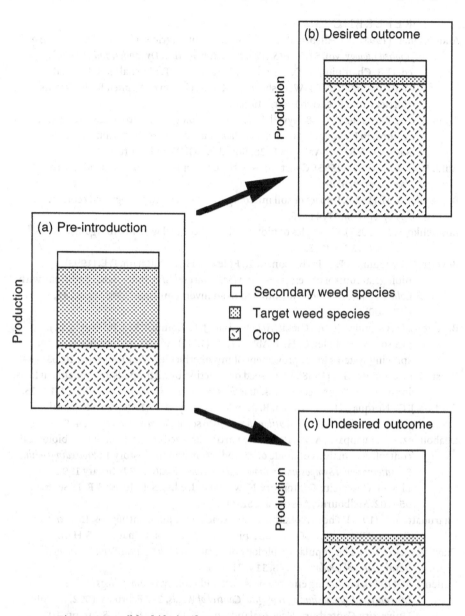

Figure 8.3 Possible shifts in plant community composition following release of a weed biocontrol agent. Before introduction of the control agent, the target weed species dominates the agroecosystem (a). Suppression of the target weed by the biocontrol agent might result in improved crop performance (b), but, alternatively, it might allow formerly minor weed species to fill the ecological gap created by removing the target weed, resulting in no improvement in crop performance (c).

REFERENCES

Andersson, L. (1998). Post-dispersal seed removal in some agricultural weeds. In *Aspects of Applied Biology*, vol. 51, *Weed Seedbanks: Determination, Dynamics, and Manipulation*, ed. G. T. Champion, A. C. Grundy, N. E. Jones, E. J. P. Marshall & R. J. Froud-Williams, pp. 159–64. Wellesborne, UK: Association of Applied Biologists and Horticultural Research International.

Andow, D. A., Ragsdale, D. W., & Nyvall, R. F. (1997). Biological control in cool temperate regions. In *Ecological Interactions and Biological Control*, ed. D. A. Andow, D. W. Ragsdale & R. F. Nyvall, pp. 1–28. Boulder, CO: Westview Press.

Auld, B. A., & Morin, L. (1995). Constraints in the development of bioherbicides. *Weed Technology*, 9, 638–52.

Boyetchko, S. M. (1996). Impact of soil microorganisms on weed biology and ecology. *Phytoprotection*, 77, 41–56.

Boyetchko, S. M. (1997). Principles of biological weed control with microorganisms. *HortScience*, 32, 201–5.

Boyette, C. D., Quimby, P. C. Jr., Bryson, C. T., Egley, G. H., & Fulgham, F. E. (1993). Biological control of hemp sesbania (*Sesbania exaltata*) under field conditions with *Colletotrichum truncatum* formulated in an invert emulsion. *Weed Science*, 41, 497–500.

Boyette, C. D., Quimby, P. C. Jr., Caesar, A. J., Birdsall, J. L., Connick, W. J. Jr., Daigle, D. J., Jackson, M. A., Egley, G. H., & Abbas, H. K. (1996). Adjuvants, formulations, and spraying systems for improvement of mycoherbicides. *Weed Technology*, 10, 637–44.

Brust, G. E., & House, G. J. (1988). Weed seed destruction by arthropods and rodents in low-input soybean agroecosystems. *American Journal of Alternative Agriculture*, 3, 19–25.

Cardina, J. C., Norquay, H. M., Stinner, B. R., & McCartney, D. A. (1996). Postdispersal predation of velvetleaf (*Abutilon theophrasti*) seeds. *Weed Science*, 44, 534–9.

Chaboudez, P., & Sheppard, A. W. (1995). Are particular weeds more amenable to biological control? A reanalysis of mode of reproduction and life history. In *Proceedings of the 8th International Symposium on Biological Control of Weeds*, 2–7 February 1992, Lincoln University, Canterbury, New Zealand, ed. E. S. Delfosse & R. R. Scott, pp. 95–102. Melbourne, Australia: CSIRO.

Charudattan, R. (1991). The mycoherbicide approach with plant pathogens. In *Microbial Control of Weeds*, ed. D. O. TeBeest, pp. 24–57. New York: Chapman & Hall.

Crawley, M. J. (1986). The population biology of invaders. *Philosophical Transactions of the Royal Society of London, Series B*, 314, 711–31.

Cullen, J. M. (1995). Predicting effectiveness: fact and fantasy. In *Proceedings of the 8th International Symposium on Biological Control of Weeds*, 2–7 February 1992, Lincoln University, Canterbury, New Zealand, ed. E. S. Delfosse & R. R. Scott, pp. 103–9. Melbourne, Australia: CSIRO.

Derksen, D. A., Blackshaw, R. E., & Boyetchko, S. M. (1996). Sustainability, conservation tillage and weeds in Canada. *Canadian Journal of Plant Science*, 76, 651–9.

DiTommaso, A., Watson, A. K., & Hallett, S. G. (1996). Infection by the fungal pathogen *Colletotrichum coccodes* affects velvetleaf (*Abutilon theophrasti*) – soybean competition in the field. *Weed Science*, 44, 924–33.

Fellows, G. M., & Roeth, F. W. (1992). Factors influencing shattercane (*Sorghum bicolor*) seed survival. *Weed Science*, 40, 434–40.

Gealy, D. R., Gurusiddaiah, S., Ogg, A. G. Jr., & Kennedy, A. C. (1996). Metabolites from *Pseudomonas fluorescens* strain D7 inhibit downy brome (*Bromus tectorum*) seedling growth. *Weed Technology*, **10**, 282–7.

Grossman, J. (1989). Biological control of weeds: what's happening, what's needed? *IPM Practitioner*, **11**(6 & 7), 1–11.

Groves, R. H., & Williams, J. D. (1975). Growth of skeleton weed (*Chondrilla juncea* L.) as affected by growth of subterranean clover (*Trifolium subterraneum* L.) and infection by *Puccinia chondrillina* Bubak & Syd. *Australian Journal of Agricultural Research*, **26**, 975–83.

Gurusiddaiah, S., Gealy, D. R., Kennedy, A. C., & Ogg, A. G. Jr. (1994). Isolation and characterization of metabolites from *Pseudomonas fluorescens*-D7 for control of downy brome (*Bromus tectorum*). *Weed Science*, **42**, 492–501.

Hallett, S. G., Paul, N. D., & Ayres, P. G. (1990). *Botrytis cinerea* kills groundsel (*Senecio vulgaris*) infected by rust (*Puccinia lagenophorae*). *New Phytologist*, **114**, 105–10.

Harper, J. L. (1977). *Population Biology of Plants*. London: Academic Press.

Harris, P. (1997). Monitoring and impact of weed biological control agents. In *Ecological Interactions and Biological Control*, ed. D. A. Andow, D. W. Ragsdale & R. F. Nyvall, pp. 215–23. Boulder, CO: Westview Press.

Harris, P., Dunn, P. H., Schroeder, D., & Vonmoos, R. (1985). Biological control of leafy spurge in North America. In *Leafy Spurge*, ed. A. K. Watson, pp. 79–92. Champaign, IL: Weed Science Society of America.

Hartke, A., Drummond, F. A., & Liebman, M. (1998). Seed feeding, seed caching, and burrowing behaviors of *Harpalus rufipes* De Geer larvae (Coleoptera: Carabidae) in the Maine potato agroecosystem. *Biological Control*, **13**, 91–100.

Hartzler, R. G. (1996). Velvetleaf (*Abutilon theophrasti*) population dynamics following a single year's seed rain. *Weed Technology*, **10**, 581–6.

Hasan, S., & Ayres, P. G. (1990). The control of weeds through fungi: principles and prospects. *New Phytologist*, **115**, 201–22.

Hodgson, R. H., Wymore, L. A., Watson, A. K., Snyder, R. H., & Collette, A. (1988). Efficacy of *Colletotrichum coccodes* and thiadiazuron for velvetleaf (*Abutilon theophrasti*) control in soybean (*Glycine max*). *Weed Technology*, **2**, 473–80.

Hoffman, J. H., & Moran, V. C. (1995). Localized failure of a weed biocontrol agent attributed to insecticide drift. *Agriculture, Ecosystems and Environment*, **52**, 197–203.

Hokkanen, H. M. T. (1986). Success in classical biological control. *CRC Critical Reviews in Plant Sciences*, **3**, 35–72.

Jackson, M. A., Schisler, D. A., Slininger, P. J., Boyette, C. D., Silman, R. W., & Bothast, R. J. (1996). Fermentation strategies for improving the fitness of a bioherbicide. *Weed Technology*, **10**, 645–50.

Jacobs, J. S., Sheley, R. L., & Maxwell, B. D. (1996). Effect of *Sclerotinia sclerotiorum* on the interference between bluebunch wheatgrass (*Agropyron spicatum*) and spotted knapweed (*Centaurea maculosa*). *Weed Technology*, **10**, 13–21.

Julien, M. H., & Griffiths, M. W. (eds) (1998). *Biological Control of Weeds: A World Catalogue of Agents and Their Target Weeds*, 4th edn. Wallingford, UK: CAB International.

Julien, M. H., Kerr, J. D., & Chan, R. R. (1984). Biological control of weeds: an evaluation. *Protection Ecology*, **7**, 3–25.

Kempenaar, C., Horsten, P. J. F. M., & Scheepens, P. C. (1996). Growth and competitiveness of common lambsquarters (*Chenopodium album*) after foliar application of *Ascochyta caulina* as a mycoherbicide. *Weed Science*, **44**, 609–14.

Kennedy, A. C. (1997). Deleterious rhizobacteria and weed biocontrol. In *Ecological Interactions and Biological Control*, ed. D. A. Andow, D. W. Ragsdale & R. F. Nyvall, pp. 164–77. Boulder, CO: Westview Press.

Kennedy, A. C., Elliot, L. F., Young, F. L., & Douglas, C. L. (1991). Rhizobacteria suppressive to the weed downy brome. *Soil Science Society of America Journal*, **55**, 722–7.

Kennedy, A. C., & Kremer, R. J. (1996). Microorganisms in weed control strategies. *Journal of Production Agriculture*, **9**, 480–5.

Klein, T. A., & Auld, B. A. (1996). Wounding can improve efficacy of *Colletotrichum orbiculare* as a mycoherbicide for Bathurst burr. *Australian Journal of Experimental Agriculture*, **36**, 185–7.

Kremer, R. J. (1993). Management of weed seed banks with microorganisms. *Ecological Applications*, **3**, 42–52.

Kremer, R. J., & Kennedy, A. C. (1996). Rhizobacteria as biocontrol agents of weeds. *Weed Technology*, **10**, 601–9.

Kremer, R. J., & Spencer, N. R. (1989). Impact of a seed-feeding insect and microorganisms on velvetleaf (*Abutilon theophrasti*) seed viability. *Weed Science*, **37**, 211–16.

Lawton, J. H. (1988). Biological control of bracken in Britain: constraints and opportunities. *Philosophical Transactions of Royal Society of London, Series B*, **318**, 335–55.

Liebman, M., & Gallandt, E. R. (1997). Many little hammers: ecological management of crop–weed interactions. In *Ecology in Agriculture*, ed. L. E. Jackson, pp. 291–343. San Diego, CA: Academic Press.

Lindquist, J. L., Maxwell, B. D., Buhler, D. D., & Gunsolus, J. L. (1995a). Velvetleaf (*Abutilon theophrasti*) recruitment, survival, seed production, and interference in soybean (*Glycine max*). *Weed Science*, **43**, 226–32.

Lindquist, J. L., Maxwell, B. D., Buhler, D. D., & Gunsolus, J. L. (1995b). Modeling the population dynamics and economics of velvetleaf (*Abutilon theophrasti*) control in a corn (*Zea mays*)–soybean (*Glycine max*) rotation. *Weed Science*, **43**, 269–75.

Louda, S. M., Kendall, K., Connor, J., & Simberloff, D. (1997). Ecological effects of an insect introduced for the biological control of weeds. *Science*, **277**, 1088–90.

Lund, R. D., & Turpin, F. T. (1977). Carabid damage to weed seeds found in Indiana corn fields. *Environmental Entomology*, **6**, 695–8.

Mabbayad, M. O., & Watson, A. K. (1995). Biological control of gooseweed (*Sphenoclea zeylandica* Gaertn.) with an *Alternaria* sp. *Crop Protection*, **14**, 429–33.

Makowski, R. (1997). Foliar pathogens in weed biocontrol: ecological and regulatory constraints. In *Ecological Interactions and Biological Control*, ed. D. A. Andow, D. W. Ragsdale & R. F. Nyvall, pp. 87–103. Boulder, CO: Westview Press.

McEvoy, P., Cox, C., & Coombs, E. (1991). Successful biological control of ragwort, *Senecio jacobaea*, by introduced insects in Oregon. *Ecological Applications*, **1**, 430–42.

McEvoy, P. B., Rudd, N. T., Cox, C. S., & Huso, M. (1993). Disturbance, competition, and herbivory effects on ragwort *Senecio jacobaea* populations. *Ecological Monographs*, **63**, 55–75.

McFayden, R. E. C. (1998). Biological control of weeds. *Annual Review of Entomology*, **43**, 369–93.

Moran, V. C. (1995). Plant/insect interactions in farmland habitats: the utility of seed-reducing insects in the suppression of alien, woody weeds. In *Ecology and Integrated Farming Systems*, ed. D. M. Glen, M. P. Greaves & H. M. Anderson, pp. 103–15. Chichester, UK: John Wiley.

Muller-Scharer, H., & Schroeder, D. (1993). The biological control of *Centaurea* spp. in North America: do insects solve the problem? *Pesticide Science*, 37, 343–53.

Newman, R., Thompson, D., & Richman, D. (1998). Conservation strategies for control of weeds. In *Conservation Biological Control*, ed. P. Barbosa, pp. 371–96. San Diego, CA: Academic Press.

Norris, R. F. (1997). Impact of leaf mining on the growth of *Portulaca oleracea* (common purslane) and its competitive interaction with *Beta vulgaris* (sugarbeet). *Journal of Applied Ecology*, 34, 349–62.

Paul, N. D., & Ayres, P. G. (1987). Effects of rust infection of *Senecio vulgaris* on competition with lettuce. *Weed Research*, 27, 431–41.

Perfecto, I. (1994). The transformation of Cuban agriculture after the cold war. *American Journal of Alternative Agriculture*, 9, 98–108.

Pfiffner, L., & Niggli, U. (1996). Effects of biodynamic, organic, and conventional farming on ground beetles (Col. Carabidae) and other epigaeic arthropods in winter wheat. *Biological Agriculture and Horticulture*, 12, 353–64.

Phatak, S. C., Callaway, M. B., & Vavrina, C. S. (1987). Biological control and its integration in weed management systems for purple and yellow nutsedge (*Cyperus rotundus* and *C. esculentus*). *Weed Technology*, 1, 84–91.

Pitty, A., Staniforth, D. W., & Tiffany, L. H. (1987). Fungi associated with caryopses of *Setaria* species from field-harvested seeds and from soil under two tillage systems. *Weed Science*, 35, 319–23.

Quimby, P. C. Jr., & Walker, H. L. (1982). Pathogens as mechanisms for integrated weed management. *Weed Science*, 30 (Supplement), 30–4.

Randall, J. M. (1996). Weed control for the preservation of biological diversity. *Weed Technology*, 10, 370–83.

Simberloff, D., & Stiling, P. (1996). How risky is biological control? *Ecology*, 77, 1965–74.

Skipper, H. D., Ogg, A. G. Jr., & Kennedy, A. C. (1996). Root biology of grasses and ecology of rhizobacteria for biological control. *Weed Technology*, 10, 610–20.

Smith, H. S., & DeBach, P. (1942). The measurement of the effect of entomophagous insects on population densities of their hosts. *Journal of Economic Entomology*, 35, 845–9.

Strong, D. R. (1997). Fear no weevil? *Science*, 277, 1058–9.

Stroo, H. F., Elliot, L. F., & Papendick, R. I. (1988). Growth, survival and toxin production of root-inhibitory pseudomonads on crop residues. *Soil Biology and Biochemistry*, 20, 201–7.

Swiadon, L., Drlik, T., & Woo, I. (1998). Integrated control of leafy spurge. *IPM Practitioner*, 20(7), 1–11.

TeBeest, D. O. (1991). Ecology and epidemiology of fungal plant pathogens studied as biological control agents of weeds. In *Microbial Control of Weeds*, ed. D. O. TeBeest, pp. 97–114. New York: Chapman & Hall.

TeBeest, D. O. (1996). Biological control of weeds with plant pathogens and microbial pesticides. *Advances in Agronomy*, 56, 115–37.

Tisdell, C. A., Auld, B. A., & Menz, K. M. (1984). On assessing the value of biological control of weeds. *Protection Ecology*, 6, 169–79.

Tranel, P. J., Gealy, D. R., & Kennedy, A. C. (1993). Inhibition of downy brome (*Bromus tectorum*) root growth by a phytotoxin from *Pseudomonas fluorescens* strain D7. *Weed Technology*, 7, 134–9.

Watson, A. K. (1989). Current advances in bioherbicide research. In *Proceedings of the Brighton Crop Protection Conference – Weeds*, pp. 987–96. Farnham, UK: British Crop Protection Council.

Watson, A. K. (1991). The classical approach with plant pathogens. In *Microbial Control of Weeds*, ed. D. O. TeBeest, pp. 3–23. New York: Chapman & Hall.

Weidemann, G. J. (1991). Host-range testing: safety and science. In *Microbial Control of Weeds*, ed. D. O. TeBeest, pp. 83–96. New York: Chapman & Hall.

Wratten, S. D., & van Emden, H. F. (1995). Habitat management for enhanced activity of natural enemies of insect pests. In *Ecology and Integrated Farming Systems*, ed. D. M. Glen, M. P. Greaves & H. M. Anderson, pp. 117–45. Chichester, UK: John Wiley.

Zhang, J. X. (1993). Biology of *Harpalus rufipes* (Coleoptera: Carabidae) in Maine and dynamics of seed predation. MSc thesis, University of Maine, Orono, ME.

CHARLES P. STAVER

9

Livestock grazing for weed management

Introduction

Cattle, sheep, goats, and other domesticated vertebrates graze more than 50% of the earth's total land area, 20% in managed pastures and 30% in rangelands (Snaydon, 1981). Animal production and cropland management are also frequently linked. Animals graze the herbaceous understory in tree crops and feed on residues and remnant vegetation in annual crop fields. Animal manures are applied to croplands, and pastures and forage crops are rotated with annual crops.

Domesticated herbivores can accentuate weed problems for humans. They disperse weed seeds (Chapter 2). They graze preferred species heavily, but leave unpalatable species to grow and reproduce. They compact soil around watering holes, at resting sites, and along trails, which fosters grazing- and trampling-tolerant unpalatable weedy vegetation. Introduced forage species naturalize to become weedy invaders (Low, 1997).

However, through managed grazing animals can also reduce weedy vegetation and promote desirable forage species. This chapter illustrates three principles for the use of livestock to reduce weeds in annual and perennial crops and on grazing lands:

1. *A weed's susceptibility to control by grazing depends on its growth habit, its life cycle stage and the growing conditions at the time of grazing, and its palatability to different herbivore species.* The identification of a weed's particular vulnerabilities to grazing contributes to understanding why it has become a problem or might become a problem. Whether a weed is vulnerable to control by grazing also depends on the other plant species in the same grazed area and their ability to tolerate and avoid grazing. In fact, a weed in one context may be a primary forage species in another.

2. *To apply grazing pressure when weed vulnerability is greatest, farmers must routinely observe and analyze the floristic makeup of the forage and weed biomass.* Timely management decisions vary with weather conditions, changing economic factors, and with the species, number, and condition of the animals that the farmer has at the moment. As the botanical composition of a grazing area fluctuates or changes, farmers must adjust their practices.
3. *To manipulate grazing pressure for weed management, farmers must have access to the appropriate species and number of animals and the means, such as fencing, to confine them in the indicated grazing area.* Animal grazing pressure must be increased or reduced to coincide with weed vulnerabilities and to maintain the productivity of desirable crop or forage species. At the same time the farmer must also insure the animals' nutritional needs. Animals with low nutritional needs may be best suited for heavy grazing of weedy fields, but weed problems can also be prevented with highly productive animals by adjusting grazing season and intensity.

Matching grazing strategies with weed problems

Grazing animals are useful for three types of weed control. First, grazing can reduce the total biomass of possible competing vegetation, for example, in timber tree plantations or orchards. Second, grazing can be directed at the biomass of a single species or groups of species. Geese grazing for grass seedling control in horticultural crops and goat grazing for shrub control on sheep or cattle pastures illustrate this case. Third, grazing can be used to reduce weed seed production and survival as, for example, when animals graze weeds and crop residues during fallow periods.

A grazing plan to reduce specific weed problems should consider differences in animal grazing habits, vulnerability of weed species to grazing, and the plant community response to grazing.

Grazing species differences

Three sets of characteristics are particularly important for understanding the effect of different types of grazing animals on weeds and using animal species differences in grazing management for weed reduction.

First, grazing animal species have different grazing actions and dietary preferences. A goose's beak is suitable for small, precise bites and very shallow furrowing action in the soil. Pigs dig to much greater depths in search of perennial plant storage organs. Based on their dietary preferences, grazers, browsers, and intermediates can be distinguished (Vallentine, 1990, pp.

220–42). The grazers, which include cattle, horses, and bison, principally consume grasses, although they may shift to forbs and shrubs low in volatile oils when grass is less available. Bison and horses are even less flexible than cattle in their grazing preferences and rarely consume more than 15% forbs. Goats consume a high percentages of forbs and shrubs and use plant materials high in volatile oils more effectively than grazers. Sheep are intermediate feeders. They use grasses, forbs, and shrubs, depending on availability. Fox & Seaney (1984) observed goats, cattle, and sheep grazing together in northeastern USA. They found that goats consumed a greater number of plant species and obtained only 34% of their diet from grasses, whereas sheep and cattle obtained 78% and 90%, respectively, from grasses.

A second important characteristic distinguishing different groups of grazing animals is their susceptibility to plant chemical compounds. A plant that is poisonous for one herbivore may not be poisonous for another species (Launchbaugh, 1996). *Centaurea solstitialis* is poisonous to horses, but the pre-spiny stages can be consumed by sheep and cattle (Thomsen et al., 1993). Cattle avoid *Euphorbia esula* and may avoid palatable plants within *E. esula* infestations, whereas sheep can be managed to control this weed with minimal detrimental effects (Lorenz & Dewey, 1988). A lethal dose of *Senecio jacobaea* for cattle or horses is 3% to 7% of body weight, whereas for sheep and goats a lethal dose is 200% to 300% of body weight (Sharrow, Ueckert & Johnson, 1988).

Lastly, treading impacts vary among animal species, a result of differences in size relative to total hoof area. Sheep range from 0.7 to 0.9 kg of body weight per square centimeter of hoof area, whereas cattle, with 1.3 to 2.8 kg cm^{-2} of hoof area, have a much greater impact (Spedding, 1971, p. 115). The treading impact of geese, with their large foot area and low body weight, is much lower. Treading impact can be increased by the deliberate movement of large groups of animals, preferably cattle due to their size, in an agitated fashion to break down unused standing forage for faster decomposition, to trample more fragile weed species, and to open up areas of dense shrub growth for grazing by smaller species (Savory, 1988, pp. 263–72; Harris, 1990).

The role of plant palatability, architecture, and life cycle in herbivory

Whether grazing can be used to reduce a weed problem depends on the ability of each plant species in the vegetation complex to avoid herbivory and to recover afterwards. Briske (1996) referred to these as avoidance and tolerance mechanisms that together determine a plant's ability to grow and reproduce under grazing.

Whether a plant is grazed at all and the extent of defoliation are influenced by positive and negative plant quality factors. Herbivores avoid or graze less on plants with low nutritional quality (protein, digestible dry matter), high secondary plant metabolites (tannins, alkaloids, nitrates, and oxalates), and physical defenses (thorns and spines) (Minson, 1981; Norton, 1982; Launchbaugh, 1996). These factors vary not only by plant species, but also by plant growth stage. For example, sheep and cattle consume the tender shoots of *Cirsium arvense*, but not the mature, spiny foliage (Hartley, Lyttle & Popay, 1984). Only those weeds that are readily grazed without harm to the herbivore can be controlled by grazing management (Scifres, 1991). Weeds that are unpalatable or toxic are not susceptible to control by selective grazing, although they may be weakened by deliberate trampling.

In addition to mechanisms to avoid grazing, plants have compensatory mechanisms to speed recovery from grazing (Crawley, 1983, pp. 86–110; Whithman *et al*., 1991). This ability to tolerate grazing depends on morphological and physiological characteristics that come into play depending on the severity of defoliation. Regrowth will be most rapid after limited defoliation if leaf growing points are still intact, with cell elongation occurring from leaf meristems. Regrowth after more severe defoliation is slower, since axillary buds must first differentiate meristems and then root mass and surface area must recuperate. Radiotracer studies of forage species after total defoliation have shown that new growth becomes independent of root reserves within 10 days for alfalfa and 36–48 hours for perennial ryegrass (Smetham, 1990). Such studies on weed species are uncommon.

The more complete and more frequent is the defoliation of a weed species, the more likely that grazing will be a useful control measure. In a worst-case scenario, cattle and wildlife do not consume the foliage of the subtropical weed *Solanum viarum* and thus cannot be used to reduce weed vigor, but ingest the fruit, spreading the seed in nutrient-rich bare spots (Mullahey *et al*., 1998). This herbivory pattern has accelerated the spread of this recently introduced weed throughout the southeastern USA.

Both weed and desirable pasture and range species can be classified morphologically by the location of their apical meristematic tissue (Clements, 1989). Many weeds species have buds that are inaccessible to common grazing herbivores and consequently recover rapidly from grazing. These include *Imperata cilindrica*, *Sorghum halepense*, *Cirsium arvense*, and *Euphorbia esula*, all of which spread by growth of roots or rhizomes. The growing points of stoloniferous and prostrate species are right at the ground surface and are partially consumed only with the most severe grazing. Rosette growth in the first year protects certain biennial and winter annual weeds during part of their life cycle. When grazing was used to destroy the bolting flowerhead of *Centaurea solstitialis*, a winter annual weed of California rangelands, plants resprouted from basal and axillary buds and produced new inflorescences, often in greater number than without grazing (Thomsen *et al.*, 1993). Annual, erect-growing broadleaf species are poorly defended against herbivory, because they produce few buds which are highly accessible at the base of leaves or branches. Shrubs and some perennial forbs, both as weeds and forage components in rangelands, regrow from aerial axillary buds when the apical bud is removed. Some species sprout from basal buds or roots. Depending on shrub architecture, many or few buds may be available for regrowth after grazing (Orshan, 1989).

The physiological mechanisms for herbivory tolerance include increased photosynthetic rate after tissue removal, temporary reallocation of photosynthates among shoots and roots, and use of accumulated carbohydrates (Briske & Richards, 1994). These mechanisms have been studied principally in range species that have different degrees of grazing tolerance. Richards (1984) compared the response of *Agropyron cristatum* and *A. spicatum* to grazing and attributed *A. cristatum*'s greater defoliation tolerance to the temporary reallocation of photosynthates from root growth to shoot growth. *Agropyron cristatum* had 50% lower root growth during the recovery period compared with *A. spicatum*.

Plants in different growth stages differ in their ability to resist grazing or tissue loss. Weather conditions also affect a plant's recovery from tissue loss. Seedlings are often the most vulnerable stage and may be damaged either by grazing or trampling. Survival of seedlings of the perennial range grasses *Agropyron cristatum* and *A. desertorum* was only 0.4% when grazed or trampled compared to 11.6% when cattle were excluded in a semiarid Utah environment (Saliki & Norton, 1987). However, seedling survival of *Macroptilium atropurpureum*, a leguminous forage in tropical Australia, increased from 0%–1% to approximately 14% when grazing was increased from 1.1 to 1.7 animals ha^{-1} (Jones & Bunch, 1987). Trampling and grazing losses were offset by the gains from improved seedling growth with increased grazing of an associated perennial grass.

Perennial plants, particularly in climates with cold winters or extended dry seasons, have regular annual cycles of accumulation and use of carbohydrate and other reserves. These fluctuations influence their response to herbivory or tissue loss (Caldwell, 1984). Cyclic seasonal fluctuations in carbohydrate reserves were key to understanding the response of the perennial pasture weed *Pteridium aquilinum* to burning which destroys plant tissue similar to grazing or trampling (Preest & Cranswick, 1978). A midsummer burn right

Table 9.1. *Grazing management by plant growth stage to favor or disfavor the growth of specific forages and weeds*

Growth stage	Actions to favor relative plant performance	Actions to disfavor relative plant performance
Germination and seedling establishment	· graze to create gaps in sward · graze to reduce growth of vegetation around seedling · reduce grazing pressure after seedling emergence	· maintain complete plant cover · trample seedlings · graze seedlings heavily
Vegetative growth	· allow sufficient leaf area for regrowth · graze to limit excessive self-shading of leaves · remove animals to allow recovery of leaf area and roots	· defoliate completely · graze to eliminate buds · repeat defoliation at short intervals
Flowering and seed production	· in annuals, graze to encourage seeding · in tillering perennials, graze to avoid flowering to promote tiller vigor and number	· graze to reduce flowering and seeding · reduce grazing to promote flowering to senesce plant
Accumulation of reserves	· reduce grazing when reserves cannot be restored · reduce grazing when growing conditions are unfavorable	· graze when reserves are low · graze when regrowth will deplete reserves

after the completion of frond expansion, when rhizome carbohydrate levels were at their low point, caused decreased bracken biomass production for the following two years. Burning in the spring or fall when carbohydrate reserves were high increased bracken biomass in the following years.

Excessive herbivory, either in frequency or intensity or when growing conditions do not favor plant regrowth, reduces root growth, retards bud and tiller development, and may lead to plant death. Table 9.1 contrasts the use of herbivory in different growth stages to either weaken a plant or to strengthen its position relative to neighboring plants. This format can be applied to specific weeds to identify and exploit their relative weaknesses.

Weeds in a plant community

Any land unit used for grazing is occupied by plant species in different proportions with different life histories, growth habits, and ecophysiologies (Figure 9.1). At any point in time, each species consists of a dispersed quantity of leaves, stems, buds, roots, and seeds (Figure 9.1). Fluctuations and changes in this vegetation complex in response to grazing, trampling, and

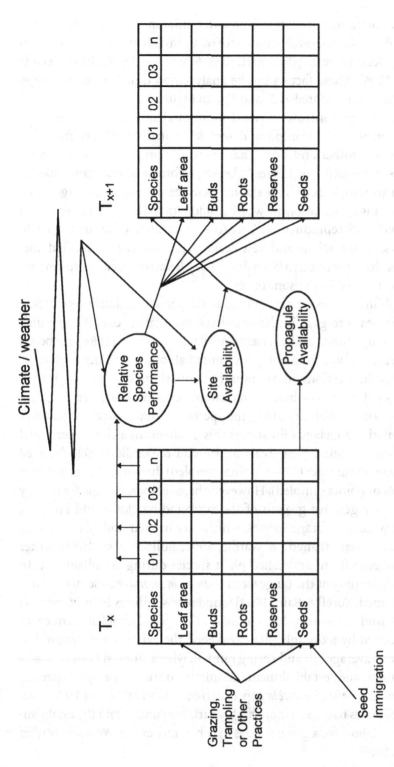

Figure 9.1 Grazing, trampling, or other practices alter the plant components, which converts to a new array of plant components through relative species performance, site availability, and propagule availability. Weather variability affects relative plant performance.

other planned management derive from differential species performance, the availability of sites for the establishment of new individuals, and the availability of propagules (Figure 9.1) (Pickett, Collins & Armesto, 1987; Sheley, Svejcar & Maxwell, 1996). These factors can be analyzed on time intervals of days, months, or years, represented as T_x and T_{x+1} in Figure 9.1.

Relative species performance is based on the mechanisms of resistance and tolerance to grazing and trampling described earlier. These vary from one plant species to another and drive changes in the proportions of different plant species in a pasture or range. Relative performance may cause species composition to remain relatively stable, cycle seasonally, or undergo large cumulative changes. Variation in weather adds unpredictable variability to species growth and reproduction. Year-to-year weather variations, even in humid areas such as England and New Zealand, produce two- to six-fold fluctuations in herbage production (Snaydon, 1981). Relative species performance also affects soil protection (Olson, 1999).

The availability of sites for new individuals depends on relative species performance that leads to gaps in the sward and on weather conditions which control soil temperature and moisture availability. Wiens (1984) proposed that ecosystems are on a gradient from limited abiotic variability with large effects of biotic interactions on community composition to high abiotic variability with small effects of biotic interactions on community composition. The nature of site availability along the spectrum from temperate pastures through semiarid rangelands illustrates this gradient. In a clover–perennial ryegrass sward in humid New Zealand, Panetta & Wardle (1992) detected more small gaps in summer (43% of points sampled) than in spring or autumn (18% and 28% of points sampled). However, these temporary gaps frequently were closed by vegetative growth of the surrounding clover and ryegrass. Seedlings of weed and forage species, which vary in their ability to colonize gaps of different sizes (Panetta & Wardle, 1992; Bullock *et al.*, 1995), suffer severe interference from established plant species during establishment. In contrast, in their study of the range weed *Gutierrizia sarothrae* in semiarid New Mexico, McDaniel, Torell & Bain (1993) found new seedlings in only one year of an 11-year study, primarily due to rainfall. The special establishment event was characterized by a drought that weakened the existing vegetation, followed by above-average fall and spring rainfall, which allowed both *G. sarothrae* germination and establishment. Similarly, during a 45-year period, significant recruitment of *Astrebla* spp. occurred only in 1945 and 1987 (Roe, 1987). Arid climates have even more highly variable rainfall, and the establishment of a new cohort for a given species may be a rare event (Westoby, Walker & Noy-Meir, 1989).

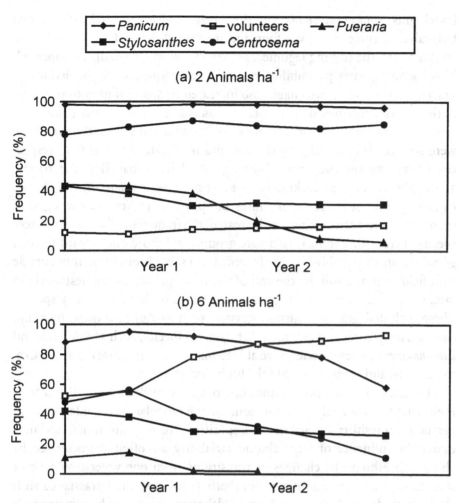

Figure 9.2 Effects of two stocking rates on relative grass and legume species proportions and the invasion of weedy volunteer species in a pasture in Malaysia. (Adapted from Eng, Kerridge & t'Mannetje, 1978.)

Propagule availability derives from the persistence of the soil seed bank and the dispersal of seeds into a site (Chapter 2). Special combinations of weather conditions may also affect propagule availability. Without adequate moisture or other germination cues, it might be argued that seeds are not available, as in the case of *G. sarothrae* in the previous paragraph. Propagule availability is related primarily to germination conditions, whereas site availability involves the seedling establishment phase.

The interactive effects of differential species performance, site availability, and seed availability are illustrated by results of a pasture experiment in Malaysia (Figure 9.2) (Eng, Kerridge & t'Mannetje, 1978). Paddocks with the bunch grass *Panicum maximum* and three legumes were stocked with weaned bull calves at three rates (two, four, and six animals ha^{-1}). Even at the lowest stocking rate, the trailing legume, *Pueraria phaseoloides*, virtually disappeared. This species' greater palatability increased the frequency with which it was grazed. Its trailing growth habit also increased its susceptibility to treading. Both factors contributed to slower *P. phaseoloides* regrowth than for other species. The other two legumes, *Centrosema pubescens* and *Stylosanthes guianensis*, were favored differentially by the low and moderate stocking rates respectively. *Centrosema pubescens* has a climbing growth habit that allowed it to grow above grasses at lower stocking rates. Erect-growing *S. guianensis* was shaded by taller grasses at lower grazing rates, but prospered when grass was grazed more heavily. As the stocking rate increased, the frequency of weedy volunteer species increased. The increased bare ground and more open sward favored germination and establishment of weeds from seeds brought in from outside each field or in the soil. By the end of the third year at the highest stocking rate, the planted grass and legumes were being replaced by weedy species. These included grazing-resistant grasses, such as *Paspalum conjugatum* and *Axonopus compressus*, and unpalatable broadleaf species, such as *Sida acuta* and *Lantana cinerea*. These species were able to maintain or increase their leaf area, roots, buds, and seeds, in spite of the high stocking rate.

The states and transitions framework, proposed by Westoby, Walker & Noy-Meir (1989) for the analysis of non-equilibrium rangelands, provides a similar perspective on shifts in vegetation composition. Non-equilibrium rangelands under the influence of large abiotic variability are often characterized by abrupt, threshold-like changes or transitions from one vegetation type or state to another. For example, in southern Texas, perennial grasslands shift abruptly to dense *Prosopis* shrublands, although there may be intermediate states such as grasslands infested with shrub seedlings (Archer, 1989). Westoby, Walker & Noy-Meir (1989) suggested that even when transitions are gradual and linear (e.g., in grasslands that receive more abundant and regular precipitation), the states and transitions framework can usefully organize information on vegetation composition and factors contributing to the change from one state to another.

A states-and-transitions diagram defines the different possible alternative persisting states for the vegetation in a grazing region (Figure 9.3). Dominant forage and weed species present, vegetation structure, or soil condition may be used to define states. The different states are the result of grazing management, fertilization, overseeding of forages, and discrete events like fires. Variability in weather frequently plays a critical role. A change in vegetation represents a different state, if it alters livestock production or signals an

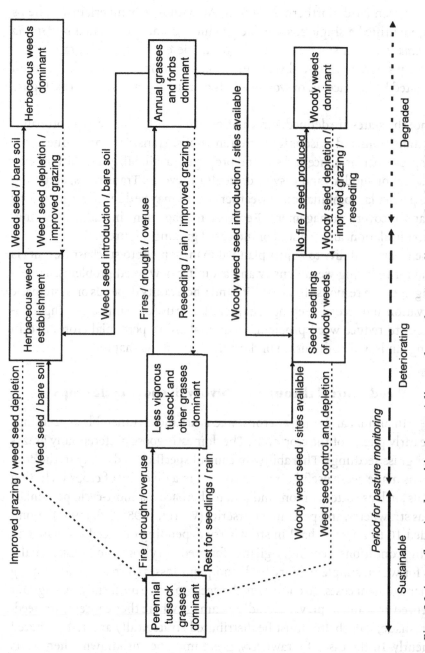

Figure 9.3 A state (boxes) and transition (arrows) diagram for northern Australian grasslands. Solid lines represent degradation factors and broken lines represent recovery forces. (Adapted from McArthur, Chamberlain & Phelps, 1994.)

incipient future state of recovery or degradation (Bellamy & Brown, 1994). Seasonal fluctuations repeated year after year are not considered different states. In semiarid northern Australia, McArthur, Chamberlain & Phelps (1994) identified a single sustainable productive state dominated by *Astrebla* and *Dichanthium*, a state with less vigorous perennial tussock grasses, two transitory stages with initial weed invasion, and three states of degradation dominated by herbaceous or woody weeds or annual grasses and forbs (Figure 9.3).

Once the states are defined, information can be assembled on the probability of transitions and reasons for such transitions. Transitions occur due to relative species performance and site and propagule availability. Transitions may be gradual or sudden, and easy or difficult to reverse. Transitions, occurring with a particular combination of weather events, may be due to overgrazing or fires and recovery management. Recovery management includes reseeding, fuel accumulation, and reduced or increased grazing (Figure 9.3).

The farmer's ability to adjust planned management to stochastic events is critical for reducing weed density and avoiding new weed problems. Timely management in response to variability and uncertainty depends on systematic observation and record-keeping (Chapter 3). The use of planned grazing management to reduce weed problems in annual crops, perennial crops, and in grazing lands will be discussed in the remainder of the chapter.

Weed control through herbivory in short-cycle crops

In annual and nursery crops, weeds are most vulnerable as seedlings in the early phases of the crop cycle. The domestic goose preferentially grazes tender grass seedlings. This ability to control specific weeds early in the crop cycle has made geese useful for weed control in a diversity of crops including annuals such as cotton, onion, and potato, herbaceous short-cycle perennials such as strawberry, and perennial nurseries (Wurtz, 1995). Doll (1981) recommended 10 to 20 geese ha^{-1} in strawberry depending on weed pressure and growing conditions. Fencing, vigilance for predatory dogs and foxes, installations for water, supplementary feed, and plant losses to treading are among the management costs. To encourage the geese to graze uniformly throughout the fenced area and to prevent treading damage where they concentrate, feed, water, shade, and shelter must be distributed strategically and redistributed frequently. In the case of strawberry, geese must be withdrawn when fruits begin to ripen.

In a two-year study Wurtz (1995) compared geese alone and in combination with hoeing and herbicides for weed control in seedlings of the unpalatable

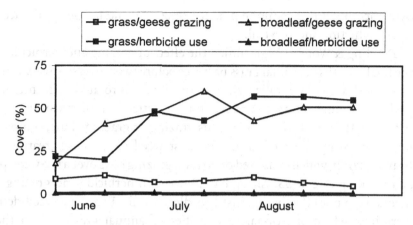

Figure 9.4 Effects of geese and herbicides for control of grasses and unpalatable weeds in the second year of an ornamental tree nursery. (Adapted from Wurtz, 1995.)

evergreen *Picea glauca*. In the first year geese were placed on the plots when palatable weed cover exceeded 5%. The lack of sufficient forage contributed to trampling losses in the crop of greater than 30%. In the second year, the weed cover threshold was increased to 15% and tree mortality declined by half to levels similar to herbicide treatments. By the end of the second year, weed floristic composition varied significantly among treatments. With geese only, unpalatable broadleaf species such as *Matricaria matricarioides*, *Polygonum aviculare*, and *Tripleurospermum phaeocephalum* had reached 50% cover and grasses were unimportant. With a recommended herbicide for Christmas trees, grasses such as *Agropyron repens* and *Hordeum jubatum* reached 50% cover and the unpalatable broadleaf species were not present (Figure 9.4). By the end of the second year in both the geese and herbicide treatments the build-up of uncontrolled weeds resulted in decreased evergreen stem diameters.

An integrated strategy to control both grass and broadleaf weeds might consist of larger flocks of geese herded slowly by workers hoeing ungrazed weeds. This would reduce treading damage and management costs such as fencing and vigilance against predators, and provide more complete weed control (Wurtz, 1995).

Aftermath and fallow grazing for weed control

Grazing fields in fallow or after the cropping period takes advantage of the forage value of weeds and crop residues and accelerates nutrient cycling by converting vegetation to manures with more concentrated and more readily available nutrients. Although there are few formal studies of weed dynamics under this common practice, grazing could be used to reduce weed seed production and survival.

Dowling & Wong (1993) studied the effect of grazing and herbicide use in the final months of annual grass pastures before two successive wheat crops in New South Wales, Australia. Wheat in this region is rotated with annual grass pastures. Annual grass seed and seedling densities in the first wheat crop were reduced 91%–99% by herbicides plus grazing or grazing alone, compared to no pre-planting weed control. The most weed suppressive treatment was heavy grazing, which consisted of 10 total grazing days over a six-week period at a stocking rate of 533 sheep ha^{-1}. Preseason herbicide and grazing treatments were not effective against broadleaf weeds. Broadleaf weed densities were inversely proportional to densities of annual grasses, with the no-grazing–no-herbicide treatment having the most annual grasses and the fewest broadleafs. The proportion of *Bromus* and *Vulpia* spp. declined into the second crop cycle, whereas the proportion of *Lolium rigidum* increased, due to less effective control by preseason treatments, less effective control in the wheat, and greater seed dormancy. Wheat yields were higher in treatments with preseason vegetation management for both years of the two-year wheat sequence.

A number of factors contributed to the effectiveness of grazing in reducing weed densities in this experiment. First, most weed seed production occurred at the end of the pasture cycle when wheat was absent from the field. In many annual crop systems, weed seed production occurs in the last part of the crop cycle before crop harvest, when grazing is not feasible. Second, the seeds of the principal annual grass weeds, except *Lolium rigidum*, maintained little viability in the soil seed bank, and were therefore highly affected by reductions in current-year seed production. Seed banks are much longer-lived in some other annual crop systems (Chapter 2). Third, the weeds to be controlled were palatable to sheep. Fourth, large numbers of livestock with maintenance-level nutrition requirements were available to be concentrated in small areas. These factors make clear that the successful use of grazing during fallow periods for reducing weed seed production will require a careful match-up of grazing rates with crop cycles and weed seed production periods.

On smallholder farms, weeds themselves may make up an important part of the available forage in aftermath and fallow grazing (Humphreys, 1991, pp. 6–12). Farmers may weed selectively during the cropping period with the express purpose of increasing forage availability during the fallow period. In coastal Ecuador Nuwanyakpa *et al.* (1983) found that seven weeds had higher digestibility and crude protein in the dry season than improved forages. Cattle readily consumed *Alternanthera gullensis*, but not other weeds such as *Sida* spp.

Grazing for weed control in tree crops

Many tree crops, including coconuts, oil palm, and most fruit trees, are palatable to domestic herbivores. Unless the young trees are protected, orchards and plantations cannot be grazed until the tree foliage is out of the reach of grazing animals. This occurs a year or two earlier for sheep than for cattle. For needleleaf timber trees, however, only buds and succulent new growth are browsed, and grazing can begin shortly after tree planting. With adequate management, needleleaf timber plantations can be grazed each year even before new growth has hardened off (Krueger, 1985).

After the first few years, the options for employing herbivory for weed control depend on differences in canopy shape and size, planting density, and planting arrangement. In oil palm the major opportunity for grazing is from years three to eight (Payne, 1985), since weed growth is minimal under the dense canopy of mature plantations. In fruit tree plantings, individual tree canopies do not overlap. Shade is heavy at the base of individual trees with full sun outside the canopy. In established coconut, canopy cover is complete, but light transmission in mature plantations may be as high as 50% to 60% (Watson & Whiteman, 1981). For a silvopastoral timber plantation in New Zealand, after year three once trees are established, grazing can be increased to reduce vegetation competition with trees and to increase animal production (Figure 9.5). The carrying capacity declines as the tree canopy increases and reduces herbage growth, but then increases again when the plantation is thinned in year six. From year eight onward, depending on initial planting density, the carrying capacity for grazing herbivores declines.

While initially livestock may have been viewed as occasional weed or brush consumers in tree plantations, grazing for weed control has evolved into dual-purpose silvopastoral management. The land is managed routinely for both tree and animal products.

Weed control in evergreen plantations

Some form of weed control such as grazing is necessary in young needleleaf plantations to reduce shading in more humid environments and moisture stress in dry environments. In southwestern Oregon either grazing or herbicides to control planted forages resulted in sufficient moisture for three additional weeks of Douglas fir (*Pseudotsuga menziesii*) growth compared with plots with no understory vegetation control (Krueger, 1985). In another study, a mixed planting had greater height and diameter at breast height with cattle grazing than without (Krueger, 1985). Cattle grazing increased height and diameter of ponderosa pine (*Pinus ponderosa*) 9% and 13% and of western larch

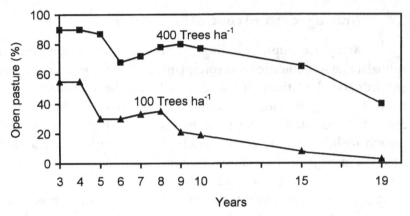

Figure 9.5 Predicted effects of timber tree age and planting density on livestock carrying capacity. (Adapted from Knowles, 1991.)

(*Larix occidentalis*) 61% and 38%, respectively. In addition to reducing moisture stress, grazing accelerates nutrient cycling. Spring and summer sheep grazing increased tree growth equivalent to the application of 100 and 50 kg ha^{-1} of ammonium sulfate respectively in the Coast Range of Oregon (Sharrow & Leininger, 1983).

Two separate economic comparisons demonstrated the effectiveness of grazing for vegetation management in needleleaf tree plantations (Krueger, 1985). In one case, grazing was found to be 25%–33% of the cost of chemical or manual control. In the other case in which grazing was being used on 240 000 hectares, chemical grass or brush control cost $75–150 ha^{-1}, whereas grazing produced a slight profit of $2.50 ha^{-1}. In both studies plantation managers thought trees grew better when grazing was used for vegetation management than when herbicides were used, although no data were presented.

Damage to trees from browsing or treading can be highly variable depending on available forage, livestock management, and age of the trees. Under the management described below, tree mortality was 1% to 5% (Currie, Edminster & Knott, 1978; Krueger, 1985; Thomas, 1985), which is usually only a small part of total mortality caused by factors such as wild herbivores and droughts. Damage levels to young plants from herbicide use were similar to or greater than the levels due to grazing (Krueger, 1985). Young plants can tolerate some defoliation, loss of lateral branches, and bark scraping (Hughes, 1976; Lewis, 1980). Trampling damage is more likely with cattle, whereas browsing damage occurs more with sheep (Sharrow & Leininger, 1983).

Grazing for vegetation control varies in effectiveness depending on the palatability of available foliage. On more favorable sites, improved forages may be planted. Krueger (1985) found that seeded grasses reduced the presence of native grasses, forbs, and shrubs, although cattle grazing was not effective in controlling shrubs. Thomas (1985) found that grazing sheep reduced heavy brush infestation of the palatable *Ceonothus integerrimus*, but consumed very little *Arctostaphylos patula*. He concluded that to avoid unfavorable shifts in botanical composition to unpalatable species and a loss of effectiveness of grazing for vegetation control, grazing must be managed not only to avoid damage to trees, but also to promote recovery of palatable browse and herbaceous species.

Recommended management of grazing for vegetation control without damage to associated needleleaf trees follows principles already described for geese grazing. Uniform grazing, careful balancing of animal grazing with available forage, and avoiding routine animal concentration in the same areas within the plantation can be achieved by careful placement of salt and water, riders or herders to observe and manage livestock grazing and behavior, and grazing of the plantation when available forage is most palatable (Krueger, 1985). Uniform grazing with limited damage to trees was easier to achieve by grazing 25% to 600% more animals than standard practice in the area, and managing them more carefully, rather than reducing stocking rate (Monfore, 1983). If an individual herd or flock shows an inclination to browse the young needleleaf trees, it must be removed from the plantation to another range. Precipitation may vary greatly between and within years (Boyd, 1985), making careful and flexible grazing management based on observation essential to high tree survival.

On many tree plantations, grazing for vegetation control may not be applicable due to the inaccessibility of the site, lack of water, excessive slopes, presence of slash, or the type of vegetation. The best sites for tree growth are also the best for the use of grazing (Krueger, 1985).

Grazing in coconut plantations

Grazing for general weed control in established tree crop plantations is a common practice in coconuts (Payne, 1985). Grazing for vegetation management is suitable in areas with more than 2000 mm of rainfall, although even in marginal areas (1300–2000 mm rainfall) coconuts are not clean tilled (Santhirasegarum, 1966). Because of shading by coconut, plant species for grazing and plantation ground cover must be adapted to irradiance levels that are only 40%–70% of full sunlight. Grasses and legumes show different degrees of tolerance for shade (Shelton, Humphreys & Batello, 1987). C_4 species may be more affected than C_3 species. Eriksen & Whitney (1982) found that legumes, such as *Desmodium intortum* and *Leucaena leucocephala*, continued to fix nitrogen, even when shaded. Humphreys (1991, p. 13) suggested that

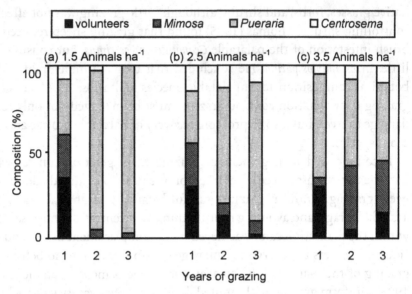

Figure 9.6 Effects of three cattle-stocking rates on floristic composition and weedy volunteer species of pastures under coconut. (Adapted from Smith & Whiteman, 1985.)

legumes were likely to dominate under grazed coconut as they are less affected by shade and may be less palatable than grasses under year-round moist conditions necessary for coconut production. Smith & Whiteman (1985) in cattle grazing trials under coconuts in the Solomon Islands found that legumes dominated at all stocking densities (Figure 9.6). With an increasing stocking rate and a diminished forage plant cover, weedy volunteer species including the legume *Mimosa pudica* also increased.

Higher coconut yields were found with grazing than without (Santhirasegarum, 1966; Rika, Nitis & Humphreys, 1981; Ferdinandez, 1973), and grazing cattle also produced milk and meat. Rika, Nitis & Humphreys (1981) found an increase from 263 to 454 nuts ha^{-1} month^{-1} with increasing stocking rate from 2.7 to 6.3 cattle ha^{-1}. This increase was attributed to more rapid nutrient cycling at higher stocking rates. Recommended stocking rates depend, however, on light transmission. In the Soloman Islands, 0.7 animals ha^{-1} were recommended with 35% transmission; 1.3 animals ha^{-1} with 50%; and 2.5 animals ha^{-1} with 80% (Humphreys, 1991, p. 15).

Grazing should be managed to maintain animal production (Humphreys, 1991, p. 15). A vigorous sward with abundant, high-quality leaf matter will also reduce the invasion of unpalatable weedy species. The grazing schedule may be arranged to coincide with nut harvesting. A short stubble right before monthly harvest aids in the collection of fallen nuts.

Grazing for weed control in pastures and rangelands

Weeds in grazing lands may be grasses, forbs, shrubs, or trees. Weed problems may result from temporary phenomena dependent on pasture age or weather, from a slow response to inadequate management, or from the invasion of exotic species.

When pastures are established in arable crop land, most weeds originate from the previous land use through the soil seed bank. While special measures are needed for their control, these primarily annual weeds are seldom an important part of the plant community for more than a short period. The increased availability of nitrogen under intensive pastures with highly productive legumes may also provide the conditions for an increase in herbaceous nitrophilous weeds, many of which are annuals (Matthews, 1981; Tothill, Mott & Gillard, 1981).

Shrub encroachment can result from either overgrazing or undergrazing. An increase of woody species on semiarid lands results from the overgrazing of the grass layer and greater infiltration of water to the subsoil. This favors the establishment and growth of shrubs which then shade out the overgrazed grass (Coppock, 1993). This situation is associated more often with cattle than sheep (Tothill, Mott & Gillard, 1981). New Zealand hill pastures established after forest clearing in wetter areas suffered shrub encroachment as soil fertility declined. Uneven grazing of low palatability forages at declining stocking rates allowed the establishment of shrubs such as *Ulex europaeus* tolerant of low fertility and acid soils (Daly, 1990). Many pastures were eventually abandoned.

An increase of low-palatability herbaceous vegetation occurs under a variety of conditions. Tothill, Mott & Gillard (1981) proposed that in drier zones more palatable species are under both grazing and environmental stress, whereas less palatable species are only under environmental stress. In the wet tropics, overgrazed planted grasses lose vigor as soil fertility declines, and are displaced by less palatable species or weeds. Grazing-tolerant grasses such as *Paspalum conjugatum*, *Paspalum notatum*, and *Axonopus compressus*, and native legumes produce a relatively stable, although not highly productive, pasture (Serrao & Toledo, 1990). However, planted pastures may also be invaded by unpalatable broadleaf and grass weeds such as *Pseudoelephantopus spicatus* or *Homolepsis aturensis* under the same conditions. With erratic manual or chemical weed control and low grazing pressure, these pastures revert to patchy shrub and secondary forest (Serrao & Toledo, 1990).

Introduced species can also become weeds in grazing lands, altering the existing vegetation composition, before monitoring and management responses can be developed (DiTomaso, 2000). *Euphorbia esula* has invaded not only cattle grazing lands in southern Canada and north-central USA, but also roadsides and parks (Swiadon, Drlik & Woo, 1998). *Solanum viarum* has invaded several hundred thousand hectares of pastures, roadsides, and perennial crops such as citrus and sugarcane in the southeastern USA, especially Florida, since it was first identified in 1987 (Mullahey *et al.*, 1998).

Weed impact on grazing land productivity

Weeds in pastures and rangelands interfere directly and indirectly with the production of animal products.

Direct interference results from either the grazing of poisonous plants, discussed earlier, or from physical damage to animals or products. For example, burrs from *Bassia birchii* in subtropical semiarid regions or *Arctium minus* in temperate humid regions contaminate wool. The sharply awned seeds of *Hordeum murinum* and *H. leporinum* perforate sheep pelts, damage meat quality, and injure young lambs, but represent minimal risks to cattle (Field & Daly, 1990).

Indirect interference by weeds is a product of their effects on the quality and quantity of forage. Kelly & Popay (1985) found *Carduus nutans* ground cover to be 2.8%, 29.7%, and 6.2% over three years in two *C. nutans*-infested pastures. They hypothesized that if animal production losses were in direct proportion to *C. nutans* ground cover, control would be profitable only in one year. *Rumex obtusifolius*, a weed of temperate pastures, has only 65% of the nutrient value of perennial ryegrass and produces only 55% to 80% as much biomass (Courtney, 1985). With three to four harvests per year, grass biomass was reduced up to 70%, depending on weed density. Grass biomass declined 1% for every 1% increase in *R. obtusifolius* cover. With five to seven cuts per year, up to 10% *R. obtusifolius* cover did not affect perennial ryegrass yields, although with greater cover, losses reached 15%. On the other hand, Meeklah & Mitchell (1985) found that 13% and 18% ground cover by *Bellis perennis* caused no yield reduction in a high-yielding sward, and 28% ground cover resulted in only a 6% decline in forage production. While *B. perennis* is highly visible in ryegrass–clover pastures, its prostrate growth habit and low growth rates compared to forage species minimize its impact on production.

Several studies have measured the effect of varying levels of weed infestations on animal production. Hartley (1983a) found that the control of *Juncus* spp. infestations of 7% to 11% ground cover resulted in a 16% to 19% increase in stock-carrying capacity. In another study, Hartley (1983b) found that over a four-month grazing period sheep live-weight gains were reduced 1.7 kg for each *Cirsium vulgare* plant m^{-2} over a range of 0.1 to 1.5 plants m^{-2}. The use of herbicides for the control of *C. vulgare* damaged clover and reduced liveweight gains by an amount equivalent to the reduction from 1.7 *C. vulgare* plants m^{-2}. In the second year of the study, live-weight gain was not affected by *C. vulgare* density.

These studies indicate that the impact of weeds on grazing land productivity depends on weed density and the level of forage use. Weeds reduce animal gains primarily when weed infestations are severe and animal use and forage availability are closely matched (Hartley, 1983b). Less palatable species do not reduce animal gains at low infestation levels and when surplus forage is available. Even when they are not grazed, weedy species still protect soil, add organic matter, and recycle nutrients. The decision to control weeds in grazing lands should incorporate not only weed control costs, but also the likelihood that additional forage will be converted to animal products and soil will not be left unprotected. Unless a weed is replaced by a more productive forage species which in turn is converted into animal products, direct weed control expenditures may be uneconomic. However, numerous experiments have shown that low-cost or income-generating practices such as changing the grazing regime, adding additional fencing, or introducing a different animal species may shift the floristic composition in a pasture or rangeland to more productive, less weedy species, reduce the severity of weed outbreaks, and maintain soil cover.

Weed control through altered grazing regimes

Simple changes in grazing rates and schedules may in some cases effectively reduce pasture weeds.

Hordeum glaucum, *H. murinum*, and *H. leporinum* are annual grasses that become unpalatable after stem elongation and flowering, and later produce seeds with spine-like awns that damage young stock (Popay & Field, 1996). Hartley *et al.* (1978) tested lax and hard grazing treatments in autumn, winter, and spring for *Hordeum* spp. control. They found that adjusting sheep numbers to maintain desirable pasture species at 2–4 cm during all grazing seasons effectively prevented flowering and seed production of *Hordeum* spp. Maintaining desirable pasture species at 2–4 cm also guaranteed adequate pasture cover in the late summer, which limited *Hordeum* spp. germination. By the third year of grazing, *Hordeum* spp. had virtually disappeared from the flexible stocking treatment which maintained sward height of 2–4 cm and from the treatment with hard spring and autumn grazing, but light summer and winter grazing. In this later treatment hard spring grazing prevented flowering and light summer grazing assured abundant pasture cover during the period of potential *Hordeum* spp. germination.

Carduus pycnocephalus and *C. tenuiflorus* are also vulnerable to an altered grazing regime (Bendall, 1973). These weeds germinate in autumn, overwinter as rosettes, and seed in the spring in Tasmania, Australia. Withholding grazing in the autumn created increased competition from desirable forage species in the pasture. Seedlings of both *Carduus* species became etiolated and less prickly compared to fall-grazed treatments in which seedlings were compact and had hardened spines. During winter or spring grazing following no grazing in autumn, sheep preferentially grazed *C. pycnocephalus* and *C. tenuiflorus*. They also consumed the growing points, which were several centimeters above the soil surface. The growing points of these weeds on autumn-grazed pastures were below ground level and inaccessible to grazing sheep. With autumn grazing *Carduus* spp. densities were 4 to 13 plants m^{-2} compared with 0.8 to 2.0 plants m^{-2} without grazing. The reduction in *Carduus* spp. numbers did not carry over to the following year. While the seed bank was not measured in the study, this was probably the source of new *Carduus* seedlings. Over time, impacts of the different grazing regimes may have become more evident.

Chrysanthemum leucanthemum is unpalatable to cattle. Olson, Wallander & Fay (1997) noted that severe infestations are the product of season-long grazing at low cattle-stocking densities and suggested that while sheep might graze *C. leucanthemum* better than cattle, most cattle ranches are not equipped to manage sheep. They proposed that intensive grazing by cattle might expose *C. leucanthemum* to non-selective grazing, trampling damage, and untimely germination and seedling mortality. In a two-year experiment comparing intensive cattle grazing to ungrazed controls, the grazed treatment had lower densities of *C. leucanthemum* in the seed bank, as seedlings, and as rosettes (Olson, Wallander & Fay, 1997). Adult *C. leucanthemum* densities were not different in grazed and ungrazed treatments, although cattle both trampled and pulled out *C. leucanthemum* stems. The investigators concluded that grazing and trampling impacts on the initial *C. leucanthemum* life stages could be expected to lead to longer-term declines in adult infestations.

Effects of paddock size and uniformity on weed control

Increasing grazing pressure when weeds are vulnerable or decreasing grazing pressure to reduce weed establishment sites can be achieved when paddocks contain relatively uniform vegetation on similar soil and slope conditions. In their reviews of the spatial heterogeneity of plant–large herbivore interactions in grazing systems, Coughenour (1991) and Bailey *et al.* (1996) concluded that the scale of large patches, corresponding to animal grazing and movement and often correlated with abiotic site factors, was the appropriate scale for the improvement of grazing management.

The difficulties in controlling grazing pressure in highly heterogeneous paddocks were demonstrated in a study by Gillen, Krueger & Miller (1984). They documented the distribution of cattle grazing within a diverse paddock using a preference index. This was calculated as the proportion of time the animals spent grazing a large relatively uniform patch divided by the relative physical space represented by that patch within the paddock. The paddocks in the study were over 3500 ha. Cattle had low preference indices for slopes greater than 20%, and spent more time on increasingly gentler slopes. They also grazed more heavily closer to water. Preference indices fell below 1.0 on areas 600 m or more from water sites. Cattle grazing distribution was not altered by the location of salt in their study. Hart *et al.* (1993) compared continuous grazing in a 24-ha paddock with rotational grazing on 20 paddocks of 7 ha. In smaller paddocks use did not vary, whereas in the large paddock forage use declined with distance from water. In large paddocks, animal grazing time was lower, animal travel time increased, and cow/calf weight gains were lower.

To facilitate timely and flexible grazing, large, internally diverse paddocks can be subdivided into smaller paddocks which are less variable internally. With smaller paddocks, grazing and other management inputs can be adjusted to the weed infestations and sward composition and vigor in each area. Animals can be concentrated in certain paddocks for short periods, while other paddocks recover.

Electric fencing has increased the practicality of permanent or temporary small paddocks. The recent development of lightweight temporary fence posts, high-tensile wire, solar-powered chargers, and portable meshes has also reduced the cost and made small paddocks feasible in areas remote from electrical installations.

A farmer's options to increase or decrease grazing animal numbers for weed control are greatly facilitated by an agile marketing system, adequate off-farm infrastructure for animal transport, and the availability of off-farm feed supplies (Medd, Kemp & Auld, 1987; Stafford Smith, 1996). For these reasons, in remote range- or pasture-based livestock areas in developing countries, farmers can not easily vary their herd or flock numbers or feed resources at the farm level in a short period. On the other hand, in temperate humid pasture regions, many options are readily available to farmers to permit short-term expansion or contraction of grazing pressure on specific paddocks.

Changing the grazing species to control weeds

The addition of an animal species specifically for weed control may require new management skills for the farmer and increased investment in facilities, fencing, and animals. However, mixed-species grazing can be effective for diminishing weed problems at low cost and under certain circumstances also produce additional income.

The addition of sheep to a bull rotational grazing system to control *Senecio jacobaea*, a weed highly toxic to cattle, but less toxic to sheep, was tested by Betteridge *et al.* (1994). Herbicides are effective for *S. jacobaea* control, but also reduce forage legume growth. Timely mowing eliminates *S. jacobaea* seed production. However, mowed plants become perennial and produce seed the following year if mowing is not continued. Young bulls were rotationally grazed at 10.5 stocking units ha^{-1} on 15 1-ha paddocks, completing the rotation every 30 to 50 days. Sheep were grazed with the bulls at 1.5 or 3.0 stocking units ha^{-1} on the same rotation or were mob-grazed for four days four times annually at two similar rates. In mob-grazing a large number of sheep are grazed in a small paddock for short periods. By the end of the first year 82% of *S. jacobaea* in the bulls-only plots were flowering, while only 32% to 36% and 0% to 5% of the *S. jacobaea* in the bulls +1.5 and 3.0 stocking units of sheep ha^{-1} were in flower. *Senecio jacobaea* plants were largest with bulls-only grazing. *Senecio jacobaea* plants grazed with bulls +3.0 stocking units of sheep were smaller than either bulls+sheep mob-grazing or bulls +1.5 stocking units of sheep. The density of new *S. jacobaea* seedlings increased when sheep were grazed rotationally with young bulls, but decreased with either level of bulls+sheep mob-grazing. Betteridge *et al.* (1994) suggested that infrequent sheep mob-grazing did not open the sward as much to new *S. jacobaea* seedling establishment, although mob-grazing was less effective in reducing the size of already established *S. jacobaea*.

The use of cattle to graze *Nardus stricta*-infested sheep pastures was studied by Grant *et al.* (1996). This weedy perennial tussock grass has a lower feed value, lower growth rate, and lower palatability than most other grasses of the Scottish hill region. It also provides poor habitat for birds compared to the heather communities that it often replaces. In the study, grasslands with more than 50% *N. stricta* cover were continuously grazed by sheep at two intensities (grasses between *N. stricta* tussocks maintained at 3.5 or 4.5 cm) and one cattle grazing routine (4.5 cm). Over a five year period, *N. stricta* cover increased 86% in the 4.5-cm sheep treatment and 72% in the 3.5-cm sheep treatment, but declined 30% in the 4.5-cm cattle treatment (Figure 9.7). *Nardus stricta* was not highly competitive, showing lower leaf expansion rates and lower tiller pro-

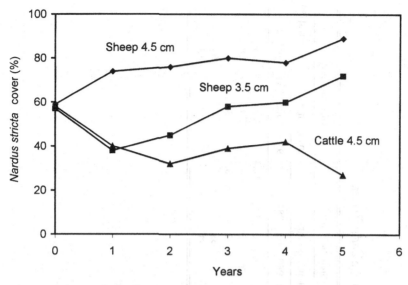

Figure 9.7 Changes in *Nardus stricta* cover in a grassland under sheep and cattle grazing treatments. (Adapted from Grant *et al.*, 1996.)

duction than other species in the sward. However, it was grazed much less by sheep than by cattle. Cattle also uprooted *N. stricta* tillers while grazing. This weedy tussock grass had lower tiller base weight and energy reserves under cattle grazing, which reduced its ability to recover after defoliation.

Goat grazing is highly effective for weed and brush control in sheep or cattle pastures (Wood, 1987; Underwood *et al.*, 1996). Rolston, Lambert & Clark (1982) and Rolston *et al.* (1981) studied effects of mixed-species grazing on a weedy pasture in New Zealand. Treatments included mob-grazing and set-grazing of sheep, goats, or sheep and goats (continuous grazing at a fixed level). The flock mix for sheep and goats was either 33%/67% or 67%/33%. The presence of goats in any proportion effectively controlled *Cirsium palustre* and *C. vulgare* (Table 9.2), but was ineffective against *C. arvense*. Goats at 100% and 66%, but not 33%, reduced *Juncus* spp. height (Table 9.2) and opened clumps to invasion by forage species. Goats at 100% and 66% decreased *Ulex europaeus* height and survival after two years of grazing (Table 9.2). Mob-grazed sheep had nearly double the annual grazing days ha^{-1} of the set-grazed sheep and proved highly effective for *U. europaeus* control. Mob-grazing was partially effective against *C. vulgare*, *C. palustre*, and *Juncus* spp. (Table 9.2). Rolston *et al.* (1981) emphasized that the experiment demonstrated the importance of the use of an appropriate mix of grazing species, small paddocks, and high stocking rates to achieve adequate grazing pressure for weed control and sward vigor.

Table 9.2 The effect of different combinations of sheep and goat grazing on weeds in a New Zealand hill pasture

Grazing species used	Juncus spp. Height (cm)		Cirsium vulgare			Cirsium palustre		Ulex europaeus		
			Height (cm)	Survival (%)		Flowers plant^{-1}		Density (m^{-1})		Survival (%)
	Feb 1980	Feb 1981	Nov 1980	Nov 1980	Apr 1981	June 1980	Feb 1981	Apr 1979	June 1981	June 1981
Sheep/goat 0%/100%	40	20	16	0	0	nd	nd	12.4	3.1	25
Sheep/goat 33%/66%	54	38	11	0	0	21	3	10.3	2.9	28
Sheep/goat 66%/33%	96	93	11	0	0	nd	2	6.3	5.3	67
Sheep/goat 100%/0%	86	105	8	53	48	68	61	7.6	5.1	67
Sheep mob-grazed	48	60	8	45	47	67	31	7.1	0.9	13

Sources: Rolston et al. (1981); Rolston, Lambert & Clark (1982).

Research directions

Livestock grazing for weed control is only one component of the ecology and management of grazing systems. In a recent comprehensive review of grazing systems covering the spectrum from intensive perennial pastures to arid rangelands, Hodgson & Illius (1996) did not cite the topic of weeds in the index, although the ecological approaches they present are readily applicable to the management of weedy grazing lands.

Future weed research in grazing systems can be considered under three general topics.

First, research to identify the ecological vulnerability of problem weed species to grazing and other practices should continue. This chapter has presented numerous examples of how timely grazing can reduce weed abundance. Concepts from sward productivity (Lemaire & Chapman, 1996) and species persistence (Marten *et al.*, 1989) can be applied to both desirable and undesirable vegetation. Management principles must continue to be developed to increase the resistance of grazing lands to weed invasion. Further application of the principles for the reduction of weed persistence, productivity, and reproduction outlined in this chapter are also needed.

Second, research on grazing for weed control must be integrated with research on other management practices, especially for the successful control of problematic invasive weeds. For weeds such as *Euphorbia esula*, *Solanum viarum*, *Centaurea solstitialis*, and *Rosa multiflora* that have invaded roadsides, pastures, and wild areas, reduction of infestations in grazing lands can only be partially remedied by livestock management. Integrated strategies for grazed and ungrazed lands which combine tactics such as biocontrol (Chapter 8), field-level monitoring for early detection, mowing, and forage reseeding need to be further explored (Sheley, Kedzie-Webb & Maxwell, 1999).

Third, research must be expanded to focus on the development of concepts and tools for improving livestock grazing for weed management on large land units. Identification of critical information needs for farmer decision-making, sampling methods for large land areas, simple predictive decision criteria in the context of early warning and risk management, management of highly variable landscapes, and contingency strategies are research themes related to the improvement of grazing and weed management at the landscape scale.

Grazing for weed control shows ecological and practical applicability in diverse situations from small, highly uniform horticultural fields to medium-sized, somewhat uniform temperate hillslope pastures to large heterogeneous rangelands. In each case the goals of grazing management are to improve vegetation quality and productivity, conserve soil, water, and biological resources, and yield animal products profitably. Essential to these goals are grazing managers who regularly monitor the state of their animals, vegetation, and natural resources (Chapter 3) (Bingham & Savory, 1990, pp.87–120; Watters, 1990; Sheath & Clark, 1996; Stafford Smith, 1996).

A practical challenge for the application of improved grazing management for weed control at the landscape scale is the development of simple sampling methods to categorize the state of a given paddock or range and to detect incipient transitions. A comparison of four sampling methods by Stohlgren, Bull & Otsuki (1998) illustrated difficulties associated with the evaluation of range vegetation. Whereas common transect methods satisfactorily estimated forage availability and soil cover of the major plant species, rare species were generally underestimated. They proposed multiscale sampling instead of transect methods. Small quantitative sample plots serve to monitor the abundance of common species, while a search of a larger surrounding area for additional species would detect rare plant species or recent invaders. The difficult task of early detection of noxious species is key to their successful management (Zamora & Thill, 1999). Temperate pasture sampling has focused on biomass or height to estimate available forage (Sheath & Clark, 1996) with little attention to weed presence. Multiscale sampling in these pastures would permit accurate estimation of available forage, while monitoring the status of less-frequent weed invaders.

The task of detecting transitions from sustainable vegetation states to deterioration (Figure 9.3) is especially challenging. Stockwell *et al.* (1994) found that farmers readily verified the states proposed by scientists for *Chrysopogon fallax* grasslands. However, identification of the factors related to transitions was more difficult, attributed by the authors to the lack of clear documentation of the nature of transitions, highly variable experiences among observers, and the circumstantial, non-systematic nature of the information. Detection or prediction of transitions requires not only the sampling of species and biomass, but also the recording of information that aids in understanding why and how the vegetation may be changing. Such information includes rainfall events, soil moisture status, grazing levels, fire, and seed production. For temperate humid grass farms, Watters (1990) and Webby & Sheath (1991) suggested daily monitoring of rainfall and soil temperature and bi-weekly estimates of pasture cover, growth rates, and grazing levels. Animal body condition should be assessed at least monthly. In their proposal, other animal performance data and soil fertility levels were monitored less frequently. For the monitoring to be used widely by scientists, technicians, and farmers, Watters (1990) recommended that pasture sampling methods be standardized and a practical notebook developed. Although not mentioned by Watters (1990), standardized sampling methods for weeds are also needed.

Methods of vegetation monitoring for improved pasture and range management can be more effective when developed through group collaboration among farmers, technicians, and scientists (Chapter 3) (Webby & Sheath, 1991; Behnke & Scoones, 1993; Clark & Filet, 1994). The analysis of transitions by farmers, extensionists, and scientists would aid in the identification of knowledge gaps and research priorities for specific grazing systems. Weed scientists can contribute to ecological weed management not only through plot research, but also by working with groups of farmers to develop simple, effective sampling methods, weed control strategies for large land units, and decision tools.

REFERENCES

Archer, S. (1989). Have southern Texas savannas been converted to woodlands in recent history? *American Naturalist*, **134**, 545–61.

Bailey, D., Gross, J., Laca, E., Rittenhouse, L., Coughenour, M., Swift, D., & Sims, P. (1996). Mechanisms that result in large herbivore grazing distribution patterns. *Journal of Range Management*, **49**, 386–400.

Behnke, R., & Scoones, I. (1993). Rethinking range ecology: implications for rangeland management in Africa. In *Range Ecology at Disequilibrium*, ed. R. Behnke, I. Scoones & C. Kerven, pp. 1–30. London: Overseas Development Institute.

Bellamy, J., & Brown, J. (1994). State and transition models for rangelands. 7. Building a state and transition for management and research on rangelands. *Tropical Grasslands*, **28**, 247–55.

Bendall, G. (1973). The control of slender thistle, *Carduus pycnocephalus* L. and *C. tenuiflorus* Curt. (Compositae), in pasture by grazing management. *Australian Journal of Agricultural Research*, **24**, 831–7.

Betteridge, K., Costall, D., Hutching, S., Devantier, B., & Liu, Y. (1994). Ragwort (*Senecio jacobaea*) control by sheep in a hill country bull beef system. *New Zealand Plant Protection Conference Proceedings*, **47**, 53–57.

Bingham, S., & Savory, A. (1990). *Holistic Resource Management Workbook*. Washington, DC: Island Press.

Boyd, R. (1985). The case for quantitative evaluation of the effects of operationally applied silvicultural weed control treatments. In *Weed Control for Forest Productivity in the Interior West*, ed. D. Baumgartner, R. Boyd, D. Breuer & D. Miller, pp. 27–8. Pullman, WA: Washington State University.

Briske, D. (1996). Strategies of plant survival in grazed systems: a functional interpretation. In *The Ecology and Management of Grazing Systems*. ed. J. Hodgson & A. Illius, pp. 37–68. Wallingford, UK: CAB International.

Briske, D., & Richards, J. (1994). Physiological responses of individual plants to grazing: current status and ecological significance. In *Ecological Implications of Livestock Herbivory in the West*, ed. M. Vavra, W. Laycock & R. Pieper, pp. 147–76. Denver, CO: Society for Range Management.

Bullock, J., Clear Hill, B., Silvertown, J., & Sutton, M. (1995). Gap colonization as a sources of grassland community change: effects of gap size and grazing on the rate and mode of colonization by different species. *Oikos*, **72**, 273–82.

Caldwell, M. (1984). Plant requirements for prudent grazing. In *Developing Strategies for Rangeland Management*, Board on Agriculture and Renewable Resources, National Research Council, National Academy of Sciences, pp. 117–52. Boulder, CO: Westview Press.

Clark, R., & Filet, P. (1994). Local best practice, participatory problem solving and benchmarking to improve rangeland management. In *Proceedings of 8th Biennial Conference, Australian Rangeland Society*, pp. 70–5. Condobolin, Australia: Australian Rangeland Society.

Clements, R. (1989). Rates of destruction of growing points of pasture legumes by grazing cattle. In *Proceedings of 16th International Grasslands Congress*, pp. 1027–8. Versailles: Association Française pour la Production Fourragère.

Coppock, D. (1993). Vegetation and pastoral dynamics in the Southern Ethiopian rangelands: implications for theory and management. In *Range Ecology at Disequilibrium*, ed. R. Behnke, I. Scoones & C. Kerven, pp. 42–61. London: Overseas Development Institute.

Coughenour, M. (1991). Spatial components of plant–herbivore interactions in pastoral, ranching, and native ungulate ecosystems. *Journal of Range Management*, **44**, 530–42.

Courtney, A. (1985). Impact and control of docks in grasslands. In *Weeds of Established Grasslands: Importance and Control*, ed. J. Brockman, pp. 120–7. Croydon, UK: British Crop Protection Council.

Crawley, M. (1983). *Herbivory: The Dynamics of Animal–Plant Interactions*. Berkeley, CA: University of California Press.

Currie, P., Edminster, C., & Knott, W. (1978). *Effects of Cattle Grazing on Ponderosa Pine Regeneration in Central Colorado*, Forest Service Research Paper RM-201. Washington, DC: US Department of Agriculture.

Daly, G. (1990). The grasslands of New Zealand. In *Pastures: Their Ecology and Management*, ed. R. Langer, pp. 1–37. Auckland, New Zealand: Oxford University Press.

DiTomaso, J. (2000). Invasive weeds in rangelands: species, impacts, and management. *Weed Science*, **48**, 255–65.

Doll, C. (1981). Geese for weeding strawberries. In *The Strawberry: Cultivars to Marketing*, ed. N. Childers, pp. 119–20. Gainesville, FL: Horticultural Publications.

Dowling, P., & Wong, P. (1993). Influence of preseason weed management and in-crop treatments on two successive wheat crops. 1. Weed seedling numbers and wheat grain yield. *Australian Journal of Experimental Science*, **33**, 167–72.

Eng, P., Kerridge, P., & t'Mannetje, L. (1978). Effects of phosphorus and stocking rate on pasture and animal production from a guineagrass–legume pasture in Johore, Malaysia. 1. Dry matter yields, botanical and chemical composition. *Tropical Grasslands*, **12**, 188–97.

Eriksen, F., & Whitney, A. (1982). Growth and nitrogen fixation of some tropical forage legumes as influenced by solar radiation regimes. *Agronomy Journal*, **74**, 703–9.

Ferdinandez, D. (1973). Utilisation of coconut lands for pasture development. *Ceylon Coconut Planter Review*, **7**, 14–19.

Field, R., & Daly, G. (1990). Weed biology and management. In *Pastures: Their Ecology and Management*, ed. R. Langer, pp. 409–47. Auckland, New Zealand: Oxford University Press.

Fox, D., & Seaney, R. (1984). Beefing up New York's abandoned farm land. *New York Food and Life Sciences Quarterly*, **15**, 7–9.

Gillen, R., Krueger, W., & Miller, R. (1984). Cattle distribution on mountain rangeland in northeastern Oregon. *Journal of Range Management*, **37**, 549–53.

Grant, S., Torvell, L., Sim, E., Small, J., & Armstrong, R. (1996). Controlled grazing studies on *Nardus* grassland: effect of between-tussock height and species of grazer on *Nardus* utilization and floristic composition in two fields in Scotland. *Journal of Applied Ecology*, **33**, 1053–64.

Harris, W. (1990). Pasture as an ecosystem. In *Pastures: Their Ecology and Management*, ed. R. Langer, pp. 75–131. Auckland, New Zealand: Oxford University Press.

Hart, R., Bissio, J., Samuel, M., & Waggoner, J. Jr. (1993). Grazing systems, pasture size, and cattle grazing behavior, distribution, and gains. *Journal of Range Management*, **46**, 81–7.

Hartley, M. (1983*a*). Effect of rushes on sheep carrying capacity. *New Zealand Weed and Pest Control Conference Proceedings*, **36**, 83–5.

Hartley, M. (1983*b*). Effect of Scotch thistles on sheep growth rates. *New Zealand Weed and Pest Control Conference Proceedings*, **36**, 86–8.

Hartley, M., Lyttle, L., & Popay, A. (1984). Control of Californian thistle by grazing management. *New Zealand Weed and Pest Control Conference Proceedings*, **37**, 24–7.

Hartley, M., Atkinson, G., Bimler, K., James, T., & Popay, A. (1978). Control of barley grass by grazing management. *New Zealand Weed and Pest Conference Proceedings*, **31**, 198–202.

Hodgson, J., & Illius, A. (1996). *The Ecology and Management of Grazing Systems*. Wallingford, UK: CAB International.

Hughes, R. (1976). Response of planted south Florida slash pine to simulated cattle damage. *Journal of Range Management*, **29**, 198–201.

Humphreys, L. (1991). *Tropical Pasture Utilization*. Cambridge, UK: Cambridge University Press.

Jones, R., & Bunch, G. (1987). The effect of stocking rate on the population dynamics of siratro in siratro (*Macroptilium atropurpureum*)–setaria (*Setaria sphacelata*) pastures in southeast Queensland. 1. Seed set, soil seed reserves, seedling recruitment and seedling survival. *Australian Journal of Agricultural Research*, **39**, 221–34.

Kelly, D., & Popay, A. (1985). Pasture production lost to unsprayed thistles at two sites. *New Zealand Weed and Pest Control Conference Proceedings*, **38**, 115–18.

Knowles, R. (1991). New Zealand experience with silvopastoril systems: a review. *Forest Ecology and Management*, **45**, 251–67.

Krueger, W. (1985). Grazing for forest weed control. In *Weed Control for Forest Productivity in the Interior West*, ed. D. Baumgartner, R. Boyd, D. Breuer & D. Miller, pp. 83–8. Pullman, WA: Washington State University.

Launchbaugh, K. (1996). Biochemical aspects of grazing behavior. In *The Ecology and Management of Grazing Systems*, ed. J. Hodgson & A. Illius, pp. 159–84. Wallingford, UK: CAB International.

Lemaire, G., & Chapman, D. (1996). Tissue flows in grazed plant communities. In *The Ecology and Management of Grazing Systems*, ed. J. Hodgson & A. Illius, pp. 3–37. Wallingford, UK: CAB International.

Lewis, C. (1980). Simulated cattle injury to planted slash pine: girdling. *Journal of Range Management*, **33**, 337–48.

Lorenz, R., & Dewey, S. (1988). Noxious weeds that are poisonous. In *The Ecology and Economic Impact of Poisonous Plants on Livestock Production*, ed. L. James, M. Ralphs & D. Nielsen, pp. 309–36. Boulder, CO: Westview Press.

Low, T. (1997). Tropical pasture plants as weeds. *Tropical Grasslands*, **31**, 337–43.

Matthews, L. (1981). Pasture weeds of New Zealand. In *Biology and Ecology of Weeds*, ed. W. Holzner & N. Numata, pp. 387–94. The Hague, Netherlands: Dr. W. Junk.

McArthur, S., Chamberlain, H., & Phelps, D. (1994). State and transition models for rangelands. 12. A general state and transition model for the mitchell grass, bluegrass-browntop and Queensland bluegrass pasture zones of northern Australia. *Tropical Grasslands*, **28**, 274–8.

McDaniel, K., Torell, L., & Bain, J. (1993). Overstory–understory relationships for broom snakeweed–blue grama grasslands. *Journal of Range Management*, **46**, 506–11.

Marten, G., Matches, A., Barnes, R., Brougham, R., Clements, R., & Sheath, G. (eds) (1989). *Persistence of Forage Legumes*. Madison, WI: American Agronomy Society.

Medd, R., Kemp, D., & Auld, B. (1987). Management of weeds in perennial pastures. In *Temperate Pastures: Their Production, Use and Management*, J. Wheeler, C. Pearson & G. Robards, pp. 253–61. Australia: Commonwealth Scientific and Industrial Research Organization (CSIRO).

Meeklah, F., & Mitchell, R. (1985). Effect of daisy removal on pasture production. *New Zealand Weed and Pest Control Conference Proceedings*, **38**, 119–21.

Minson, D. (1981). Nutritional differences between tropical and temperate pastures. In *Grazing Animals*, ed. F. Morley, pp. 143–58. New York: Elsevier.

Monfore, J. (1985). Livestock: a useful tool for vegetation control in ponderosa pine and lodgepole pine plantations. In *Forestland Grazing*, ed. B. Roche & D. Baumgartner, pp. 105–7. Pullman, WA: Washington State University.

Mullahey, J., Shilling, D., Mislevy, P., & Akanda, R. (1998). Invasion of tropical soda apple (*Solanum viarum*) into the US: lessons learned. *Weed Technology*, **12**, 733–6.

Norton, B. (1982). Differences between species in forage quality. In *Nutritional Limits to Animal Production from Pastures*, ed. J. Hacker, pp. 89–110. Farnham, UK: Commonwealth Agricultural Bureaux.

Nuwanyakpa, M., Bolsen, K., Posler, G., Diaz, M., & Rivera, F. (1983). Nutritive value of seven tropical weed species during the dry season. *Agronomy Journal*, **75**, 566–9.

Olson, B. (1999). Impacts of noxious weeds on ecologic and economic systems. In *Biology and Management of Noxious Rangeland Weeds*, ed. R. Sheley & J. Petroff, pp. 4–18. Corvallis, OR: Oregon State University Press.

Olson, B., Wallander, R., & Fay, P. (1997). Intensive cattle grazing of oxeye daisy (*Chrysanthemum leucanthemum*). *Weed Technology*, **11**, 176–81.

Orshan, G. (1989). Shrubs as a growth form. In *The Biology and Utilization of Shrubs*, ed. C. McKell, pp. 249–65. New York: Academic Press.

Panetta, F., & Wardle, D. (1992). Gap size and regeneration in a New Zealand dairy pasture. *Australian Journal of Ecology*, **17**, 169–75.

Payne, W. (1985). A review of the possibilities for integrating cattle and tree crop production systems in the tropics. *Forest Ecology and Management*, **12**, 1–36.

Pickett, S., Collins, S., & Armesto, J. (1987). Models, mechanisms, and pathways of succession. *Botanical Review*, **53**, 335–71.

Popay, I., & Field, R. (1996). Grazing animals as weed control agents. *Weed Technology*, **10**, 217–31.

Preest, D., & Cranswick, A. (1978). Burn-timing and bracken vigor. *New Zealand Weed and Pest Control Conference Proceedings*, **31**, 69–73.

Richards, J. (1984). Root growth response to defoliation in two *Agropyron* bunchgrasses: field observations with an improved root periscope. *Oecologia*, **64**, 21–5.

Rika, I., Nitis, I., & Humphreys, L. (1981). Effects of stocking rate on cattle growth, pasture production and coconut yield in Bali. *Tropical Grasslands*, **15**, 149–57.

Roe, R. (1987). Recruitment of *Astrebla* spp. in the Warrego region of southwestern Queensland. *Tropical Grasslands*, **21**, 61–2.

Rolston, M., Lambert, M., & Clark, D. (1982). Weed control options in hill country. *New Zealand Grassland Association Proceedings*, **43**, 196–203.

Rolston, M., Lambert, M., Clark, D., & Devantier, B. (1981). Control of rushes and thistles in pastures by goat and sheep grazing. *New Zealand Weed and Pest Control Conference Proceedings*, **34**, 117–21.

Saliki, D., & Norton, B. (1987). Survival of perennial grass seedlings under intensive grazing in semi-arid rangelands. *Journal of Applied Ecology*, **24**, 145–51.

Santhirasegarum, K. (1966). The effects of pastures on yield of coconut. *Journal of Agriculture (Trinidad–Tobago)*, **66**, 183–93.

Savory, A. (1988). *Holistic Resource Management*. Washington, DC: Island Press.

Scifres, A. (1991). Selective grazing as a weed control method. In *CRC Handbook of Pest Management in Agriculture*, ed. D. Pimentel, pp. 369–75. Boca Raton, FL: CRC Press.

Serrao, E., & Toledo, J. (1990). The search for sustainability in Amazonian Pastures. In *Alternatives to Deforestation*, ed. A. Anderson, pp. 195–214. New York: Columbia University Press.

Sharrow, S., & Leininger, W. (1983). Sheep as a silvicultural tool in coastal Douglas fir forest. In *Foothills for Food and Forest*, ed. D. Hannaway, pp. 219–31. Corvallis, OR: Oregon State University Press.

Sharrow, S., Ueckert, D., & Johnson, A. (1988). Ecology and toxicology of *Senecio* species with special reference to *Senecio jacobaea* and *Senecio longilobus*. In *The Ecology and Economic Impact of Poisonous Plants on Livestock Production*, ed. L. James, M. Ralphs & D. Nielsen, pp. 181–96. Boulder, CO: Westview Press.

Sheath, G., & Clark, D. (1996). Management of grazing systems: temperate pastures. In *The Ecology and Management of Grazing Systems*, ed. J. Hodgson & A. Illius, pp. 301–24. Wallingford, UK: CAB International.

Sheley, R., Kedzie-Webb, S., & Maxwell, B. (1999). Integrated weed management on rangeland. In *Biology and Management of Noxious Rangeland Weeds*, ed. R. Sheley & J. Petroff, pp. 57–68. Corvallis, OR: Oregon State University Press.

Sheley, R., Svejcar, T., & Maxwell, B. (1996). A theoretical framework for developing successional weed management strategies on rangeland. *Weed Technology*, **10**, 766–73.

Shelton, H., Humphreys, L., & Batello, C. (1987). Pastures in the plantations of Asia and the Pacific: performance and prospect. *Tropical Grasslands*, **21**, 159–68.

Smetham, M. (1990). Pasture management. In *Pastures: Their Ecology and Management*, ed. R. Langer, pp. 197–240. Auckland, New Zealand: Oxford University Press.

Smith, M., & Whiteman, P. (1985). Animal production from rotationally-grazed natural and sown pastures under coconuts at three stocking rates in the Solomon Islands. *Journal of Agricultural Science (Cambridge)*, **104**, 173–80.

Snaydon, R. (1981). The ecology of grazed pastures. In *Grazing Animals*, ed. F. Morley, pp. 13–31. New York: Elsevier.

Spedding, C. (1971). *Grassland Ecology*. Oxford, UK: Oxford University Press.

Stafford Smith, M. (1996). Management of rangelands: paradigms at their limits. In *The Ecology and Management of Grazing Systems*, ed. J. Hodgson & A. Illius, pp. 325–58. Wallingford, UK: CAB International.

Stockwell, T., Andison, R., Ash, A., Bellamy, J., & Dyer, R. (1994). State and transition models for rangelands. 9. Development of state and transition models for pastoral management of the golden beard grass and limestone grass pasture lands of NW Australia. *Tropical Grasslands*, **28**, 260–5.

Stohlgren, T., Bull, K., & Otsuki, Y. (1998). Comparison of rangeland vegetation sampling techniques in the Central Grasslands. *Journal of Range Management*, **51**, 164–72.

Swiadon, L., Drlik, T., & Woo, I. (1998). Integrated control of leafy spurge. *Integrated Pest Management Practitioner*, **20**(7), 1–11.

Thomas, D. (1985). The use of sheep to control competing vegetation in conifer plantations in the Downieville Ranger District, Tahoe National Forest: 1981–1984. In *Weed Control for Forest Productivity in the Interior West*, ed. D. Baumgartner, R. Boyd, D. Breuer & D. Miller, pp. 89–91. Pullman, WA: Washington State University.

Thomsen, C., Williams, W., Vayssieras, M., Bell, F., & George, M. (1993). Controlled grazing on annual grassland decreases yellow starthistle. *California Agriculture*, **47**, 36–40.

Tothill, J., Mott, J., & Gillard, P. (1981). Pasture weeds of the tropics and subtropics with special reference to Australia. In *Biology and Ecology of Weeds*, ed. W. Holzner & N. Numata, pp. 403–27. The Hague, Netherlands: Dr. W. Junk.

Underwood, J., Loux, M., Amrine, J. Jr., & Bryan, W. (1996). *Multiflora Rose Control*, Ohio State Extension Bulletin no. 857. Columbus, OH: Ohio State University.

Vallentine, J. (1990). *Grazing Management*. San Diego, CA: Academic Press.

Watson, S., & Whiteman, P. (1981). Grazing studies on the Guadalcanal Plains, Solomon Islands. 2. Effects of pasture mixtures and stocking rate on animal production and pasture components. *Journal of Agricultural Science (Cambridge)*, **97**, 353–64.

Watters, A. (1990). Farm monitoring as an initiator of improved farm management practices. *New Zealand Grassland Association Proceedings*, **52**, 17–19.

Webby, R., & Sheath, G. (1991). Group monitoring, a basis for decision making and technology transfer on sheep and beef farms. *New Zealand Grassland Association Proceedings*, **53**, 13–16.

Westoby, M., Walker, B., & Noy-Meir, I. (1989). Opportunistic management of rangelands not at equilibrium. *Journal of Range Management*, **42**, 266–74.

Whithman, T., Maschinski, J., Larson, K., & Paige, K. (1991). Plant responses to herbivory: the continuum from negative to positive and underlying physiological mechanisms. In *Plant–Animal Interactions*, ed. P. Price, T. Lewinsohn, G. Fernandes & W. Benson, pp. 227–56. New York: John Wiley.

Wiens, J. (1984). On understanding a non-equilibrium world: myth and reality in community patterns and processes. In *Ecological Communities: Conceptual Issues and the Evidence*, ed. D. Strong, D. Simberloff, L. Abele & A. Thistle, pp. 439–57. Princeton, NJ: Princeton University Press.

Wood, G. (1987). Animals for biological brush control. *Agronomy Journal*, 79, 319–21.

Wurtz, T. (1995). Domestic geese: biological weed control in an agricultural setting. *Ecological Applications*, 5, 570–8.

Zamora, D., & Thill, D. (1999). Early detection and eradication of new weed infestations. In *Biology and Management of Noxious Rangeland Weeds*, ed. R. Sheley & J. Petroff, pp. 73–84. Corvallis, OR: Oregon State University Press.

CHARLES L. MOHLER

10

Weed evolution and community structure

Introduction

Most weed management practices are motivated by short-term goals: reduction of weed impact on the current crop and prevention of seed production that could pose problems in succeeding crops. A slightly longer perspective may enter considerations of crop rotation and its impact on weeds (see Chapter 7), but weed management planning horizons of farmers rarely exceed five years. In contrast, important phenomena relating to weed diversity, community composition, and weed evolution affect weed communities on time scales of five years to centuries. In principle, these processes could be managed, though at present they largely are not. This chapter explains why long-term management of these phenomena may be needed, and outlines some tentative strategies.

The nature of long-term changes in weed species and communities has not been well documented and proposals for managing these changes are therefore necessarily speculative. Consequently, most of this chapter focuses on the ecological and evolutionary processes governing the changing nature of weed species and communities, with most suggestions for management reserved for the final sections. Three general points will be made.

First, evolutionary and community responses of the earth's flora to the resources available in farm fields leads to a continuous increase in the global diversity of agricultural weeds. Simultaneously, long-distance colonization events and local spread of species to new locations create a tendency toward increase in regional and local weed diversity. In the long run, *coping with the growing flexibility implied by this increasing weed diversity will require responsive management methods and may also require continual development of new weed control tactics to achieve a constant level of control.* At present, however, the global and regional increase in weed diversity is masked by a decrease in weed diversity at the farm scale that results from specialization on a few crops grown with highly standardized cultural practices.

Second, weeds are highly adaptable, as is evidenced by the locally differentiated races that have been documented in many species. Moreover, the growing problem of weed resistance to herbicides indicates that weeds adapt to management practices. In principle, weeds might adapt to ecological management methods as well. However, the selection pressure exerted by most ecological management tactics is less severe than the selection pressure from herbicides. Moreover, genetic models indicate that diversification of cropping systems and use of multiple control tactics should reduce the rate at which individual weed species evolve in response to a particular herbicide or ecological tactic. *Hence, flexible management using multiple ecological weed control tactics within a diverse cropping system may present sufficiently weak and contradictory selection pressures to avoid adaptation of weed species to management.*

Third, floristic and genetic changes in weed communities are fueled by dispersal of weeds between regions and within landscapes. Hence, *long-term management of weeds requires management of weed dispersal and early eradication of new colonies.*

The degree to which weeds are controlled depends on the balance between the characteristics of the weeds present and the management tools available to growers. Although tools for managing weeds improved greatly during the 20th century, weed communities have also changed rapidly, and floristic changes are likely to continue. The present relatively favorable balance between weed communities and management tools could be lost if attention is not given to management of weeds at landscape and regional levels, and to preservation of herbicides and ecological control tactics in the face of evolutionary responses of weeds. Management of weeds over large areas and long time scales requires an expanded perspective on weed community dynamics and weed evolution (Cardina *et al.*, 1999). It also requires institutional structures that are poorly developed at present.

Formation and management of weed communities

In previous chapters of this book, the management consequences of community properties such as weed species richness and the relative abundance of species have mostly been ignored. Although the effects of weed community properties on the success of management efforts are largely unknown, they may be substantial (Clements, Weise & Swanton, 1994). In this section, a conception of the weed community is developed, the role of colonization and extinction processes in weed communities is assessed, and some of the consequences of weed species richness are considered. Finally, tactics for management of community structure are discussed.

The composition of weed communities

The composition of a weed community is determined by multiple interacting factors. The species present in a field include (i) the species present as vegetative plants, (ii) species in the seed bank, and (iii) species that disperse into the field from neighboring habitats such as roadsides but which are incapable of surviving the agricultural practices in use in the field. This pool is periodically supplemented by medium- to long-distance dispersal from other places.

The primary factors affecting weed density, biomass, and propagule production are crop competition and the cultural conditions of the cropping system. Factors influencing the degree of crop competition include species, cultivar, density, arrangement, and planting date (see Chapter 6). Cultural practices, including tillage regime, irrigation, fertility management, herbicides, cultivation, and the timing of all of these interact with each other and with the degree of crop competition to influence the composition of the weed community (see Chapters 4 and 5). A change in any of these crop and management factors is likely to favor certain species while suppressing others, and thereby shift composition of the community. Soil and weather directly affect population dynamics of the weeds (see Chapter 5) but also indirectly influence weed populations by dictating cropping practices. Grazing livestock and natural enemies may also have effects on the relative abundance of weed species (see Chapters 8 and 9). Sometimes weed species may affect the population size of other weeds via competition or by providing shelter that facilitates establishment, particularly in species-diverse, perennial-dominated systems like pastures and tropical orchards. However, as explained below, interactions among weeds may be a small influence on community composition in most annual crops.

The various factors just discussed determine how many seeds and buds are produced and what proportion remain at the beginning of the next cropping cycle. If cultural practices keep production of propagules to negligible levels, a species may be eliminated from a field eventually. This probably takes a very long time for some species with highly persistent seed banks (see Chapter 2). However, other species may be eliminated by changed practices within a few years. Taken together, the preceding considerations indicate that composition of the weed community is not a fixed feature of a field, but rather is largely determined by management practices.

The degree to which interspecific interactions among weeds determine their presence and relative abundance is largely unexplored. Provided weed control is even moderately effective, little competition is expected between weeds of different species. First, if the weeds are reasonably well controlled, the frequency of competitive contact among weeds will be low. Second, to the extent that weed species are distributed in patches, intraspecific competition will be more common than competition between weed species. Third, in annual crops, no competition at all occurs for a period after planting, and most weed species can store sufficient resources during this time to permit at least limited reproduction. Consequently, in agricultural systems where weed biomass is normally a small fraction of crop biomass, interaction among weeds probably has little effect on weed community composition. If this is the case, then the factors affecting abundance act on each species individually, and contrary to statements commonly encountered in the literature (e.g., Clements, Weise & Swanton, 1994; Maillet & Lopez-Garcia, 2000), the presence or absence of particular species will have negligible effects on abundance of other weeds or on the invasibility of the weed community.

In some tropical smallholder cropping systems, extensive grazing systems and other situations where farmer resources are limited, crop value is low, or the crop tolerates moderate weed abundance, weed biomass may be high. When weed biomass is substantial, one weed species may competitively suppress another (Liebman *et al.*, 1996). This effect may be useful in weed management if easily controlled weeds can be used as living mulches to suppress more competitively harmful ones. The extent and importance of competition between weed species needs systematic evaluation.

Species introduction and the species richness of weed communities

Human activity extensively mixes the floras of similar bioclimatic regions through both deliberate and accidental introduction of species. For example, the state of New York has a total of 2078 native vascular plant species and 1117 introduced species (Mitchell & Tucker, 1997, p. 6). Thus, the four centuries of contact with other parts of the world has created a flora that is 35% alien. Similarly, the flora of the British Isles is 44% alien (Crawley, Harvey & Purvis, 1996). This movement of species between continents and regions acts to increase local species richness, at least in the short term. It occurs, however, at the expense of compositional differences between regions. Essentially, the human species is homogenizing the flora of the planet to the extent that the climatic limitations of species and the invasion resistance of natural communities allow.

Disturbance facilitates invasion of aliens by removing competition from established plants. In an analysis of the British flora, Crawley (1987) found that the species on waste ground, walls, and farm fields included 78%, 46%, and 37% aliens, respectively. In contrast, deciduous and pine forests had 5% and 0% aliens, and all wetland habitats had similarly low percentages.

Because the habitats that are suitable for invasion are disturbed, successful invaders are usually disturbance-adapted (Hobbs, 1991). That is, a high percentage of aliens have weedy tendencies as defined in Chapter 2. Consequently, the weed floras of many regions include a high percentage of introduced species. For example, of the 500 weeds of the northern USA discussed in Muenscher (1955), 61% have been introduced, mostly from Eurasia.

Occasionally however, native species predominate. This is the case for the weed flora of Californian rice fields which includes only 34% aliens, many occurring only in limited areas (Barrett & Seaman, 1980). The similarity of the water regime in these continuously flooded rice fields to the marshlands they replaced apparently facilitated retention of native species. Also, rice cropping is relatively recent in California, and so introduced species have not had long to accumulate (Barrett & Seaman, 1980). Nevertheless, several of the most abundant weeds in California rice are introduced (e.g., *Echinochloa crus-galli*, *Bacopa rotundifolia*).

The rate of introduction of new species into any given region must eventually decline as fewer species are left that have not already immigrated. Given the large number of alien species that have already been added to most floras, a currently low rate of introduction might therefore be postulated (McNeill, 1976). Forcella & Harvey (1983) tabulated the date of first observation of alien species by county from herbarium sheets and other sources for the five northwestern states of the USA. Their data indicate that introduction of species into this region continued throughout the 20th century, but at a declining rate (Figure 10.1a). Their data thus support McNeill's (1976) hypothesis. However, most species spread after first being sighted: the mean number of counties infested by the 188 species of weeds that were present before 1910 increased steadily throughout the century (Figure 10.1b). Thus, local arrival of introduced species is an ongoing problem.

The normally low degree of interaction among the species of a weed community implies that the species present in a field represent some proportional sample of the regional weed flora (Cornell & Lawton, 1992). Hence, an increase in the regional flora will tend to push up local weed species richness by invasion pressure (Figure 10.2a). At any given time, weed species richness is determined by the balance between local extinction and medium-range dispersal out of the regional species pool (Figure 10.2a). The rate of species immigration to a field (number of species arriving per year) declines through time (i)

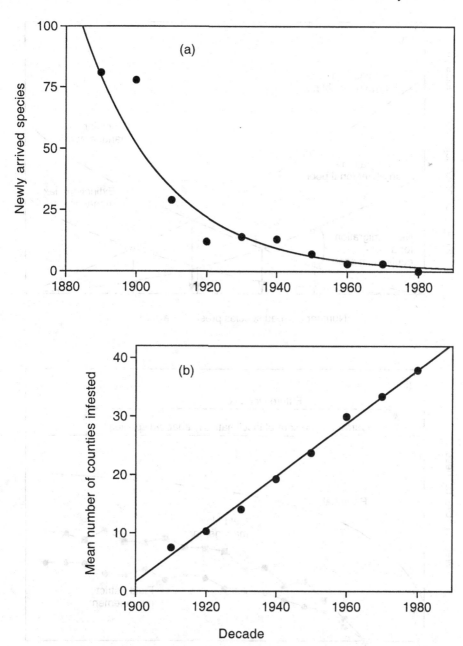

Figure 10.1 (a) Number of first records of invading species by decade in the five northwestern states of the USA. Line is a negative exponential curve fitted to the data. Data point for 1890 includes all species that had arrived by that date. Ten species that were present in only one or two counties in 1980 were not included. (b) Mean number of counties infested by the 188 invasive weeds that had established in the same region by 1910. (Plotted from data of F. Forcella & S. J. Harvey; see Forcella & Harvey, 1983.)

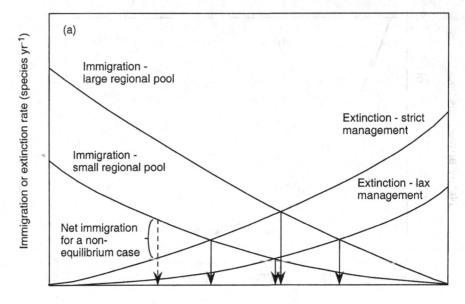

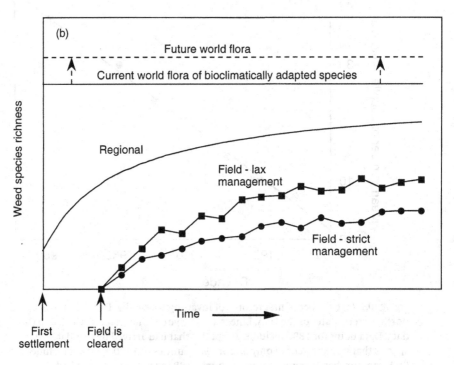

Figure 10.2 Factors affecting species richness of agricultural weeds in a field. (a) The species richness of any given field within the region is a subset of the regional species richness, and is determined at equilibrium by the balance between

because as the number of species increases, the proportion of species that have yet to immigrate declines, and (ii) because the species that are best dispersed by humans tend to arrive first. The extinction rate increases as a function of species richness for two reasons. First, the control strategies in use in a field at a given time will, on average, extinguish more species if more are present. Second, species that for any reason tend to form sparse populations will likely arrive later due to low propagule density in crop seed, manure, soil picked up by machinery, etc. (see Chapter 2), and the characteristically small populations of these late arrivers will make them more prone to local extinction. Although Figure 10.2a was inspired by island biogeography theory (MacArthur & Wilson, 1967), the rising extinction curve for islands is generated by a different mechanism, namely the smaller average population size that occurs on islands as more species partition the available resources. No evidence supports an increased partitioning of resources in species-rich communities of agricultural weeds, and the nature of these communities indicates that such partitioning is unlikely (see preceding section). Hence, the classic mechanism generating a rising extinction curve would not apply.

Fluctuations in the species richness of a field occur as changing practices eliminate suites of species or facilitate the invasion of others (Figure 10.2b). On average over long time periods, strict management should keep the extinction rate higher and therefore species richness lower than would be the case if management of the farm were lax, but species richness should still increase as the regional pool grows (Figure 10.2a). Whether the recent decrease in weed species richness at the field level associated with introduction of chemical weed management (Andreasen, Stryhn & Streibig, 1996) represents a temporary fluctuation or a shift from a lax to a strict management curve remains to be determined.

Increased species richness is likely to confer greater flexibility on a weed community. At any given time, most weed problems are usually the result of

invasion and extinction (solid arrows). Invasion rate (species yr^{-1}) is determined largely by the species richness of the regional species pool. Extinction rate (species yr^{-1}) is determined largely by management, and especially by the effectiveness of rotation between management strategies. In many locations, invasion exceeds extinction and species richness is not in equilibrium (dashed arrow). (b) Species richness in a region increases asymptotically through time following initial settlement by agriculturalists due to immigration of species from other regions and recruitment of preadapted native species from non-agricultural habitats. Worldwide evolution of new weed species increases the asymptote. As the regional species pool increases, the species richness of individual fields increases due to a shift from a lower to a higher local invasion rate.

one or two species that are particularly good at avoiding the farmer's current management practices. Innovative management may drive these problem weeds to low abundance, but if species richness of the field is high, then formerly minor species may increase in response to the changed management regime. Consequently, a more diverse local weed flora may require application of additional weed control measures. This will particularly be the case if the diversity of species includes a wide range of ecologically distinct types (e.g., grasses and broadleafs, perennials and annuals, etc.). In addition, high weed species richness may also contribute to unpredictability since weed species will respond differently to variation in weather conditions.

Species richness also allows evolutionary flexibility, since some species may adapt rapidly to a particular management practice whereas others cannot. Introduction of many species into Australia increased the probability that at least one species would be particularly competent at evolving herbicide-resistant forms. The extraordinary ability of *Lolium rigidum* to evolve resistant biotypes (Powles & Howat, 1990) would not pose a problem for Australian farmers if that species had not been introduced to the continent.

Crop diversity and weed diversity

This book advocates the use of crop diversity in the management of weeds. Although many studies show that crop rotation and intercropping help control populations of particular weed species (see Chapter 7), only a few have addressed the effects of cropping system diversity on the structure of weed communities (Liebman & Dyck, 1993).

Conceivably, diversification of crop rotations could either increase or decrease weed community diversity. If crop rotation poses a sufficient problem for particular species, then some of those species may be extirpated from the field. Alternatively, if diversifying the rotation provides opportunities for establishment of additional species, or facilitates invasion of the field by additional dispersal routes (e.g., weeds sown with forage seed), then diversification of the rotation may foster an increase in weed species richness. However, regardless of effects on the number of weed species present, the equitability among weed species should increase with the diversity of a rotation. Continuous cropping favors a very few weeds that are well adapted to that crop whereas a diverse rotation will tend to favor any given species only in certain years, and hence the relative abundance of species will tend to be more equal. Liebman & Dyck (1993) reviewed several studies in which dominance by a single problem weed occurred with continuous cropping but not with a rotation of crops. Cardina, Webster & Herms (1998) found that species richness and equitability of seed banks were greater with more complex crop rotations. Covarelli & Tei (1988) found that the number of weed species was equivalent in various maize–winter wheat rotations and continuous maize. However, equitability in the maize phase of the rotations was greater than in continuous maize.

Thus, diverse crop rotations appear to favor a mix of generalist weeds whereas continuous cropping of a single species tends to favor one or a few weeds that are well adapted to the control measures that are possible in that crop. Although generalists may pose substantial problems, they are usually easier to manage across a rotation than are specialists in a continuous culture of the crop to which they are specialized. Managing for equitable distribution in the relative abundance of weed species in a field is therefore reasonable, because high equitability generally indicates a low degree of specialization in the community.

In contrast, high weed species richness is not desirable for the reasons discussed in the preceding section. Presence of a diversity of plant species in a field often improves management of insects and disease (Risch, Andow & Altieri, 1983; Altieri & Liebman, 1986; Andow, 1991). However, better management of the pests can be accomplished with less risk through intercropping of species that are easy to control and have properties that effectively inhibit the particular pest organisms, rather than with weeds that can potentially decrease crop yield. Similarly, attempts to conserve rare weeds by increasing weed diversity in farm fields are probably misguided (Hebden *et al.*, 1998); a better approach may be to favor particular rare species in some fields by appropriate management based on their biology. Moreover, the higher weed species richness that is often observed in integrated weed management systems and organic agriculture (Clements, Weise & Swanton, 1994; Rasmussen & Ascard, 1995; Doll, 1997) is not desirable from the standpoint of weed management. Fortunately, high weed species richness may not be an inevitable consequence of these systems.

Although more work on this subject is needed, intensification of crop competition appears to offer a means for simultaneously increasing equitability while decreasing species richness. Palmer & Maurer (1997) examined weed diversity in five sole crops and the corresponding five-species intercrop, all at a single density. The intercrop had the highest weed species richness. The cause of this pattern was unclear, but it was apparently not due to different crop species favoring different sets of weeds. Crop biomass, however, had a significant negative effect on weed species richness in this experiment. Chapter 7 discusses other intercropping experiments that showed lower species richness and greater equitability as the height, density, and biomass of the crops increased. In a sole-cropping experiment, Lawson & Topham (1985) similarly found that weed species richness was inversely related to density of a pea crop. Thus, the intensity of crop competition is probably more important in controlling weed species richness and the relative abundance among weed species than is the diversity of the crops *per se*.

Other weed management measures may similarly eliminate some species from the above-ground segment of the community. However, to truly lower the weed species richness of a field requires removing species from the seed bank, which is difficult. For this, tillage regimes that stimulate germination must be combined with consistent prevention of reproduction. The species present will then be driven to extinction in the particular field roughly in reverse order of their seed longevity. Probably the best strategy for limiting species richness is to prevent dispersal of species into the field in the first place. This is addressed in Chapter 2, and in the section "Controlling the spread of new weeds" below.

Human-dominated ecosystems as an evolutionary context

The defining characteristic of the present geological era is the widespread and intense exploitation of ecosystems that results from high human population and the high per capita consumption of resources in the developed countries. The biota of the planet expresses three basic responses to this human presence. First, the majority of species retreat to ever-smaller sanctuaries of relatively undisturbed habitat. Many of these appear likely to go extinct in the near future as their final refuges are radically changed by human disturbance (Myers, 1994).

Second, a substantial number of species are becoming domesticated. This process is particularly common among higher vertebrates (e.g., the many reptiles now bred as pets) and vascular plants (e.g., cultivation of Pacific yew, *Taxus brevifolia*, for medicinal products – Piesch & Wheeler, 1993). Few quantitative data document this process. *Hortus*, however, lists the species available in the horticultural trade in North America. Succeeding editions document an increase from 12 659 species in 1930 to 18 447 in 1941 and 20 397 in 1976 (Bailey & Bailey, 1930, 1941, 1976), though the 1976 compilation included some plants that are only suitable for Hawaii and Puerto Rico. Most cultivated plant species are not yet domesticated in the genetic sense of being so dependent on human propagation that they would be incapable of successfully maintaining populations in their original native habitat (de Wet, 1975). However, because their native habitats are disappearing, many of these species may eventually become ecologically dependent on humans for their continued existence. This includes not only the growing number of species that humans consider useful, but also species that are maintained in cultivation specifically because they are rare or extinct in the wild (e.g., the franklintree, *Gordonia alatamaha* – Harrar & Harrar, 1962, pp. 521–2).

Finally, a growing segment of the biota is taking advantage of the habitats created by human activity. These species exploit humans, our domestic species, and the ecosystems we create. Examples of species making this transition currently or in the recent past include the Colorado potato beetle, human immunodeficiency virus, and the raccoon. Weeds with an apparently recent adaptation to agriculture include *Diodia teres* (Jordan, 1989a, 1989b), *Echinochloa microstachya*, and *Oryza punctata* (Barrett, 1983). As human civilization becomes the only game in town, a growing number of species will come to play.

Unless human civilization changes radically, most species that are not either domesticated or weedy will experience shrinking populations as presently wild habitats are managed with increasing intensity or converted to urban and agricultural uses. Many of these species are probably doomed to eventual extinction (Quammen, 1998). The future of the non-domesticated flora of the earth may thus depend largely on the evolution of increasing weediness. In addition to active agricultural land, arenas for the evolution of weeds include cities, roadsides, mine spoils, degraded and intensively managed forests, and agricultural areas abandoned due to salinization, erosion, desertification, shifting economic conditions, and warfare. Fortunately, only a small fraction of the world's flora currently thrives as agricultural weeds. However, about 11% of the terrestrial area of the earth is cultivated land and another 26% is used for permanent pasture (World Resources Institute, 1998, p. 298). Most of this land is well suited for plant growth. The great increase in fitness associated with characteristics that allow ruderal species to adapt into agricultural niches can be expected to increase the diversity of agricultural weeds during the coming millennia.

Origins of weeds

The preceding section hypothesized that evolution of new weed species is an ongoing process. This section provides evidence in support of that view. Currently, preventing the evolution of new weeds may not be possible. However, an increasing diversity of agricultural weed species is probably undesirable, and the processes of weed origination are discussed here in hopes of directing attention to this potential problem.

New weed races and species continue to evolve from wild and domesticated plants via several pathways: (i) directly from wild plants by selection of races adapted to human-dominated ecosystems, (ii) by reversion of cultivated species to weedy forms, (iii) by hybridization, (iv) by evolution of new forms within existing weed species, and (v) by speciation of geographically isolated populations. These pathways are not always distinct, and complex genetic interchanges have been hypothesized within crop–weed–wild assemblages of closely related species (Zohary, 1965). Nevertheless, these pathways provide a structure for organizing an understanding of weed origins. Unless a species can reproduce in agricultural fields, selection for traits that further adapt the species as an agricultural weed is unlikely. Thus, a species must be preadapted to agricultural conditions to some extent if it is to become an agricultural weed.

Preadaptation for weediness

For most weed species, truly wild populations are known from the taxon's center of origin. Typically, these inhabit naturally open or disturbed habitats: stream margins, marshlands, beaches, dunes, cliffs, scree, exposed or high-elevation sites, and animal-disturbed areas (Godwin, 1960; Baker, 1974). Such habitats provide the limited competition required for persistence of these species. Which types of habitats are most likely to contribute agricultural weeds has not been assessed for any agricultural weed flora. In related work, Marks (1983) found that most of the plants invading abandoned agricultural fields in the northeastern USA originated in wetlands and on cliffs.

Characteristics that adapt plant species to fertile disturbed habitats, namely rapid growth, early maturity, high allocation to reproduction, resistance to trampling, and resilience following shoot burial and damage to the root system, preadapt species to thrive in sites disturbed by agriculture. From the perspective of a potential weed, bare fields created by plowing represent a bonanza of resources, and species that are preadapted to exploit such conditions experience a great increase in fitness when agriculture is introduced to a region.

Because potential weeds existed in the landscape at the time humans first began deliberate cultivation of plants, even the earliest crops were probably infested with weeds. Thus, for example, Garfinkel, Kislev & Zohary (1988) reported that a store of carbonized lentils from Israel radiocarbon-dated at 8800 BP contained seeds of *Galium tricornutum*. This date is close to the earliest records for Middle Eastern agriculture. *Galium tricornutum* still infests lentil fields of non-industrial Middle Eastern farmers today. Similarly, strata deposited from 6300 BP to 4800 BP in old oxbows of the Vistula River near Krakow, Poland showed marked correlation between the abundance of cereal fossils and the abundance of weed pollen, particularly *Plantago major* (or *P. paucifolia*)

Table 10.1. *Weeds present in interglacial, full glacial, and late glacial deposits in Britain*[a]

Species	Interglacial	Full glacial	Late glacial
Aethusa cynapium	M		
Carduus nutans	M		(M)
Chenopodium album	M		M
Cirsium vulgare	M		
Heracleum sphondylium	M		
Pastinaca sativa	M	(P)	P
Plantago (*major* or *media*)	(P)	P	P
Valerianella (3 species)	M		
Artemisia spp.	(P)	(P)	(P)
Linum catharticum		P	
Trifolium spp.		(P)	
Barbarea vulgaris			M
Centaurea cyanus			P
Centaurea nigra			P
Cerastium vulgatum			M
Daucus carota			M
Galeopsis tetrahit			M
Galium aparine			M
Linaria vulgaris			M
Polygonum aviculare			M
Rumex acetosa			M
Rumex crispus			M
Sonchus arvensis			M
Taraxacum officinale			M
Urtica spp.			(P)

Notes:
[a] M indicates macroscopic remains (chiefly seeds or fruits) and P indicates pollen. Where material from a period was only identified to genus, the entry is given in parentheses.
Source: Summarized from Godwin (1960).

and *Polygonum aviculare* (Wasylikowa *et al.*, 1985). Iversen (1941) and Godwin (1960) similarly noted the presence of weed pollen in association with the arrival of agriculture in Denmark and Britain. The presence of plant remains from interglacial and glacial deposits in Britain indicates that many weed species were already present when agriculture first arrived (Table 10.1) (Godwin, 1960). Thus, as agriculture has spread across the globe, species native to each region have been added to the total pool of weeds.

This process continues as agriculture penetrates the remaining non-agricultural regions of the world. For example, Conn & DeLapp (1983) noted that several native species could invade and persist in grain fields recently established in the Alaskan interior. Similarly, the forest gap species *Solanum crinitum* and *Vismia guianensis* have become pasture weeds as permanent pasture has replaced forest in Amazonia (Dias Filho, 1994, pp. 7–8).

Movement of species into agriculture is also facilitated by introduction of a new cropping system into a region. For example, the advent of the native Californian marsh grass *Echinochloa microstachya* as a weed presumably postdates the initiation of rice cultivation in that state during the period 1912–15 (Barrett & Seaman, 1980; Barrett, 1988). This species has subsequently spread to rice-growing areas in Australia (Barrett, 1988). Similarly, the native wild rice *Oryza punctata* was first noted in rice fields in Swaziland in the mid 1950s. By the 1970s the species had forced abandonment of mechanized rice cultivation over large areas (Barrett, 1983). Throughout the eastern USA, the advent of no-till planting has resulted in colonization of fields by robust native perennial species like *Solidago altissima* and *Rhus typhina* that were previously not associated with row crop agriculture.

Evolution of weed races from wild populations

Although the initial movement of early successional species from natural habitats into human disturbances involves simply the exploitation of somewhat similar habitats by preadapted genotypes, subsequent selection can be expected to modify weedy populations substantially. Tilled fields differ from the ancestral habitats of weeds in several important respects. First, the temporal pattern of soil disturbance by tillage is typically much more predictable than natural disturbance by animals, shifting dunes, or erosional/depositional events along streams and cliffs. Second, the increase in competition as the crop matures is also more predictable than in a natural succession. Third, the selective pressures from weed pulling, cultivation, and, more recently, herbicides differ from those exerted by natural mortality factors in wild habitats. Finally, with the exception of some animal-caused disturbances, the nitrogen fertility of most naturally open habitats is low; cliffs, dunes, beaches, and gravel bars lack the organic matter necessary for generation of a pulse of nitrate following soil disturbance. Thus, human disturbance selects from the wild not only preadapted species but particular preadapted genotypes. These then become the ancestors of weedy races, and ultimately, species.

Because only certain genotypes of a potential agrestal weed species may be capable of founding a weed race, the invasion of cultivated fields by wild species is probably an ongoing process partially determined by chance events. Due to the diversity of relatively innocuous weeds that are often present at low abundance in a given field, the new arrival of a weedy strain of a wild species usually goes unnoticed until it spreads. By then it is usually too late to document the time and place of the weed's origin. Consequently, few studies address the transformation of wild species into weeds (but see Jordan, 1989a, 1989b discussed in section "Ecotype formation in weeds" below).

Baker (1991) described a series of events that indicate the sort of processes that may be involved in the creation of new weeds. He noted that in the gardens around Berkeley, California occurs a weedy, tetraploid race of *Viola alba*, a species that is normally diploid with a chromosome number of $2n = 20$. He observed in his own garden a particularly vigorous cytological mutant with 34 chromosomes. Although the mutant was sterile, it spread aggressively by runners and perhaps could have become a significant weed had it not died out during the drought of 1977–78. A large number of nonweedy and ruderal species "test" genotypes in fields and gardens around the world each year. Though most, as with Baker's *Viola*, are ultimately unsuccessful, the few that do persist and thrive contribute to the weed problems of succeeding centuries. Which species are most likely to evolve an agrestal habit is difficult to predict. Presumably, these species must have some agrestal characteristics to begin with (see Chapter 2), but, for example, are populations that spin off new weed races most likely to be sparse or dense, continuous or patchy, outbreeding or inbreeding?

Weedy races of crop species

A second avenue for the development of weeds is via the evolution of weedy races of crop species. Such weeds can be extremely troublesome; by mimicking key crop attributes, such as herbicide tolerance and seedling coloration, they prevent the use of selective control measures (Barrett, 1983). The genetic changes that allow development of a weed race from a crop may come about through mutation or through acquisition of characters from conspecifics and congeners.

For grains, often all that is required to initiate a potential weed race is a mutation that causes the inflorescence to shatter. Although additional characteristics, such as nonsynchronous germination, early seed maturation, or increased seed dormancy may be required to make the new form fully effective as a weed (Baker 1965, de Wet 1975), the development of a shattering form initiates the process of evolution to weediness. Baum (1969) and Scholz (1986) provided evidence for the mutant origin of shattering races of *Avena sativa* and *Hordeum vulgare*. Harlan (1975) noted that although some of the African weed sorghums disarticulate via formation of an abscission layer like the wild progenitor, others disarticulate by breakage of the rachis. Domesticated sorghum has lost the abscission layer but has a strong rachis and thus retains seeds on the plant until harvest. The rachis-shattering weed races likely originated from domesticated sorghum, since rachis breakage would be pointless prior to loss of the abscission layer. The rachis-shattering types are more widespread, and all weedy *Sorghum* in North American disperse by rachis breakage (Harlan, 1975).

Cavers & Bough (1985) analyzed populations of weedy proso millet (*Panicum miliaceum*) of several forms. Although the most aggressive, a black-seeded type, was probably introduced from Europe, the others resembled cultivars and probably evolved in North America directly from the domesticate. Weedy *Panicum miliaceum* has only become a problem in North America since 1970 (Cavers & Bough, 1985), so these crop-derived weeds may be of recent origin.

Creation of new weeds by hybridization

Domesticates can also become weeds through acquisition of traits from wild or weed relatives via hybridization. Most domesticated plants are part of a wild–weed–domesticate complex within which genetic exchange occurs regularly (de Wet, 1975; de Wet & Harlan, 1975; Harlan, 1982).

For example, a weedy race of domesticated radish, *Raphanus sativus*, has developed in California through introgression from *R. raphanistrum* (Baker, 1965). In contrast with the domestic radish, the weed form develops a deep taproot with many lateral roots, flowers rapidly, and has fruits that dehisce from the plant. Moreover, the harder fruit coat on the weed race confers some dormancy. Because the weedy *R. sativus* is better adapted than *R. raphanistrum* to the coastal region where radishes are commonly grown, it is becoming a worse problem for growers than the more widespread *R. raphanistrum* (Baker, 1991).

Similarly, populations of a weed beet (*Beta vulgaris*) expanded rapidly in the sugar-beet-producing regions of Europe during the 1970s. Analysis of mitochondrial and chloroplast DNA by Boudry *et al.* (1993) demonstrated that the cytoplasm of the weed race was derived from the domesticated beet. In contrast, the gene for annual flowering habit was nuclear and apparently derived from wild beet populations (*Beta vulgaris* ssp. *maritima*) growing along the French coast.

This recent creation of a new weed is especially interesting given the probable origins of domestic beets. Domesticated leaf beets were apparently derived by selection from populations of weedy beets infesting Near Eastern grain fields about 3500 to 4000 years ago (King, 1966). Following introduction into Europe in the 1st century AD, the leaf beet increased in sugar content by hybridization with wild *B. vulgaris* ssp. *maritima* (Zossimovich, 1939). The large-rooted table beet was developed from the leaf beet around the 5th or 6th century AD in the region of modern Iraq or Iran, and subsequently spread into

Europe as well. Hybridization of table beet with the sweet form of leaf beet in Europe followed by selection led to the modern sugar beet. Thus, the recent evolution of a new weed race continues the long history of genetic interaction between wild, domesticated, and weedy races of *B. vulgaris*.

Hybridization can also create new weeds without involvement of a cultivated species. A well-studied case is the British hexaploid species *Senecio cambrensis* which arose by chromosome doubling in hybrids between diploid *S. vulgaris* and tetraploid *S. squallidus* (Ashton & Abbott, 1992; Harris & Ingram, 1992). Additional examples of the generation of new weeds by hybridization are discussed in Stebbins (1965) and Sun & Corke (1992).

Evolution of new biotypes within existing weed species

Among weed scientists, the term *biotype* refers to a group of individuals that share some distinctive heritable trait or suite of traits. The concept is not completely distinct from that of the *ecotype*, which refers to a race that is adapted to particular ecological conditions (Lincoln, Boxshall & Clark, 1998). An ecotype, however, consists of one or more populations, whereas multiple biotypes are frequently present within a single population. Because most weeds reproduce regularly by selfing, apomixis, or vegetative propagation, well-adapted genotypes are easily preserved across generations as relatively stable biotypes.

New weed biotypes may act as distinct problems if they respond to management differently than the parental type. For example, many herbicide-resistant biotypes have evolved in response to herbicide application (Holt & LeBaron, 1990; Shaner, 1995). Usually these resistant biotypes require changes in management strategy, and may complicate weed management considerably.

Other sorts of biotypes with apparently recent origin and significant consequences for management have been identified. For example, Abbas, Pantone & Paul (1999) reported that a biotype of *Xanthium strumarium* with up to 25 seeds per bur has appeared recently in central Texas. Additional examples are discussed in the section "Ecotype formation in weeds" below.

Speciation of geographically isolated populations

In addition to the forces acting to create new weed biotypes, races, and species discussed above, speciation by geographic isolation will likely add weed species to the world flora over the coming millennia. Many agricultural weed species have now been introduced onto dozens of different landmasses (Table 10.2). To some extent populations on separate continents have already diverged genetically (see section "Ecotype formation in weeds" below) and can

Table 10.2. *Number of major landmasses on which the world's 18 worst weeds occur*

Species	Continents	Major islands[a]	Total
Cyperus rotundus	6	16	22
Cynodon dactylon	6	15	21
Echinochloa crus-galli	6	12	18
Echinochloa colonum	6	15	21
Eleusine indica	6	14	20
Sorghum halepense	6	11	19
Imperata cylindrica	6	13	19
Eichhornia crassipes	6	15	21
Portulaca oleracea	6	12	18
Chenopodium album	6	9	15
Digitaria sanguinalis	6	12	18
Convolvulus arvensis	6	9	15
Avena fatua	6	9	15
Amaranthus hybridus	6	6	12
Amaranthus spinosus	5	12	17
Cyperus esculentus	6	5	11
Paspalum conjugatum	5	12	17
Rottboellia exaltata	5	6	11

Notes:

[a] Islands larger in area than Puerto Rico (8860 km^2). The number shown is a minimum due to (i) gaps in knowledge during construction of the distribution maps (Holm *et al.*, 1977), (ii) combining of some major islands with the parent country on a neighboring continent (e.g., Sardinia and Sicily with Italy), and (iii) grouping of several islands in an archipelago when reporting distribution (e.g., Japan, the Philippines).

Source: compiled from distribution maps in Holm *et al.* (1977).

be expected to continue to do so, despite some gene flow from secondary introductions. Geographical isolation is a powerful predisposing condition for speciation (Mayr, 1963, pp. 481–515; Grant, 1981, pp. 149–69). Humans have created that condition for hundreds of weed species, and more introductions into geographically isolated areas occur each year.

Effects of weed evolution on weed diversity

In terms of the conceptual model in Figure 10.2b, not only is the regional species richness of agrestal weeds rising asymptotically in ecological time due to immigration, but the asymptote is itself rising due to the several mechanisms discussed above. The relative rate at which agricultural weeds are added to regional pools by immigration versus creation of new weed species is unknown, though presumably immigration is the faster process.

Like global climate change, mass extinction of species, and many other environmental problems, the evolution of new weeds occurs so slowly and over such large areas that an effective response by society is difficult. Although the short planning horizons of modern societies preclude significant attempts to block most of the routes for development of new weeds discussed above, the first step is recognition that the problem exists. Perhaps future cultures will learn how to prevent further increase in the weed flora, even as they cope with the weeds that are currently developing in farm fields today. A tentative, partial approach to the problem is proposed in the section "Controlling the spread of new weeds" below.

Weed genecology

This section discusses several aspects of weed genecology (genetical ecology) that are useful in understanding the evolutionary response of weeds to management. Evolutionary changes in weeds are driven by a variety of factors. The two most important are (i) selection exerted by control efforts and other aspects of cropping systems, and (ii) genetic bottlenecks associated with colonization of new locations. Response to selection is a function of the genetic variability in the population and the intensity of the selection pressure. To a large extent, genetic variability is determined by the weed's breeding system, and by losses of variability that occur during colonization events. Selection acting on this genetic variability commonly leads to differentiation of ecotypes specialized for particular physical environments and cropping systems. However, most weeds have a high degree of phenotypic plasticity that allows them to survive and reproduce in a wide range of environmental conditions even without adaptive change in the genome. Each of these issues is discussed further below.

Genetic variability and breeding systems

During the establishment of new populations, weeds are faced with the need to reproduce even though potential mates may be few and scattered, or absent altogether. Consequently, most weeds are either self-compatible, apomictic (able to set seeds asexually), or propagate vegetatively.

The correlation between life history and breeding system is strong. Most annual and stationary perennial weeds are self-compatible or apomictic whereas most wandering perennials regularly outcross. For example, in a study of 64 Canadian weeds, all 33 annuals and 21 of 23 stationary perennials set seed when inflorescences were bagged to prevent cross-pollination. In contrast, none of the eight wandering perennials set seeds when bagged (Mulligan & Findlay, 1970).

Outbreeding species tend to have high heterozygosity and rapid generation of new gene combinations (Table 10.3) (Clegg & Brown, 1983) due to

Table 10.3. *Breeding system, genetic variation, and life history*

Breeding system	Heterozygosity	Formation of new gene combinations	Life history
Outbreeding	High	Fast	Mostly wandering perennials
Inbreeding	Low	Slow	Mostly annuals and stationary perennials
Asexual	High	Very slow	Wandering perennials and apomicts

Source: Modified from Clegg & Brown (1983).

random assortment of chromosomes obtained from different parents and crossing-over between chromosomes bearing different alleles. In weeds, outbreeding may allow rapid differentiation of invading populations in response to newly encountered conditions (Warwick, Thompson & Black, 1987). A variety of mechanisms have evolved to insure a high rate of outbreeding in plants. These include self-sterility genes (e.g., *Trifolium repens* – Atwood, 1940, 1942), differences in the timing of pollen shed and style receptivity (e.g., *Daucus carota* – Bell, 1971), and dioeciousness (e.g., *Rumex acetosella* – Baker, 1962), though the last mechanism is rare in weeds.

Repeated selfing tends to increase random loss of variability. Selfing decreases heterozygosity (Solbrig & Solbrig, 1979, pp. 168–9) and this increases the chance that rare alleles will be lost. Moreover, crossing-over during meiosis in a selfing event rarely creates new gene combinations, because the chromosomes involved are usually identical.

An important consequence of selfing is that well-adapted gene combinations are protected against disruption during reproduction (Stebbins, 1957). This and the ability to propagate at low density have probably selected for selfing in many weed species that lack vegetative reproduction.

Asexual plants have only limited ability to generate new genotypes. Somatic mutation may occasionally add variability to the population, and even basically asexual species like *Taraxacum officinale* outcross occasionally (Richards, 1970). Nevertheless, generation of new genotypes in most asexual species is probably very slow. However, the principal process that creates sterile plant populations, namely interspecific hybridization, necessarily results in high levels of heterozygosity. This can confer competitive ability and broad adaptation via heterosis if the two genomes are compatible with respect to vegetative characters. Examples of highly successful asexual weeds include *Oxalis pes-caprae*, *Taraxacum officinale*, and *Erigeron annuus* (Baker, 1965; Solbrig, 1971; Stratton, 1991). Genetic evidence indicates that *Cyperus rotundus* is also essentially asexual, with single clones covering large regions (Okoli *et al.*, 1997).

Baker (1965) proposed that many asexual weed species possess a general-purpose genotype that allows them to thrive under a variety of conditions even without local adaptation. However, populations of some asexual species have substantial genetic variation, presumably from rare outcrossing events, and are capable of adaptation to local conditions (Gadgil & Solbrig, 1972; Stratton, 1991).

Although high levels of homozygosity and low variability are common in self-compatible annual species (Stebbins, 1957), many successful self-compatible weeds have surprisingly high genetic variability (Allard, 1965). For example, coefficients of variability for some quantitative characters in *Avena fatua*, a highly self-pollinating species, are as great as for *Lolium multiflorum*, a self-incompatible species (Allard, 1965). Allard (1965) explored the processes maintaining variability in self-compatible species with a series of experiments and simulations. Using two strains of lima bean differentiated for a series of morphological markers as a model system, he found that although outcrossing was only 5%, most individuals in the population were of hybrid origin after six generations. In this experiment, apparently heterozygotes had a selective advantage, since the rate of increase of the hybrids was too fast to be explained by outcrossing alone. Heterozygote advantage is apparently common in weeds (Allard, 1965). Using computer simulations Allard & Hansche (1964) showed that even in highly inbred species, recombination can generate many genotypes within just a few generations and that low rates of outcrossing can allow rates of evolutionary change that approach those of panmictic outcrossers (Figure 10.3). Thus, self-compatible weeds appear to have a highly flexible genetic system in which adapted genotypes can be perpetuated, but genetic variability is available when needed.

In some self-compatible weeds the outcrossing rate varies among populations (e.g., *Avena fatua* populations in California – Allard, 1965). Variation in outcrossing rate suggests that it may be modified by selection (Jain & Martins, 1979). Allard (1963) has demonstrated that crossover rate can also be altered by selection in predominantly selfing populations. Conceivably, after the invasion of a new region or a change in cropping system is implemented, the most fit individuals may be products of the occasional outcrossing event. In that case, outcrossing itself might be selected for in addition to the particular traits that are adaptive in the new environment. This would then generate additional variability allowing further adaptation. In contrast, during periods of relative stability, highly fit individuals that do not outcross will likely leave more offspring that retain their highly fit characteristics than will equally fit individuals that do outcross. This could then favor a decline in outcrossing, with consequent decrease in variability. Thus, the breeding system of some

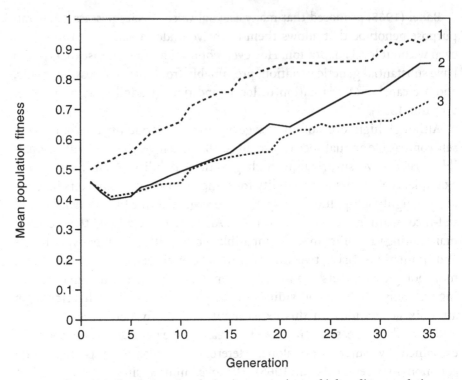

Figure 10.3 Simulation of evolution in outcrossing and inbreeding populations. Even a low rate of outcrossing allows rapid and sustained adaptation to selection pressure. (1) Random mating and moderate directional selection without heterozygote advantage. (2) 95% selfing and moderate directional selection with heterozygote advantage. (3) 95% selfing and moderate directional selection without heterozygote advantage. (Redrawn from Allard & Hansche, 1964.)

self-compatible species may evolve to changing environmental conditions in ways that make control more difficult. This hypothesis requires further investigation.

Similar adjustments in variability occur even more easily in perennial weeds that spread vegetatively. The process is illustrated by *Solidago altissima* in abandoned agricultural fields of various ages in central New York (Table 10.4). Variability was low in the youngest field due to the small number of early colonizers. Variability in a somewhat older field was higher, probably because additional individuals had time to arrive. Finally, variability was low in older populations as the competitively superior genotypes replaced others (Table 10.4). Probably a similar sequence occurs in many perennial weed species as populations establish and develop. In cases where ongoing dispersal into a field is limited, variability would still be expected to increase in the short run as recombination mixed the genetic material of the original founders. Then,

Table 10.4. *Genetic diversity of* Solidago altissima *in fields of several ages in central New York*

Field age[a] (yr)	Total ramets sampled	Number of plots	Genotypes per 0.75 m² plot
1	30	14	~1.5
5	244	3	9.3
~20	131	3	3.3
~35	165	3	2.0

Notes:
[a] Field age for the two older fields was approximate.
Source: Maddox et al. (1989).

once highly adapted genotypes have been created they will increase in relative abundance by vegetative propagation.

Genetic variability following colonization events

Dispersal to a new location represents a genetic bottleneck for several reasons. First, only a sample of the genetic diversity of the parental population is transported to the new site. Second, some variability may be lost by chance while the population is still small (genetic drift). Third, some variation may be lost via selection in the new habitat. Several cases of substantially depleted genetic variation, as indicated by a reduced frequency of isozyme polymorphisms, have been documented following long-range dispersal of weed species. These include *Striga asiatica* in the southeastern USA (Werth, Riopel & Gillespie, 1984), and *Chondrilla juncea* and *Emex spinosa* in Australia (Burdon, Marshall & Groves, 1980; Marshall & Weiss, 1982). In the case of *E. spinosa*, although individual populations are apparently genetically uniform or nearly so, variability exists among populations, probably due to separate introductions from the source region (Marshall & Weiss, 1982). Thus, this species has potential for rapid increase in variability in Australia if populations spread into sympatry and interbreed.

In contrast to the previous examples, some outbreeding species have retained high genetic diversity even after transoceanic colonization. Notable examples include *Echium plantagineum* in Australia (Brown & Burdon, 1983), *Apera spica-venti* in Canada (Warwick, Thompson & Black, 1987), and *Centaurea solstitialis* in the western USA (Sun, 1997). All three species show high levels of isozyme heterozygosity. Highly heterozygous founders together with high rates of outcrossing in these species quickly regenerate a diversity of genotypes following colonization. The high genetic diversity in *A. spica-venti* led

Warwick, Thompson & Black (1987) to predict rapid evolution of a race locally adapted to Canadian conditions.

Few studies have attempted to measure the effects of genetic diversity of the founders on colony success. In an unusual study, Martins & Jain (1979) sowed seeds of a newly introduced Eurasian forage legume, *Trifolium hirtum*, along roadsides in northern California. The seed sources for these colonies had low, medium, or high genetic polymorphism for eight morphological and allozyme markers. In the first year following sowing, 14 of 135 colonies established and success was uncorrelated with degree of polymorphism. The following year, however, seven additional colonies established from dormant seed, all from sowings with medium and high polymorphism. Combined, the two years data showed a weakly significant relation between polymorphism and establishment success ($p \sim 0.1$). In addition, however, Martins & Jain (1979) noted that the genotypes emerging the second year were different from those emerging the first year, further indicating the importance of genetic diversity for colonizing success. An implication of this study is that even if preventive measures cannot eliminate dispersal of all weed seeds, reducing the number of individuals reaching new locations may limit the frequency of successful establishment by lowering the genetic diversity of new populations.

Somatic variation

Somatic variation includes both somatic polymorphism and plastic developmental response to environmental conditions (Dekker, 1997). An example of the former is the non-genetic seed polymorphism that governs the variable germination requirements of some species (e.g., *Chenopodium album* – Williams & Harper, 1965; *Xanthium strumarium* – Weaver & Lechowicz, 1983). Examples of developmental plasticity include the great range in size and seed production of many annual weeds when subjected to a range of light, competition, or soil fertility conditions (Table 10.5) (Moran, Marshall & Müller, 1981; Rice & Mack, 1991; Sultan & Bazzaz, 1993a; Hermanutz & Weaver, 1996). In essence, somatic variation increases the size of a species' potential niche.

The degree of plasticity in a population is itself an evolved trait that sometimes responds to the selective forces of agriculture (Bradshaw, 1965). For example, *Camelina sativa* var. *sativa* is a widely distributed weed whose highly plastic morphology makes it an effective competitor in a variety of crops. In contrast, *C. sativa* var. *linicola* is a flax mimic; it grows tall, with a narrow crown, like the crop to which it is specialized, and has lost the flexibility of its more generally adapted ancestor (Stebbins, 1950; Bradshaw, 1965). Knowing

Table 10.5. *Phenotypic plasticity in* Echinochloa crus-galli *var.* crus-galli[a]

Character	Fertilized[b]	Density stressed[c]
Above-ground biomass (g)	219	0.026
Time to flowering (days)	54	95
Tiller production	17.6	1.0
Seed production	17 900	1.9
Harvest index (%)	13.4	10.4

Notes:
[a] Values are the means of plants in 10 pots.
[b] Plants in the fertilized treatment were grown singly in large (31 860 cm^3) pots and fertilized with N–P–K.
[c] Plants in the density stressed treatment were grown in small (88 cm^3) pots at a starting density of 100 plants.
Source: Barrett (1992).

to what extent weed races have increased or decreased phenotypic plasticity relative to their wild progenitors could yield insight into the evolutionary processes that create weeds.

Theory indicates that the evolution of plasticity for a character should be favored over genetic polymorphism in situations where environmental conditions vary greatly in short time spans or within the space occupied by an individual plant (Levins, 1963, 1968, pp. 10–22, 66–72). For example, if the light environment varies between high light and shade during the day or over the course of the growing season, then plasticity in response to light level should be favored rather than differentiation of sun- and shade-adapted genotypes (Gross, 1984; Sultan & Bazzaz, 1993b). Even if genetic variability for a character exists in a population, plasticity may prevent evolutionary change by allowing the various genotypes to assume similar phenotypes (Bradshaw, 1965; Sultan, 1987).

Plasticity and adaptation are largely complementary processes. High levels of plasticity in many weed characteristics allow weeds to avoid "dead-end" adaptation to temporary or local conditions. However, even highly plastic characteristics may shift under selection if the environment favors forms that are beyond the range of the plasticity response or, as in the case of *Camelina sativa* var. *linicola* cited above, the plasticity response itself is selectively penalized.

Ecotype formation in weeds

Well-differentiated ecotypes, races, and subspecies are the norm in weed species. Colonization with small founding populations and episodic reduction in population size of established populations by disturbance and weed control efforts create periods of extreme inbreeding and genetic drift. As discussed previously, selfing, apomixis, and vegetative reproduction then replicate the surviving genotypes as the population subsequently expands. More importantly, divergent selection pressure exerted by spatial variation in soil and climate conditions, cropping systems, and weed control methods act to differentiate weed populations. Great variation in the ecology of weed species can thus be generated rapidly and sometimes over short distances.

Ecotypic variation across the geographic range of a weed species can have important consequences for management. For example, *Hypericum perforatum* grows in both natural and disturbed habitats in Europe, but it is rarely a pest there (Pritchard, 1960). In contrast, it has been an aggressive and economically serious weed of grazing lands in Australia, California, and South Africa. Common garden studies have shown that individuals from Australia and California are substantially taller than individuals from Britain and other places where the species is not a pest (Pritchard, 1960). To some extent, its status as a weed in recently colonized portions of its range has been due to release from herbivory (Huffaker, 1957). Blossey & Nötzold (1995) hypothesized that release from herbivore pressure in invasive species like *H. perforatum* selects for reallocation of resources from defense to vegetative growth and seed production.

Weeds also differentiate in the process of range extension into new climatic zones. Between 1926 and the early 1970s *Sorghum halepense* spread northward in eastern North America from 38° to 43° latitude (Warwick, Phillips & Andrews, 1986), a range extension into an unfavorable climatic zone of about 550 km. In most of its range, this tropical grass is a perennial. However, most northern populations are annual. Probably these populations acquired the annual habit and associated characters by hybridization with the domesticated *S. bicolor* (Warwick, Thompson & Black, 1984). The shift to an annual habit in *S. halepense* has been accompanied by changes in several quantitative characters, including increased seed weight, increased percentage seedling emergence, faster seedling growth, earlier flowering, and greater allocation to reproduction (Warwick, Thompson & Black, 1984). Although the annual habit and associated characters are apparently adaptive in the northern populations, nothing obviously precludes the same characters from being adaptive in more southerly climates as well. Will the annual race spread south, or will gene flow from the more common perennial form prevent this? Ecotypic differentiation associated with range extension has also been documented for *Datura stramonium* and *Abutilon theophrasti* (Weaver, Dirks & Warwick, 1985; Warwick & Black, 1986; Warwick, 1990).

This book advocates management of weeds by manipulation of cropping systems, and it is therefore worthwhile asking to what extent weeds vary genetically in response to cropping system characteristics. Unfortunately, few relevant data exist. Naylor & Jana (1976) compared seed dormancy of *Avena fatua* in fields with different cropping histories. One site had been cropped in a rotation with summer fallow in one year out of three. A second site had only experienced occasional summer fallows, none of them recent. Naylor & Jana (1976) established that percentage dormancy was a heritable characteristic in these populations, and that dormancy was much more frequent in seeds of plants from the first site, probably because of strong selection against plants that emerged during the summer fallow years. In a follow-up experiment, artificial populations established from mixtures of dormant and non-dormant lines were subjected to either continuous cropping or a rotation with summer fallow in alternate years (Jana & Thai, 1987). After seven years (three summer fallows) the continuously cropped fields still had an equitable balance of dormant and non-dormant types whereas most individuals in fields with summer fallow were dormant. Similarly, Wilkes (1977) observed a population of weedy *Zea mexicana* with strong seed dormancy in a location where maize was consistently rotated with pasture. Elsewhere in its range, seeds of *Z. mexicana* apparently do not persist beyond the next growing season.

Several studies have compared weed with nonweed races, or compared weed populations subjected to different disturbance regimes. These give some idea of the type and magnitude of adaptive response that might be expected from changes in cropping systems or weed management practices. For example, Hodgson (1970) compared responses of 10 populations of *Cirsium arvense* to frequent cultivation with goosefoot shovels. After seven cultivations, some ecotypes were nearly eliminated, but one retained 32% of its original shoot density.

Theaker & Briggs (1993) analyzed several life history and morphological characteristics of *Senecio vulgaris* in a common garden experiment to test the hypothesis that frequent intensive weeding in botanical gardens selects for precocious development. As hypothesized, plants from botanical gardens flowered and set fruit earlier than plants from field edges or seminatural habitats (Table 10.6a). Also, plants from the botanical gardens were shorter and senesced earlier than plants from other habitats (Table 10.6a), indicating a trade-off between early fruiting and stature. Exposure to frequent intensive weeding had similar effects on plant size and rate of development in *Arabidopsis thaliana* (Jones, 1971) and *Stellaria media* (Sobey, 1987; Briggs, Hodkinson & Block, 1991).

Theaker & Briggs (1993) also found that a field population of *Senecio vulgaris*

Table 10.6. *Mean values of Senecio vulgaris traits from different habitats, observed in a common garden*

Habitat	Days to first anthesis	Days to first fruiting	Days from first anthesis to first fruiting	Plant height (mm)	Total capitula at first fruiting	Proportion dead by day 103
Mean over three populations from each of five habitat types						
Botanic gardens	37.5	51.7	14.2	203	39.1	0.81
Field margins	42.9	60.3	17.4	281	39.6	0.12
Inland sand	43.5	61.3	17.8	219	42.3	0.16
Coastal sand	43.2	60.9	17.6	237	38.0	0.18
Shingle beaches	47.2	67.1	19.9	213	32.6	0.06
Mean values for two nearby populations from contrasting habitats						
Field margin	42.7	59.7	17.0	290	47.5	0.13
Shingle beach	47.8	68.4	20.6	170	33.7	0.13

Source: Theaker & Briggs (1993).

and one on a shingle beach differed in most of the characters analyzed even though they were separated by just 50 m and an embankment (Table 10.6b). This implies that the strength of the contrasting selective pressures in the two populations was sufficient to maintain differences despite evidence of gene flow.

Jordan (1989a) compared populations of *Diodia teres* from a coastal dune and a soybean field. The weed population emerged two days earlier after planting and rapidly produced more leaves and meristems than the nonweed population. In the absence of competition, the weed grew 38% larger than the nonweed; in the presence of a soybean crop, the weed grew twice as large as the nonweed. This difference in final size was due solely to rapid early growth, since the weed grew slower than the nonweed after the crop became competitive. Thus, the weed race had apparently evolved to avoid competition by completing more growth early in the season. Unlike the previously cited studies in which the selective regime favored precocious flowering, the weed population of *Diodia teres* had plenty of time to complete development, hence selection for early growth led to larger, not smaller, mature plants. In further analysis, Jordan (1989b) showed that the genetic variability of the nonweed population was sufficient to allow rapid evolution of growth characteristics in response to selection by presence of the crop. Moreover, selection by the crop on the nonweed population was predicted to select for the phenological shift observed in the weed.

A troublesome form of weed adaptation involves mimicry of crop characteristics. Mimicry of vegetative characteristics occurs primarily in agricultural systems where hand-weeding is practiced regularly. It is most common in grass weeds, probably because of the superficial resemblance of many grass species to grain crops, especially in the seedling stage when hand-weeding is most likely (Barrett, 1983). Weed taxa closely related to the crop pose a special problem as mimics because they can acquire characteristics from the crop by introgressive hybridization. Apparently due to a combination of introgression and selection, weedy varieties of *Sorghum* and *Zea mexicana* mimic variation in the appearance of the related crop across regions of Africa and Mesoamerica, respectively (de Wet, Harlan & Price, 1976; Wilkes, 1977). Langevin, Clay & Grace (1990) found F_1 hybrids between weedy *Oryza sativa* and all six cultivars of domesticated rice they tested. However, because all of the F_1 hybrids flowered late, F_1 plants resulting from crosses with early-season cultivars would be destroyed during the harvest before they could shed seed. This would prevent backcrossing of hybrids with the weed. Thus, use of short-season varieties in this case offers a potential means for limiting convergence of the weed with the crop.

A large number of weeds have evolved races in which the size and shape of the propagules cause them to be threshed out with the crop seed, particularly with traditional seed-cleaning techniques (Barrett, 1983). Many seed crops have such mimics. The problem of seed mimics has been greatly reduced in the developed countries by modern seed-cleaning procedures and seed certification programs. Thus, for example, the once common weed of winter wheat, *Agrostemma githago*, is now rarely seen on industrialized wheat farms. In such cases, technology advanced faster than the adaptive response of the weeds. Presumably, if improvements in seed cleaning had taken place more slowly, the size and shape of the weed seeds would have become even more closely matched to those of their model crops, and seed-cleaning techniques that are presently effective would be insufficient. Although greatly reduced, the problem of seed mimic weeds has not been eliminated. Thus, for example, mimetic *Vicia sativa* is still found in lentil fields in the USA (Erskine, Smartt & Muehlbauer, 1994), and *Cardiospermum halicacabum* has become a problem for soybean seed producers in parts of the southern USA (Johnston, Murray & Williams, 1979; Bridges, 1992).

Variation among weed populations of a species implies that best management practices will sometimes vary between locations. For example, Cavers (1985) summarized data on seven biotypes of *Panicum miliacium* found in Canada. 'Black' was extremely distasteful to birds, whereas most other biotypes were attractive to birds and could be expected to suffer high seed predation if left on the soil surface through the winter. On the other hand, 'black' was sensitive to some herbicides to which other biotypes were tolerant. Also, collection of seed during crop harvest could in principle control the several non-shattering biotypes, whereas this option is not available for 'black' which shatters easily. As weed management shifts from reliance on herbicides to tactics that target specific ecological characteristics of weeds, variation in those characteristics will need to be taken into account.

Managing the adaptation of weed populations

Will weeds adapt to ecological management tactics?

The studies discussed above reveal several points about the differentiation of weed populations. First, substantial spatial variation in ecologically significant characteristics occurs in many weed species. This variation involves dormancy, phenology, stature, growth rate, and other characters that affect the efficacy of various ecological weed management tactics. Second, since much of the variation in such traits is heritable (Naylor & Jana, 1976; Jordan,

1989*b*; Theaker & Briggs, 1993; Thomas & Bazzaz, 1993), management practices can presumably select for forms more resistant to the practices. Third, differentiation between populations can occur rapidly. Apparently, a few years to a few decades are often sufficient to create detectable phenotypic divergence between populations. Fourth, ecotypes are found in a wide range of weed species: annuals and perennials, grasses and broadleafs, and species with and without seed banks. Thus, the potential for evolution of management-adapted weed races is a general problem that is not restricted to a few taxa or ecological categories.

Whether weed adaptation to management factors other than herbicides is a cause of the ongoing crop losses to weeds is presently unknown (Jordan & Jannink, 1997). Much of the ongoing crop losses are clearly attributable to shifts in species composition (e.g., the increase in *Senna obtusifolia* in soybean in the southern USA – Webster & Coble, 1997) rather than to genetic changes within species. Moreover, if the rate of adaptation to a weed control practice is slow, farmers and weed scientists have little reason to take preventive action since tactics and technology change through time and a particular adaptation may be irrelevant to future management practices. However, some species appear to evolve resistance rapidly to some classes of herbicides (Gill, 1995), and as discussed in the preceding section, adaptation to other management practices can be identified.

The fact that there are relatively few documented cases of management-adapted weeds other than crop mimics and herbicide-resistant biotypes appears to argue against the importance of weed evolution as a general management consideration. However, many ecotypes adapted to characteristics of particular cropping systems may be cryptic. First, subtle but important adaptive variation between farms in characters like seed longevity, the timing of emergence, or degree of dormancy could easily remain unnoticed by both farmers and weed scientists. Second, cropping systems vary within and between regions. To what extent are geographic weed races the result of management, including factors associated with the types of crops grown, rather than just a response to soil and climate as is usually assumed?

Obviously, the rate and magnitude of adaptive response to ecological management practices are unclear. Some possible responses to specific practices discussed in other chapters of this book are listed in Table 10.7, but whether these adaptations actually occur in specific weeds in response to control measures remains to be determined. Some idea of the selection pressure exerted on weed populations by these practices can be gathered from percentage mortality data (Table 10.7). However, additional data on the genetic variability and heritability of the characters involved, and

Table 10.7. *Percentage mortality from selected ecological weed management practices and adaptations that may be selected for by the practices*

Practice	Mortality (%)	Species	Reference	May select for
Rotary hoeing	53–90	*Chenopodium album* *Amaranthus* spp. *Setaria faberi*	Buhler, Gunsolus & Ralston (1992)	Ability to emerge from greater depth
Late planting of spring crops (delay of ~1 month)	43 63–85	*Setaria faberi* *Chenopodium album* *Amaranthus retroflexus* *Setaria faberi*	Forcella, Eradat-Oskoui & Wagner (1993) Gunsolus (1990)	Delayed germination; higher temperature for induction of secondary dormancy
Collection of weed seeds during grain harvest	91–99	*Datura ferox*	Ballaré *et al.* (1987)	Early maturation

information on how changes in a character affect exposure to the mortality factor would be needed to predict whether a population is likely to evolve resistance to a given practice. Gathering such data would require considerable effort. However, since ecological management practices are primarily useful within an integrated program, the potential of a weed to adapt to a single tactic may be irrelevant. A reasonable first step in assessing the ability of weeds to adapt to integrated ecological management would be to use common garden or reciprocal planting techniques to compare the ecological characteristics of weed populations from several pairs of adjacent conventional and organic farms. This would at least reveal whether significant divergence in response to cropping practices appears to occur over time scales relevant to management.

Due to doubts as to the importance of weed adaptation and the dearth of research on management of adaptation to control factors other than herbicides, the following sections provide few prescriptions. Instead, they are intended to indicate factors that may impact management of weed adaptation should such management prove worthwhile.

Management of weed adaptation: basic concepts

Jordan & Jannink (1997) identified three general levels of management of weed evolution. First, efforts may be directed toward limiting adaptation to specific weed control tactics. Most of the literature on management of weed evolution deals with this subject. Second, attempts can be made to prevent adaptation to systems of weed management. Such adaptation may involve evolution of several specific traits that provide simultaneous resistance to each of the component weed management tactics in the system. In this case, the management goal is to create selective regimes that make the co-occurrence of all the necessary traits in a single genotype unlikely. Alternatively, highly plastic, general-purpose genotypes may allow weeds to be successful within integrated weed management systems. Avoiding selection for plasticity in diversified cropping systems may be difficult. Third, weeds exist as metapopulations, systems of semi-isolated subpopulations that can facilitate rapid evolutionary change (Wade & Goodnight, 1991; Gould, 1993; Hastings & Harrison, 1994). Controlling evolutionary processes driven by metapopulation structure is likely to involve managing gene flow between subpopulations. For example, work by Paulson & Gould reported in Gould (1993) showed that increase of adaptive recessive genes could not occur if gene flow between subpopulations exceeded a threshold value. In other situations, gene flow may facilitate adaptive evolution of weeds (see section "Controlling the spread of new weeds" below). The evolutionary consequences of metapopulation structure in weed populations require exploration through modeling, model systems, and case studies.

Discussion in the weed science literature of methods for managing the adaptation of weeds to specific control measures has focused primarily on prevention of herbicide resistance. Although some analogies may be made between control of herbicide resistance and control of adaptation to ecological management practices, such analogies are limited by differences in the mode of inheritance. Resistance to herbicides and other anthropogenic toxins is usually controlled by one to a few major genes (Macnair, 1991; Jasieniuk, Brûlé-Babel & Morrison, 1996). First, because herbicides usually target a single enzyme, a small but highly specific change in the genome is sufficient to confer resistance. Second, mortality from an effective herbicide is high (often 95% to 99%), and consequently the amount of phenotypic change required in a single generation usually exceeds the available additive genetic variation for tolerance in the population. Intermediate forms cannot survive, and hence one to three genes with large effects usually control herbicide resistance. However, the degree of herbicide tolerance to low doses or in naturally tolerant species is usually controlled by many genes (Putwain & Collin, 1989). In contrast to herbicide resistance, characters like growth rate, stature, seed mass, seed longevity, and phenology of germination that might change during adaptation to ecological management practices are quantitative characters controlled by many genes.

Several strategies for slowing or preventing the evolution of herbicide resistance have been suggested based on genetic models. These include (i) slowing the rate of increase in the frequency of resistant genes by rotating use of a given herbicide with other chemical and nonchemical controls (Gressel & Segel, 1990; Gorddard, Pannel & Hertzler, 1996; Jasieniuk, Brûlé-Babel & Morrison, 1996), (ii) decreasing the probability of resistant individuals in the population by use of multiple herbicides simultaneously (Gressel & Segel, 1990), (iii) decreasing selection pressure for resistance by using rates that allow some escapes (Maxwell, Roush & Radosevich, 1990), and (iv) leaving untreated areas to provide flow of susceptible genes into the population (Maxwell, Roush & Radosevich, 1990).

With the exception of rotating control methods, these tactics are largely irrelevant to ecological weed management. Gene flow occurs over such short distances in most weed species that the last approach would usually require leaving unacceptable numbers of uncontrolled individuals (Jasieniuk, Brûlé-Babel & Morrison, 1996; but see discussion of spatially complex cropping systems in the next section). The second and third approaches apply specifically to adaptation via major genes. In particular, decreasing selection pressure for a resistance trait controlled by multiple genes from a high to a moderate level tends to increase rather than decrease the rate of adaptation (Cousens & Mortimer, 1995, p. 280). With characters controlled by multiple genes, lower selection pressure allows more escapes and thus more possibilities for resistant forms to arise by recombination. Hard selection that results in low population size is thus a defense against resistance controlled by multiple genes.

Cropping system diversity and the management of weed adaptation

Regardless of whether the response to a practice is via polygenes or a single major gene, alternation of methods between years provides an important approach for limiting adaptation to control measures. This is particularly the case if fitness of genotypes adapted to some practice, A, is less than that of unadapted genotypes in years when some other practice, B, is used. In such cases, the frequency of A-adapted types will decline in years when practice B is used. On the other hand, if practice B is neutral with respect to A-adapted types then the rate of adaptation to practice A is proportional to how often it is used (Gressel & Segel, 1990; Jasieniuk, Brûlé-Babel & Morrison, 1996). Alternation between the practices then slows adaptation to practice A, but does not prevent evolution of a high level of resistance eventually.

Continuous monocultures appear to favor special adaptations like crop mimicry and herbicide resistance (Gressel, 1991). In contrast, diverse crop rotations may favor genetic variability since variation in selection pressures between years would tend to slow fixation of particular alleles. This hypothesis requires testing. To the extent that the several crops and corresponding cultural practices of a rotation put different selection pressures on a weed population, diverse crop rotations might even disrupt multigene adaptations to particular control practices that may be used in several crops (e.g., cultivation, mulch).

Fine-grained spatial variation in the cropping system probably also tends to slow weed adaptation. Growing different crops repeatedly in separate fields creates a selective regime potentially favoring genetic divergence in the several subpopulations. In contrast, when different crops are grown together in polycultures, then differences in the competitive and cultural environments of the several crops act as disruptive selection on many potentially adaptive traits of the weeds.

In actual practice, diverse crop rotations and intercropping systems usually do not repeat the exact sequence and arrangement of crops, but rather respond flexibly to changes in market conditions, weather, weeds, and other pests. Use of cropping system diversity to limit weed adaptation will likely work best if farmers and crop advisors learn to recognize incipient adaptive responses in weed populations and know how to respond appropriately.

Although diversification of cropping systems should in general slow the adaptive response of weeds to specific crops and control tactics, weeds can potentially cope with complex rotations and polycultures by several means. First, long gaps between years with a particular class of crop should select for species and genotypes that have great seed longevity. A testable corollary is that annual species whose seed store well in the soil should make up a greater proportion of the seed bank in fields that have been regularly rotated into sod crops. Moreover, populations of particular species from such fields should have greater seed longevity than populations from fields that have been in continuous monoculture of an annual crop. Second, weeds may evolve additional mechanisms for cueing germination to particular cropping conditions, and conversely, mechanisms for remaining dormant through unfavorable periods. Third, complex cropping systems may select for greater plasticity and polymorphism. Biotypes that can survive at low stature and abundance through several unfavorable years but produce abundant seed in the occasional good year should be favored. Similarly, a diverse cropping system may select for increased polymorphism, especially in germination characteristics.

Is a trade-off between selection for general adaptations by diverse cropping systems and selection for specific adaptations by continuous monoculture inevitable? In principle, if multiple tactics that attack the weed population at several life stages keep population densities perennially low, then evolution of both specific and general adaptations should be slow due to the limited number of genotypes available for selection and the chance loss of adapted types when they do occur. Additional strategies for limiting evolution of general adaptations within the context of diversified cropping systems need to be identified.

Seed banks and the management of weed adaptation

Presence of a seed bank tends to prevent genetic drift by increasing effective population size during years in which the above-ground population is small (Epling, Lewis & Ball, 1960). More importantly, a seed bank can prevent loss of genetic variability due to transient selection in a fluctuating environment by shielding a subset of individuals from the selection pressure (Templeton & Levin, 1979). A seed bank thereby slows adaptation to management.

Due to more rapid turnover of seed banks in reduced tillage systems relative to conventional tillage, reduced tillage, and especially no-till, probably has greater risk of generating management-resistant races of species with seed banks. For example, Gressel (1991) noted that *Senecio vulgaris* first evolved triazine resistance in orchards, nurseries, and roadsides rather than in maize fields. Few *S. vulgaris* seeds persist longer than one year on the soil surface (Popay & Roberts, 1970), whereas many remain viable for several years when incorporated into the soil (Watson, Mortimer & Putwain, 1987).

A seed bank also biases selection toward traits that are favored during years with high seed production (Templeton & Levin, 1979). Conversely, selection for traits that are favored during years when few seeds are produced has little effect on the population, because seeds carrying those traits are few relative to those from years of high seed production. Consequently, when a weed with a persistent seed bank first begins to show resistance to a control tactic, use of additional measures that reduce seed production will slow the rate of adaptation. Analogously with seed production, selection during years when emergence from the seed bank is great is more effective in changing the genetic makeup of a population than selection in years when emergence rate is low. Essentially, a seed bank provides a mechanism whereby evolution is partly governed by the absolute quality of the environment as measured by seed production and percentage emergence. One consequence is that adaptation to control tactics that occasionally fail is likely to be slow for weeds with seed banks.

Controlling the spread of new weeds

A key part of any general strategy for management of weed communities and the adaptation of weed populations is prevention of dispersal between fields and regions, and rapid response to new forms once they are present. This requires a mode of operation that goes beyond the usual prescriptive approach to the management of weeds with herbicides and even beyond the integrative approach advocated throughout this book. Rather, managing the spread of new weed species and biotypes in a region requires an approach analogous to that used to control communicable diseases, with emphasis on epidemiological analysis, education, and societal response (Green, 1990, pp. 254–85).

Preventing the spread of weeds between fields and regions is a critical and much neglected aspect of weed science. Blocking movement of weeds clearly reduces the rate of increase in the number of infested fields for an alien species that is spreading within a region. Less obviously, it also reduces the total number of fields infested at equilibrium by shifting the balance between colonization and local extinction (see section "Species introduction and the species richness of weed communities" above).

Moreover, reducing dispersal between fields restricts the entry of new genetic variability into weed populations. The utility of not allowing immigration of herbicide-resistant biotypes into populations is well understood (Darmency, 1994; Morris, Kareiva & Raymer, 1994; Jasieniuk, Brûlé-Babel & Morrison, 1996), but entry of other sorts of genetic variability is often undesirable as well since it may provide the raw material for future adaptation of the population. Many self-compatible weed species have low within-population isozyme polymorphism but substantial variation among populations (Barrett, 1988; Wang, Wendel & Dekker, 1995a, 1995b). As discussed previously, this probably reflects past genetic bottlenecks during colonization. Assuming the bottlenecks have had similar effects on adaptive variation, further adaptation at the local level may depend on arrival of genetic material from outside the population.

Methods for preventing the spread of weeds in manure, on machinery, in crop seed, etc. were detailed in Chapter 2. In addition, since a large proportion of the propagules dispersing between fields in a district in any given year probably originate from a relatively few heavily infested fields, control of weeds at those locations will reduce spread. Finally, if the newly arrived species or problem biotype can be recognized soon after arrival, it can be eradicated while the population is small and localized.

The problem is to implement all these methods through education and institutional response. This may be accomplished through a three-pronged approach using the participatory learning-for-action model discussed in Chapter 3. In some respects the proposed approach resembles quarantine and eradication programs used against particularly noxious weeds (e.g., *Striga asiatica* in the USA – Sand, 1987). However, for most weeds such extreme measures are not cost-effective. Rather, mechanisms are needed for co-ordinating the collective good sense of the agricultural community for regional management of emerging weed problems.

First, the epidemiology of major weeds and potentially serious new weeds and weed races in a region needs to be elucidated. In particular, how is each taxon spreading among farms? Are additional mechanisms promoting spread among fields within farms? What is the relation between the number of propagules in an inoculum and the probability of establishing a population? What is the distribution of infested fields in the region? Where are the fields with extreme infestations located? These questions need to be answered first for those species and biotypes that are most potentially damaging to the agriculture of the region. With time, databases for additional species can be added. Understanding the mode of spread and effective inoculum size will generally require a focused, but not necessarily extensive, research effort. Information on distribution of problem weeds could be collected by extension agencies with cooperation from crop consultants and farmers. Well-advertised telephone numbers and electronic mail addresses for reporting suspected infestations could be used to help gather information. Taxonomically trained local volunteers could then check putative sightings. Standardized database templates and clearly designated lines for reporting locally collected information to regional and national institutions will facilitate assembly of information into a useful form.

Education forms a second critical component of programs for preventing emergence and spread of new weeds. Farmers, extension agents, and crop consultants need to be better trained in weed identification. This can be accomplished through workshops, interactive web sites, preparation of inexpensive guides available through extension agencies, and attractive, easy-to-use weed identification handbooks (e.g., Stucky, Monaco & Worsham, 1981; Uva, Neal & DiTomaso, 1997). When research or extension institutions become aware that a particular problem species or biotype is spreading in a region, information on identification of the weed needs to be disseminated to farmers via fliers, newsletters, and grower meetings.

Education is also critical for informing farmers about the most effective methods for preventing dispersal of weeds onto and within their farms, the value of early eradication of new infestations, and the importance of controlling severe infestations on fallow land and other areas where weeds may not have an immediate economic impact. Finally, education efforts will be required to convince growers of the value of their support for the institutional responses to weed spread discussed below. Education of farmers by farmers is particularly important in developing support for prevention programs.

The third component of a comprehensive program for the prevention of new weed problems is institutional and farmer action. Some institutional actions can be taken to limit weed dispersal. These include certification of seed and feed, inspection of produce crossing national borders, for example, by the Animal and Plant Health Inspection Service (APHIS) in the USA, and manure handling standards that encourage composting (e.g., Northeast Organic Farming Association of New York, 1995). Institutional responses to new or severe infestations could include government cost-sharing of control efforts, and advice and logistic support for farmer–extension teams trained to encourage effective management of weed problems in the community. Most often, however, all that will be required to eradicate a new infestation is recognition by the farmer that new weed infestations represent a threat, an effective control strategy, and farmer persistence.

The critical first step in developing a proactive approach to the management of new weed problems is commitment by researchers to address the problem. The scientific community is better situated than farmers and extensionists to take a long view of weed problems, and to suggest to government and the other segments of the agricultural community general strategies for their solution. In particular, researchers need (i) to construct theories and models that predict how and why new weed problems develop, (ii) to monitor the introduction of new crops, technologies, and cropping systems for the emergence of novel weed problems, and (iii) to consider how regional problems can be managed before rather than after they have a major economic impact. The public health style approach suggested here for the control of new weed problems is a substantial extension of conventional weed science. It nevertheless represents only a small first step toward long-term management of the slow but persistent increase in local, regional, and global weed diversity.

Conclusions

As argued in the preceding sections, agricultural ecosystems continuously generate new weed problems. Recent advances in control measures appear to have reduced losses due to weeds relative to historical levels (Pimentel et al., 1978). However, economic losses from weeds continue to be large (Bridges, 1992). Continuous advances in both weed biology and weed control technology may be necessary just to maintain the current levels of loss in the face of the continuing spread of introduced species, shifts in the composition of weed communities, and the evolution of resistance to herbicides and other management practices.

Two overall strategies for management of changes in weed communities are apparent. One is to continue the present essentially responsive approach in which shifts in weed composition and development of herbicide resistance are attacked with newly developed herbicides and complex mixtures of existing materials. This approach guarantees a continuing market for new chemical technologies, but leaves the grower with a generally increasing bill for weed control.

The alternative is to take a more methodical approach in which principles are elucidated that predict the response of weeds to control measures, and strategies are developed to intercept problems before they become severe. The growing interest among weed scientists in modeling the dynamics and genetic composition of weed populations is a first step in implementing this alternative approach to the management of incipient weed problems. However, new categories of higher-level models are needed to understand and predict phenomena like species shifts, the spread of weeds within and between regions, and the evolution of herbicide resistance in taxa that are currently susceptible. Such phenomena occur at spatial and temporal scales that exceed the boundaries of farms and the attention span of individual growers. Consequently, the extension of human understanding of weeds into larger scales will make management decisions at the community, regional, and national levels both practical and desirable. Developing a higher-level theory of weeds probably represents the greatest challenge for weed science in the coming century. Implementation of that understanding by farmers, communities, and government agencies may prove equally challenging.

REFERENCES

Abbas, H. K., Pantone, D. J., & Paul, R. N. (1999). Characteristics of multiple-seeded cocklebur: a biotype of common cocklebur (*Xanthium strumarium* L.). *Weed Technology*, **13**, 257–63.

Allard, R. W. (1963). Evidence for genetic restriction of recombination in the lima bean. *Genetics*, **48**, 1389–95.

Allard, R. W. (1965). Genetic systems associated with colonizing ability in predominantly self-pollinated species. In *The Genetics of Colonizing Species, Proceedings of the 1st International Union of Biological Sciences Symposia on General Biology*, ed. H. G. Baker & G. L. Stebbins, pp. 49–75. New York: Academic Press.

Allard, R. W., & Hansche, P. E. (1964). Some parameters of population variability and their implications in plant breeding. *Advances in Agronomy*, **16**, 281–325.

Altieri, M. A., & Liebman, M. (1986). Insect, weed, and plant disease management in multiple cropping systems. In *Multiple Cropping Systems*, ed. C. A. Francis, pp. 182–218. New York: Macmillan.

Andow, D. A. (1991). Vegetational diversity and arthropod population response. *Annual Review of Entomology*, **36**, 561–86.

Andreasen, C., Stryhn, H., & Streibig, J. C. (1996). Decline of the flora in Danish arable fields. *Journal of Applied Ecology*, **33**, 619–26.

Ashton, P. A., & Abbott, R. J. (1992). Multiple origins and genetic diversity in the newly arisen allopolyploid species, *Senecio cambrensis* Rosser (Compositae). *Heredity*, **68**, 25–32.

Atwood, S. S. (1940). Genetics of cross-incompatibility among self-incompatible plants of *Trifolium repens*. *Journal of the American Society of Agronomy*, **32**, 955–68.

Atwood, S. S. (1942). Oppositional alleles causing cross-incompatibility in *Trifolium repens*. *Genetics*, **27**, 333–8.

Bailey, L. H., & Bailey, E. Z. (1930). *Hortus: A Concise Dictionary of Gardening, General Horticulture and Cultivated Plants in North America*. New York: Macmillan.

Bailey, L. H., & Bailey, E. Z. (1941). *Hortus Second: A Concise Dictionary of Gardening, General Horticulture and Cultivated Plants in North America*. New York: Macmillan.

Bailey, L. H., & Bailey, E. Z., revised and expanded by the staff of the Liberty Hyde Bailey Hortorium (1976). *Hortus Third: A Concise Dictionary of Gardening, General Horticulture and Cultivated Plants in North America*. New York: Macmillan.

Baker, H. G. (1962). Weeds: native and introduced. *Journal of the California Horticultural Society*, **23**, 97–104.

Baker, H. G. (1965). Characteristics and modes of origin of weeds. In *The Genetics of Colonizing Species, Proceedings of the 1st International Union of Biological Sciences*

Symposia on General Biology, ed. H. G. Baker & G. L. Stebbins, pp. 147–68. New York: Academic Press.

Baker, H. G. (1974). The evolution of weeds. *Annual Review of Ecology and Systematics*, **5**, 1–24.

Baker, H. G. (1991). The continuing evolution of weeds. *Economic Botany*, **45**, 445–9.

Ballaré, C. L., Scopel, A. L., Ghersa, C. M., & Sánchez, R. A. (1987). The demography of *Datura ferox* (L.) in soybean crops. *Weed Research*, **27**, 91–102.

Barrett, S. C. H. (1992). Genetic variation in weeds. In *Biological Control of Weeds with Plant Pathogens*, ed. R. Charudattan & H. L. Walker, pp. 73–98. New York: John Wiley.

Barrett, S. C. H. (1983). Crop mimicry in weeds. *Economic Botany*, **37**, 255–82.

Barrett, S. C. H. (1988). Genetics and evolution of agricultural weeds. In *Weed Management in Agroecosystems: Ecological Approaches*, ed. M. A. Altieri & M. Liebman, pp. 57–75. Boca Raton, FL: CRC Press.

Barrett, S. C. H., & Seaman, D. E. (1980). The weed flora of Californian rice fields. *Aquatic Botany*, **9**, 351–76.

Baum, B. R. (1969). The use of lodicule type in assessing the origin of *Avena* fatuoids. *Canadian Journal of Botany*, **47**, 931–44.

Bell, C. R. (1971). Breeding systems and floral biology of the Umbelliferae or evidence for specialization in unspecialized flowers. In *Biology and Chemistry of the Umbelliferae*, ed. V. H. Heyward, pp. 93–107. New York: Academic Press.

Blossey, B., & Nötzold, R. (1995). Evolution of increased competitive ability in invasive nonindigenous plants: a hypothesis. *Journal of Ecology*, **83**, 887–9.

Boudry, P., Mörchen, M., Saumitou-Laprade, P., Vernet, P., & Van Dijk, H. (1993). The origin and evolution of weed beets: consequences for the breeding and release of herbicide-resistant transgenic sugar beets. *Theoretical and Applied Genetics*, **87**, 471–8.

Bradshaw, A. D. (1965). Evolutionary significance of phenotypic plasticity in plants. *Advances in Genetics*, **13**, 115–55.

Briggs, D., Hodkinson, H., & Block, M. (1991). Precociously developing individuals in populations of chickweed [*Stellaria media* (L.) Vill.] from different habitat types, with special reference to the effects of weed control measures. *New Phytologist*, **117**, 153–64.

Bridges, D. C. (ed.) (1992). *Crop Losses Due to Weeds in the United States: 1992*. Champaign, IL: Weed Science Society of America.

Brown, A. H. D., & Burdon, J. J. (1983). Multilocus diversity in an outbreeding weed, *Echium plantagineum* L. *Australian Journal of Biological Sciences*, **36**, 503–9.

Buhler, D. D., Gunsolus, J. L., & Ralston, D. F. (1992). Integrated weed management techniques to reduce herbicide inputs in soybean. *Agronomy Journal*, **84**, 973–8.

Burdon, J. J., Marshall, D. R., & Groves, R. H. (1980). Isozyme variation in *Chondrilla juncea* L. in Australia. *Australian Journal of Botany*, **28**, 193–8.

Cardina, J., Webster, T. M., & Herms, C. P. (1998). Long-term tillage and rotation effects on soil seedbank characteristics. In *Weed Seedbanks: Determination, Dynamics and Manipulation* eds. G. T. Champion, A. C. Grundy, N. E. Jones, E. J. P. Marshall & R. J. Froud-Williams, pp. 212–20. Wellsbourne, UK: Association of Applied Biologists.

Cardina, J., Webster, T. M., Herms, C. P., & Regnier, E. E. (1999). Development of weed IPM: levels of integration for weed management. *Journal of Crop Production*, **2**, 239–67.

Cavers, P. B. (1985). Intractable weeds: intraspecific variation must be considered in formulating control measures. In *Proceedings of the 1985 British Crop Protection Conference – Weeds*, vol. 1, pp. 367–76. Croydon, UK: British Crop Protection Council.

Cavers, P. B., & Bough, M. A. (1985). Proso millet (*Panicum miliaceum* L.): a crop and a weed. In *Studies on Plant Demography*, ed. J. White, pp. 143–55. London: Academic Press.

Clegg, M. T., & Brown, A. H. D. (1983). The founding of plant populations. In *Genetics and Conservation*, ed. C. M. Schonewald-Cox, S. M. Chambers, B. MacBryde & L. Thomas, pp. 216–28. Menlo Park, CA: Benjamin/Cummings.

Clements, D. R., Weise, S. F., & Swanton, C. J. (1994). Integrated weed management and weed species diversity. *Phytoprotection*, **75**, 1–18.

Conn, J. S., & DeLapp, J. A. (1983). Weed species shifts with increasing field age in Alaska. *Weed Science*, **31**, 520–4.

Cornell, H. V., & Lawton, J. H. (1992). Species interactions, local and regional processes, and limits to the richness of ecological communities: a theoretical perspective. *Journal of Animal Ecology*, **61**, 1–12.

Cousens, R., & Mortimer, M. (1995). *Dynamics of Weed Populations*. Cambridge, UK: Cambridge University Press.

Covarelli, G., & Tei, F. (1988). Effet de la rotation culturale sur la flore adventice du mais. In *7iéme Colloque International sur la Biologie, l'Ecologie et la Systematique des Mauvaises Herbes*, vol. 2, pp. 477–84. Paris, France: Comité Français de Lutte contre les Mauvaises Herbes.

Crawley, M. J. (1987). What makes a community invasible? In *Colonization, Succession, and Stability: 26th Symposium of the British Ecological Society*, ed. A. J. Gray, M. J. Crawley & P. J. Edwards, pp. 429–53. Oxford, UK: Blackwell.

Crawley, M. J., Harvey, P. H., & Purvis, A. (1996). Comparative ecology of the native and alien floras of the British Isles. *Philosophical Transactions of the Royal Society of London*, series B, **351**, 1251–9.

Darmency, H. (1994). The impact of hybrids between genetically modified crop plants and their related species: introgression and weediness. *Molecular Ecology*, **3**, 37–40.

Dekker, J. (1997). Weed diversity and weed management. *Weed Science*, **45**, 357–63.

de Wet, J. M. J. (1975). Evolutionary dynamics of cereal domestication. *Bulletin of the Torrey Botanical Club*, **102**, 307–12.

de Wet, J. M. J., & Harlan, J. R. (1975). Weeds and domesticates: evolution in the man-made habitat. *Economic Botany*, **29**, 99–107.

de Wet, J. M. J., Harlan, J. R., & Price, E. G. (1976). Variability in *Sorghum bicolor*. In *Origins of African Plant Domestication*, ed. J. R. Harlan, J. M. J. de Wet & A. B. I. Stemler, pp. 453–63. The Hague, Netherlands: Mouton.

Dias-Filho, M. B. (1994). Ecophysiological studies of four Amazonian weedy species: implications for their invasive potential. PhD thesis, Cornell University, Ithaca, NY.

Doll, H. (1997). The ability of barley to compete with weeds. *Biological Agriculture and Horticulture*, **14**, 43–51.

Epling, C., Lewis, H., & Ball, F. M. (1960). The breeding group and seed storage: a study in population dynamics. *Evolution*, **14**, 238–55.

Erskine, W., Smartt, J., & Muehlbauer, F. J. (1994). Mimicry of lentil and the domestication of common vetch and grass pea. *Economic Botany*, **48**, 326–32.

Forcella, F., & Harvey, S. J. (1983). Relative abundance in an alien weed flora. *Oecologia (Berlin)*, **59**, 292–5.

Forcella, F., Eradat-Oskoui, K., & Wagner, S. W. (1993). Application of weed seedbank ecology to low-input crop management. *Ecological Applications*, **3**, 74–83.

Gadgil, M., & Solbrig, O. T. (1972). The concept of r- and K-selection: evidence from wild flowers and some theoretical considerations. *American Naturalist*, **106**, 14–31.

Garfinkel, Y., Kislev, M. E., & Zohary, D. (1988). Lentil in the Pre-Pottery Neolithic B Yiftah'el: additional evidence of its early domestication. *Israel Journal of Botany*, **37**, 49–51.

Gill, G. S. (1995). Development of herbicide resistance in annual ryegrass populations (*Lolium rigidum* Gaud.) in the cropping belt of Western Australia. *Australian Journal of Experimental Agriculture*, **35**, 67–72.

Godwin, H. (1960). The history of weeds in Britain. In *The Biology of Weeds: A Symposium of the British Ecological Society*, ed. J. L. Harper, pp. 1–10. Oxford, UK: Blackwell.

Gorddard, R. J., Pannel, D. J., & Hertzler, G. (1996). Economic evaluation of strategies for management of herbicide resistance. *Agricultural Systems*, **51**, 281–98.

Gould, F. (1993). The spatial scale of genetic variation in insect populations. In *Evolution of Insect Pests.*, ed. K. E. C. Kim & B. A. McPherson, pp. 67–85. New York: John Wiley.

Grant, V. (1981). *Plant Speciation*, 2nd edn. New York: Columbia University Press.

Green, L. W. (1990). *Community Health*, 6th edn. Saint Louis, MO: Times Mirror/Mosby College Publishing.

Gressel, J. (1991). Why get resistance? It can be prevented or delayed. In *Herbicide Resistance in Weeds and Crops*, ed. J. C. Caseley, G. W. Cussans & P. K. Atkins, pp. 1–26. Oxford, UK: Butterworth-Heinemann.

Gressel, J., & Segel, L. (1990). Modelling the effectiveness of herbicide rotations and mixtures as strategies to delay or preclude resistance. *Weed Technology*, **4**, 186–98.

Gross, L. J. (1984). On the phenotypic plasticity of leaf photosynthetic capacity. In *Mathematical Ecology, Proceedings of the Autumn Course (Research Seminars) Held at the International Center for Theoretical Physics*, 29 November–1 December, 1982, Miramare-Trieste, Italy, ed. S. A. Levin & T. G. Hallam, pp. 2–14. Berlin: Springer Verlag.

Gunsolus, J. L. (1990). Mechanical and cultural weed control in corn and soybeans. *American Journal of Alternative Agriculture*, **5**, 114–19.

Harlan, J. R. (1975). *Crops and Man*. Madison, WI: American Society of Agronomy.

Harlan, J. R. (1982). Relationships between weeds and crops. In *Biology and Ecology of Weeds*, ed. W. Holzner & N. Numata, pp. 91–6. Hague, Netherlands: Dr. W. Junk.

Harrar, E. S., & Harrar, J. G. (1962). *Guide to Southern Trees*, 2nd edn. New York: Dover.

Harris, S. A., & Ingram, R. (1992). Molecular systematics of the genus *Senecio* L. 1. Hybridization in a British polyploid complex. *Heredity*, **69**, 1–10.

Hastings, A., & Harrison, S. (1994). Metapopulation dynamics and genetics. *Annual Review of Ecology and Systematics*, **25**, 167–88.

Hebden, P. M., Rodger, S. J., Wright, G., & Squire, G. (1998). Effects of rotation and cropping system on dynamics of seedbank species diversity. In *Weed Seedbanks: Determination, Dynamics and Manipulation*, eds. G. T. Champion, A. C. Grundy, N. E. Jones, E. J. P. Marshall & R. J. Froud-Williams, pp. 243–8. Wellsbourne, UK: Association of Applied Biologists.

Hermanutz, L. A., & Weaver, S. E. (1996). Agroecotypes or phenotypic plasticity? Comparison of agrestal and ruderal populations of the weed *Solanum ptycanthum*. *Oecologia*, 105, 271–80.

Hobbs, R. J. (1991). Disturbance a precursor to weed invasion in native vegetation. *Plant Protection Quarterly*, 6, 99–104.

Hodgson, J. M. (1970). The response of Canada thistle ecotypes to 2,4-D, amitrole, and intensive cultivation. *Weed Science*, 18, 253–5.

Holm, L. G., Plucknett, D. L., Pancho, J. V., & Herberger, J. P. (1977). *The World's Worst Weeds*. Honolulu, HI: University Press of Hawaii.

Holt, J. S., & LeBaron, H. M. (1990). Significance and distribution of herbicide resistance. *Weed Technology*, 4, 141–9.

Huffaker, C. B. (1957). Fundamentals of biological control of weeds. *Hilgardia*, 27, 101–57.

Iversen, J. (1941). Land occupation in Denmark's Stone Age. *Danmarks Geologiske Undersogelse II Raekke*, 66, 1–68.

Jain, S. K., & Martins, P. S. (1979). Ecological genetics of the colonizing ability of rose clover (*Trifolium hirtum* All.). *American Journal of Botany*, 66, 361–6.

Jana, S., & Thai, K. M. (1987). Patterns of changes of dormant genotypes in *Avena fatua* populations under different agricultural conditions. *Canadian Journal of Botany*, 65, 1741–5.

Jasieniuk, M., Brûlé-Babel, A. L., & Morrison, I. N. (1996). The evolution and genetics of herbicide resistance in weeds. *Weed Science*, 44, 176–93.

Johnston, S. K., Murray, D. S., & Williams, J. C. (1979). Germination and emergence of balloonvine (*Cardiospermum halicacabum*). *Weed Science*, 27, 73–6.

Jones, M. E. (1971). The population genetics of *Arabidopsis thaliana*. *Heredity*, 27, 51–8.

Jordan, N. (1989a). Path analysis of growth differences between weed and nonweed populations of poorjoe (*Diodia teres*) in competition with soybean (*Glycine max*). *Weed Science*, 37, 129–36.

Jordan, N. (1989b). Predicted evolutionary response to selection for tolerance of soybean (*Glycine max*) and intraspecific competition in a nonweed population of poorjoe (*Diodia teres*). *Weed Science*, 37, 451–7.

Jordan, N., & Jannink, J. L. (1997). Assessing the practical importance of weed evolution: a research agenda. *Weed Research*, 37, 237–46.

King, L. J. (1966). *Weeds of the World*. New York: Interscience.

Langevin, S. A., Clay, K., & Grace, J. B. (1990). The incidence and effects of hybridization between cultivated rice and its related weed red rice (*Oryza sativa* L.). *Evolution*, 44, 1000–8.

Lawson, H. M., & Topham, P. B. (1985). Competition between annual weeds and vining peas grown at a range of population densities: effects on the weeds. *Weed Research*, 25, 221–9.

Levins, R. (1963). Theory of fitness in a heterogeneous environment. II. Developmental flexibility and niche selection. *American Naturalist*, 97, 75–90.

Levins, R. (1968). *Evolution in Changing Environments*. Princeton, NJ: Princeton University Press.

Liebman, M., & Dyck, E. (1993). Crop rotation and intercropping strategies for weed management. *Ecological Applications*, 3, 92–122.

Liebman, M., Drummond, F. A., Corson, S., & Zhang, J. (1996). Tillage and rotation crop effects on weed dynamics in potato production systems. *Agronomy Journal*, **88**, 18–26.

Lincoln, R., Boxshall, G., & Clark, P. (1998). *A Dictionary of Ecology, Evolution and Systematics*, 2nd edn. Cambridge, UK: Cambridge University Press.

MacArthur, R. H., & Wilson, E. O. (1967). *The Theory of Island Biogeography*. Princeton, NJ: Princeton University Press.

Macnair, M. R. (1991). Why the evolution of resistance to anthropogenic toxins normally involves major gene changes: the limits to natural selection. *Genetica*, **84**, 212–19.

Maddox, G. D., Cook, R. E., Wimberger, P. H., & Gardescu, S. (1989). Clone structure in four *Solidago altissima* (Asteraceae) populations: rhizome connections within genotypes. *American Journal of Botany*, **76**, 318–26.

Maillet, J., & Lopez-Garcia, C. (2000). What criteria are relevant for predicting the invasive capacity of a new agricultural weed: the case of invasive American species in France. *Weed Research*, **40**, 11–26.

Marks, P. L. (1983). On the origin of the field plants of the northeastern United States. *American Naturalist*, **122**, 210–28.

Marshall, D. R., & Weiss, P. W. (1982). Isozyme variation within and among Australian populations of *Emex spinosa* (L.) Campd. *Australian Journal of Biological Sciences*, **35**, 327–32.

Martins, P. S., & Jain, S. K. (1979). Role of genetic variation in the colonizing ability of rose clover (*Trifolium hirtum* All.). *American Naturalist*, **114**, 591–5.

Maxwell, B. D., Roush, M. L., & Radosevich, S. R. (1990). Predicting the evolution and dynamics of herbicide resistance in weed populations. *Weed Technology*, **4**, 2–13.

Mayr, E. (1963). *Animal Species and Evolution*. Cambridge, MA: Harvard University Press.

McNeill, J. (1976). The taxonomy and evolution of weeds. *Weed Research*, **16**, 399–413.

Mitchell, R. S., & Tucker, G. C. (1997). *Revised Checklist of New York State Plants*, New York State Museum Bulletin no. 490. Albany, NY: New York State Museum.

Moran, G. F., Marshall, D. R., & Müller, W. J. (1981). Phenotypic variation and plasticity in the colonizing species *Xanthium strumarium* L. (Noogoora Burr). *Australian Journal of Biological Sciences*, **34**, 639–48.

Morris, W. F., Kareiva, P. M., & Raymer, P. L. (1994). Do barren zones and pollen traps reduce gene escape from transgenic crops? *Ecological Applications*, **4**, 157–65.

Muenscher, W. C. (1955). *Weeds*, 2nd edn. Ithaca, NY: Cornell University Press.

Mulligan, G. A., & Findlay, J. N. (1970). Reproductive systems and colonization in Canadian weeds. *Canadian Journal of Botany*, **48**, 859–60.

Myers, N. (1994). Global diversity. II. Losses. In *Principles of Conservation Biology*, ed. G. K. Meffe & C. R. Carroll, pp. 110–40. Sunderland, MA: Sinauer Associates.

Northeast Organic Farming Association of New York (1995). *Organic Farm Certification Standards and Administrative Procedures*. Port Crane, NY: Northeast Organic Farming Association of New York.

Naylor, J. M., & Jana, S. (1976). Genetic adaptation for seed dormancy in *Avena fatua*. *Canadian Journal of Botany*, **54**, 306–12.

Okoli, C. A. N., Shilling, D. G., Smith, R. L., & Bewick, T. A. (1997). Genetic diversity in purple nutsedge (*Cyperus rotundus* L.) and yellow nutsedge (*Cyperus esculentus* L.). *Biological Control*, **8**, 111–18.

Palmer, M., & Maurer, T. A. (1997). Does diversity beget diversity? A case study of crops and weeds. *Journal of Vegetation Science*, **8**, 235–40.

Piesch, R. F., and Wheeler, N. C. (1993). Intensive cultivation of *Taxus* species for the production of Taxol®: integrating research and production in a new crop plant. *Acta Horticulturae*, **344**, 219–28.

Pimentel, D., Krummel, J., Gallahan, D., Hough, J., Merrill, A., Schreiner, I., Vittum, P., Koziol, F., Back, E., Yen, D., & Fiance, S. (1978). Benefits and costs of pesticide use in US food production. *BioScience*, **28**, 772, 778–84.

Popay, A. I., & Roberts, E. H. (1970). Ecology of *Capsella bursa-pastoris* (L.) Medik. and *Senecio vulgaris* L. in relation to germination behaviour. *Journal of Ecology*, **58**, 123–39.

Powles, S. B., & Howat, P. D. (1990). Herbicide-resistant weeds in Australia. *Weed Technology*, **4**, 178–85.

Pritchard, T. (1960). Race formation in weedy species with special reference to *Euphorbia cyparissias* L. and *Hypericum perforatum* L. In *The Biology of Weeds: A Symposium of the British Ecological Society*, ed. J. L. Harper, pp. 61–6. Oxford, UK: Blackwell.

Putwain, P. D., & Collin, H. A. (1989). Mechanisms involved in the evolution of herbicide resistance in weeds. In *Herbicides and Plant Metabolism*, ed. A. D. Dodge, pp. 211–35. Cambridge, UK: Cambridge University Press.

Quammen, D. (1998). Planet of weeds. *Harper's*, October, 1998, 57–69.

Rasmussen, J., & Ascard, J. (1995). Weed control in organic farming systems. In *Ecology and Integrated Farming Systems*, ed. D. M. Glen, M. P. Greaves & H. M. Anderson, pp. 49–67. Chichester, UK: John Wiley.

Rice, K. J., & Mack, R. N. (1991). Ecological genetics of *Bromus tectorum*. II. Intraspecific variation in phenotypic plasticity. *Oecologia*, **88**, 84–99.

Richards, A. J. (1970). Eutriploid facultative agamospermy in *Taraxacum*. *New Phytologist*, **69**, 761–74.

Risch, S. J., Andow, D., & Altieri, M. A. (1983). Agroecosystem diversity and pest control: data, tentative conclusions and new research directions. *Environmental Entomology*, **12**, 625–9.

Sand, P. F. (1987). The American witchweed quarantine and eradication program. In *Parasitic Weeds in Agriculture*, vol. 1, Striga, ed. L. J. Musselman, pp. 207–23. Boca Raton, FL: CRC Press.

Scholz, H. (1986). Die Entstehung der Unkraut-Gerste *Hordeum vulgare* subsp. *agriocrithon* emend. *Botanische Jahrbucher für Systematik Pflanzengeschichte und Pflanzengeographie*, **106**, 419–26.

Shaner, D. L. (1995). Herbicide resistance: Where are we? How did we get here? Where are we going? *Weed Technology*, **9**, 850–6.

Sobey, D. G. (1987). Differences in seed production between *Stellaria media* populations from different habitat types. *Annals of Botany*, **59**, 543–9.

Solbrig, O. T. (1971). The population biology of dandelions. *American Scientist*, **59**, 686–94.

Solbrig, O. T., & Solbrig, D. J. (1979). *Introduction to Population Biology and Evolution*. Reading, MA: Addison-Wesley.

Stebbins, G. L. Jr. (1950). *Variation and Evolution in Plants*. New York: Columbia University Press.

Stebbins, G. L. (1957). Self fertilization and population variability in the higher plants. *American Naturalist*, **91**, 337–54.

Stebbins, G. L. Jr. (1965). Colonizing species of the native California flora. In *The Genetics of Colonizing Species, Proceedings of the 1st International Union of Biological Sciences Symposia on General Biology*, ed. H. G. Baker & G. L. Stebbins, pp. 173–91. New York: Academic Press.

Stratton, D. A. (1991). Life history variation within populations of an asexual plant, *Erigeron annuus* (Asteraceae). *American Journal of Botany*, **78**, 723–8.

Stucky, J. M., Monaco, T. J., & Worsham, A. D. (1981). *Identifying Seedling and Mature Weeds Common in the Southeastern United States*. Raleigh, NC: North Carolina Agricultural Research Service and North Carolina Agricultural Extension Service.

Sultan, S. E. (1987). Evolutionary implications of phenotypic plasticity in plants. *Evolutionary Biology*, **21**, 127–78.

Sultan, S. E., & Bazzaz, F. A. (1993a). Phenotypic plasticity in *Polygonum persicaria*. III. The evolution of ecological breadth for nutrient environment. *Evolution*, **47**, 1050–71.

Sultan, S. E., & Bazzaz, F. A. (1993b). Phenotypic plasticity in *Polygonum persicaria*. I. Diversity and uniformity in genotypic norms of reaction to light. *Evolution*, **47**, 1009–31.

Sun, M. (1997). Population genetic structure of yellow starthistle (*Centaurea solstitialis*), a colonizing weed in the western United States. *Canadian Journal of Botany*, **75**, 1470–8.

Sun, M., & Corke, H. (1992). Population genetics of colonizing success of weedy rye in northern California. *Theoretical and Applied Genetics*, **83**, 321–9.

Templeton, A. R., & Levin, D. A. (1979). Evolutionary consequences of seed pools. *American Naturalist*, **114**, 232–49.

Theaker, A. J., & Briggs, D. (1993). Genecological studies of groundsel (*Senecio vulgaris* L.). IV. Rate of development in plants from different habitat types. *New Phytologist*, **123**, 185–94.

Thomas, S. C., & Bazzaz, F. A. (1993). The genetic component in plant size hierarchies: norms of reaction to density in a *Polygonum* species. *Ecological Monographs*, **63**, 231–49.

Uva, R. H., Neal, J. C., & DiTomaso, J. M. (1997). *Weeds of the Northeast*. Ithaca, NY: Comstock.

Wade, M. J., & Goodnight, C. J. (1991). Wright's shifting balance theory: an experimental study. *Science*, **253**, 1015–18.

Wang, R.-L., Wendel, J. F., & Dekker, J. H. (1995a). Weed adaptation in *Setaria* spp. I. Isozyme analysis of genetic diversity and population genetic structure in *Setaria viridis*. *American Journal of Botany*, **82**, 308–17.

Wang, R.-L., Wendel, J. F., & Dekker, J. H. (1995b). Weed adaptation in *Setaria* spp. II. Genetic diversity and population genetic structure in *S. glauca*, *S. geniculata*, and *S. faberii* (Poaceae). *American Journal of Botany*, **82**, 1031–9.

Warwick, S. I. (1990). Allozyme and life history variation in five northwardly colonizing North American weed species. *Plant Systematics and Evolution*, **169**, 41–54.

Warwick, S. I., & Black, L. D. (1986). Genecological variation in recently established populations of *Abutilon theophrasti* (velvetleaf). *Canadian Journal of Botany*, **64**, 1632–43.

Warwick, S. I., Phillips. D., & Andrews, C. (1986). Rhizome depth: the critical factor in winter survival of *Sorghum halepense* (L.) Pers. (Johnson grass). *Weed Research*, **26**, 381–7.

Warwick, S. I., Thompson, B. K., & Black, L. D. (1984). Population variation in *Sorghum halepense*, Johnson grass, at the northern limits of its range. *Canadian Journal of Botany*, **62**, 1781–90.

Warwick, S. I., Thompson, B. K., & Black, L. D. (1987). Genetic variation in Canadian and European populations of the colonizing weed species *Apera spica-venti*. *New Phytologist*, **106**, 301–17.

Wasylikowa, K., Starkel, L., Niedzialkowska, E., Skiba, S., & Stworzewicz, E. (1985). Environmental changes in the Vistula Valley at Pleszów caused by Neolithic man. *Przeglad Archeologiczny*, **33**, 19–55.

Watson, D., Mortimer, A. M., & Putwain, P. D. (1987). The seed bank dynamics of triazine resistant and susceptible biotypes of *Senecio vulgaris*: implications for control strategies. In *Proceedings of the 1987 British Crop Protection Conference – Weeds*, vol. 3, pp. 917–24. Croydon, UK: British Crop Protection Council.

Weaver, S. E., & Lechowicz, M. J. (1983). The biology of Canadian weeds. 56. *Xanthium strumarium* L. *Canadian Journal of Plant Science*, **63**, 211–25.

Weaver, S. E., Dirks, V. A., & Warwick, S. I. (1985). Variation and climatic adaptation in northern populations of *Datura stramonium*. *Canadian Journal of Botany*, **63**, 1303–8.

Webster, T. M., & Coble, H. D. (1997). Changes in the weed species composition of the southern United States: 1974 to 1995. *Weed Technology*, **11**, 308–17.

Werth, C. R., Riopel, J. L., & Gillespie, N. W. (1984). Genetic uniformity in an introduced population of witchweed (*Striga asiatica*) in the United States. *Weed Science*, **32**, 645–8.

Wilkes, H. G. (1977). Hybridization of maize and teosinte, in Mexico and Guatemala and the improvement of maize. *Economic Botany*, **31**, 254–93.

Williams, J. T., & Harper, J. L. (1965). Seed polymorphism and germination. I. The influence of nitrates and low temperatures on the germination of *Chenopodium album*. *Weed Research*, **5**, 141–50.

World Resources Institute. (1998). *World Resources, 1998–1999: A Guide to the Global Environment*. New York: Oxford University Press.

Zohary, D. (1965). Colonizer species in the wheat group. In *The Genetics of Colonizing Species, Proceedings of the 1st International Union of Biological Sciences Symposia on General Biology*, ed. H. G. Baker & G. L. Stebbins, pp. 403–19. New York: Academic Press.

Zossimovich, V. P. (1939). Evolution of cultivated beet B. *vulgaris* L. *Comptes Rendus (Doklady) de l'Académie des Sciences de l'URSS*, **24**, 73–6.

CHARLES L. MOHLER, MATT LIEBMAN, AND CHARLES P. STAVER

11

Weed management: the broader context

Introduction

Biological and physical techniques that can serve as components of multitactic weed management strategies abound. Examples throughout this book illustrate how greater knowledge of ecological processes can maintain or improve crop yields while decreasing dependence on herbicides. We suggest that by reducing the need for herbicides, ecologically based weed management strategies can help farmers reduce their input costs, reduce threats to the environment and human health, and minimize selection for herbicide-resistant weeds.

Despite the potential benefits of ecological weed management, most farmers in industrialized countries continue to rely heavily on herbicides, and the use of herbicides in developing countries is increasing. Many agricultural analysts question the ability of the world's farmers to produce enough food for a burgeoning human population without continued emphasis on herbicides and other agrichemical technologies. Some analysts argue that it will be possible to protect natural habitats and wildlife only by increasing production per unit of farmland through the intensive use of pesticides, synthetic fertilizers, genetically engineered seeds, and other purchased inputs.

In this chapter we examine reasons why ecological weed management has not been widely embraced and address whether ecological weed management is indeed consistent with the goals of increasing food security and protecting nature. We then suggest ways to promote research on ecological weed management. Finally, we address ways to foster ecological weed management on farms in both industrialized and developing countries.

If ecological weed management is effective, why do farmers rely heavily on herbicides?

Although we have argued throughout this book that ecological weed management can be effective at reducing weed density, growth, and damage to crops, some of the procedures discussed are probably not economically competitive with the cheaper herbicides within the context of the present world economy. Several factors strongly reinforce reliance on herbicides for weed control and reduce the likelihood that farmers will adopt ecologically based alternatives: (i) the apparent ease and low risk of chemical weed management; (ii) the aggressive marketing of chemical solutions to weed management problems, coupled with a lack of widely available information concerning alternatives; (iii) the externalization of environmental and human health costs of agrichemical technologies; (iv) the increasing prevalence of large-scale industrial farms; and (v) government policies that foster input-intensive agricultural practices.

The apparent ease and low risk of chemical farming

Chemical management of weeds offers apparent convenience. Applying chemicals with a large boom sprayer is one of the fastest of all field operations, and applying chemicals from an airplane requires even less of a farmer's time because such work is generally done by an outside contractor. Comparisons of conventional farming systems with low-purchased-input or organic systems generally show lower labor requirements per hectare for conventional farming (Karlen, Duffy & Colvin, 1995; Lighthall, 1996: Hanson, Lichtenberg & Peters, 1997). Moreover, the level of management skill required to apply herbicides is low relative to the skill required for ecological management of weeds. With herbicides, the instructions come on the label. In contrast, ecological farming requires adaptation of diverse sources of information to the local environment of the farm through ongoing observation and experimentation.

In actual practice, however, the farmer only realizes part of the promised labor savings. Use of chemicals essentially makes pest management, including the management of weeds, a largely off-farm activity (Smith, 1992). Consequently, most of the economic returns for chemical pest management go to input suppliers rather than to the farmer. The same studies that show labor savings with chemical farming also frequently show greater net returns per hectare for ecological management (Karlen, Duffy & Colvin, 1995; Smolik, Dobbs & Rickerl, 1995; Hanson, Lichtenberg & Peters, 1997). Depending on the outlook of the farmer, the extra labor requirements of ecological management can thus be viewed either as an added cost or as a route for obtaining full on-farm employment without expanding farm size (Lighthall, 1996). Some farmers may consciously prefer to spend their workdays managing purchased inputs on a larger farm rather than working with soil, crops, and animals on a medium-sized farm. Many farmers, however, may be using chemical management without careful assessment of all the trade-offs involved.

Farmers also perceive herbicides as low-risk solutions to weed management. As all herbicide users know, applications sometimes fail to control weeds to the desired extent for a variety of reasons related to weather, soil conditions, timing of application, and equipment failure. Increasingly, however, companies guarantee control by supplying materials for reapplication if target control levels are not reached (Benbrook, 1996, pp. 47–9). For example, Owen (1998) estimated that 25% to 30% of maize in Iowa was re-treated in 1994 through herbicide guarantee programs. Since aggressive marketing gives growers the idea that weed-free fields are the only acceptable condition, many second applications are probably unnecessary. In other cases, the second application is untimely and the problem could be better solved through cultivation or use of an alternative chemical. Although respray programs are supposedly free for the user, the costs are in fact paid for in the price of the material. Nevertheless, such programs offer farmers a means for reducing one of the risks associated with crop production. To the extent that respray programs foster unnecessary use of herbicides, however, they may increase risks to human health and the environment.

Solving problems by selling products

Modern high-input mechanized agriculture co-developed with herbicide technology. A corollary of this co-development process is that most growers receive little in-depth information concerning alternative approaches. As noted in Chapter 1, information on chemical tactics has dominated the weed science literature for the last 30 years. Consequently, the intellectual underpinnings of modern weed management are closely linked and largely limited to herbicide technology (Wyse, 1992). Additionally, large corporations use well-financed advertising campaigns and extensive networks of sales personnel to promote herbicides. In developing countries, promotional campaigns include the free distribution of small quantities of herbicides and other pesticides. Pesticides are advertised heavily in farm magazines (Benbrook, 1996, pp. 164–5), whereas relatively few publications inform farmers about how to manage weeds through the integrated use of cover crops, crop rotation, grazing practices, biocontrol agents, cultivation, and other ecologically based tactics.

In contrast to the highly commercial nature of chemical weed management, ecological weed management relies on information that is free or inexpensive. Since few companies profit from ecological weed management methods, they are not advertised in the media and few people are paid specifically to promote their use. The only public agencies in a good position to even partially redress this imbalance are extension services. However, extension educators are often resistant to the concepts of sustainable agriculture and skeptical about the value of ecologically based approaches to pest management (Agunga, 1995; Paulson, 1995; Conner & Kolodinsky, 1997). Training in sustainable agriculture is now mandatory for extension agents in the USA (Hoag & Pasour, 1992; Schaller, 1992). Since training programs vary in content and effectiveness, however, the impact of this mandate is likely to be spotty. Both in the USA and elsewhere, additional training of extension agents is still needed to strengthen the link between ecology and agricultural production. If ecological weed management is to be implemented on a broad scale, the public sector needs to place emphasis on improving farmer management of ecological processes, rather than on programs to promote input use.

Cost externalization

Many costs associated with the production and use of chemical technologies, including herbicides, are not included in the price of the products. These include purchase of alternative water sources, cost of medical treatment, lost productivity due to illness and injury, and expenses for clean-up of pollutants not borne by manufacturers and applicators (Pearce & Tinch, 1998; see also Chapter 1). Exclusion of these "external" costs from the price of herbicides makes chemical weed management appear more cost-effective than it actually is for society as a whole. Only when external costs are integrated into price are they readily visible to all participants in the marketplace (Hawken, 1993, p. 83).

External costs are controversial because they can be hard to quantify and because damage to ecosystems and human health may become apparent only after a substantial time lag. Nonetheless, failure to estimate and anticipate external costs can lead to situations where toxic compounds are widely used for many years, only to be banned after sufficient proof of harm eventually accumulates. The regulatory history of the herbicide dinoseb illustrates this situation. It was first registered for use in the USA in 1945. More than 40 years later, the United States Environmental Protection Agency (USEPA) completed a review of the compound's acute and chronic toxicity and declared it an "imminent hazard" to agricultural workers and the environment (Haskell, 1991). At the time dinoseb was prohibited in the mid 1980s, 3 to 5 million kg of the material were applied annually to a variety of American crops (Gorney, 1987).

The Nordic countries have pursued cost internalization through taxes on pesticides. These are intended to reduce pesticide use and encourage farmers to develop alternative strategies. Finnish and Norwegian farmers pay 2.5% and 11% taxes, respectively, on pesticides; Swedish farmers pay a 20% "regulation charge" on top of price, a 10% "environmental tax" on price, and a charge equivalent to $5.65 per hectare per pesticide application (Pearce & Tinch, 1998). As discussed later in this chapter, imposition of such taxes, initiation of educational programs, and greater reliance on newer low-rate products led to a 54% reduction in herbicide use (on a kg ha^{-1} basis) in Sweden during the 1980s (Bellinder, Gummesson & Karlsson, 1994).

In principle, a program for internalizing the external costs of pesticide use could tax more harmful materials at a higher rate. However, reducing the multidimensional impacts of any particular pesticide to a single number may be impossible using purely objective procedures (Dushoff, Caldwell & Mohler, 1994). A general tax on pesticides aims to internalize some unknown fraction of the costs of chemically based agriculture as a way to encourage an alternative approach. In contrast, a tax that varies with the specific pesticide aims to shift use from more harmful to less harmful materials. In principle, specific taxes may be fairer. However, a general tax on pesticides may better internalize costs to ecosystems and communities that result from interaction of pesticide use with other aspects of farm management like farm size and degree of specialization. A general tax is also probably more effective for changing the overall direction of a nation's agriculture.

Some ecological weed management measures can also have external costs if used improperly. In particular, tillage and cultivation for weed control can lead to soil erosion with consequent damage to long-term farm productivity and aquatic ecosystems. However, erosion is not an inevitable consequence of tillage. Many methods are available to minimize erosion, including rotation with sod crops, terracing, strip cropping, sod water ways, cover crops, mulches, wind breaks, and use of conservation tillage equipment (Brady, 1984, pp. 534–69). Insufficient use of these practices results more from economic conditions that force farmers to maximize short-term profits in order to remain solvent than from ecological weed management. Careful use of soil conservation practices allows low-purchased-input and organic farms to produce crops on potentially erodable land with low rates of soil loss (Reganold, Elliott & Unger, 1987; National Research Council, 1989, p. 269).

Farm size and industrial farm management

Another factor favoring heavy reliance on herbicides is the growing dominance of industrial farm management. In recent decades, farms in both developed and developing countries have tended to become larger, more standardized in their production practices, more specialized in the crops and livestock they produce, and more reliant on capital-intensive technologies (MacCannell, 1988; Kirschenmann, 1991). Although the number of small farms (<20 ha) in the USA has recently risen after falling for many years, middle-sized farms have increasingly been incorporated into large farms. Most American farmland is now controlled by farms greater than 400 ha (Stanton, 1993). Farms are smaller in most other parts of the world, but aggregation is occurring rapidly in many countries (Ehrensaft et al., 1984; Maunder, 1984; Thiesenhusen & Melmed-Sanjak, 1990). By the mid 1990s, the concentration of American farming in larger farms had proceeded to the point where 6% of the nation's farms sold 60% of the farm products (United States Department of Agriculture, 1998). The largest, most industrialized farms often convert inputs into products with lower efficiency than smaller farms, but dominate the marketplace for other reasons: improved access to credit, price breaks on inputs from suppliers, better ability to insure markets through contracts with purchasers, and greater subsidies and tax breaks from government programs (Strange, 1988, pp. 78–103; National Research Council, 1989, pp. 76–7; United States Department of Agriculture, 1998).

As farms increase in size, they become less suited for ecological weed management and more prone to intensive use of herbicides (Lighthall, 1996; Welsh & Lyson, 1997). Certain ecological weed management practices, such as the use of high crop densities and optimal planting patterns, are insensitive to farm scale and specialization. Other more systemic procedures, such as long rotations, intercropping, and crop–livestock integration, tend to become rare as farms grow larger and more specialized (Buttel, 1984; Strange, 1988, pp. 111–14, 131–4; see also Chapter 7). Timely cultivation for weed control may also be precluded by large farm size and concentrated production of only one or two crops. Weed management with herbicides fits well with an industrial mode of farming because it can be conducted almost uniformly over large areas in a narrow time window. Herbicides reduce management complexity and the need for local adaptation.

Although farm scale can affect reliance on herbicides and opportunities for ecological weed management, it is not their sole determinant. Matheson *et al.* (1991) described farms from 400 to 1100 ha in the northwestern USA and western Canada on which weeds are managed with rotations, intercropping, cultivation, adjustments in seeding rates, and other practices, but without herbicides. Conversely, small farms may rely heavily on herbicides for weed control. Smallholder Asian cereal producers have become increasingly reliant on herbicides as expanded off-farm employment opportunities have driven up agricultural wages and increased the cost of hand-weeding (Pingali & Gerpacio, 1997).

Bigger obstacles to the adoption of ecological weed management than large scale *per se* are the decision-making processes, relationships with adjacent communities, and patterns of natural resource and labor management that characterize industrial farming systems. As farms become highly industrialized, decision-making increasingly resides in the hands of office-bound executives who by predilection, training, and feedback are oriented more toward the growth of the farm enterprise than its long-term integration with the ecosystem that supports it (Kirschenmann, 1991). In such situations, information feedback from the fields to the manager may be minimal and environmental problems will tend to elicit short-term fixes based on additional purchased inputs rather than reorganization of the production process. Rural communities dominated by industrial farms tend to have higher levels of poverty, lower quality and quantity of social and commercial services, lower education levels, and weak local governments (MacCannell, 1988; United States Department of Agriculture, 1998). In addition to reducing economic opportunities and retarding rural development, these social impacts of industrial agriculture are likely to inhibit effective feedback from adjacent communities to farm managers about environmental problems such as off-site herbicide drift and water contamination.

Agricultural policy

Throughout the world, farm industrialization has resulted in oversupply relative to market demand and long periods of low prices for a wide range of commodities (Buttel, 1990; Pretty, 1995, p. 55). In response, governments in many nations attempt to raise farm income by providing farmers with price supports, subsidizing input costs, and exporting surpluses. All of these practices affect the likelihood of implementing ecological weed management and other components of sustainable agricultural systems.

Although cost savings associated with reduced use of purchased inputs tends to favor adoption of sustainable farming practices, widespread adoption is unlikely where government regulations, tax and subsidy policies, and research and extension activities favor increases in farm scale, greater capital investment, and continued dependence on agrichemical inputs. Until government policies integrate social, environmental, and production goals, industrial farms with sufficient credit and capital to ride out periods of low prices will continue to acquire the land and facilities of financially stressed smaller-scale farms (Strange, 1988, p. 104). Rather than cutting use of all inputs, cost containment on industrial farms is likely to involve reducing labor expenses and increasing reliance on purchased inputs, including herbicides.

When government programs subsidize only a few crops and link payments to yield levels and land areas planted with these "program crops," diverse rotations become economically unattractive to farmers (National Research Council, 1989, pp. 235–40). This in turn increases the uniformity of a region's cropland, exacerbates weed problems, and provokes greater reliance on herbicides (see Chapter 8). Pretty, Vorley & Keeney (1998) noted a strong positive correlation between crop subsidy levels and pesticide use (kg ha^{-1}) in Japan, the European Union, the USA, and Brazil.

Government subsidies that lower pesticide costs to farmers directly favor high rates of pesticide use (Pearce & Tinch, 1998), largely to the exclusion of other pest management tactics. Repetto (1985) surveyed nine developing countries in Asia, Latin America, and Africa in the mid 1980s and found that government pesticide subsidies ranged from 19% to 89% of full retail costs. Until such subsidies are removed, farmers have little incentive to adopt ecologically based strategies for managing weeds and other pests.

To improve farm income in developed countries, government programs often export crop surpluses to developing countries, thereby increasing supply and reducing crop prices in those countries. Though urban consumers in developing countries may pay less for food in the near term, local farmers lose income and incentives to maintain production (World Commission on Environment and Development, 1987, pp. 122–3, 129; McMichael, 1998). In effect, this type of agricultural globalization creates dependency on food imports in developing countries and reinforces farm industrialization in food exporting countries. When foreign exchange rates become unfavorable or funds for purchasing food shrink, food dependency may manifest itself as food shortages.

Given the pervasive effects of current government policies on farming practices and financial returns, politicians and policy makers must carefully analyze the differences between ecological and conventional weed management systems. Would lower weed densities be observed if crop rotations lengthened in response to policy changes? Would ecological weed management appear more favorable financially if conventional systems had to include costs of surface and ground water contamination? In the long run, depletion of non-renewable resources, particularly petroleum, and cumulative damage to ecosystems and human health by resource degradation and pollution are likely to precipitate substantial changes in the nature of economic activity (Brown, Flavin & Postel, 1991; Hawken, 1993; Daly & Cobb, 1994). Under optimistic scenarios, the need for restoring balance between human economic activity and ecosystem processes will provide incentives for agriculture, industry, and government to move toward an economy that fosters sustainable agricultural systems. At that point, ecological weed management will present a feasible option for managing weeds in an efficient, environmentally harmonious manner. We discuss needed changes in policy in the final section of this chapter.

Feeding a growing human population

Policies favoring industrial agriculture are justified as an antidote to hunger and malnutrition in a rapidly growing human population and to the increasing destruction of forest and wetland habitats. Advocates of industrial farming present herbicides as one of the modern tools necessary to feed the world and protect the environment (Schneiderman & Carpenter, 1990; Avery, 1995).

If farmers relied on ecologically based weed management strategies and greatly reduced their use of herbicides, would they produce enough food? Would more land be required to match the output from conventional farming, thereby reducing wildlife habitat and further threatening biological diversity? We believe that the answer to the first question is "yes," and that the answer to the second is "no." Moreover, the second question falsely poses a dichotomy between agriculture and the natural world.

Food security

Recent studies indicate that intensive use of chemical inputs is not the only path to increasing the productivity and efficiency of farms. In both industrialized and developing countries, farming systems operated with minimal use of herbicides, other pesticides, and synthetic fertilizers can produce high yields if farmers manage ecological processes intelligently (National Research Council, 1989; Stanhill, 1990; United Nations Development Programme, 1992; Pretty, 1995, pp. 204–37). Pretty (1995, pp. 19, 206) characterized such systems as "regenerative" and summarized their performance by noting:

> (1) "In the industrialized agricultural systems, a transition to sustainable agriculture could mean a fall in per hectare yields of 10 to 20 percent in

the short term, but with better levels of financial returns to farmers" and "substantial environmental improvements."

(2) In the Green Revolution lands of developing countries, in which irrigation and modern crop varieties are used routinely, farmers adopting regenerative technologies have maintained yields and gained environmental benefits, while substantially reducing inputs.

(3) "In the diverse, complex, and 'resource-poor' lands of the Third World, farmers adopting regenerative technologies have doubled or trebled crop yields, often with little or no use of external inputs."

Food security is not synonymous with increased production. People go hungry not because of inadequate world or regional food production, but because they lack the means to buy food or grow it (Pretty, Vorley & Keeney, 1998). For both rural and urban residents of three regions of Brazil, caloric intake was found to be positively correlated with income (Murdoch, 1980, p. 100). In Bangladesh, 75% of rural families own little or no land and 11% of the country's families own half the land. Average caloric and protein intake of Bangladeshi peasants owning less than 0.2 ha was found to be 19% and 23% less than for those owning at least 1.2 ha (Brown & Jacobson, 1986). In a food-short year, the death rate of landless Bangladeshi peasants was three times higher than that of peasants owning at least 1.2 ha (Eckholm, 1979). In a range of countries, family size has been noted to be negatively correlated with income, literacy, years in school, and equality of income levels (Murdoch, 1980, pp. 15–83; Roodman, 1999). Thus, policies addressing hunger, health, and human population growth must focus on patterns of land tenure and social, economic, and political development.

Food security is also related to how crops are used. More than 40% of the herbicides sold worldwide are used for maize and soybean production (Jutsum, 1988). Although both of these crops can be eaten directly by people, most of the production is used as livestock feed. In Iowa, where farmers annually apply more than 20 million kg of herbicides to maize and soybean, almost 70% of maize and soybean production is used for feed; an additional 14% is used for sweeteners for soft drinks and other processed foods; 9% is used to make ethanol; and the remaining 8% is used for used for seed, industrial products, and food for humans (Mayerfeld *et al.*, 1996). The principal beneficiaries of industrial agriculture are not the poor and hungry who subsist primarily on grain, but consumers who have cash to buy processed foods and products from grain-fed animals. The world's cropland currently produces enough calories and protein to support 10 billion human beings on a vegetarian diet (Waggoner, 1994). Even if ecological farming methods led to slightly lower yields in the developed countries, production would meet human needs. In

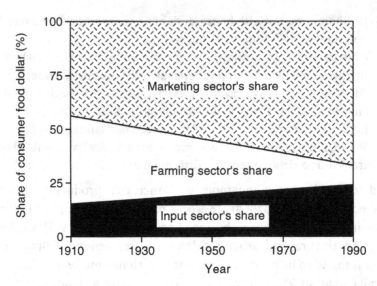

Figure 11.1 Changes in the proportions of American consumers' food dollars going to farm input suppliers, the farm sector, and the agricultural marketing sector (including distribution), 1910–90. (Redrawn from Smith, 1992.)

the developing countries, where most of the increase in human population occurs, more equitable distribution of resources, improvements in local food production and distribution systems, and strengthened community resource management skills offer the best opportunities for reducing hunger (World Commission on Environment and Development, 1987, p. 129).

Some critics have suggested that use of ecological weed management and other sustainable agriculture strategies would lead to an unacceptable rise in food prices. Smith's (1992) analysis of the American food system suggests that this argument is wrong where agriculture follows an industrial model that emphasizes food processing, long-range distribution, and consolidated retailing. As shown in Figure 11.1, the farm sector in the USA now receives less than 9% of each dollar a consumer spends on food. That is, only 9 cents of each dollar goes for capital consumption, property taxes, wages for farm labor, and net farm income. Input costs for producing crops and livestock (e.g., pesticides, fertilizers, fuel, etc.) take more than 24% of each consumer food dollar, and 67% of each food dollar goes to processors, packagers, advertisers, and sellers. Since input costs are less in sustainable agricultural systems (Chase & Duffy, 1991; Nguyen & Haynes, 1995; Hanson, Lichtenberg & Peters, 1997), any increase in food prices resulting from a shift toward such systems would presumably be due to higher costs in the farm sector. Even in the unlikely event that ecological management methods doubled farm sector costs, costs were fully passed on to consumers, and no savings occurred through reduced purchase of external inputs, food prices in the USA would increase only 9% (24 + 67 + [2 × 9] = 109).

In developing countries, productivity gains associated with ecological management (see above) will often tend to decrease rather than increase food prices. Nevertheless, farmers also need to decrease input costs and increase production efficiency. Where labor is inexpensive relative to land and chemical inputs, intensive intercropping, multiple cropping, and agroforestry systems can be used to increase yields per unit area, minimize soil erosion, recycle nutrients, and suppress weeds (Beets, 1982; Steiner, 1984; Nair, 1993). In regions where labor is expensive compared with herbicides, ecologically based weed management strategies may be similar to those used in the developed countries. Reductions in herbicide use without corresponding increases in labor requirements are possible through changes in tillage systems, improved soil moisture management, better use of cover cropping and intercropping practices, and development of weed-suppressive and weed-tolerant crop varieties (Pingali & Gerpacio, 1997; see Chapters 4, 5, 6, and 7). Throughout the developing countries, access to land, development of technologies based on local resources, and improvement in the managerial abilities of small-scale farmers are key elements in implementing sustainable agricultural practices and maintaining an abundant, affordable food supply.

Habitat protection and wildlife conservation

Advocates of chemically intensive agriculture maintain that it protects land for wildlife habitat because farmers harvest more yield from a smaller area (Avery, 1995). This argument presumes that low levels of chemical inputs result in low yields, and that wildlife and agriculture are inherently incompatible. As noted earlier, studies in many countries show that low-external-input farming systems can produce high and stable yields. Regarding the second assumption, wildlife indeed fares poorly on industrial farms where monoculture, high use of pesticides, and a lack of natural vegetation are the norm (Papendick, Elliot & Dahlgren, 1986). However, many species thrive in diverse agricultural ecosystems subjected to minimal applications of pesticides.

Studies conducted in Europe and the USA have shown that bird densities may be up to 20 times greater on organic farms than on conventionally managed farms, an effect attributed to greater habitat diversity and additional food sources (Lampkin, 1990, pp. 574–9). Similarly, the diversity of wild plant species around field edges can be higher when the use of herbicides and other agrichemicals are minimized (Boutin & Jobin, 1998). Elimination of herbicide and insecticide use in 3-m margins around cereal fields can increase the abundance of partridges and other game birds (Jahn & Schenck, 1991). Recent studies of Sumatran agroforestry systems indicate they can contain over 50% of the total regional pool of resident tropical forest bird species, most of the mammals, and about 70% of the plants (Leakey, 1999). In Central American coffee orchards with shade trees, populations of birds, bats, invertebrates, and other taxa are more abundant and diverse than in conventionally managed orchards without shade trees, and more closely resemble the fauna of the surrounding forest (Perfecto et al., 1996; Greenberg et al., 1997).

These and other examples lead to the conclusion that agricultural landscapes with crop diversity, patches of natural vegetation, and limited use of toxic materials can support a broad diversity of wildlife. For weed scientists, this management approach requires a shift in perspective from narrow categories of crops and weeds toward a wider view of vegetation and landscape management.

Developing an environment for research on ecological weed management

Ecological weed management is information intensive. Its further development therefore requires greater investment in research and shifts in research priorities. Basic ecological data are lacking for many weed species. For some important species, no published data on seed longevity in the soil are available, and the effects of soil conditions on seed survival are understood for only a handful of species. Information on the response of weed growth rate to environmental factors is equally spotty. Yet species-specific ecological management requires precisely these sorts of information. Furthermore, substantial technological advances are needed in a number of areas. For example, breeding crops for improved ability to compete against weeds has barely begun (see Chapter 6). Methods for preserving residue at the soil surface during tillage are similarly in their infancy (see Chapter 5). Testing soil samples for weed seeds could improve weed management by allowing better targeting of control efforts (King et al., 1986; Lybecker, Schweizer & King, 1991), but equipment that can rapidly separate and automatically identify weed seeds is not yet available (Buhler & Maxwell, 1993). Throughout this book, we have indicated many other important unanswered questions in ecological weed management, and other areas for research that we believe could increase the sustainability of agricultural systems. Developing this potential requires further commitment from society.

Part of the need for increased investment in research focused on ecological weed management stems from the chronic understaffing of weed science programs. In 1995 the top 15 programs in entomology and plant pathology in the USA had an average of 15 and 12 teaching faculty, respectively, whereas the top 15 weed science programs had an average of only three (Kuhns & Harpster, 1997). Note that these are the top schools; most agricultural universities have only one or two weed scientists. Due to understaffing, coping with the pressing needs of farmers for information on how to use the growing array of herbicides against ever-changing weed communities leaves most weed scientists with little time for long-term research or basic research on weed ecology. Consequently, the present imbalance among different pest management disciplines needs to be redressed if the promise of ecological weed management is to be fully realized.

Because research on ecological weed management mostly results in procedures rather than products, development of this field depends largely on public sector support rather than private sector funds. Industry has little incentive to develop the use of cover crops for weed management, study the weed suppressive effects of crop rotation, or determine the best ways to control weeds with livestock. For the foreseeable future, most of that type of research will have to come from government funded programs, with a minor additional contribution from private foundations. This is a problem at a time when government support for agricultural research is at best increasing only slowly (Westendorf, Zimbelman & Pray, 1995).

As government funding stalls or disappears, small industry grants leverage larger public sector funds. This occurs because faculty in agricultural colleges need to obtain outside funding to run their research programs, obtain tenure, and advance their status within their institution and profession (Strange, 1988, pp. 216–20). Consequently, small (usually less than $10 000) chemical company grants for product-related research can redirect faculty and technician salaries, land, facilities, equipment, and government research grants toward investigations of direct benefit to the company. Through this process, publicly funded weed science tends to be co-opted into herbicide-oriented research.

A search made in September 1998 of the Current Research Information System database of the United States Department of Agriculture (USDA) funded projects illustrates this process. Forty-seven projects focused on the development of herbicide-resistant maize, soybean, and cotton, and cropping systems based on these resistant genotypes. In contrast, only nine projects focused on development or use of maize, soybean, and cotton varieties with improved ability to tolerate or competitively suppress weeds. Although both technologies are economically useful and could reduce negative environmental impacts of crop production, current government funded research clearly emphasizes the chemical approach rather than the ecological one. Many factors are involved, but we believe that this emphasis largely reflects the ability of chemical company funding to direct researchers toward issues and problems favorable to the industry, and away from technologies that could potentially reduce herbicide sales.

Those who argue that the relative attention paid to herbicide tolerant and weed tolerant crops in USDA sponsored research results from their relative merit might consider the following scenario. Suppose that half of the chemical company grants to public sector weed scientists that currently support research on herbicide-tolerant crops went instead for research on weed-tolerant crops. Would the number of USDA projects still be so skewed toward research on herbicide-tolerant crops?

How can support for research on ecological weed management be improved? First, government research programs need to target more funding for work on sustainable agricultural systems and ecological pest management. Emphasis needs to be placed on incorporating research priorities for sustainable, ecologically based agriculture into all funding programs for agricultural research, rather than pigeonholing them into a few small programs. Otherwise, research on sustainable agricultural systems will not enter the mainstream and will be vulnerable to cuts in specific programs, and most funds will be directed elsewhere (Batie & Swinton, 1994).

Second, funding programs need to support a balance between research on farming systems and component research on sustainable agriculture. Some advances in sustainable agriculture will only come through understanding how farming systems can be better designed to achieve efficiencies of integration. Much remains to be learned, however, about individual system components, and in particular, the ecology of weeds and their interaction with crops. Often this is best studied with small, controlled experiments rather than in a farming system context. Funding programs must be designed to allow investigators to ask a wide range of questions using a wide range of research approaches, including mixed approaches in which multiple small component studies feed into an evolving farming system comparison.

Third, mechanisms need to be developed for funding long-term agricultural research projects. Many fundamental questions in agriculture, particularly those involving crop rotation, changes in soil properties, and evolution of pests and pest communities are difficult to approach with the two- to five-year grants currently available in most countries. These processes are important in all agricultural systems, but understanding their operation is especially critical for continued development of sustainable agriculture and ecological weed management. The United States National Science Foundation's Long Term

Ecological Research Site program might serve as a model if substantial modifications were made to meet the particular needs of agricultural research.

Fourth, researchers can foster better funding for ecological weed management simply by making it their own research priority. Writing competitive grant proposals is considerably more work than collecting funding from herbicide manufacturers for spray trials. Over the long run, however, funding levels for government programs reflect the quantity and quality of the proposals submitted. If weed scientists do not work for competitive funding, the funding will go to other areas. Moreover, researchers must resist the tendency for herbicide money to pull resources out of government sponsored programs. Ideally, industry money should be supporting overhead on government sponsored research on ecological weed management, not vice versa. One way to create greater awareness of the research priorities and constituencies for ecological weed management would be to develop a sabbatical program that linked researchers and extensionists with farmers who are particularly successful in the management of sustainable agricultural systems (Vorley & Keeney, 1998). Such a program might partially counterbalance the ideological influence of herbicide company funding.

Fifth, the lack of weed science positions in academic institutions, particularly positions focused on the development of ecological weed management strategies, could be compensated for by improving links between weed scientists and entomologists, plant pathologists, soil scientists, crop breeders, agricultural engineers, modelers, "basic" plant ecologists, and others. As weed scientists develop a more holistic perspective on farming systems and agroecosystem management, opportunities for collaborative research should increase.

Sixth, researchers and administrators could expand the funding base and resources available for ecological weed management by developing links with private foundations and nongovernmental organizations (NGOs). Until recently, funding by private foundations for agricultural research outside of the international centers was meager, perhaps due to the belief that national governments covered all the needs. This appears to be changing (Viederman, 1990). Emphasis on links between sustainable agriculture, environmental quality, and community development could encourage funding from foundations that previously have not funded agricultural research. Additional opportunities exist with NGOs, which have proven increasingly effective in facilitating interactions between farmers and researchers seeking improved pest management systems (Thrupp, 1996). Strategies to develop the number and scope of NGO–farmer–researcher collaborations should be pursued.

Finally, researchers can improve support for ecological weed management by forming alliances with industries having shared interests, e.g., cultivator manufacturers, and organic food processors and distributors. At present, industry support in weed science is essentially synonymous with support by herbicide manufacturers. Industries with a vested interest in sustainable agriculture must be made to realize that the changes sweeping through the agricultural sector can to some extent be directed by strategic placement of research money. The chemical companies are masters at this game, but the rapid growth of the organic food industry opens the possibility of other players. Unlocking this potential source of funding will require careful identification of how proposed research would reduce the wholesale price of organic products needed by food processors.

Implementing ecological weed management

Because the market forces presently acting on agriculture appear likely to compromise the well-being of future generations, we believe it is appropriate for government agencies to proactively protect the environment and promote economically and socially viable rural communities. Given that the structure of agriculture contributes to heavy reliance on herbicides, the widespread implementation of ecological weed management requires changes in government policies and a reworking of relationships among farmers, extensionists and researchers. We believe five types of changes are particularly necessary.

First, tax laws and government financial policies need to favor medium-sized owner-worked farms and investment in sustainable technologies rather than large industrial farms and the acquisition of land, buildings, and machinery (Strange, 1988, pp. 134–64, 262–4). Although environmental and economic problems in farming are not precluded by an agricultural system based on family farms, the forces creating those problems tend to be less virulent when the family is the primary source of management and labor. Diminishing the trend toward larger farms by removing biases against small and mid-sized farms would increase opportunities for crop diversification, crop–livestock integration, and timely cultivation, all of which would facilitate weed management with greater emphasis on ecological processes and less use of chemicals. If small and mid-scale farms are going to be viable without substantial dependence on non-farm employment or government subsidies, then policy makers need to foster processes whereby farmers can cost-effectively produce food and fiber in an ecologically sound manner and retain a greater share of agricultural profits on their farms.

Second, legislative initiatives promoting soil and water conservation

(United States Department of Agriculture, 1990; Kuch & Ogg, 1996) should include additional practices that improve environmental quality such as ecological weed management. To the extent commodity support programs are actually phased out in the USA and elsewhere, new methods will be needed to encourage environmentally sound farming practices (Tweeten, 1996). These could consist of a mix of direct subsidies, new tax incentives, and refocused regulations. To be effective, these measures need to be accompanied by changes in international trade policies that protect farmers using sustainable practices from unfair competition by offshore polluters. Prices should reflect the full cost of production, including damage to human health, the environment, and the socioeconomic condition of communities. Accordingly, adjustment of market prices through import taxes should be allowed in cases where commodities are produced with destructive practices.

Third, global society must resolve contradictions among policies that set agricultural goals (e.g., increased production), environmental goals (e.g., soil and wildlife conservation, protection of water quality), and social goals (e.g., food security, adequate farm income, improved economic opportunities for small and mid-scale farmers). National and international agencies have recognized these contradictions (World Commission on Environment and Development, 1987, pp. 118–46; National Research Council, 1989, pp. 65–84; United States Department of Agriculture, 1998), but they persist in most regions, largely because crop and livestock output continues to take precedence over rural development and protection of natural resources. Agriculture based on ecological concepts, as illustrated here by ecological weed management, offers a unified social, environmental, and production paradigm.

Fourth, a tax on agrichemicals should be used to generate increased research and extension funding for sustainable agriculture, including ecological weed management. In the state of Iowa, the Leopold Center for Sustainable Agriculture supports research, education, and demonstration projects focused on weed management and a range of other issues with funds derived from fees on fertilizer and pesticide sales, as well as appropriations from the Iowa legislature (Keeney, 1998). Nationally in the USA, where herbicide sales are about 6.3 billion dollars annually (Aspelin, 1997), a 0.5% surcharge on herbicides would generate 31.5 million dollars each year. By comparison, during the late 1990s, funding for USDA's entire program for Sustainable Agriculture Research and Education has been at the level of about 8 million dollars per year. Given the substantial amount of publicly funded research on herbicides (see the preceding section), a 0.5% tax would probably still leave the industry with a net public subsidy.

For the grower, a 0.5% surcharge on pesticides would not be excessively burdensome. Farmers who spent $10 000 per year on herbicides would have to pay an additional $50. Moreover, if the tax were used to fund the development of ecologically benign alternatives to pesticides, growers would recoup some of this expense in the form of reduced input and health costs. To some extent, a pesticide surcharge would probably be passed on to consumers as a slight increase in food prices. However, if the tax paid for development of ecological methods of pest management, consumers would also spend less on general taxes for the treatment of environmental and health problems caused by pesticide use. As long as the external costs of pesticides are denied and ignored, funding research with a tax on pesticides will appear politically impossible. Recognition of those external costs, however, would allow this type of approach to develop rapidly. Weed scientists need to realize that their discipline would be a major beneficiary of such a tax, although their allegiance to herbicide technology would have to change.

Finally, governments need to assist farmers in obtaining information and management skills necessary for ecologically based strategies. The success of farmer-first, information-intensive approaches can be seen in the implementation of ecologically based strategies for managing insect pests in Asian rice production systems. When governments in Indonesia and six other Asian nations reduced pesticide subsidies and organized season-long "farmer field schools" stressing an ecosystem approach to crop health, participating farmers reduced insecticide use, increased yields, and improved net returns (Pretty, 1995, p. 227; Thrupp, 1996, p.7; Pretty, Vorley & Keeney, 1998). Key to this success was the development of "schools without walls" in rice fields where farmers learned new principles, concepts, and terms relating to crop, pest, and natural enemy management. Farmers learned to make observations in their own fields and present their observations and management decisions to other farmers and members of research and extension teams working with them. Farmers used dyes in their knapsack sprayers to observe where the pesticides actually were deposited. "Insect zoos" were developed to increase farmer knowledge of pest life cycles, and predation and parasitism of pests by natural enemies. Surveys conducted in Indonesia to measure the impact of this training showed that rice yields increased an average of 0.5 Mg ha^{-1}, whereas the average number of insecticide applications fell from 2.9 to 1.1 per season. About a quarter of the 110 000 Indonesian farmers who completed the program by 1993 applied no pesticides thereafter (Pretty, 1995, p. 227).

Industrialized countries can make similarly large changes in agricultural practices. As a result of government policy initiatives, shifts in research and extension priorities, and attention to farmers' concerns, Sweden's annual use of herbicide active ingredients fell from 3829 Mg during the baseline period of 1981–85 to 1743 Mg in 1990 (Bellinder, Gummesson & Karlsson, 1994). To accomplish this reduction, the Swedish government first convened a panel of scientists to identify ways to decrease herbicide use by 50%, while maintaining crop yields. Energy requirements for crop production were to be maintained or reduced. During the period of strategy development, attention was directed toward cooperation and dialogue among grower groups, agricultural scientists, environmentalists, and policy makers. The Swedish Farmers Association actively supported the program and mounted a marketing campaign for "the world's most environmentally friendly agriculture."

The Swedish government substantially increased research funding, with emphasis given to improving knowledge of basic weed ecology and nonchemical control methods, increasing crop competitiveness, better understanding of crop rotation effects on weed dynamics, and developing methods to reduce weed seed dispersal during crop harvests (Bellinder, Gummesson & Karlsson, 1994). Programs also reduced herbicide use through promotion of surfactants and better timing of applications. Research on newer low-dose herbicides was encouraged.

The Swedish extension service was expanded to better deliver information to farmers. Emphasis was placed on impressing farmers that "less is best." Farmers were provided with information about when herbicides and other pesticides were actually needed to protect crops economically, how application rates and frequencies could be reduced, and what alternatives were available (Bellinder, Gummesson & Karlsson, 1994). Information was developed for product labels that related herbicide dosage and efficacy to weed population densities causing crop yield reductions, rather than to the number of weeds surviving treatment. Researchers and extensionists demonstrated that lower rates of herbicides could provide acceptable weed control while increasing cereal yields and profits, due to lower toxicity effects on crops. Agricultural scientists developed mobile units that helped farmers calibrate sprayers to improve precision and performance. A government funded grant program to assist Swedish farmers to convert to organic farming was established in 1989 (Matteson, 1995).

Sweden's success in lowering herbicide application on a mass basis (kg ha^{-1} active ingredients) has been criticized as a "phantom reduction," since much of the change resulted from a switch to low-dose products (e.g., sulfonylurea compounds) and more efficient application technologies (Matteson, 1995). Adoption of alternative management strategies was limited. Success with reduced doses may have been possible because previous herbicide use reduced weed seed banks to low levels (Bellinder, Gummesson & Karlsson, 1994).

Nonetheless, a further 50% reduction in pesticide use was mandated by the Swedish parliament in 1990 (Matteson, 1995). This drop will require much greater attention to farming system redesign and weed ecology, since adjustments in herbicide management have already been made.

These experiences in Asia and Sweden offer successful models that could be adapted to other farming systems around the world. Other specific programs may work equally well. The important message from these examples is that government policy makers, researchers, extensionists, and farmers working together can rapidly lay the groundwork for broadscale implementation of ecological weed management. We hope that readers of this book will help with that task and aid the development of sustainable agricultural systems wherever they live and work.

REFERENCES

Agunga, R. A. (1995). What Ohio extension agents say about sustainable agriculture. *Journal of Sustainable Agriculture*, **5**, 169–87.

Aspelin, A. L. (1997). *Pesticide industry sales and usage: 1994 and 1995 market estimates*, Office of Prevention, Pesticides, and Toxic Substances Publication no. 733-R-97-002. Washington, DC: US Environmental Protection Agency.

Avery, D. (1995). *Saving the Planet with Pesticides and Plastic: The Environmental Triumph of High-Yield Farming*. Indianapolis, IN: Hudson Institute.

Batie, S. S., & Swinton, S. M. (1994). Institutional issues and strategies for sustainable agriculture: view from within the land-grant university. *American Journal of Alternative Agriculture*, **9**, 23–7.

Beets, W. C. (1982). *Multiple Cropping and Tropical Farming Systems*. Boulder, CO: Westview Press.

Bellinder, R. R., Gummesson, G., & Karlsson, C. (1994). Percentage-driven government mandates for pesticide reduction: the Swedish model. *Weed Technology*, **8**, 350–9.

Benbrook, C. M. (1996). *Pest Management at the Crossroads*. Yonkers, NY: Consumers Union.

Boutin, C., & Jobin, B. (1998). Intensity of agricultural practices and effects on adjacent habitats. *Ecological Applications*, **8**, 544–57.

Brady, N. C. (1984). *The Nature and Properties of Soils*, 9th edn. New York: Macmillan.

Brown, L. R., & Jacobson, J. L. (1986). *Our Demographically Divided World*, Worldwatch Paper no. 74. Washington, DC: Worldwatch Institute.

Brown, L. R., Flavin, C., & Postel, S. (1991). Vision of a sustainable world. In *The World Watch Reader on Global Environmental Issues*, ed. L. R. Brown, pp. 299–315. New York: W. W. Norton.

Buhler, D. D., & Maxwell, B. D. (1993). Seed separation and enumeration from soil using K_2CO_3-centrifugation and image analysis. *Weed Science*, **41**, 298–302.

Buttel, F. H. (1984). Socioeconomic equity and environmental quality in North American agriculture: alternative trajectories for future development. In *Agricultural Sustainability in a Changing World Order*, ed. G. K. Douglass, pp. 89–106. Boulder, CO: Westview Press.

Buttel, F. H. (1990). Social relations and the growth of modern agriculture. In *Agroecology*, ed. C. R. Carroll, J. H. Vandermeer & P. M. Rosset, pp. 113–45. New York: McGraw-Hill.

Chase, C., & Duffy, M. (1991). An economic comparison of conventional and reduced-chemical farming systems in Iowa. *American Journal of Alternative Agriculture*, **6**, 168–73.

Conner, D., & Kolodinsky, J. (1997). Can you teach an old dog new tricks? An evaluation of extension training in sustainable agriculture. *Journal of Sustainable Agriculture*, **10**, 5–20.

Daly, H. E., & Cobb, J. B. Jr. (1994). *For the Common Good*, 2nd edn. Boston, MA: Beacon Press.

Dushoff, J. Caldwell, B., & Mohler, C. L. (1994). Evaluating the environmental effect of pesticides: a critique of the environmental impact quotient. *American Entomologist*, **40**, 180–4.

Eckholm, E. (1979). *The Dispossessed of the Earth: Land Reform and Sustainable Development*, Worldwatch Paper no. 30. Washington, DC: Worldwatch Institute.

Ehrensaft, P., LaRamée, P., Bollman, R. D., & Buttel, F. H. (1984). The microdynamics of farm structural change in North America: the Canadian experience and Canada–USA comparisons. *American Journal of Agricultural Economics*, **66**, 823–8.

Gorney, C. (1987). Banning a weedkiller: no middle ground. *Washington Post*, 3 August 1987, A1 & A10.

Greenberg, R., Bichier, P., Angon, A. C., & Reitsma, R. (1997). Bird populations in shade and sun coffee plantations in central Guatemala. *Conservation Biology*, **11**, 448–59.

Hanson, J. C., Lichtenberg, E., & Peters, S. E. (1997). Organic versus conventional grain production in the mid-Atlantic: an economic and farming system overview. *American Journal of Alternative Agriculture*, **12**, 2–9.

Haskell, D. E. (1991). A stroll down herbicides' memory lane. In *Proceedings of the 43rd Annual California Weed Conference*, 21–3 January 1991, Santa Barbara, CA, pp. 154–9.

Hawken, P. (1993). *The Ecology of Commerce*. New York: HarperBusiness.

Hoag, D. L., & Pasour, E. C. Jr. (1992). It's a dilemma. *Choices*, **7**(1), 32 & 34.

Jahn, L. R., & Schenck, E. W. (1991). What sustainable agriculture means for fish and wildlife. *Journal of Soil and Water Conservation*, **46**, 251–4.

Jutsum, A. R. (1988). Commercial application of biological control: status and prospects. *Philosophical Transactions of the Royal Society of London, Series B*, **318**, 357–73.

Karlen, D. L., Duffy, M. D., & Colvin, T. S. (1995). Nutrient, labor, energy, and economic evaluations of two farming systems in Iowa. *Journal of Production Agriculture*, **8**, 540–6.

Keeney, D. (1998). Sustainable technologies and management for Iowa agriculture: the role of the Leopold Center for Sustainable Agriculture. In *Concept Papers for the World Bank Workshop on Sustainable Intensification of Agricultural Systems: Linking Policy, Institutions, and Technology*, ed. R. Kanwar, pp. 38–47. Ames, IA: College of Agriculture, Iowa State University.

King, R. P., Lybecker, D. W., Schweizer, E. E., & Zimdahl, R. L. (1986). Bioeconomic modeling to simulate weed control strategies for continuous corn (*Zea mays*). *Weed Science*, **34**, 972–9.

Kirschenmann, F. (1991). Fundamental fallacies of building agricultural sustainability. *Journal of Soil and Water Conservation*, **46**, 165–8.

Kuch, P. J., & Ogg, C. W. (1996). The 1995 Farm Bill and natural resource conservation: major new opportunities. *American Journal of Agricultural Economics*, **78**, 1207–14.

Kuhns, L. J., & Harpster, T. L. (1997). Agriculture needs more weed scientists. *Proceedings of the Northeastern Weed Science Society*, **51**, 192–7.

Lampkin, N. (1990). *Organic Farming*. Ipswich, UK: Farming Press.

Leakey, R. R. B. (1999). Agroforestry for biodiversity in farming systems. In *Biodiversity in Agroecosystems*, ed. W. W. Collins & C. O. Qualset, pp. 127–46. Boca Raton, FL: CRC Press.

Lighthall, D. R. (1996). Sustainable agriculture in the corn belt: production-side progress and demand-side constraints. *American Journal of Alternative Agriculture*, **11**, 168–74.

Lybecker, D. W., Schweizer, E. E., & King, R. P. (1991). Weed management decisions in corn based on bioeconomic modeling. *Weed Science*, **39**, 124–9.

MacCannell, D. (1988). Industrial agriculture and rural community degradation. In *Agriculture and Community Change in the US, The Congressional Research Reports*, ed. L. E. Swanson, pp. 15–75. Boulder, CO: Westview Press.

Matheson, N., Rusmore, B., Sims, J. R., Spengler, M., & Michalson, E. L. (1991). *Cereal–Legume Cropping Systems: Nine Farm Case Studies in the Dryland Northern Plains, Canadian Prairies and Intermountain Northwest*. Helena, MT: Alternative Energy Resources Organization.

Matteson, P. (1995). The "50% pesticide cuts" in Europe: a glimpse of our future? *American Entomologist*, **41**, 210–20.

Maunder, A. H. (1984). Land tenure and structural change in the European Economic Community. *Oxford Agrarian Studies*, **13**, 103–22.

Mayerfeld, D. B., Hallberg, G. R., Miller, G. A., Wintersteen, W. K., Hartzler, R. G., Brown, S. S., Duffy, M. D., & DeWitt, J. R. (1996). *Pest Management in Iowa: Planning for the Future*, Publication IFM 17. Ames, IA: Iowa State University Extension.

McMichael, P. (1998). Global food politics. *Monthly Review*, **50**(3), 97–111.

Murdoch, W. W. (1980). *The Poverty of Nations: The Political Economy of Hunger and Population*. Baltimore, MD: Johns Hopkins University Press.

Nair, P. K. R. (1993). *An Introduction to Agroforestry*. Dordrecht, Netherlands: Kluwer.

National Research Council (1989). *Alternative Agriculture*. Washington, DC: National Academy Press.

Nguyen, M. J., & Haynes, R. J. (1995). Energy and labour efficiency for three pairs of conventional and alternative mixed cropping (pasture–arable) farms in Canterbury, New Zealand. *Agriculture, Ecosystems and Environment*, **52**, 163–72.

Owen, M. D. K. (1998). Producer attitudes and weed management. In *Integrated Weed and Soil Management*, ed. J. L. Hatfield, D. D. Buhler & B. A. Stewart, pp. 43–59. Chelsea, MI: Ann Arbor Press.

Papendick, R. I., Elliot, L. F., & Dahlgren, R. B. (1986). Environmental consequences of modern production agriculture: how can alternative agriculture address these concerns? *American Journal of Alternative Agriculture*, **1**, 3–10.

Paulson, D. D. (1995). Minnesota extension agents' knowledge and views of alternative agriculture. *American Journal of Alternative Agriculture*, **10**, 122–8.

Pearce, D., & Tinch, R. (1998). The true price of pesticides. In *Bugs in the System: Redesigning the Pesticide Industry for Sustainable Agriculture*, ed. W. Vorley & D. Keeney, pp. 50–93. London: Earthscan Publications.

Perfecto, I., Rice, R. A., Greenberg, R., & Van Der Voort, M. (1996). Shade coffee: a disappearing refuge for biodiversity. *BioScience*, **46**, 598–608.

Pingali, P. L., & Gerpacio, R. V. (1997). *Towards Reduced Pesticide Use for Cereal Crops in Asia*,

International Maize and Wheat Improvement Center (CIMMYT) Economics Working Paper no. 97–04. Mexico, DF: CIMMYT.
Pretty, J. N. (1995). *Regenerating Agriculture: Policies and Practice for Sustainability and Self-Reliance*. Washington, DC: Joseph Henry Press.
Pretty, J. N., Vorley, W., & Keeney, D. (1998). Pesticides in world agriculture: causes, consequences, and alternative courses. In *Bugs in the System: Redesigning the Pesticide Industry for Sustainable Agriculture*, ed. W. Vorley & D. Keeney, pp. 17–49. London: Earthscan Publications.
Reganold, J. P., Elliott, L. F., & Unger, Y. L. (1987). Long-term effects of organic and conventional farming on soil erosion. *Nature*, **330**, 370–2.
Repetto, R. (1985). *Paying the Price: Pesticide Subsidies in Developing Countries*. Washington, DC: World Resources Institute.
Roodman, D. M. (1999). Building a sustainable society. In *State of the World, 1999*, ed. L. R. Brown, C. Flavin & H. French, pp. 169–88. New York: W. W. Norton.
Schaller, N. (1992). It's an opportunity. *Choices*, **7**(1), 33.
Schneiderman, H. A., & Carpenter, W. D. (1990). Planetary patriotism: sustainable agriculture for the future. *Environmental Science and Technology*, **24**, 466–73.
Smith, S. (1992). Farming activities and family farms: getting the concepts right. In *Symposium: Agricultural Industrialization and Family Farms – The Role of Federal Policy*, pp. 117–33. Washington, DC: Joint Economic Committee, Congress of the United States.
Smolik, J. D., Dobbs, T. L., & Rickerl, D. H. (1995). The relative sustainability of alternative, conventional, and reduced-till farming systems. *American Journal of Alternative Agriculture*, **10**, 25–35.
Stanhill, G. (1990). The comparative productivity of organic agriculture. *Agriculture, Ecosystems and Environment*, **30**, 1–26.
Stanton, B. F. (1993). Changes in farm size and structure in American agriculture in the twentieth century. In *Size, Structure, and the Changing Face of American Agriculture*, ed. A. Hallam, pp. 42–70. Boulder, CO: Westview Press.
Steiner, K. G. (1984). *Intercropping in Tropical Smallholder Agriculture with Special Reference to West Africa*, 2nd edn. Eschborn, Germany: Deutsche Gesellschaft fur Technische Zusammenarbeit (GTZ).
Strange, M. (1988). *Family Farming: A New Economic Vision*. Lincoln, NE: University of Nebraska Press.
Thiesenhusen, W. C., & Melmed-Sanjak, J. (1990). Brazil's agrarian structure: changes from 1970 through 1980. *World Development*, **18**, 393–415.
Thrupp, L. A. (ed.) (1996). *New Partnerships for Sustainable Agriculture*. Washington, DC: World Resources Institute.
Tweeten, L. (1996). The environment and the 1995 farm bill: discussion. *American Journal of Agricultural Economics*, **78**, 1217–18.
United States Department of Agriculture (1990). *Farm Bill, Public Law 101-624-Nov. 28, 1990; 104 STAT. 3359; 101st Congress*. Washington, DC: US Government Printing Office.
United States Department of Agriculture (1998). *A Time to Act: A Report of the USDA National Commission on Small Farms*. Washington, DC: US Department of Agriculture.
United Nations Development Programme (1992). *Benefits of Diversity: An Incentive Towards Sustainable Agriculture*. New York: UN Development Programme.
Viederman, S. (1990). A matter of commitment. In *Sustainable Agricultural Systems*, ed. C. A.

Edwards, R. Lal, P. Madden, R. H. Miller & G. House, pp. 666–73. Ankeny, IA: Soil and Water Conservation Society.

Vorley, W., & Keeney, D. (1998). Solving for pattern. In *Bugs in the System: Redesigning the Pesticide Industry for Sustainable Agriculture*, ed. W. Vorley & D. Keeney, pp. 193–215. London: Earthscan Publications.

Waggoner, P. E. (1994). *How Much Land Can 10 Billion People Spare for Nature?* Task Force Report no. 121. Ames, IA: Council for Agricultural Science & Technology.

Welsh, R., & Lyson, T. A. (1997). Farm structure, market structure and agricultural sustainability goals: the case of New York State dairying. *American Journal of Alternative Agriculture*, **12**, 14–19.

Westendorf, M. L., Zimbelman, R. G., & Pray, C. E. (1995). Science and agriculture policy at land grant institutions. *Journal of Animal Science*, **73**, 1628–38.

World Commission on Environment and Development (1987). *Our Common Future*. Oxford, UK: Oxford University Press.

Wyse, D. L. (1992). Future of weed science research. *Weed Technology*, **6**, 162–5.

Taxonomic index

Abutilon theophrasti, 51, 54, 57, 61, 64–5, 68–9, 79, 161, 243, 247–8, 303, 334, 377–8, 380, 396–8, 400, 470
Achillea millefolium, 146, 148
Achromobacter, 397
Acioa barteri, 357–8
Aegilops cylindrica, 223, 397
Aeschynomene virginica, 393
Aethusa cynapium, 329, 457
Agapeta zoegana, 382, 384
Ageratina riparia, 387
Agropyron cristatum, 413
Agropyron desertorum, 413
Agropyron repens, 45, 115, 142, 145–9, 335, 348–9, 421
Agropyron spicatum, 413
Agrostemma githago, 102, 474
Agrostis gigantea, 146, 148
Agrostis spp., 147
Alcaligenes, 397
alfalfa, 21–3, 289, 334–5
Alisma triviale, 216
Allium vineale, 145–6, 329
almond, 359
Alopecurus myosuroides, 10, 51, 83, 152, 154, 164, 304
Alternaria, 380
Alternaria spp., 393–4, 396
Alternaria cassiae, 393
Alternaria macrospora, 396
Alternathera gullensis, 422
Amaranthus, 1
Amaranthus spp., 49, 214, 397
Amaranthus albus, 76
Amaranthus hybridus, 462
Amaranthus retroflexus, 47, 51–2, 61, 64–6, 68–9, 73–4, 158, 219, 222, 225–6, 243, 247, 296, 333, 351
Amaranthus spinosus, 462
Ambrosia artemisiifolia, 47, 61, 351
Anoda cristata, 396
Anthomyiidae, 382–3
Anthoxanthum odoratum, 398

Apera spica-venti, 304, 467–8
Aphanes arvensis, 102
Aphthona spp., 382–3
Apocynum cannabinium, 147
Arabidopsis thaliana, 47, 471
Arachis pintoi, 359
Archlagocheirus funeatus, 385
Arctium lappa, 76–7
Arctium minus, 428
Arctostaphylos patula, 425
Artemisia spp., 457
Arthrobacter, 397
Asclepias spp., 76
Asclepias syriaca, 68, 144
Ascochyta caulina, 393
Asteraceae, 361
Astrebla spp., 416, 420
Atriplex patula, 48
Avena fatua, 56, 58, 61, 161–2, 163, 168, 221, 282, 291–2, 334–5, 462, 465, 471
Avena sativa, 459
Avena sterilis ssp. *ludoviciana* (*A. ludoviciana*), 54, 58, 152
Axonopus compressus, 418, 427

Bangasternus fausti, 384
Barbarea vulgaris, 457
barley, 271, 277, 284, 289, 291–2, 301–2, 345–6, 348–9, 351
 cv. 'Fergus', 292
Bassia birchii, 428
bean, 227, 271, 284, 289, 290, 300, 324, 329, 341
beetles, 382, 383, 384, 385
Bellis perennis, 428
berseem clover, 232
Beta vulgaris, 460
 ssp. *maritima*, 460–1
Bidens alba, 361
Bilderdykia convolvulus, see *Polygonum convolvulus*
blackgram, 349–50
bluegrass, Kentucky, 271
Botanophila semecoella, 386
Botrytis cinerea, 395

Brassica, 1
Brassica arvensis, 51
Brassica hirta, 220–1, 227–8, 293, 345, 346
Brassica kaber, 61, 293, 334–5, 351, 378
Brassica napus, 181, 278, 304
bromegrass, smooth, 334
Bromus spp., 422
Bromus diandrus, 58
Bromus interruptus, 77
Bromus japonicum, 397
Bromus rigidus, 57
Bromus secalinus, 58, 74–5, 152
Bromus sterilis, 54, 58, 77
Bromus tectorum, 47, 79, 223, 225, 331, 397
bugs, 385
Buprestidae, 384–5

cabbage, 271, 300
Cactoblastis cactorum, 385, 391
cactus, 385, 391
Calliandra calothyrsus, 356–7
Camelina sativa
    var. *linicola*, 468–9
    var. *sativa*, 468
Canavalia ensiformis, 116–17
Capsella bursa-pastoris, 47, 48, 55
carabid beetles, 377–80
Cardaria draba, 102
Cardiospermum halicacabum, 474
Carduus, 390
Carduus nutans, 428, 457
Carduus pycnocephalus, 402, 430
Carduus tenuiflorus, 430
carrot, 289
cassava, 324, 354–5
Cassia obtusifolia, see *Senna obtusifolia*
Cassia occidentalis, 393
Cassia siamea, 356
Cecidomyiidae, 382–3
Celosia spp., 350–1
Cenchrus incertus, 76–7
Centaurea cyanus, 457
Centaurea maculosa, 344, 382
Centaurea nigra, 457
Centaurea repens, 147
Centaurea solstitialis, 411, 435, 467
Centrosema pubescens, 418
Ceonothus integerrimus, 425
Cerambycidae, 382–3
Cerastium vulgatum, 457
Chaetorellia acolophi, 384
Chamaesphecia spp., 383
Chelinidea tabulata, 385
Chenopodium spp., 1, 397
Chenopodium album, 47–8, 50–1, 53, 55, 61, 64, 68–9, 71, 158, 172, 186, 222, 244–8, 378, 393–4, 457, 462
Chondrilla juncea, 387–8, 398, 467
Chromolaena odorata, 104
Chrysanthemum leucanthemum, 430
Chrysomelidae, 382–3, 387

Circium canescens, 390–1
Cirsium arvense, 45, 47, 75, 147, 150, 334–5, 412, 471
Cirsium palustre, 433–4
Cirsium vulgare, 428–9, 433–4, 457
Citrobacter, 397
Citrus spp., 361
Cladosporium, 380
Clusia rosea, 362
coconut, 423, 425–6
coffee, 112, 129–31, 293, 359–62
Colletotrichum coccodes, 398, 400
Colletotrichum gloeosporioides, 393, 396
Colletotrichum orbiculare, 398
Colletotrichum truncatum, 395, 396
Commelina difusa, 360
Commelina erecta, 361
Commelinaceae, 360–1
Conservula cinisigna, 389
Convolvulus arvensis, 54, 63, 79, 144, 146, 147, 161, 181, 462
cotton, 271, 284, 289
cowpea, 271, 289, 337, 339–40, 344–5, 348
crimson clover, 231, 232, 245–7
Crotalaria spectabilis, 393
cucumber, 271, 293, 300
Cucurbita texama, 396
Curculionidae, 384–5, 387
Cuscuta spp., 393
Cynodon dactylon, 73, 149, 215, 462
Cyperus spp., 142, 213, 350
Cyperus esculentus, 75, 393, 398, 462
Cyperus rotundus, 144, 147, 149–50, 214, 357, 462, 464
Cyphocleonus achates, 382, 384

Dactylis glomerata, 70, 398
Dactylopius spp., 385
Dacylopiidae, 387
Datura ferox, 51, 52, 74
Datura stramonium, 65, 397, 470
Daucus carota, 457
Desmodium adscendens, 237–8
Desmodium intortum, 425
Desmodium ovalifolium, 355–6, 358, 359
Dichanthium, 420
Digitaria spp., 104, 350
Digitaria sanguinalis, 68, 71, 241–3, 462
Diodia teres, 455
Drymaria cordata, 360

Echinochloa spp., 301
Echinochloa crus-galli, 58, 73, 213, 217–18, 224, 304, 333, 378, 462
    var. *crus-galli*, 469
Echinochloa microstachya, 455, 458
Echinochloa oryzoides, 217
Echinochloa phyllopogon, 217
Echium plantagineum, 390, 467
egusi melon, 358
Eichhornia crassipes, 462

Eleusine indica, 462
Elytrigia repens, see *Agropyron repens*
Emex australis, 392
Emex spinosa, 467
Enterobacter, 397
Entyloma ageratinae, 387
Epilobium spp., 76
Equisetum arvensis, 102
Erigeron annnuus, 464
Erwinia, 380, 397
Erysimum cheiranthoides, 51
Erythrina spp., 352
Erythrina poeppigiana, 357
Euphorbia esula, 75, 77, 382, 383, 411, 412, 428, 435
Euphorbia heliscopia, 55

Fallopia convolvulus, 226
fava bean, 341–3
Festuca arundinacea, 289, 359, 398
Flavobacterium, 380, 397
flax, 271, 278, 284
flea beetles, 382, 383, 386–7, 399–400
Flemingia congesta, 356
Flemingia macrophylla, 356
Fumaria officinalis, 48, 55, 77
Fusarium, 380
Fusarium laterium, 39
Fusarium oxysporum, 215, 396
Fusarium solani f.sp. *cucurbitae*, 396

Galeopsis tetrahit, 48, 226, 457
Galinsoga ciliata, 71, 152
Galinsoga parviflora, 102
Galinsoga quadriradiata, 361
Galium aparine, 46, 56, 278, 335, 457
Galium tricornutum, 456
gallflies, 382, 383, 384
Gelechiidae, 384–5
Geometridae, 382–3
Gliricidia sepium, 356–8, 362
Gordonia alatamaha, 455
guineagrass, 289, 292
Gutierrizia sarothrae, 416

hairy vetch, 232, 237–8, 237–9, 239, 254
Helianthus annuus, 161–2
Heracleum sphondylium, 457
Holcus lanatus, 70, 398
Holcus mollis, 148
Homolepsis aturensis, 427
Hordeum glaucum, 429
Hordeum jubatum, 58, 421
Hordeum leporinum, 428, 429
Hordeum murinum, 428, 429
Hordeum vulgare, 459
Hylemia semecoella, see *Botanophila semecoella*
Hyles euphorbiae, 382–3
Hypericum perforatum, 402, 470
Hypurus bertrandi, 378–9

Imperata cylindrica, 28–9, 146, 356, 358, 412, 462
Indigofera cordifolia, 1
Inga spp., 352, 361
Inga edulis, 355–6, 358
Inga paterna, 362
Ipomoea spp., 214, 397
Ipomoea hederacea, 61
Ipomoea purpurea, 51, 59

Juncus spp., 428, 433–4

Klebsiella, 397
Kochia scoparia, 56–7

Lamium amplexicaule, 48
Lamium purpureum, 53, 56, 152
Lantana cinerea, 418
Larinus spp., 384
Larix occidentalis, 423–4
leek, 301
lentil, 271
Lepidium draba, 147
Lepidium perfoliatum, 47
Leptochloa chinensis, 217
lettuce, 167, 301, 341, 394
Leucaena leucocephala, 356–8, 425
Leucaena spp., 104
Linaria vulgaris, 457
Linum catharticum, 457
Lobesia euphorbiana, 382–3
Lolium multiflorum, 51, 56, 221, 465
Lolium rigidum, 9–10, 227, 422, 452
Lolium temulentum, 102
Longitarsus jacobaeae, 386–7, 399–400
lupin, 271, 284, 289

Macroptilium atropurpureum, 413
maize, 19–20, 131–2, 152–3, 240, 271, 284, 287, 289, 303, 324, 332–3, 341, 347–8, 350, 354–5, 393–4
Malva pusilla, 393
Matricaria matricarioides, 55, 421
Matricaria perforata, 378
Medicago littoralis, 237
Medicago lupulina, 47, 55
Melampodium microcephalum, 361
Mentha arvensis, 148
Metzneria paucipunctella, 382, 384
midges, 382, 383
millet, 1, 284, 350–1; see also *Panicum*, *Sorghum*
Mimosa pudica, 426
Minoa murinata, 383
mollusks, 69, 241–4
monarch butterfly, 68–9
Morrenia odorata, 393
moths, 382, 384, 385, 386-7, 389, 391
Mucana spp., 131–2, 255
mungbean, 289, 290, 344, 348
Musa spp., 361
mustard, white, 335

Nardus stricta, 432–3
Neodiplogrammus quadrivittatus, 385
Niesthrea louisianica, 380

oat, 271, 282, 284, 287, 289, 293, 459
Oberea erythrocephala, 382–3
Oidaematophorus beneficus, 387
onion, 271
Onopordum, 390
Oplismenus burmani, 360
Opuntia spp., 385, 391
Opuntia spinosissima, 391
Opuntia triacantha, 391
Oryza punctata, 455, 458
Oryza sativa, 473
Oxalis pes-caprae, 464
Oxalis stricta, 77

Panicum, 104
Panicum capillare, 76, 247, 248
Panicum dichotomiflorum, 50
Panicum maximum, 418
Panicum miliaceum, 83, 285, 293, 303, 460, 474
Panicum repens, 81
Panotima spp., 389
Papaver rhoeas, 55
Paspalum conjugatum, 418, 427, 462
Paspalum notatum, 427
Pastinaca sativa, 457
'Paterson's curse', 390
pea, 271–2, 289, 341, 351
    cv. 'Alaska', 345
    cv. 'Century', 345
peach, 359
peanut, 272, 284, 289, 350, 350–1
pearl millet, 1, 284
Pegomya spp., 383
Pelochrista medullana, 384
Phacelia tanacetifolia, 181
Phaseolus vulgaris, 227
Phomopsis emicis, 392
Phragmidium violaceum, 387
Phytophthora palmivora, 393
Picea glauca, 420–1
pigeonpea, 272, 284, 289, 337–9, 341–2, 344
Pinus ponderosa, 423–4
Plantago lanceolata, 50–1, 69–70
Plantago major, 47, 51, 456, 457
plantain, 354–5, 358
Poa annua, 48, 51, 55, 58, 63, 69–70, 145, 213, 329
Polygonum spp., 397
Polygonum aviculare, 48, 55, 65, 421, 457
Polygonum convolvulus, 48, 49, 55, 79, 278, 378
Polygonum lapathifolium, 378
Polygonum pennsylvanicum, 49
Polygonum persicaria, 15–16
Portulaca oleracea, 47, 51, 68–9, 71, 152, 378–9, 462
potato, 244–5, 252, 253, 289, 290, 293–6, 302
    cv. 'Green Mountain', 293–6
    cv. 'Katahdin', 293–6
    cv. 'Norchip', 294–5
    cv. 'Sebago', 294–5
Pseudoelephantopus spicatus, 427
Pseudomonas, 244, 380, 397
Pseudomonas fluorescens strain D7, 397
Pseudotsuga menziesii, 423–4
Pteridium aquilinum, 388–90, 413–14
Pterolonche inspersa, 384
Pterolonchidae, 384–5
Puccinia spp., 396
Puccinia canaliculata, 393, 398
Puccinia chondrillina, 387–8, 398
Puccinia lagenophorae, 394–5
Pueraria phaseoloides, 237–8, 358, 418
Pyralidae, 387
Pythium spp., 244

Ranunculus arvensis, 102
rape, see *Brassica napus*
rapeseed, 15–16, 168, 244–5, 272, 284, 289, 331
Raphanus raphanistrum, 460
Raphanus sativus, 460
red clover, 234, 238, 348–9
Rhinocyllus conicus, 390
rhizobacteria, 397–8
Rhizoctonia, 244
rhizoma peanut, 272
Rhus typhina, 458
Rhyssomatus marginatus, 385
rice, 289, 291–2, 293, 300, 301, 330, 349–50, 393, 448
Rosa multiflora, 435
Rottboellia cochinchinensis, 132, 191
Rottboellia exaltata, 357, 462
Rubus constrictus, 387
Rubus ulmifolius, 387
Rumex acetosa, 457
Rumex acetosella, 464
Rumex crispus, 50, 51, 149, 239, 248–9, 457
Rumex obtusifolius, 149, 239, 428
rust fungi, 387–8
rye, 230–2, 234–6, 248–50, 304, 335
ryegrass, 222, 272, 289, 292, 345, 350

safflower, 270, 272, 274, 278, 284, 289, 290
Salsola iberica, 73
'Salvation Jane', 390
sawfly, 378–9
scale insects, 385
Schizocerella pilicornis, 378–9
Scirpus maritimus, 330
Senecio cambrensis, 461
Senecio jacobaea, 386–7, 398–400, 402, 411, 432
Senecio squallidus, 461
Senecio vulgaris, 48, 55, 186, 394–5, 461, 471–3
Senna obtusifolia, 59, 65, 393, 475
sesame, 1
Sesbania aculeata, 237
Sesbania exaltata, 395–6
Sesbania punicea, 385–6

Sesiidae, 382–3
Setaria spp., 58
Setaria faberi, 49, 71, 296, 333, 380
Setaria glauca, see *Setaria pumila*
Setaria pumila, 304
Setaria viridis, 47, 49, 222, 270, 274, 278, 304, 380
Sida spp., 422
Sida acuta, 418
Silybum, 390
Simarouba glauca, 362
Sinapis alba, 186, 277
Sinapis arvensis, 47, 224
Sisymbrium altissimum, 76
Smallanthus maculatus, 361
snap bean, 271, 284–5
Solanum spp., 77
Solanum crinitum, 457–8
Solanum viarum, 53, 412, 428, 435
Solidago altissima, 458, 466
Sonchus arvensis, 51, 145–6, 148, 457
Sonchus asper, 278
Sophia multifida, 47
sorghum, 234, 272, 284, 289, 330, 337, 338–9, 345, 460, 473
Sorghum bicolor, 378
Sorghum halepense, 51, 147, 216, 239, 412, 462, 470
soybean, 15–16, 19–20, 64, 152–3, 240, 272, 284, 289–90, 293, 303, 330, 333, 344, 395
   cv. 'Evans', 296
   cv. 'Gnome', 296
Spergula arvensis, 48, 55
Sphenoclea zeylanica, 393–4
Sphenoptera jugoslavica, 384
Spurgia capitigena, 382, 383
Spurgia esulae, 382, 383
squash, 289
Stellaria media, 48, 51, 53, 55, 71, 102, 181, 186, 226, 248–9, 471
strawberry, 420
Striga asiatica, 467, 482
Stylosanthes guianensis, 418
subterranean clover, 227, 232, 292, 347–8, 398–9
sudangrass, 244–5
sugar beet, 299, 329, 393–4
sugarcane, 289
sunflower, 15–16, 64, 284, 293, 341
sweet corn, 245–7, 284–5, 303
sweet potato, 272, 293

Taraxacum officinale, 47, 66, 142, 248–9, 334, 457, 464
taro, 356
Taxus brevifolia, 454
Tephritidae, 384–5, 387

Terellia virens, 384
Thlaspi arvense, 55, 226, 335, 378
timothy, 272
Tingidae, 387
Tithonia diversifolia, 116
tomato, 127–8, 301
Tortricidae, 382–4
Toxicodendron radicans, 77
Trianthema portulacastrum, 214
Trichapion lativentre, 385–6
Trifolium spp., 457
Trifolium hirtum, 468
Trifolium repens, 464
Tripleurospermum maritimum ssp. *inodorum*, 55
Tripleurospermum phaeocephalum, 421
Tripogandra serrulata, 361
Tyria jacobaeae, 386–7, 399

Ulex europaeus, 427, 433–4
urdbean, 344
Urophora spp., 382, 384
Urtica spp., 457
Urtica urens, 48, 55

Valerianella spp., 457
velvetbean, 131–2, 255
Veronica hederifolia, 47, 48, 55, 73, 224
Veronica persica, 48, 55, 56, 102, 152, 226
Verticillium spp., 378
Verticillium dahliae, 215
Vicia hirsuta, 55
Vicia sativa, 474
Viola alba, 459
Viola arvensis, 55–6
Vismia guianensis, 458
Vulpia spp., 422

weevils, 378–9, 382–6, 390
wheat, 232, 272–3, 284, 289, 291–2, 304, 331–3, 342–3, 397
white mustard, 335
winter barley, 301–2
winter rye, 230, 231
winter wheat, 331–3, 397

Xanthium canadense, 397
Xanthium spinosum, 398
Xanthium strumarium, 54, 61, 63–5, 303, 461

Zea mays, see maize
Zea mexicana, 471, 473
Zizania palustris, 216

Subject index

adaptive weed management, 112–13
    and farmer-extensionist-scientist interactions, 117–18
    in indigenous agriculture, 116–17
    mechanized ecological cropping, 115–16
    non-mechanized ecological cropping, 116
    precision agriculture, 114–15, 117
    see also participatory learning for action
aftermath grazing, 421–2
'agrestal' weeds, 40
agrichemicals
    historical use of, 111–12
    precision agriculture, 114–15, 117
    see also fertilizers; herbicides; pesticides
agricultural machinery
    for intercropping systems, 363–4
    weed dispersal by, 80–1
    see also cultivators; tillage implements
agricultural policies
    herbicide use, 500–1
    international, 511
    promotion of crop diversity, 364
agricultural science, 8, 102–3
agricultural subsidies, 501
agroecosystem redesign, 27–9
agroforestry, 322–3, 324, 351–62
    alley cropping, 353, 356–8
    benefits to farmer, 351–2
    and crop diversity, 322–3, 324
    forest fallow systems, 353–6
    temporal patterns of weed infestation, 352–3
    understory cover crops, 358–60
    use of multilayered canopies, 360–2
alachlor, 12, 14, 15, 18
aldicarb, 379
alien weed species, 68, 447–52
    biological control of, 381–8
    and disturbance, 447–8
    rate of introduction, 448–9
    and weed species richness, 449–52
allelochemicals, 233–4, 245
allelopathy
    and crop genotype, 292–3

    of crop residues, 233–6, 330
alley cropping, 353, 356–8
ALLOCATE model, 222–3
American northern plains, 28
ammonium nitrogen, 225–6
animal-drawn implements, 169
annual weeds, 42–4
    ridge tillage, 164–7
    seed production, 71–3
ants, 77
aquatic life guidelines, 14
asexual weeds, 464–5
atrazine, 12–14, 14, 15, 18

basket weeders, 177
beet knives, 176
2(3H)-benzoxazolinone (BOA), 234, 235
biological control, 375–6
    accessibility to farmers, 401
    conservation methods, 377–80
    inoculation methods, 380–91
        for exotic weeds, 381–8
        for native weeds, 388–9
        and non-target species, 390–1
        social and ecological impacts, 389–90
    interactions of agents, 402
    inundation methods, 391–8
        deleterious rhizobacteria, 397–8
        mycoherbicides, 392–6
    levels of success achieved, 376
    plant community changes, 402–3
    principles, 376–7
    production and culture methods, 398, 401
    use of multiple stressors, 398–400
biotypes, 461
bipyridilium herbicides, 17
birds, 75–6, 77, 505–6
birth defects, 18
BOA see 2(3H)-benzoxazolinone
breeding systems, 463–7
brush control, 433–4
brush weeders, 177–9
bulbs, 144

Subject index

cancer, 17–18
carbon dioxide, 52
carcinogens, 17
case studies
   Californian tomato cropping, 127–8
   cover crops in Central America, 131–2
   ground cover in coffee, 129–31
   Iowa grain cropping, 128–9
cattle grazing, 411, 422, 426, 432–3
chisel plow, 141, 142
chloroacetamide herbicides, 14–15, 17
chlorsulfuron, 15–16
coastal pollution, 14
coconut plantations, 423, 425–6
colonization
   distribution of major weeds, 462
   ecotype formation, 469–70
   genetic variability following, 467–8
   see also weed disperal
combine harvester, 74, 81
competition
   and biocontrol efficacy, 398–400
   weed–weed, 446–7
   see also crop–weed competition
competitive effect, 287–8
competitive response, 288
composting, 79
cost–price squeeze, 19–21
cover crops, 28–9, 229–33
   in agroforestry systems, 358–60
   case study, 131–2
   killing, 230–3
   planting and establishment, 230, 231
   in rotation systems, 335–6
   see also crop residues
crop breeding programs, 296–7
crop competitiveness, 287–9
   and crop density responses, 279–81
   and genotype, 287–96
   see also crop–weed competition
crop density, 270–81
   and crop yield, 271–3, 276–8
   in intercropping systems, 339, 341–4
   and weed biomass, 270–6
   and weed suppression, 278–81
crop diversification
   farmer education, 364
   machinery/technology for, 363–4
   obstacles and opportunities, 363–4
   principles of, 325–6
   weed adaptation
      management of, 479–81
      potential for, 471–4
   see also agroforestry; crop rotations; intercropping
crop diversity
   decline in, 323–4
   in organic systems, 325
   and weed diversity, 452–4
crop genotype
   breeding programs, 296–7
   and competitive ability, 287–9
   allelopathy, 292–3
   below ground, 292
   in cereal crops, 291–2
   cultivar screening, 288, 290
   in forage crops, 292
   multiple factors, 293–6
   in row crops, 290–1
crop height, 278, 283, 285, 290–1
crop phenology
   weed-free period, 297–9
   weed infestation period, 299–301
crop planting date, 49, 151–3
   and crop–weed competition, 302–4
   in rotations, 327–9
   and time of weed emergence, 151–3
crop planting depth, 218, 220
crop residues, 229–50, 298–9
   allelopathy of, 233–6
   green manures and mulches, 115–16, 229–33
   and herbivores, 241–4
   light extinction, 240–2
   mechanical management, 254–6
   nutrient release, 236–8
   and pathogens, 244, 398
   soil moisture, 239–40
   soil temperature, 238–40
   use in crop rotations, 330
crop rotation, 28, 322
   cover crops, 335–6
   and ecotypic variation, 471
   herbicides used in, 330–3
   increasing profitability of, 21–3
   perennial forage crops, 333–5
   potential benefits, 326–7
   soil conditions, 330
   timing of crop management practices, 49, 327–9
   and weed diversity, 452–3
crop spatial arrangement, 281–7
   in intercropping systems, 344
   planting patterns, 281–2
   row orientation, 286–7
   row spacing, 282–6
crop–weed competition, 1
   below ground, 66–7, 221–3, 292
   'competitive effect', 287–8
   'competitive response', 288
   and crop density, 270–81
   and crop genotype, 287–96
   crop initial size, 63–5, 301–2
   and crop planting date, 302–4
   and intercrops, 338–9, 340
   legume/non-legume, 227–9
   for light, 65–6, 278, 290–1
   and mycoherbicides, 393–4
   vegetative growth and, 62–7
   weed infestation period, 299–301
   weed-free period, 297–9
   see also allelopathy
crops
   growth parameters, 63–5
   herbicide resistance, 10–11

life-history characteristics, 41
surpluses, 501
use of, 503
value of, 23–4
cultivation, 139
crop row spacing, 285–6
dark, 171–3
depth, 52–3
guidance systems, 188–9
timing, 173
cultivators, 169, 173–4
engineering of, 193–4
field, 142
full-field tools, 179–82
guidance systems, 188–9, 194
in-row tools, 182–6
inter-row, 174–6
near-row tools, 176–9
rolling, 174, 176
shovel, 174–5, 184
used in mulch systems, 255–6
cyanazine, 12, 14

2,4-D, 16, 18
damping-off fungi, 244
dark cultivation, 171–3
deep tine, 156
deleterious rhizobacteria, 397–8
desiccation, 149–50
developing countries
crop diversity, 324
farm profitability, 19, 21
food imports, 501
food security, 503–5
herbicide use, 6, 8
high input agriculture in, 111–12
transition to ecological weed management, 25–6
dhurrin, 234
4-dihydroxy-1, 4(2H)-benzoxazin-3-one (DIBOA), 234, 235
dinitroaniline herbicides, 17
disk hillers, 176
disk tillage implements, 142, 155, 156
dispersal *see* weed dispersal
disturbance, 447–8, 456
diversity *see* crop diversity; weed diversity; wild plant diversity
domestication, of plants, 454–5
dormancy, 46–9, 59
drip irrigation, 218–19
drought, 66, 68

earthworms, 243
ecological weed management, 30–1
adaption to weed variability, 114–16
benefits of, 25
farmer education, 512, 513
future research, 506–10
implementation, 510–14
objectives, 2–5
obstacles to uptake, 499–500
training/education, 497
transition to, 25–9
weed adaptation management, 477–9
see also farmer–extensionist–scientist interactions; participatory action for learning
ecotype formation, 469–74
electric weeders, 188
eliasomes, 77
ensiling of forage, 79
environmental taxes, 498, 511–12
ethylene, 51–2, 213
evergreen plantations, 420–1, 423–5
exotic weeds *see* alien weeds
extensionist, 105–8, 497; *see also* farmer–extensionist–scientist interactions; participatory learning for action

facilitative interactions, 338
fallow cultivation, 146, 147
fallow grazing, 421–2
false seedbed, 53, 168–9
farm profitability, 19–24
cost–price squeeze, 19–21
crop value, 23–4
reducing production costs, 21–3
farm size, 499–500, 510
farmer, 100
as adviser, 127–8
decision-making, 25, 100
communication networks, 108, 124–7, 131–2
education, 483, 512, 513
health effects of pesticides, 17–18
perceived benefits of herbicides, 495–6
perspectives on weeds, 105–8
role in agricultural research, 118–21
role in technology development, 101–2, 104
weed management requirements, 4
weed observation and measurement, 122–4
see also farmer–extensionist–scientist interactions; participatory learning for action
farmer–extensionist–scientists interactions, 99–133
adaptive management, 117–18
historical perspective, 100–4
improving weed management, 109–12
participatory learning, 118–33
fertilizers
banding, 223–4, 283
differential crop–weed response, 220–3
nitrogen source, 225–7
timing of application, 224–5
use in intercropping, 346
ferulic acid, 236
field cultivator, 142
field margins, 380, 505–6
Finland, pesticide taxes, 498
flame weeders, 186–7
flame weeding, 167, 191, 192
flooding, 216

Subject index

food prices, 504–5
food security, 502–5
forage, weed seed contamination, 79
forage crops, 21–3, 292, 333–5
forest fallow systems, 353–6
freezing damage, 150
full-field cultivation tools, 179–82
fungi
    as biocontrol agents, 387–8, 392–6, 400
    soil-borne, 244, 378

genetic variation *see* crop genotype; weed genetic variation
geographic isolation, 461–2
glacial deposits, 457
glyphosate, 9, 16
goat grazing, 411, 433–4
goose grazing, 410, 420–1
'goosefoot' shovels, 174–5
grass strips, 380
grass weeds
    annual, 164
    in forage crops, 335
    seedling survival, 69–71, 413
grazing *see* livestock grazing
grazing tolerance, 411–14
green manures, 115–16, 229–33
groundwater contamination, 14–15

habitat protection, 505–6
hand weeding, 116
harrows, 180–2
health advisory levels, 12
hedgerow intercropping *see* alley cropping
herbicide resistance, 8–11, 478–9, 481
herbicides, 5
    comparison with mechanical weeding, 190–2
    cost of development/registration, 10
    cost externalization, 497–9
    in crop rotations, 330–3
    development of, 103
    factors promoting use, 6–8, 496–7
    global sales, 5–6
    global use, 6–8
    increasing efficiency of use, 26, 114–15
    perceived benefits, 495–6
    pollution from, 11–16
    reducing use of, 498, 512–14
    substitution, 26–7
    see also named herbicides
herbivory, 68–9
    as biocontrol
        introduced species, 381–91
        native species, 377–80
    effects of crop residues, 241–4
    seedlings, 68–9, 241, 243, 378–9
    seeds, 57–9, 243–4, 377–8, 379
    see also livestock grazing
hill pastures, 432–4
Hodgkin's lymphoma, 18
Huastec agroforestry, 324
human health
    acute effects of herbicides, 16–17
    chronic effects of herbicides, 17–18
human-dominated ecosystems, 454–5
hybridization, 460–1

ice-encasement, 216
Illinois River, 13
imidiazolinone compounds, 15
immunotoxicity, 18
in-row cultivation tools, 179, 182–6
inbreeding, 464–6
indigenous agriculture, 116–17, 233, 324
industrialized agriculture, 111–12
    beneficiaries of, 503–4
    crop diversity, 323–4
    farms, 499–500
infertile soils, 66–7
inoculative biocontrol
    exotic weeds, 380–8
    native weeds, 388–91
    non-target species, 390–1
    social/ecolocial impacts, 389–91
input substitution, 26–7
insect herbivores
    introduced, 381–91
    native, 57–9, 68–9, 377–80
integrated pest management (IPM), 24–5
inter-row cultivation, 174–6, 191, 286
INTERCOM model, 222–3
intercropping, 336–51
    benefits to farmer, 336–8
    crop density, 341–4
    crop diversity, 339–42
    crop genotype, 345
    crop spatial arrangement, 344
    living mulches, 346–8
    machinery/technology for, 363–4
    resource use and weed suppression, 338–9, 340
    smother crops, 348–50
    soil fertility, 346
    weed community composition, 350–1, 453–4
inundative biological control, 391–2
    mycoherbicides, 392–6
    rhizobacteria, 397–8
IPM *see* integrated pest management
irrigation
    selective placement, 218–20
    weed dispersal, 81–2
isothiocyanates, 245

kebuntalun agroforestry, 324
ketones, 236
knowledge communities, 105–9

land equivalent ratio (LER), 336–7
large-seeded weeds *see* seed size
leaf mulch, 357, 362
legumes, 327
    competition with non-legumes, 227–9
    cover crops, 28–9, 131–2, 230–3
    forage crops, 21–3, 333–5
    in pastures, 67

LER *see* land equivalent ratio
life span, 71
life-history strategies, 41–5
  breeding systems, 463–7
  dispersal, 75–83
  dormancy/germination, 46–53
  life span, 71
  seed longevity, 53–60
  seed production, 71–5
  seedling establishment, 61–2
  survival to maturity, 67–71
  vegetative growth, 62–7
light
  and crop row orientation, 286–7
  crop–weed competition, 65–6
  and seed germination, 51, 52
  transmission through mulches, 240–2
light flash 161
light quality, 52, 241
livestock grazing
  aftermath and fallow, 421–2
  grazing species differences, 410–11
  invasive weeds, 435
  large land units, 435
  management principles, 409–10
  pastures and rangelands, 427–8
    altering grazing regime, 429–30
    animal productivity, 428–9
    changing grazing species, 432–4
    paddock size/uniformity, 430–2
  plant community changes, 414–20
  plant resistance, 411–14
  in short-cycle crops, 420–1
  in tree crops, 420–1, 423–6
  vegetation monitoring, 436–7
living mulches, 346–8

manure
  addition to soil, 251–3
  weed dispersal in, 78–9
  see also green manures
MCPA *see* 2-methyl-4-chlorophenoxyacetic acid
mechanical weed management, 62
  comparison with herbicides, 190–2
  crop stand losses, 185–6, 192–3
  future research directions, 192–4
  in mulch systems, 254–6
  principles, 169–73
  systems for, 189–90
  see also cultivators; tillage
2-methyl-4-chlorophenoxyacetic acid (MCPA), 18
metolachlor, 14
metribuzin, 14, 18, 252
Mexico, Gulf of, 14
mimicry, 473–4
mineral elements, 220
Mississippi River basin, 11–14
Missouri River, 13
modular growth, 61–2
moldboard plow, 141–2, 155, 156–7
mollusks, 69, 241–4
monocultures, 323–4, 327, 452

mowing, 255
mulches
  from tree crops, 357, 362
  living, 346–50
  polythene, 212–15
  see also crop residues
mycoherbicides, 392–6
  formulations, 395–6
  and plant competition, 400
  production and culture, 396

napropamide, 218–19
near-row tools, 176–9
needleleaf plantations *see* evergreen plantations
nitrate, 50–1, 225–6
nitrogen, 220
nitrogen fertilizers
  banding 223–4, 283
  effect of nitrogen source, 225–7
  in intercropping systems, 346
  legume/non-legume competition, 227–9
no-till systems, 141, 254–5, 380–1
non-dormant seeds, 46, 47
non-Hodgkin's lymphoma, 17–18
Norway, pesticide taxes, 498

Ohio River, 13
orchards, 360–2
organic amendments, 251–4
organic farming
  crop diversity, 325
  crop values, 23–4
outbreeding, 463–6
oxygen, 50–1

paraquat, 18, 252
parasitic weeds, 45–6
participatory learning for action, 100, 119
  case studies, 127–32
    Californian tomato cropping, 127–8
    cover crops in Central America, 131–2
    ground cover in coffee, 129–31
    Iowa grain cropping, 128–9
  and farmer decision-making, 121
  farmer–farmer communications, 124–7
  potential rewards of, 124, 126
  role of farmer groups, 118–21
  weed monitoring methods, 122–4
pastures
  biological weed control, 386–90
  grazing regimes, 429–30
  infertile, 66–7
  invasive weeds, 435
  paddock size/uniformity, 430–1
  seedling survival, 69–71, 413
  unpalatable species, 432–3
  vegetation composition changes, 414–20
  vegetation monitoring, 436–7
  weed impact on productivity, 428–9
pebulate, 218–19

perennial weeds
  breeding systems, 463–7
  effects of fallow cultivation, 146, 147
  life-history strategies, 42–5
  mechanical control, 45, 141–51
    by chopping and burying roots/rhizomes, 146, 148–9
    of established weeds, 141–5
    by exposure of roots/rhizomes, 149–51
    by fallow cultivation, 146, 147
    timing of, 145–6
  seed production, 73
  survival after emergence, 69–71
pesticides
  cost of development/registration, 10
  poisoning, 16–17
  taxes on, 498, 511–12
  see also herbicides
phenotypic plasticity, 468–9
phenoxy herbicides, 18
pig grazing, 410
plant community composition
  and biological weed control, 402–3
  livestock grazing regimes, 414–20
  see also weed community composition
plant morphology
  and grazing tolerance, 412–13
  and light competition, 65–6
plantations see agroforestry; tree crops
plastic mulching, 212–15
Platte River, 13
poisonous plants, 386–7, 388–9, 411, 432
pollution
  coastal ecosystems, 14
  groundwater, 14–15
  spray drift, 15–16
  surface waters, 11–14
potassium uptake, 292
Practical Farmers of Iowa, 128–9
precision agriculture, 26, 114–15, 117
program crops, 501

rachis-shattering, 459–60
range extension, 470
rangeland, 382–5
rapeseed green manure, 244–5
recombinant DNA technology, 10
relative growth rate, 41, 62–5
relay intercropping, 322, 350
resource complementarity, 338
rhizobacteria see deleterious rhizobacteria
rhizomes see roots and rhizomes
rice production
  crop rotations, 330
  water management, 216–18
  weed flora, 216–17
ridge tillage, 27, 164–7
rigid tine, 157
rolling cultivators, 174, 176
root competition, 66–7, 221–3, 292
roots and rhizomes
  chopping/burying, 146, 148–9
  effects of tillage implements/practices, 141–5
  exposure to desiccation/freezing, 149–51
  and fallow cultivation, 146, 147
  removal from field, 150–1
rotary hoe, 179
rotary hoeing, 62, 166, 179–80
rotary tiller, 142, 156, 176
row crops
  competitive ability, 290–1
  harrowing, 182
  planting date, 303
rubber-finger weeder, 185
ruderal weeds, 40

scarification, 161
science see agricultural science
scientist
  networks, 108
  perspective on weeds, 105–8
  see also farmer–extensionist–scientist interactions; participatory learning for action
seed banks
  management of weed adaptation, 480–1
  seed persistence in, 53–9
  tillage effects
    depletion, 158–62
    germination, 50–3
    redistribution of seeds, 154–7
seed cleaning, 78
seed germination
  and dark cultivation, 171–3
  and depth of burial, 56–7
  effects of soil fungi, 378
  large seeds, 53, 54
  seasonality of, 46–9, 327–8
  and tillage, 50–3, 158–62
seed mortality, 56–8
  predation, 57–8, 59, 243, 377–8, 379, 380
  solarization, 213
  tillage, 158–62
  see also seed survival
seed productivity, 71–5
  effects on future crops, 74
  herbicides affecting, 15–16
  measures to reduce, 74–5
  and seed size, 42
seed size, 41–2
  of crops, 61, 301–2
  and emergence through mulch, 249
  and establishment, 60–1
  and germination, 53, 54
  and growth parameters, 63–5
  mortality, 57
  and persistence, 59
  and seed productivity, 42
  and susceptibility to allelopathy, 234–5
seed survival, 53–6
  consequences for weed management, 59–60
  in forage crops, 334
  patterns among species, 58–9
  and tillage, 158–62

see also seed mortality
seedbed, 171
  false, 53, 168–9
  stale, 167
  see also tillage
seedling density
  and timing of tillage, 151–3
  and type of tillage, 153–4
seedling emergence
  and mulches, 247–50
  periodicity of, 46–9, 327–8
  and tillage, 155–8
seedling establishment, 60–2
seedling survival, 67–71, 241, 243–4, 378
self-compatibility, 464–6
shade trees, 360–2
shading, 65–6, 270, 286–7
shallow tine, 156
sheep grazing, 411, 425, 432–4
short-cycle crops, 420–1
shovel cultivator, 174–5, 184
simazine, 14
slash-and-burn systems, 212
slash–mulch systems, 233
slash–mulch–fallow rotations, 116–17
small-seeded species *see* seed size
smother crops, 347, 348–50
soil amendments, 251–4
soil drainage, 171, 180
soil erosion, 140, 498–9
soil fertility
  crop–weed competition, 66–7, 221–3, 292
  and germination, 50, 51
  in intercrops, 346
  legume/non-legume competition, 227–9
  nitrogen source, 225–7
  spatial variation, 223–4
  temporal variation, 224–5
soil gas exchange, 50–2
soil management
  integration with weed management, 250–4
  principles, 211
  temperature, 212–15, 238–40
  see also crop residues; soil fertility; soil water management
soil pathogens, 215, 244
soil temperature, 212–15, 238–40
soil tilth, 140, 157–8, 171
soil water, 66, 68, 171, 180, 239–40
soil water management
  flooding, 216
  in rice production, 216–18, 230
  selective irrigation, 218–20
solarization, 212–15
somatic variation, 468–9
sorgoleone, 234
speciation, 461–2
spinners, 183–5
spray drift, 15–16
spring hoes, 182, 183
spring-tine harrows, 180
spyders, 176–7

stale seedbed, 167
states-and-transitions framework, 418–20
stationary perennials, 42–4, 73
storage reserves
  chopping/burying of storage organs, 146–9
  and fallow cultivation, 146, 147
  and grazing/tissue loss, 413–14
  and tillage timing, 145–6
  see also rhizomes; roots
stubble burning, 212
sulfonylurea herbicides, 9, 15
Sweden
  pesticide taxes, 498
  reduction of herbicide use, 512–14
sweep plow, 142
sweeps, 174–5

tapado systems, 233
taproots, 144–5
technology *see* agricultural technology
thermal weed suppression, 212–15
thermal weeders, 186–8
tillage
  in crop rotations, 327–9
  in dark, 171–3
  depth of, 52–3
  minimal, 141, 254–5, 480–1
  in mulch systems, 254–5
  perennial weed control, 45, 141–51
    chopping and burying roots/rhizomes, 146, 148–9
    established plants, 141–5
    exposure of roots/rhizomes, 149–51
    fallow cultivation, 146, 147
    removal of roots/rhizomes, 150–1
    timing of and weed growth, 145–6
  principles, 139–40
  pros and cons, 140–1
  ridge, 27, 164–7
  seed bank management, 162–4
  seed germination, 50–3
  seed redistribution, 154–5
  seed survival, 53, 158–62
  seedling emergence, 155, 157–8
  stale/false seedbed, 168–9
  timing of, 140, 151–3
  type of, 153-4
  weed dispersal during, 80–1
tillage implements, 141–2, 155, 156–7
tine implements, 156–7
torsion weeders, 182, 183
transplants, 301
trash wheels, 255
tree crops
  coconut, 425–6
  evergreen, 420–1, 423–5
  livestock grazing under, 420–1, 423–6
  selective weeding, 360
  understorey cover crops, 358–9
  see also agroforestry
tree prunings, 357
triazine herbicides, 14–15, 17

trifluralin, 18
tropical agriculture 111–12, 212, 233
tubers, 75, 142, 143, 144

United States Environmental Protection Agency (USEPA), 12, 17, 497–8

vegetable crops, 293–6, 301–2
vegetable knives, 176
vegetation monitoring, 122–4, 436–7
vegetative growth, 62–7
vegetative reproduction, 75, 466–7

wandering perennials, 43–5
   dispersal, 80
   genetic variability, 464, 466–7
   seed production, 73
   seed survival, 58–9
   vegetative reproduction, 75, 466–7
water contamination, 11–15
weather, 173
weed community composition, 3, 4, 445–7
   and alien species, 447–52
   and crop diversification, 350–1, 452–4
   and nitrogen source, 225–7
weed dispersal
   adaptative mechanisms, 76–8
   in bulk commodities, 80
   by machinery, 80–1
   crop seed contamination, 78
   genetic variation following, 467–8
   in irrigation water, 81–2
   in manure and forage, 78–9
   prevention, 82, 481–4
   spatial patterns, 82–3, 109–10
   spatial scales, 75–6
weed distribution patterns
   management approaches to, 112–17
   patchiness, 82–3, 109–10
   uncertainty, 110–11
weed diversity
   and crop diversity, 452–4
   and species introduction, 447–52
   and weed evolution, 462–3
weed ecology, development, 103–4
weed eradication, 30
weed evolution, 455–6
   biotypes, 461
   and disturbance, 447–8
   from crop species, 459–60
   from wild populations, 458–9
   herbicide resistance, 8–11, 478–9
   and human-dominated ecosystems, 454–5
   hybridization, 460
   mimicry, 473–4
   preadaptation for weediness, 456–8
   speciation, 461–2
   and weed diversity, 462–3
   weed–nonweed comparisons, 471–3
weed genetic variation
   after colonization, 467–8
   and breeding systems, 463–7
   ecotype formation, 469–74
   somatic variation, 468–9
weed infestation period, 299–301
weed monitoring, 122–4, 436–7
weed seed *see* seed
weed seedling *see* seedling
weed-free crop trials, 269
weed-free period, 297–9
weed–weed interactions, 446–7
weeding, 116, 471; *see also* mechanical weeding
weeds, value of, 1–2
well contamination, 14–15
wetland crops, 216–18, 330
White River, 13
wild plant diversity, 505–6
wildlife conservation, 505–6
wind-dispersed seed, 77–8
winter annuals, 46, 46–9
woody perennials, 43–5
wool, raw, 80